Case Studies in
CANCER

Case Studies in
CANCER

Richard J. Lee
Massachusetts General Hospital
Harvard Medical School

Jeremy S. Abramson
Massachusetts General Hospital
Harvard Medical School

Richard A. Goldsby
Amherst College

W. W. NORTON & COMPANY
NEW YORK • LONDON

W. W. Norton & Company has been independent since its founding in 1923, when William Warder Norton and Mary D. Herter Norton first published lectures delivered at the People's Institute, the adult education division of New York City's Cooper Union. The firm soon expanded its program beyond the Institute, publishing books by celebrated academics from America and abroad. By midcentury, the two major pillars of Norton's publishing program—trade books and college texts—were firmly established. In the 1950s, the Norton family transferred control of the company to its employees, and today—with a staff of four hundred and a comparable number of trade, college, and professional titles published each year—W. W. Norton & Company stands as the largest and oldest publishing house owned wholly by its employees.

Editors: Betsy Twitchell and Summers Scholl

Associate Editor: Katie Callahan

Editorial Consultant: Denise Schanck

Senior Associate Managing Editor, College:
    Carla L. Talmadge

Editorial Assistant: Danny Vargo

Managing Editor, College: Marian Johnson

Managing Editor, College Digital Media: Kim Yi

Production Manager: Elizabeth Marotta

Media Editor: Kate Brayton

Associate Media Editor: Gina Forsythe

Media Project Editor: Jesse Newkirk

Media Editorial Assistant: Katie Daloia

Ebook Production Manager: Michael Hicks

Marketing Manager, Biology: Stacy Loyal

Content Development Specialist: Todd Pearson

Director of College Permissions: Megan Schindel

Permissions Clearer: Sheri Gilbert

Composition: Six Red Marbles

Illustrations: Nigel Orme

Manufacturing: TC–Transcontinental
    Printing

Permission to use copyrighted material is included alongside the appropriate content.

Library of Congress Cataloging-in-Publication Data

Names: Lee, Richard (Richard Jae Bong) 1972- author. | Abramson, Jeremy,
  author. | Goldsby, Richard A., author.
Title: Case studies in cancer / Richard Lee, Jeremy Abramson, Richard Goldsby.
Description: New York : W.W. Norton & Company, [2019] | Includes index.
Identifiers: LCCN 2018039133 | ISBN 9780393679519 (pbk.)
Subjects: LCSH: Cancer—Research. | Cancer—Case studies.
Classification: LCC RC267 .L434 2019 | DDC 362.19699/40072—dc23 LC record available at
https://lccn.loc.gov/2018039133

W. W. Norton & Company, Inc., 500 Fifth Avenue, New York, NY 10110

wwnorton.com

W. W. Norton & Company Ltd., 15 Carlisle Street, London W1D 3BS

1 2 3 4 5 6 7 8 9 0

Dedication

To our better halves: Young Oh, Dave Kroll, and Barbara Osborne, for their love and support

To our students

To our patients and their families, for their limitless courage and inspiration

Brief Contents

Preface   xvii

About the Authors   xxi

CHAPTER 1   Overview of Cancer and Clinical Oncology   1

CHAPTER 2   Introduction to Cancer Therapy   13

CHAPTER 3   Cytotoxic Chemotherapy   39

CHAPTER 4   Protein Kinase Inhibition   61

CHAPTER 5   Inhibition of the PI3-Kinase/Akt/mTOR Pathway   103

CHAPTER 6   Hormone Therapy   131

CHAPTER 7   Differentiation Therapy   159

CHAPTER 8   Epigenetic Therapy   171

CHAPTER 9   Targeting Protein Degradation and DNA Repair   193

CHAPTER 10   Interference with the Tumor Microenvironment   215

CHAPTER 11   Immunological Strategies: Monoclonal Antibodies   239

CHAPTER 12   Immunological Strategies: Vaccination and Adoptive T-Cell Transfer   269

Glossary   G-1

Index   I-1

Contents

Preface xvii

About the Authors xxi

CHAPTER 1

Overview of Cancer and Clinical Oncology 1

Introduction 1

Cancer facts and figures 2

Clinical oncology 3

The hallmarks of cancer 4

The tumor microenvironment 6

Diagnosis, workup, and staging 8

Role of cancer screening 8

Biomarkers 10

Development of cancer therapies 11

The future of clinical oncology 11

Chapter discussion questions 12

Selected references 12

CHAPTER 2

Introduction to Cancer Therapy 13

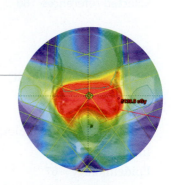

Introduction 13

Surgery 13

Radiation therapy 17

Medical cancer therapies 19

Genomics and oncology 23

Cytogenic approaches 23

Microarrays 25

Sequencing 27

NGS technologies 31

Genomics 32

Genomic assays of circulating tumor cells and tumor DNA 34

Pharmacogenomics and cancer therapy 35

Summary and future prospects 35

Chapter discussion questions 37

Selected references 38

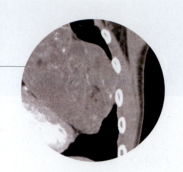

CHAPTER 3
Cytotoxic Chemotherapy   39

Introduction   39

CASE 3-1 Testicular Cancer   41

Introduction   41

Epidemiology and etiology   42

Diagnosis, workup, and staging   42

Management of stage I and II testicular cancer   45

Management of stage III testicular cancer   46

The case of Michael LaPorta, a healthy young man with a painless scrotal mass   47

Targeted therapy and testicular cancer   48

Future prospects and challenges in metastatic testicular cancer   49

Discussion questions   49

CASE 3-2 Pancreatic Cancer   50

Introduction   50

Epidemiology and etiology   50

Diagnosis, workup, and staging   51

Management of localized pancreatic cancer   53

Management of metastatic pancreatic cancer   54

The case of Jeanne Brantley, a real estate broker with painless jaundice   55

Targeted therapy and pancreatic cancer   57

Future prospects and challenges in metastatic pancreatic adenocarcinoma   58

Discussion questions   59

Chapter summary   59

Chapter discussion questions   59

Selected references   60

CHAPTER 4
Protein Kinase Inhibition   61

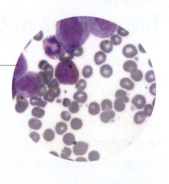

Introduction   61

CASE 4-1 Chronic Myelogenous Leukemia   66

Introduction   66

Epidemiology   67

Diagnosis, workup, and staging   67

Management of CML in chronic phase   70

The case of Jeffrey Paul, a man with a twisted ankle and an elevated white blood cell count   71

Management of accelerated- and blast-phase CML   73

Management of CML when initial TKI therapy fails   74

Future prospects and challenges in CML   74

Discussion questions   75

CASE 4-2 Chronic Lymphocytic Leukemia   75

Introduction   75

Epidemiology   75

Diagnosis, workup, and staging   76

Principles of CLL management   78

The case of Patricia O'Halloran, a 68-year-old professor of German history with
an elevated WBC count   79

Treatment of relapsed CLL   80

Future prospects and challenges in CLL   81

Discussion questions   82

CASE 4-3 Lung Cancer   82

Introduction and epidemiology   82

Diagnosis, workup, and staging   83

Management of localized lung cancer   85

Management of metastatic lung cancer   85

Genetic alterations and NSCLC   86

The case of Alex Washington, a nonsmoker with a lingering cough   86

EML4-ALK translocation-positive NSCLC   87

Future prospects and challenges in lung cancer   89

Discussion questions   89

CASE 4-4 Metastatic Melanoma   90

Introduction and epidemiology   90

Diagnosis, workup, and staging   90

Management of localized melanoma   93

Management of metastatic melanoma   94

Genetic alterations and melanoma   94

The case of Tom Schaeffer, a retired machinist with an irregular brown skin lesion   95

Kinase inhibition and melanoma   96

Future prospects and challenges in metastatic melanoma   99

Discussion questions   100

Chapter summary   100

Chapter discussion questions   101

Selected references   102

CHAPTER 5
Inhibition of the PI3-Kinase/Akt/mTOR Pathway 103

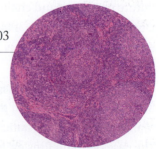

Introduction   103

CASE 5-1 Renal Cell Carcinoma   108

Introduction   108

Diagnosis, workup, and staging   109

Management of localized RCC   111

Management of metastatic RCC   111

Molecular alterations and RCC   113

The case of Mary Carlton, a woman with an incidental kidney mass found during
evaluation of abdominal pain   114

Targeted therapies and metastatic RCC   115

Future prospects and challenges in metastatic RCC   116

Discussion questions   117

CASE 5-2 Follicular Lymphoma   118

Introduction   118

Epidemiology   120

Diagnosis, workup, and staging   120

Principles of FL management   122

The case of Stephen Kim, a community organizer with kidney stones   124

PI3-kinase inhibition in follicular lymphoma   125

Future prospects and challenges in follicular lymphoma   126

Discussion questions   127

Chapter summary   128

Chapter discussion questions   129

Selected references   130

CHAPTER 6

Hormone Therapy   131

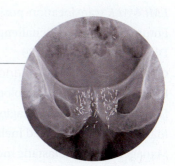

Introduction   131

Inhibition of gonadotropin release   133

Steroidogenesis inhibition   134

Inhibition of conversion of steroids in peripheral tissues   135

Steroid hormone receptor antagonists   135

CASE 6-1 Breast Cancer   136

Introduction   136

Etiology   136

Diagnosis, workup, and staging   137

Breast cancer subtypes   139

Management of localized breast cancer   140

Management of metastatic breast cancer   141

The case of Brenda Nathan, a paramedic with a breast mass   142

Hormone therapy and metastatic breast cancer   143

Future prospects and challenges in metastatic breast cancer   143

Discussion questions   144

CASE 6-2 Prostate Cancer   145

Introduction   145

Diagnosis, workup, and staging   146

Management of localized prostate cancer   149

Management of metastatic prostate cancer   151

The case of Adam Presley, a retired high school principal with hip pain   152

AR targeting and prostate cancer: beyond ADT   153

Future prospects and challenges in metastatic prostate cancer   155

Discussion questions   156

Chapter summary   156

Chapter discussion questions   157

Selected references   157

CHAPTER 7

Differentiation Therapy 159

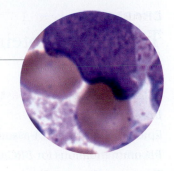

Introduction   159

CASE 7-1 Acute Promyelocytic Leukemia   161

Introduction   161

Epidemiology and etiology   163

Diagnosis, workup, and staging   164

Initial treatment of APL   165

The case of Mark Gable, a young man with bleeding gums following dental cleaning   166

Management of relapsed APL   167

Discussion questions   168

Chapter summary   168

Chapter discussion questions   168

Selected references   169

CHAPTER 8

Epigenetic Therapy 171

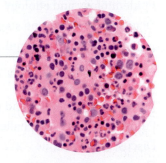

Introduction   171

Histone modification   171

DNA methylation   173

MicroRNA (miRNA) targeting   174

Long noncoding RNAs (lncRNAs)   175

CASE 8-1 Myelodysplastic Syndromes   175

Introduction and epidemiology   175

Diagnosis and workup   176

Prognosis   178

The case of Hank Johnson, an older male with dyspnea on exertion and progressive cytopenias   178

Treatment   180

Future prospects and challenges in MDS   182

Discussion questions   182

CASE 8-2 Cutaneous T-cell Lymphoma   183

Introduction   183

Epidemiology   184

Diagnosis, workup, and staging   184

Principles of MF and SS management   186

The case of Jon Carter, a 62-year-old man with skin lesions   187

HDAC inhibition in cutaneous MF and SS   188

Future prospects and challenges in cutaneous T-cell lymphomas   189

Discussion questions   190

Chapter summary   190

Chapter discussion questions   191

Selected references   191

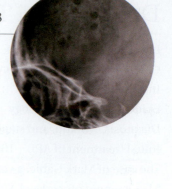

CHAPTER 9

Targeting Protein Degradation and DNA Repair 193

Introduction   193

CASE 9-1 Hereditary Breast Cancer   197

Introduction   197

Epidemiology of *BRCA*-associated cancers   200

Recommendations for *BRCA1/2* asymptomatic carriers   200

The case of Yvonne Charles, a young mother with a breast lump   201

Neoadjuvant therapy for breast cancer   203

Management of localized *BRCA1/2*-related breast cancer   203

Management of metastatic *BRCA1/2*-related breast cancer   203

Future prospects and challenges in *BRCA*-mutant cancers   204

Discussion questions   205

CASE 9-2 Multiple Myeloma   205

Introduction   205

Epidemiology and etiology   205

Diagnosis, workup, and staging   206

Initial management of multiple myeloma   208

The case of Malik Jackson, a 39-year-old man who presented with anemia and multiple lytic bone lesions   208

Proteasome inhibition and multiple myeloma   209

Future prospects and challenges in multiple myeloma   211

Discussion questions   212

Chapter summary   212

Chapter discussion questions   214

Selected references   214

CHAPTER 10

Interference with the Tumor Microenvironment 215

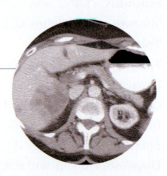

Introduction   215

The immunosuppressive TME   217

Angiogenesis in the TME   217

Targeting and disturbing the TME   218

CASE 10-1 Colorectal Cancer   219

Introduction   219

Molecular alterations and CRC   220

Epidemiology and etiology   221

Diagnosis, workup, and staging   222

Management of localized CRC   223

Management of metastatic CRC   225

The case of Mary Maspin, a woman with blood in her stool and an unintentional 20-pound weight loss   225

Targeted therapies and metastatic CRC   226

Future prospects and challenges in metastatic CRC treatment   227

Discussion questions   228

CASE 10-2 Multiple Myeloma   228

Introduction   228

Principles of management for transplant-ineligible patients with newly diagnosed MM   229

The role of IMiDs in multiple myeloma therapy   230

The case of Tom Brown, a 70-year-old male with a history of monoclonal gammopathy and severe low back pain   230

The bone marrow microenvironment in multiple myeloma   231

Bone metabolism   232

Immunomodulatory drugs and the bone marrow microenvironment   233

Future prospects and challenges in treating multiple myeloma   235

Discussion questions   235

Chapter summary   236

Chapter discussion questions   238

Selected references   238

CHAPTER 11

Immunological Strategies: Monoclonal Antibodies 239

Introduction   239

CASE 11-1 Non-Hodgkin Lymphoma   248

Introduction   248

Epidemiology   251

Diagnosis, workup, and staging   251

Principles of DLBCL management   253

The case of Linda White, a 61-year-old school teacher with pain in her shoulder   253

Management of relapsed DLBCL   254

Future prospects and challenges in DLBCL   254

Discussion questions   256

CASE 11-2 Hodgkin Lymphoma   256

Introduction   256

Epidemiology   258

Diagnosis, workup, and staging   258

Principles of CHL management   259

The case of Jessica Romero, a 22-year-old graduate student with shortness of breath   260

Treatment of relapsed Hodgkin lymphoma   261

Future prospects and challenges in CHL   262

Discussion questions   263

CASE 11-3 Breast Cancer   263

The case of Julie Rosen, a 53-year-old woman with a growing breast mass and a family history of breast cancer   264

Future prospects and challenges in breast cancer   266

Discussion questions   267

Chapter summary   267

Chapter discussion questions   268

Selected references   268

CHAPTER 12

Immunological Strategies: Vaccination and Adoptive T-Cell Transfer 269

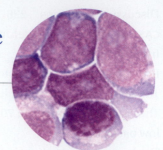

Introduction 269

Cancer vaccines 272

Adoptive T-cell therapies 273

Blockade of immune checkpoints 277

Other immunotherapies 279

CASE 12-1 Cervical Cancer 282

Introduction 282

Epidemiology and etiology 283

Diagnosis, workup, and staging 284

Cell biology of HPV 285

Management of localized cervical cancer 287

Management of advanced cervical cancer 287

The case of Pamela Johnson, a 38-year-old woman with abnormal Pap smear findings 288

HPV vaccination and prevention of cervical cancer 288

Future prospects and challenges in HPV vaccines 290

Discussion questions 291

CASE 12-2 Acute Lymphoblastic Leukemia 291

Introduction and epidemiology 291

Diagnosis, workup, and staging 292

Management of ALL 293

The case of Roger Gonzales, a 5-year-old boy with fatigue, pallor, and bruising 294

Treatment with CAR T cells 296

Discussion questions 298

CASE 12-3 Metastatic Melanoma 299

Introduction 299

The case of Adam Stephenson, a 66-year-old man with a growing lesion on his left shoulder 299

Immunotherapy and melanoma 300

Future prospects and challenges in metastatic melanoma 302

Discussion questions 303

Chapter summary 303

Chapter discussion questions 305

Selected references 305

Glossary G-1

Index I-1

Preface

We are in the midst of a precision medicine revolution. In no disease is this more evident than cancer, where previous standards of care employed cytotoxic chemotherapies that kill rapidly growing cells indiscriminately. Drugs were combined that might act synergistically, unfortunately compounding toxicities at maximum tolerable doses, and with few real victories. Molecular biology enabled an era of discovery in cancer biology with the promise of one day translating into new therapies. That day arrived in 1997 when for the first time the FDA approved a targeted therapy: rituximab (Rituxan), a monoclonal antibody for the treatment of non-Hodgkin lymphoma. The subsequent development of other targeted therapies based on cancer biology has transformed the care of oncology patients.

Despite their clear links, the fields of cancer biology and clinical oncology have been largely taught as separate entities, leaving a gap of knowledge separating the basic scientist and the clinician. This book was conceived to bridge that gap by demonstrating how cancer biology *drives* clinical oncology, and how clinical realities can inform and pose research questions. As targeted approaches have accelerated and achieved clinical reality for an increasing number of malignancies, the time is ripe to write about the themes revealed in the emerging therapies.

Our goals are to educate and inspire the ongoing dialog and cross-fertilization of ideas between cancer biologists and clinical oncologists. Each chapter tackles a targeted approach, with salient examples from clinical practice. For our scientist readers, our overviews of a range of cancers offer insight into evaluation and management, with illustrative cases of targeted therapies in action. Since many cancer researchers do not have training in oncology, a well-chosen and clearly explained set of cases will provide them an opportunity to view cancer as it is encountered in the clinic. For our clinical readers, we describe the cancer biology concepts and discoveries that have helped bring therapeutic targets into focus. Neither the science nor the oncology we present is encyclopedic or exhaustive; our readers are directed to other sources to extend their education and exploration.

Note that all cases are based on actual patients but the names, demographics, and other histories for the described patients are fictitious. No identification with actual persons (living or deceased) is intended or should be inferred.

Features

- This book stands at the intersection of cancer biology and clinical oncology and is designed to offer an integrated view of the two disciplines.

- Readers without clinical backgrounds will learn the fundamentals of clinical oncology.

- The chapters are centered on themes in cancer therapy spanning different diseases, as opposed to just specific diseases and their individual molecular targets, because the future of clinical oncology will likely be driven by therapeutic targets as opposed to organ of origin.

RESOURCES FOR INSTRUCTORS AND STUDENTS

Ebook   Norton Ebooks give students and instructors an enhanced reading experience at a fraction of the cost of a print textbook. Students are able to have an active reading experience and can take notes, bookmark, search, highlight, and even read offline. As an instructor, you can even add your own notes for students to see as they read the text. Norton Ebooks can be viewed on—and synced among—all computers and mobile devices.

Image files   All of the images in the book are provided in both JPEG and PowerPoint formats for instructor use.

Acknowledgments

The authors acknowledge the inspiration and critical feedback of Dr. Robert A. Weinberg during this book's development, the brilliance of scientific colleagues hell-bent on discovery, the determination of clinical colleagues to test new therapies, the funders of our research, and, most important, the bravery of patients in the face of the most humbling of diseases.

We also would like to acknowledge the many contributions of our colleagues in the creation of this text. In particular, we wish to give special thanks to the following doctors, whose research, clinical work, and expertise were instrumental to this book:

Lauren Dias, Massachusetts General Hospital
Amir Fathi, Massachusetts General Hospital
Michaela Higgins, The Mater Hospital, Dublin
Elizabeth O'Donnell, Massachusetts General Hospital
Noopur Raje, Massachusetts General Hospital
Aliyah R. Sohani, Massachusetts General Hospital
Andrew J. Yee, Massachusetts General Hospital

In addition to these contributors, we also received detailed reviews from the following faculty and scientists, and we thank them for their feedback:

Ancha Baranova, George Mason University
Michael J. Barber, University of South Florida
Jan Blancato, Georgetown University
Nicole Bournias-Vardiabasis, California State University, San Bernardino
Joana Desterro, University of Lisbon
Paul Edwards, University of Cambridge
Evi Farazi, University of Nicosia
Robert Goldin, Imperial College, London
Raj K. Gopal, Harvard Medical School
Ward Kirlin, Morehouse School of Medicine
Carita Lanner, Northern Ontario School of Medicine
Michael Lipscomb, Howard University
Sylvia Lopez-Vetrone, Whittier College
Stuart McBain, Keele University, Newcastle-under-Lyme, UK
Matthew Meyerson, Harvard University
Patricia D. Murphy, GeneWISE

Asha Nayak-Kapor, Augusta University

Barbara A. Osborne, University of Massachusetts–Amherst

Ryan C. Owyang, Phillips Academy

Michael E. Pacold, Massachusetts Institute of Technology

Angel Pellicer, New York University School of Medicine

Helen Remotti, Columbia University

Jeffrey Segall, Albert Einstein College of Medicine

Andreas Seyfang, University of South Florida

David Virshup, Duke–NUS Medical School, Singapore

Danny Welch, University of Kansas Medical Center

Claudia Wellbrock, University of Manchester

Ellen Wheeler, Massachusetts General Hospital Cancer Center

Meng Zhang, Sanofi U.S.

Finally, we want to thank the long list of people at Garland Science and W. W. Norton & Company who labored toward publication of this book. Summers Scholl, our editor at Garland Science, and Claudia Acevedo-Quiñones, our assistant editor, managed this project at its inception. Our developmental editor Susan Teahan's keen readerly eye helped us produce a working manuscript, and our copy editor, Richard K. Mickey, ensured that the text was stylistically consistent and error-free. We would also like to thank Marian Provenzano for getting this book through the last stages of copy editing and figure drawing. For welcoming our book to W. W. Norton and bringing this edition to print, we thank our editor Betsy Twitchell. Denise Schanck deserves extra special gratitude for overseeing the transition from Garland Science to Norton. In addition, we would like to thank Danny Vargo for his editorial assistance on this project. We are grateful to content development specialist Todd Pearson, media editor Kate Brayton, associate media editor Gina Forsythe, and media editorial assistant Katie Daloia for their help in the creation of ancillary instructor resources. We are thankful for marketing manager Stacy Loyal's passionate support for our work. Megan Schindel, Sheri Gilbert, Stacey Stambaugh, and Ted Szczepanski are all owed a huge debt of gratitude for clearing the permissions for this textbook. Last but not least, Elizabeth Marotta's able management of production and Carla Talmadge's attention to detail made the book you hold in your hands today a reality.

Richard J. Lee
Jeremy S. Abramson
Richard A. Goldsby

About the Authors

RICHARD J. LEE After completing a bachelor's and a master's degree at Harvard, Dr. Lee then completed his medical school (M.D.) and graduate school (Ph.D.) training at the Albert Einstein College of Medicine, Bronx, New York. He came to Massachusetts General Hospital for internal medicine residency training and then completed medical oncology fellowship training through the Dana–Farber/Harvard Cancer Center joint program. After a postdoc with Dr. Robert A. Weinberg at the Whitehead Institute, he joined the medical oncology faculty at the Massachusetts General Hospital Cancer Center. Dr. Lee is currently Assistant Professor of Medicine at Harvard Medical School.

Courtesy Mass General Hospital Cancer Center

JEREMY S. ABRAMSON is Director of the Lymphoma Program and holds the Jon and Jo Ann Hagler Chair in Lymphoma at the Massachusetts General Hospital Cancer Center. He is also Assistant Professor of Medicine at Harvard Medical School. Dr. Abramson earned his medical degree from the Mount Sinai School of Medicine in New York and a Master of Medical Science degree from Harvard Medical School. His research focuses on the development of novel targeted therapies and cellular immunotherapy for lymphoid malignancies.

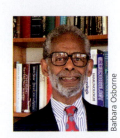

Courtesy Mass General Hospital Cancer Center

RICHARD A. GOLDSBY received a Ph.D. in chemistry at the University of California–Berkeley. He is the Thomas B. Walton Jr. Memorial Professor of Biology Emeritus at Amherst College, coauthored three editions of a leading immunology textbook, and was a cofounder of Hematech. He is also Adjunct Professor of Veterinary and Animal Sciences at the University of Massachusetts–Amherst and Visiting Scientist at the Whitehead Institute, Cambridge, Massachusetts.

Barbara Osborne

Case Studies in

CANCER

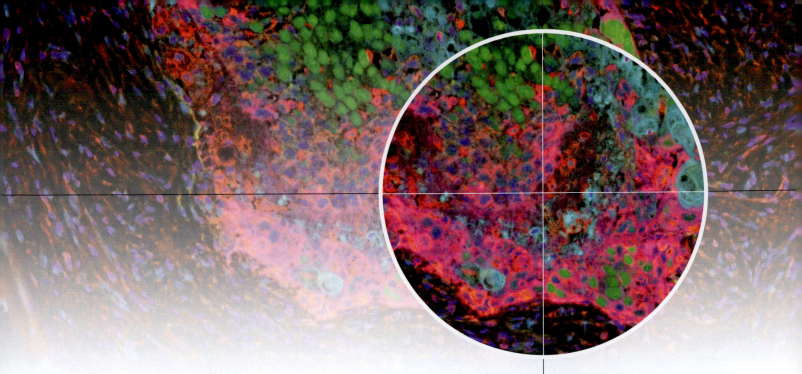

Shannon L. Stott, Ph.D., Center for Cancer Research, Massachusetts General Hospital and Harvard Medical School and Associate Member, Broad Institute of Harvard and MIT/João Paulo Oliveria-Costa, Ph.D., Center for Cancer Research, Massachusetts General Hospital and Harvard Medical School.

Chapter 1

Overview of Cancer and Clinical Oncology

Introduction

There are many types and subtypes of cancer. Cancers are broadly categorized according to their cell or tissue of origin. The major categories are:

1. **Carcinomas** (>90% of all human cancers), arising from epithelial cells and including cancers of the lung, breast, prostate, colon, cervix, bladder, non-melanoma skin cancers, and others.

2. Hematopoietic cancers (~5% of cancers), arising from blood or blood-forming cells and including **leukemias**, **lymphomas**, **myelomas**, and others.

3. **Sarcomas** (~1% of cancers), arising from cells of muscle, bone, and connective tissues and including fibrosarcomas, osteosarcomas, and others.

4. Neuroectodermal cancers (~1% of all cancers), arising from cells of neuronal tissues and including gliomas, neuroblastomas, schwannomas, **melanomas**, and others.

While all types of cancer share the property of uncontrolled cell proliferation, cancers diverge in many ways, and effective treatment of different cancers requires different therapies.

estimated new cases

males

prostate	164,690	19%
lung and bronchus	121,680	14%
colon and rectum	75,610	9%
urinary bladder	62,380	7%
melanoma of the skin	55,150	6%
kidney and renal pelvis	42,680	5%
non-Hodgkin lymphoma	41,730	5%
oral cavity and pharynx	37,160	4%
leukemia	35,030	4%
liver and intrahepatic bile duct	30,610	4%
ALL SITES	856,370	100%

females

breast	266,120	30%
lung and bronchus	112,350	13%
colon and rectum	64,640	7%
uterine corpus	63,230	7%
thyroid	40,900	5%
melanoma of the skin	36,120	4%
non-Hodgkin lymphoma	32,950	4%
pancreas	26,240	3%
leukemia	25,270	3%
kidney and renal pelvis	22,660	3%
ALL SITES	878,980	100%

estimated deaths

males

lung and bronchus	83,550	26%
prostate	29,430	9%
colon and rectum	27,390	8%
pancreas	23,020	7%
liver and intrahepatic bile duct	20,540	6%
leukemia	14,270	4%
esophagus	12,850	4%
urinary bladder	12,520	4%
non-Hodgkin lymphoma	11,510	4%
kidney and renal pelvis	10,010	3%
ALL SITES	323,630	100%

females

lung and bronchus	70,500	25%
breast	40,920	14%
colon and rectum	23,240	8%
pancreas	21,310	7%
ovary	14,070	5%
uterine corpus	11,350	4%
leukemia	10,100	4%
liver and intrahepatic bile duct	9,660	3%
non-Hodgkin lymphoma	8,400	3%
brain and other nervous system	7,340	3%
ALL SITES	286,010	100%

Figure 1.1. The ten leading types of cancer with estimated diagnoses and deaths by gender, United States, 2018. (Adapted from R.L. Siegel, K.D. Miller and A. Jemal, *CA Cancer J. Clin.* 68:7–30, 2018. With permission from John Wiley & Sons.)

Cancer facts and figures

Cancer is a significant medical burden in the United States, with one in four deaths attributed to cancer. Over the course of a lifetime, the probability of being diagnosed with invasive cancer is 40% for American men and 38% for American women. Over 1.7 million new cases of cancer, known as the **incidence**, are estimated to be diagnosed in 2018, with over 600,000 deaths. Cancers of the prostate, breast, lung, and colon/rectum are the top four types of cancer diagnosed, and of all the types of cancer, they are responsible for the most cancer-related deaths of men and women in the United States (Figure 1.1). Globally, the annual incidence of cancer has been estimated at 14 million new cases as of 2012, with over 8.2 million cancer-related deaths. The leading causes of cancer death worldwide are lung, liver, stomach, colon/rectum, breast, and esophageal cancers. Important differences—including genetic predisposition, endemic infections (such as viral hepatitis), and cultural habits (such as smoking and diet), among many others—may explain the discrepancy among leading causes of cancer incidence and deaths in the United States compared with the rest of the world.

Although the annual incidence of cancer has climbed, death rates for men and women in the United States have fallen since a peak in the 1990s. The increase in incidence is largely attributed to more cancers detected by screening, with the spike in male incidence in the early 1990s due to the introduction of the prostate-specific antigen (PSA) blood test for prostate cancer. Other screenings include colonoscopy for colorectal cancer and mammography for breast cancer. Although cancer occurs in all age groups,

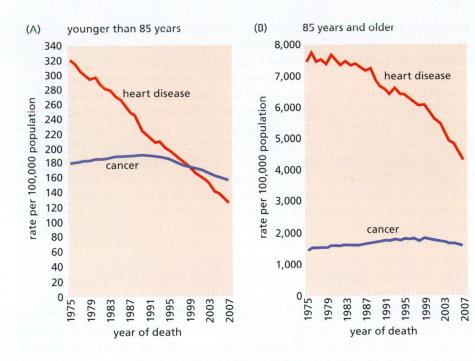

Figure 1.2. Death rates per 100,000 people from heart disease (red curve) and cancer (blue curve) per year in the U.S. population under age 85 **(A)** and 85 years and above **(B)**. (Adapted from R.L. Siegel, D. Naishadham and J. Ahmedin, *CA Cancer J. Clin.* 63:11–30, 2013. With permission from John Wiley & Sons.)

the incidence rises with age; for the segment of the population under 85, cancer is now the leading cause of death, as death rates from heart disease have fallen steadily due to improvements in treating that disease (Figure 1.2). The death rate from heart disease for patients over 85 has similarly diminished, although heart disease remains the number one cause of death for this segment of the population, with cancer coming in at a distant second. Because people are surviving longer despite heart disease and other co-morbid illnesses, they are more apt to develop cancer. Thus, innovation and improvements in cancer treatments could not come at a more opportune time.

Clinical oncology

Clinical oncology, the care of patients with cancer, is a constantly evolving art. Care of the cancer patient begins with diagnosis and then moves to staging, prognostication, and selection and administration of the optimal treatment and necessary supportive care. Specific treatment choices are informed by outcomes of clinical trials, which detail the benefits and toxicities of specific therapies in a given cancer. A chosen treatment is then tailored to the individual patient based on the patient's disease characteristics and overall health. Care of the patient with cancer is a coordinated, multidisciplinary approach, involving, as might be expected, medical oncologists, radiation oncologists, and surgical oncologists, with the collaborative and supportive care of nurses. In addition, a variety of other specialists, such as genetic counselors, nutritionists, and social workers, make essential contributions to the overall landscape of clinical oncology.

Clinical oncology remains heavily dependent on surgery, chemotherapy, and radiation—the traditional triad of cancer treatments. However, the translation of basic research from laboratory benches to patient bedsides and doctors' offices has added new and powerful drugs to the treatment of cancer. There is considerable optimism that these drugs, based on a detailed molecular and cellular understanding of the tumors they target, are the vanguard of many more cancer drugs, some already in clinical trials and others yet to be discovered.

Beginning in the 1970s and rapidly accelerating in the decades since, the growth in the understanding of cancer at the molecular and cellular levels has been the basis of insights that have led to more precise diagnostic strategies and to the development and deployment of more focused and effective cancer therapies. We now understand that cancer is a genetic disease, affecting a relatively small number of genes, and ultimately traceable to **"driver" mutations**—known as such because they confer a selective growth advantage for cancer cells. The progressive acquisition of a small set of these driver mutations by a normal cell, a multi-step process taking many years, transforms it into a cancer cell that develops into one of the myriad types of cancer, many of which are discussed in the chapters that follow. Comprehensive genome-wide sequencing has compared cancer cells, including those of the breast, lung, prostate, and colon and a variety of leukemias and lymphomas, with their normal counterparts. These studies have shown that cancer cells typically contain many mutations, scores in some, hundreds in others, and even thousands in a few types of cancer. Strikingly, any specific type of cancer cell possesses only a few driver mutations, typically between two and eight. Careful search of the thousands of distinct mutations observed in genomic studies of cancer has identified only a relatively small number (~150) of driver mutations. The other mutations do not impact the physiology of the cell and are referred to as **"passenger" mutations**. The practice of contemporary oncology requires knowledge of the constantly evolving therapeutic treatment options and increasingly requires integration of the emerging molecular and cellular insights from cancer biology with diagnosis and management of the patient.

Despite the broad diversity of cancers confronted in clinical oncology, it is now possible to identify key attributes, or hallmarks, that provide a set of principles for understanding neoplastic disease. These hallmarks collaborate to enable the uncontrolled growth and metastatic spread that sets cancer cells apart from their normal counterparts, and they play a major role in causing the pathology that clinical oncology strives to treat.

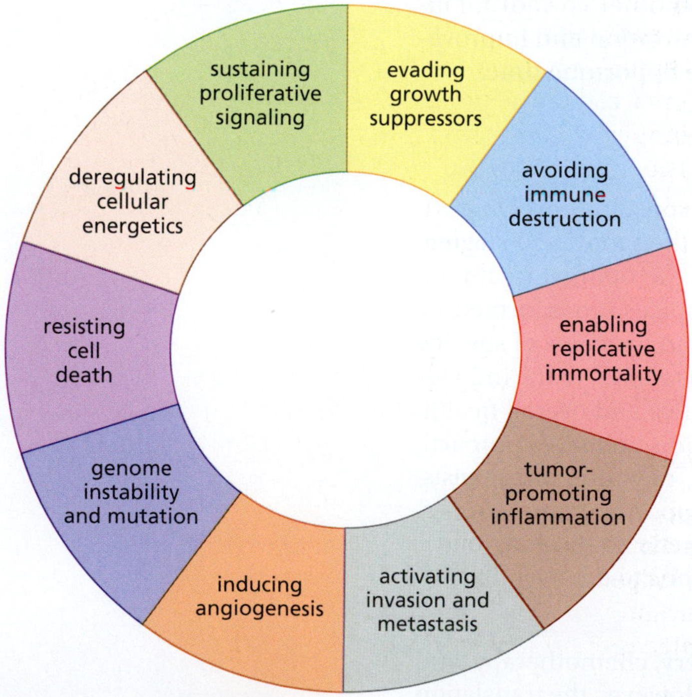

The hallmarks of cancer

Cancer includes an enormous diversity of neoplastic diseases, including the liquid tumors such as leukemia and the enormous variety of solid tumors that appear as cancers of the epithelium, connective tissue, bone, and brain. The variety of diseases classified as cancer is vast, and pathologists recognize hundreds of distinct types and subtypes of cancers. Despite a bewildering diversity of manifestations, Douglas Hanahan and Robert Weinberg have identified a set of capabilities they dubbed the "hallmarks of cancer" that underlie and enable cancer growth. The hallmarks of cancer, grouped together with an auxiliary set of emerging hallmarks they identified, are illustrated in Figure 1.3 and described below.

Figure 1.3. The hallmarks of cancer. The figure provides a summary of the core attributes and capabilities manifested by cancers. (Adapted from D. Hanahan and R.A. Weinberg, *Cell* 144(5):646–674, 2011. With permission from Elsevier.)

1. *Sustaining proliferative signaling.* Cancer cells divide chronically, inappropriately, and without limit—if the required nutrients are available. This contrasts with the limited and tightly regulated division of normal cells. Cancer

cells employ any one or a combination of many different strategies to sustain proliferative signaling, such as producing growth factors, upregulating the receptors for growth ligands that stimulate proliferative signaling, modifying proliferative signaling pathways to signal constitutively, and making cell division independent of growth factor ligands and their receptors.

2. *Resisting cell death*. Normal cells have an intact signaling circuitry that can be activated to induce their death, most commonly by **apoptosis**. Life or death is determined by external and intrinsic death signaling pathways counterbalanced by pathways promoting survival. Damage sensors, with the **tumor suppressor** p53 preeminent among them, can trigger apoptotic pathways while anti-apoptotic proteins, such as Bcl-2 and some of its relatives, oppose this programmed cell death.

3. *Evading growth suppressors*. Normal cells are endowed with a variety of molecules collectively called tumor suppressors that suppress cell proliferation. For example, the retinoblastoma protein (pRB) presides over signaling circuitry that restricts the entry of cells into the cell cycle. Cancer cells must lose, disable, or circumvent the action of tumor suppressors that would otherwise limit proliferation and survival.

4. *Avoiding immune destruction*. Immune responses can inhibit tumor growth or even eliminate tumors. A number of strategies including inhibition of immune responses and the loss of tumor antigens are employed by cancer cells to evade immune responses.

5. *Enabling replicative immortality*. **Telomeres**, multiple tandem hexanucleotide repeats, cap the ends of chromosomes and are essential for the maintenance of chromosomal integrity. Telomeres shorten during chromosomal replication and cell division. Telomere integrity is maintained by the enzyme **telomerase**, which is normally present in stem cells but is not expressed in most normal cells. Consequently, the absence of telomerase means that after a limited number of cell divisions, normal cells have greatly reduced telomeres, eventually suffer irreversible chromosomal damage, and die. Cancer cells can increase telomerase expression, preventing telomere shortening and cell death.

6. *Tumor-promoting inflammation*. Cancers often display an inflammatory response that supports tumor growth and is mediated by cells of the immune system, a paradox in light of the tumor-inhibiting action of hallmark number 4 above. The tumor-promoting effects of inflammation include production of growth factors and survival factors, pro-angiogenic factors, and extracellular matrix (ECM) modifying molecules, especially certain proteases that facilitate invasion and metastasis.

7. *Activating invasion and metastasis*. Primary tumors not only disseminate cancer cells to the immediate local environment but distribute them through blood and lymph to distant organs. Once lodged in the distant environment, some of these emigrant cancer cells surmount the challenges of adaptation to the new environment and proliferate, establishing metastatic foci of the primary tumor.

8. *Inducing angiogenesis*. Although diffusion can accommodate the needs of small tumors (<1 mm in diameter) to import oxygen and nutrients and export carbon dioxide and metabolic wastes, growth beyond the diameter of a millimeter or so requires the acquisition of a network of blood vessels, a vasculature. Cancers activate a developmental program known as the **angiogenic switch** to continuously grow and expand a vasculature

network as the tumor grows. Activation of the angiogenic switch is complex and involves increased production of proteins, such as VEGF-A (vascular endothelial growth factor A), that stimulate vascular growth, and decreased expression of inhibitors, such as the protein thrombospondin-1 (TSP-1), that discourage vascular expansion.

9. *Genome instability and mutation.* A cell's acquisition of mutations that confer a growth or survival advantage can enable its differential growth. The multi-step progression to cancer involves the successive accumulation of mutations that enable cell proliferation and disable growth suppressors. Consequently, factors that increase genome instability will facilitate this multi-step progression to cancer. Genomic instability can also allow cancer cells to continually acquire additional growth-promoting properties and to develop resistance to therapies and treatment.

10. *Deregulating cellular energetics.* Although energy generation via glycolysis is inherently less efficient than ATP production by mitochondria-based oxygen-dependent oxidative phosphorylation, cancer cells typically adjust their metabolism to increase the fraction of energy generated by glycolysis. Cancer cells that feature increased glucose utilization via glycolysis produce lactate and other glycolytic intermediates that can be diverted to biosynthetic activities, such as the production of amino acids and nucleosides necessary for the synthesis of the macromolecules necessary for cell growth and proliferation. In some cases, cancer cells further adapt to their needs for biosynthetic intermediates by increasing their metabolism of glutamine.

The tumor microenvironment

Detailed cytologic analysis of tumors reveals that they are composed of both cancer cells and several types of noncancer cells, coexisting in a complex and dynamically diverse community of cells known as the **tumor microenvironment** (TME). The TME is essential for a cancer's growth and maintenance. The cellular inventory of the TME includes endothelial cells, pericytes, cancer-associated fibroblasts (CAFs), and various immune cells (Figure 1.4).

Figure 1.4. The tumor microenvironment (TME). The TME is shown as a tissue- or organ-like community of cancer cells and many other cell types. Some of these support the maintenance, growth, survival, and, in some cases, metastatic spread of cancer cells.

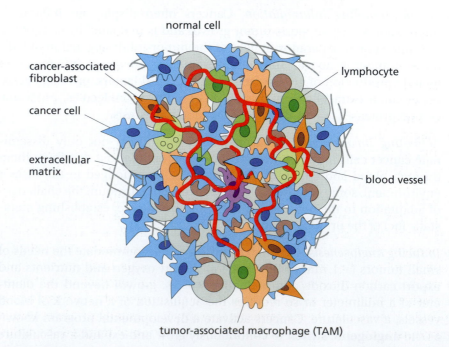

Each type of noncancer cell impacts the growth, proliferation, and survival of the resident cancer cells. Endothelial cells are major components of arteries and veins and compose the tumor vasculature essential for tumor growth and maintenance. Pericytes are related to smooth muscle cells and wrap the vasculature in fingerlike projections, buttressing it against the hydrostatic pressure of blood flow. CAFs enhance cell proliferation and angiogenesis and secrete a variety of extracellular matrix components. Cells of the immune system include macrophages, neutrophils, immature myeloid cells, NK cells, T cells, and other types of immune cells. Some of these immune cells—including an immunosuppressive subtype of macrophage, immune suppressing regulatory T cells (T_{reg}s), and immature myeloid cells—aid tumor growth. On the other hand, cytotoxic T (T_C) cells, NK cells, and a subtype of macrophage that supports cytotoxic immune responses inhibit tumor growth by killing cancer cells. The key cell type in the TME is the cancer cell. Although cancer cells are dependent upon the many types of noncancer cells that populate the TME, they are the ultimate instigators and sustainers of this vital tumor niche.

The hallmarks of cancer and the growing understanding of the indispensable supporting roles played by the TME reveal that an elaborate and dynamic network of processes, some intrinsic to cancer cells and others resident in normal cells, interact to produce malignancy. While daunting in their still-unfolding complexity, these many pathways and cells provide an abundance of opportunities for mounting mechanism-based therapeutic attacks. The massive efforts to design and develop pharmaceutical agents that target the mechanisms underlying cancer maintenance, growth, proliferation, and survival have met with increasing success, and several successful examples of molecularly targeted therapies (or simply **targeted therapies**) are already approved and in clinical use. Many more can be expected during the coming years. Targeted therapies zero in on the molecular mechanisms exploited by cancer cells or key cell types of the TME and selectively destroy or change the behavior of those particular cells—distinguishing them from the traditional cytotoxic chemotherapies, which indiscriminately kill rapidly growing cells, whether normal or diseased. It is the indiscriminate nature of conventional cytotoxic chemotherapy that leads to its commonly seen side effects, such as hair loss, skin sensitivity, shedding of the gastrointestinal lining, and suppression of blood-forming cells in the bone marrow.

Sequencing studies have shown that even among cancers of the same type, significant genetic differences exist, as well as a high level of mutational heterogeneity, explaining why some cancers of the same type respond to a targeted therapy while others may harbor genetic alterations that make them insensitive to the drug. Even a targeted therapy that yields initial and perhaps dramatic clinical responses is likely to eventually become ineffective due to the emergence in the tumor of resistant cancer cells—clones of the original cancer cell but with genetic alterations that bypass the targeted therapy's mechanism of action. The genetic heterogeneity that develops within a cancer provides an increasingly diverse repertoire of mutations from which variants resistant to even initially highly effective target therapies arise. Furthermore, under the selective pressure of a therapy effectively targeting a given hallmark, cells within the tumor heavily dependent on that hallmark may switch their dependence to another. A cancer cell's ability to circumvent mechanisms while under attack make it unlikely that an individual patient's cancer will succumb to a therapy exclusively targeting a single hallmark. Therefore, even when using a highly specific, mechanism-based targeted therapy, it is usually prudent, and in many instances necessary, to deploy a combination of targeted therapies, enabling the simultaneous attack on multiple hallmarks. However, the key elements in the design of a therapeutic program are the diagnosis, workup, and staging of a patient's cancer.

Diagnosis, workup, and staging

The diagnosis of cancer may be prompted by symptoms described by a patient or by screening tests in otherwise asymptomatic patients. Occasionally, cancers may be detected incidentally in laboratory or radiology tests performed for unrelated reasons. Given the heterogeneity of cancers, there is no standard workup for all malignancies. The diagnostic workup and evidence for a benefit to screening will be discussed in the following chapters pertaining to specific diseases.

For all cancers, regardless of whether they are of solid organ or hematologic origin, a tissue sample is critical to establish the diagnosis. Tissue samples are often acquired via surgical resection or needle biopsy, depending on the location of the tissue. The refinement of imaging tests and equipment has allowed for scan-guided (computed tomography [CT] or ultrasound) biopsies to be safely performed, even in deep locations that were previously difficult to access with minimally invasive techniques. The features of malignant tissue under the microscope (for example, abnormal appearance of cells, such as nuclear atypia or an unusual nucleus/cytoplasm ratio) in conjunction with immunohistochemical stains for specific proteins are usually sufficient for a pathologist to conclusively diagnose the majority of cancers. Detection of molecular alterations in cancer cells is a more recent addition to the diagnostic armamentarium and is playing an increasingly important role in both diagnosis and treatment selection.

Following a diagnosis of cancer, patients undergo a staging workup. Staging is the process of determining exactly where in the body cancer can be detected, information that is critical to determine prognosis and optimal therapy. Staging relies primarily on imaging studies but may also include tests of the peripheral blood or biopsies of the bone marrow in select diseases. The optimal staging workup considers the underlying cancer because of known propensities for patterns of spread in a given disease. Imaging modalities most commonly used in staging workups are described in Figure 1.5 and Table 1.1.

Role of cancer screening

Cancer screening tests must be proven in large studies of healthy patients to improve the survival of those individuals who receive the diagnosis. General guiding principles of screening tests for cancer include the following:

- The screening test must be for a disease that is common and associated with a high **morbidity** or mortality.

- A treatment must be available that reduces the morbidity and the mortality rates for the disease the screening test is designed to detect.

- The screening tests must be safe, ethically acceptable to society, and relatively inexpensive.

Examples of effective screening tests include mammography for breast cancer, Papanicolaou (Pap) smears for cervical cancer, and colonoscopy for colorectal cancer, each of which satisfies all three guiding principles for screening tests. Screening tests for other cancers are controversial. For example, the value of

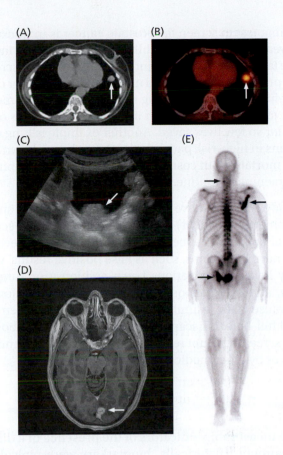

Figure 1.5. Imaging modalities for staging. Arrows depict tumors. (A) CT scan of lungs indicates a tumor in the lung periphery. (B) PET scan of the same tumor fused with the CT scan shows increased metabolic activity within the tumor. (C) Ultrasound of the bladder shows a tumor along the wall; urine in the bladder is dark, and the solid mass is gray. (D) MRI shows a brain tumor. (E) Bone scan shows areas of active bone metabolism where prostate cancer cells have spread.

Table 1.1. Imaging modalities in common use for staging cancer.

Imaging modality	Energy type	Description
Computed tomography (CT)	X-rays	Combines X-rays from many angles and creates cross-sectional images with computer assistance. The CT scan is much more powerful than plain X-ray films. Tissues are distinguished by density. Intravenous or oral dyes (known as contrast) can be administered before imaging to highlight tissue for better contrast in the CT scan.
Positron-emission tomography (PET)	Radioactive tracer	Uses a tracer, typically ^{18}F-labeled fluorodeoxyglucose (FDG), which is taken up preferentially by cancer cells; often combined with CT scans (as a "fused PET/CT scan") to localize the radioactivity to the abnormal area on standard CT images. FDG is a glucose analog. Normal tissues that also avidly take up FDG include brain, heart, and kidneys. Inflammatory reactions or infections will also take up FDG and thus can hinder interpretation.
Ultrasound (US)	Ultrasound waves	Also referred to as sonography; uses high-frequency sound waves that pass through the body and bounce off tissues and organs with "echoes" transmitted as visual images on a screen; useful to distinguish solid masses from cystic (fluid-filled) lesions.
Magnetic resonance imaging (MRI)	Nuclear magnetic resonance	A powerful magnetic field that causes the protons (hydrogen nuclei) of water molecules within the body to become aligned in the direction of the field. Activation of a radio-frequency current produces an electromagnetic field that flips the spins of the protons. When the current is turned off, the spins "relax" and become realigned; this relaxation generates a radio-frequency signal that can be captured. Protons in different tissues relax at different rates, and this can be used to construct anatomic images.
Bone scan	Radioactive ^{99m}Tc (technetium-99m)-methylene diphosphonate (MDP)	A tracer injected intravenously that localizes to areas of bone turnover. Cancers that cause bone formation in addition to destruction, e.g., prostate cancer, will show increased uptake; while those that are exclusively destructive, or lytic, e.g., multiple myeloma, will not.

the PSA screening test to screen for prostate cancer is in question despite its more than 20-year history, because while effective at identifying prostate cancer, at least one large clinical trial has indicated that the PSA screening test has not altered the prostate cancer-related mortality rate. Another clinical trial did indicate a benefit to the PSA screening test, but nearly 1000 men would need to be screened to save one life, which in the view of some is overtesting and overtreatment for so few lives saved. Another example is lung cancer screening with CT scans in high-risk patients, which has been shown to reduce lung cancer-related mortality, but cost considerations have prohibited implementation of the screening as routine.

Routine full-body CT scans have been proposed for detection of subclinical (not yet symptomatic or evident) disease. While appealing because they offer the possibility of detecting cancer that is very early in its development, full-body CT scans have limitations, such as false positive results, which can lead to unnecessary invasive evaluations (such as biopsy or surgery), needless patient anxiety, and overdiagnosis of clinically indolent (slow progressing) conditions. Because of these limitations, and a lack of data demonstrating the effectiveness of full-body CT scans in detecting cancer, full-body imaging for general cancer screening is not recommended by the Food and Drug Administration (FDA) or by other medical organizations.

Biomarkers

Biomarkers are measurable indicators of the presence of a disease or of the effects of a treatment. Historically, biomarkers were typically physiological indicators, such as heart rate or blood pressure. A biomarker in clinical oncology typically refers to proteins detected in peripheral blood that are expressed at higher levels by cancerous compared with normal cells. However, other biomarkers in clinical oncology refer to proteins on tumor cells, circulating tumor cells, antibodies, nucleic acids circulating in peripheral blood, and proteins in urine, saliva, or cerebrospinal fluid.

To be clinically useful, biomarkers must be readily obtained, rapidly processed, and accurate. For screening tests, a biomarker in the abnormal range should have a high rate of detecting disease, whereas the same biomarker in the normal range should have a low chance at missing a diagnosis (false negative rate). The determination of the threshold of normal versus abnormal range is therefore an important one. In the case of PSA screening described earlier, the key issue is accuracy, because benign prostatic hypertrophy (enlargement of the prostate, a common benign problem in aging men) can raise the PSA above the threshold that is considered normal (often resulting in a false positive reading), and because some aggressive prostate cancers express little PSA and therefore the serum PSA is well within the "normal" range.

By contrast with screening, some molecular biomarkers can be very useful after a diagnosis, during treatment. As stated, PSA is not a reliable screening test in identifying prostate cancer, but it can be very useful when prostate cancer has been diagnosed, as the test is a relative measure of the burden of the disease during treatment. In the modern era of targeted therapies in clinical oncology, **predictive biomarkers** can be identified in subpopulations of cancer patients and thereby indicate who among those patients is most likely to respond to a specific targeted therapy. Such biomarkers may be the very targets of the therapy. For example, the biomarker human epidermal growth factor 2 (HER2), which is overexpressed in breast cancers, predicts response to the anti-HER2 antibody trastuzumab (see Case 11-3 in Chapter 11).

Table 1.2. Phases in drug development.

Phase	Population tested	Purpose
Pre-clinical	Cell lines (*in vitro*) or animal studies (typically two species)	Demonstrate a drug's cytotoxicity, effective concentrations, and safety (including carcinogenic effects)
I	Human subjects, typically advanced cancers (unselected)	Screen a drug for safety in humans; establish a maximal tolerable dose; identify side effects; perform pharmacodynamic and pharmacokinetic studies
II	Human subjects, typically from a selected cancer type	Evaluate effectiveness and safety of a drug in a larger group of subjects
III	Human subjects, from a selected cancer type with similar clinical features	Evaluate effectiveness of the drug compared with standard-of-care treatments (if available) or placebo; typically a randomized, controlled trial
IV	Human subjects, from a selected cancer type with similar clinical features	Evaluate further the optimal use of the drug as well as safety; occurs after FDA approval of the drug (also referred to as a postmarketing or postapproval study)

Development of cancer therapies

Cancer therapies undergo rigorous trial testing, as summarized in Table 1.2. To be considered for FDA approval, new therapies typically must pass through each phase of a trial and ultimately show a benefit in survival, disease control, or symptom control, with tolerable side effect profiles, for patients in Phase III trials.

The effects of standard cytotoxic chemotherapy drugs (agents) are often initially evaluated in patients who have been previously treated with standard treatment regimens but whose disease has progressed. Thus, it is likely that the patient's cancer cells have evolved resistance to the prior agents and, consequently, the disease is more difficult to treat. To test the effects of targeted therapy drugs, which are often used when standard cytotoxic chemotherapy agents fail, verification of a mutation in cancer cells may be required before a trial begins. Furthermore, pinpointing the mutation that is causing cancer cells to resist treatment helps to accelerate targeted therapy drug development.

The future of clinical oncology

The care of cancer patients continues to evolve. While our molecular and cellular understanding of the disease has accelerated with the advent of high-throughput technologies (and their shrinking price tags), the application of this knowledge to clinical care is only now poised to truly revolutionize the care of the cancer patient. The optimal combination of therapies targeting multiple hallmarks of the disease carries our best hope to control advanced cases. The aim of this book is to inspire a greater understanding of the nexus between science (cancer biology) and medicine (clinical oncology).

Take-home points

✓ Cancer remains a significant burden in the United States with lifetime incidence of 40% and 38% for men and women, respectively.

✓ The practice of oncology is evolving. Pathology and clinical factors remain key in diagnosing cancer; however, molecular testing is playing an increasingly important role in cancer diagnoses.

✓ Staging, the determination of extent of disease, is integral in the prognosis and the selection of an optimal treatment for a cancer patient.

✓ Although therapies with defined molecular "targets" (targeted therapies) are in development, traditional nonselective cytotoxic chemotherapies, surgery, and radiation will be the mainstay components of treatment for the foreseeable future.

Chapter discussion questions

1. Why has cancer overtaken heart disease as the number one cause of death in Americans under 85 years of age?

2. Why might targeting just one hallmark of cancer be a faulty strategy to control metastatic cancer?

3. Identify and discuss the distinctions between cancer cells and the tumors in which they reside.

4. Explain why some cancer therapies target noncancer cells.

5. Why is screening for breast cancer, colon cancer, and cervical cancer a routine part of primary care recommendations, but not prostate cancer?

Selected references

DeVita VT, Lawrence TS & Rosenberg SA (2015) DeVita, Hellman, and Rosenberg's Cancer: Principles & Practice of Oncology, 10th Ed. Philadelphia: Wolters Kluwer.

Hanahan D & Weinberg RA (2011) Hallmarks of cancer: the next generation. *Cell* 144:646–674.

Siegel RL, Miller KD & Jemal A. (2018) Cancer statistics, 2018. *CA Cancer J. Clin.* 68:7–30.

Siegel R, Ward E, Brawley O & Jemal A (2011) Cancer statistics, 2011. *CA Cancer J. Clin.* 61:212–236.

Steward BW & Wild CP (eds) (2014) World Cancer Report 2014. Lyon, France: International Agency for Research on Cancer.

Vogelstein B, Papadopoulos N, Velculescu VE et al. (2013) Cancer genome landscapes. *Science* 339:1546–1558.

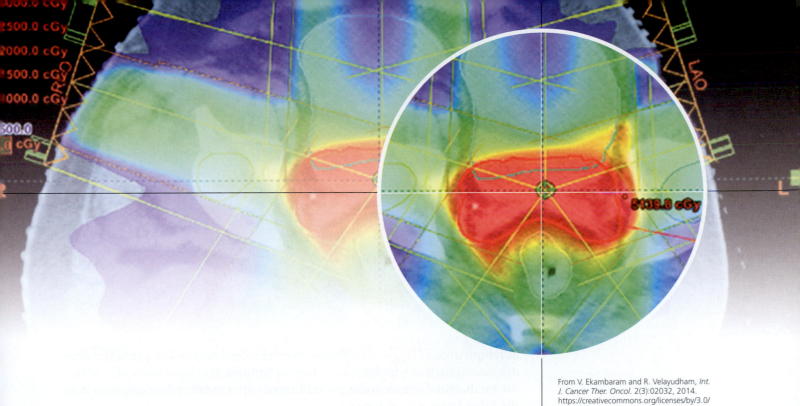

Chapter 2

Introduction to Cancer Therapy

Introduction

Cancer treatment frequently requires a multidisciplinary approach, involving specialists in surgical oncology, radiation oncology, and medical oncology. Each discipline has seen significant improvements in cancer treatments and cancer patient outcomes over the past few decades. In medical oncology, the recent genomic revolution has spurred a greater understanding of cancer biology and laid the groundwork for molecularly targeted therapies to treat cancer.

Surgery

Surgery, or what is also referred to as **resection**, is often the initial procedure performed in treating cancer. Broadly, the objectives of surgery in cancer patients can be diagnostic, curative, reconstructive, preventative, debulking, or palliative. When a patient presents with a suspicious lesion, surgery often involves no more than a small excision to obtain a tissue sample for diagnosis (see Chapter 1). The term **primary tumor** describes the anatomical site where the cancer cells originated; **secondary tumors** indicate other organs where the cancer has spread, or metastasized. Diagnostic surgical procedures can include biopsies of a presumed primary or secondary tumor, to establish a tissue diagnosis and possibly to define whether the cancer has spread. For example, when a mass is suspected in or between the lungs, a mediastinoscopy procedure, which employs a small camera (mediastinoscope) inserted into a small incision in the chest, allows for direct examination of the mediastinum (middle of the chest cavity) as well as biopsy of suspicious sites such

13

Figure 2.1. An example of a diagnostic procedure: mediastinoscopy.
A mediastinoscope is a thin tube with an affixed camera that is inserted into the chest through an incision above the breastbone on the left side of the chest. The camera allows for direct visualization of the structures in the middle of the lung cavity. Biopsies may be taken from suspicious- or normal-appearing lesions that can yield a diagnosis of a primary or secondary tumor.

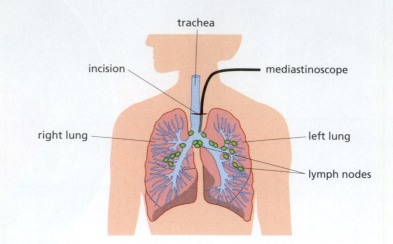

as lymph nodes (Figure 2.1). This procedure might thus yield a primary tumor diagnosis (such as lymphoma, a cancer of lymphocytes) or might indicate that the mediastinal lymph nodes are infiltrated with cancer cells originating from the lungs (a secondary tumor).

Curative surgical procedures represent the majority of cancer operations. In the case of diagnosed cancer, when a malignant tumor is known to be arising from a solid organ, an oncological surgeon is frequently consulted to ascertain whether resection of the primary tumor is necessary and beneficial. Depending on the extent of the diseased tissue and the involved organ, cancer surgery requires either excising the tumor along with a "rim" of surrounding normal tissue but leaving the rest of the normal organ intact, or cutting out the entire organ wherein the tumor resides, with or without removal of nearby tissue.

Figure 2.2. An example of draining lymph nodes: the breast.
Lymph normally flows from breast tissue along lymphatic ducts toward the regional lymph nodes. Cancer cells that have the capacity to spread may be found in these lymph nodes at the time of primary tumor resection. The axillary nodes are the most common site for early spread of breast cancer. (From A.T. Skarin, Dana-Farber Cancer Institute Atlas of Diagnostic Oncology, 4th ed. Mosby Elsevier, 2010. With permission from Elsevier.)

Local structures may be resected along with the primary tumor for cancers that either have a known propensity to spread or have imaging characteristics that are suspicious for early spread. For example, **lymph nodes**, which consist of B lymphocytes and T lymphocytes enclosed in fibrous capsules, are commonly resected during primary tumor surgery. The lymphatic system captures and returns plasma that has been lost to surrounding tissues through capillary leakage back to the bloodstream (Figure 2.2). The lymphatic vessels channel the fluid, termed lymph, through successive lymph nodes before emptying the lymph into the subclavian vein, a blood vessel lying beneath the collar bone, and into the general blood circulation. Lymph nodes also play a critical role as the site where lymphocytes encounter antigens under circumstances likely to generate immune responses. Because the lymphatic system drains fluid from all major organs into local lymph nodes, cancer cells that metastasize from their original location may be detected first in the regional lymph nodes draining the primary organ. Further, since the lymphatic system runs parallel to the blood circulatory system, metastasizing cancer cells can travel along lymphatic channels and grow in lymph nodes quite distant from the regional "lymphatic drainage basin."

Surgical approaches to cancer have evolved over time, leading to improved cancer control, decreased complication rates, or both. Breast cancer and rectal cancer surgeries exemplify the evolution of surgical oncology over the past decades. In the case of breast cancer, radical mastectomy was first performed by Dr. William Halsted in 1882 and

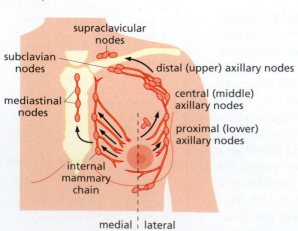

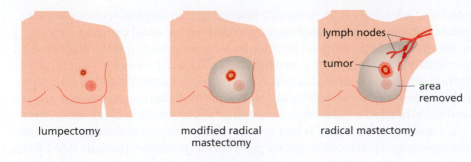

lumpectomy   modified radical   radical mastectomy
                    mastectomy

Figure 2.3. Evolution of breast cancer surgical management. Radical mastectomy was the original standard procedure, and included removal of the breast, underlying chest muscles, and lymph nodes from the axilla, or armpit. The modified radical mastectomy involves removal of the breast and axillary lymph nodes, but not the muscles. A lumpectomy, also termed "partial mastectomy" and "breast-conserving surgery," removes only the tumor and a rim of normal surrounding tissue. An evaluation of sentinel lymph nodes is also routinely performed.

included removal of the breast, underlying chest muscles (pectoralis major and minor), and lymph nodes from the axilla (armpit) (Figure 2.3). The modified radical mastectomy involves removal of the breast and axillary lymph nodes, but not the muscles. A lumpectomy, also termed "partial mastectomy" and "breast-conserving surgery," removes only the tumor and a rim of normal surrounding tissue and is appropriate for small tumors that can be completely excised. An evaluation of sentinel lymph nodes is also routinely performed (see Cases 6-1 and 4-4, and Figure 4.26). Appropriate limited surgery with or without additional treatment has decreased complication rates from surgery but has not negatively affected outcomes. Because of the potentially deformative outcomes of breast cancer surgery and consequent body-image issues that breast cancer patients may face, the ability to perform limited surgeries without compromise in cancer cure rates has had a significant positive impact for the long-term care (also called **survivorship**) in this population. **Reconstructive surgery** is also frequently employed to restore the cosmetic appearance of the breasts to a more normal state, using other tissues such as muscle or other implants, with improvement in the psychological aspects of coping with body changes related to cancer treatment (see Case 6-1 for more about surgical management of localized breast cancer).

In contrast to the visible differences seen during the evolution of breast cancer surgery techniques, rectal cancer surgery has continued to use traditional incisions but has evolved in terms of dissection technique (Figure 2.4). Total mesorectal excision (TME) is associated with improved cancer control, less local recurrence (which is often painful), and improved overall survival. Thus, these improvements are less obvious to patients, family, and caregivers, but clearly critical for cancer care.

Figure 2.4. Evolution of rectal cancer therapy: total mesorectal excision (TME). Curative removal of a rectal tumor requires wide excision around the tumor, including removal of the mesorectum as well as regional lymph nodes. The mesorectum is a part of the peritoneum, a membrane that attaches the abdominal organs to the body wall. Whereas prior surgeries used "blunt" dissection (using fingers or blunt instruments) to separate the mesorectum, the TME uses sharp, meticulous dissection (denoted by dashed lines) and is associated with significant improvements in local control of disease and survival rates. The TME became a standard of care in the 1990s. Later improvements for rectal cancer care outcomes included perioperative chemotherapy and radiation. (From A.T. Skarin, Dana-Farber Cancer Institute Atlas of Diagnostic Oncology, 4th ed. Philadelphia: Elsevier Science Ltd., 2003.)

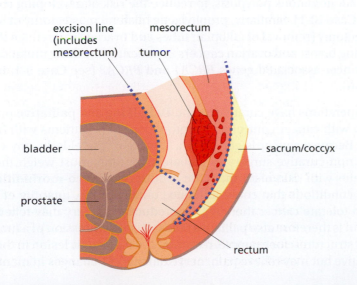

Modern technology is often rapidly integrated into surgical techniques. Powerful optics and improved software have allowed for small cameras ("scopes") to be integrated into surgery; techniques include mediastinoscopy (described previously) and laparoscopy, in which the camera (laparoscope) offers visualization into the abdominal cavity. A laparoscope is constructed of fiber-optic cable with one end connecting to a telescopic or digital lens and video camera and the other end to a halogen or xenon gas light source and television monitor. Laparoscopic surgery is less invasive than conventional surgery. In laparoscopic surgery, the oncologic surgeon cuts a "porthole" (usually 0.5–1.5 cm in diameter) at a strategic area on the body relatively distant from the location of the cancer and feeds the laparoscope through the porthole. Using the camera to guide other instruments inserted through other portholes, the surgeon holds the instruments outside the abdominal cavity and performs essentially the same operation as the traditional "open" procedure. A further innovation on laparoscopic surgery includes the use of robotic assistance, in which the laparoscope and other instruments are held by a robot that is controlled remotely by the surgeon using a console elsewhere in the operating room. An advantage of laparoscopic (or robotic-assisted laparoscopic) surgery over conventional surgery is the faster recovery time and potentially less blood loss. A disadvantage, especially when using the robot, is higher cost. It should be noted that most comparisons of open versus laparoscopic surgeries, such as prostate removal for prostate cancer, have not revealed an improvement in cancer control or cure rates (see Chapter 6 and Figure 6.11 for robotic-assisted laparoscopic prostatectomy for prostate cancer). Thus, technology and technical advances for cancer care do not uniformly lead to improved cancer control.

Prophylactic, or preventative, surgery is performed in some patients who have a high likelihood of developing cancer due to an inherited mutant gene. In these cases, a family member has typically been diagnosed with cancer (the "**index patient**"), and genetic testing of the index patient has been performed in consideration of young age at diagnosis, a spectrum of other cancers in the index patient or close relatives that could indicate an inherited predisposition for malignancy, or pathologic features of the cancer under the microscope that might suggest a genetic link. If a mutation is found, close family members may then be advised to pursue similar genetic testing, and if they have a matching mutation, they may face an important decision about prophylactic surgery. For example, prophylactic colectomy (removal of the entire colon) may be strongly considered for patients with *APC* mutations, associated with familial adenomatous polyposis, to reduce the risk of developing colon cancer (see Case 10-1). Similarly, prophylactic bilateral mastectomy or salpingo-oophorectomy (removal of fallopian tubes and ovaries) can reduce the risk of developing breast and ovarian cancers in patients inheriting mutations in the breast cancer-associated genes *BRCA1* and *BRCA2* (see Case 9-1, including Figure 9.6).

Not all operations have curative intent. **Debulking** and **palliative** operations can help with cancer control or symptom control in patients with incurable disease. Because cancer is generally a disease of the elderly, any consideration of non-curative surgery to palliate symptoms must weigh the potential benefits with risks of harm, in the context of other **co-morbidities** (other medical conditions that could influence quality of life, quantity of life, and ability to tolerate cancer therapy). Cytoreductive surgery may relieve symptoms (and is therefore also palliative) caused by compression of a tumor mass on other structures; for example, resection of a metastatic lesion in the brain is not curative but may relieve pain, nerve damage, or changes in mental status (confusion, cognition). In ovarian cancer, cytoreductive surgery is used to minimize metastatic disease in the abdomen and pelvis prior to chemotherapy. In select diseases such as kidney cancer and melanoma, cytoreduction can occasionally cause stability or even regression of other metastatic sites (see Case 5-1), though this is considered the exception and not the rule. Other palliative procedures can include diversion of intestine in a patient with a bowel obstruction that has resulted from a growing primary or secondary tumor. The diversion can help with the pain typically associated with a bowel obstruction, which can improve quality but not quantity of life.

For some cancers with substantial risk of recurrence, clinical trials have established a role for perioperative treatment, which refers to additional therapy near the time of surgery to decrease the risk of recurrence and increase the chance for cure. When given after surgery, perioperative treatment is referred to as **adjuvant therapy**, whereas before surgery it is known as **neo-adjuvant therapy**. Radiation and chemotherapy—either or both—are the cornerstones of perioperative treatment, with a potentially growing role for molecularly targeted therapies in adjuvant treatment.

Radiation therapy

Radiation in medicine was first described in 1895 in Germany by Dr. Wilhelm Röntgen, who dubbed a newly discovered type of electromagnetic radiation "X-rays." This finding led to uses of X-rays for diagnostic purposes, and in 1901, Dr. Röntgen received the Nobel Prize in Physics for this discovery. The 1898 discovery of radium by the Polish-born French physicist and chemist Marie Curie eventually led to the discovery that radiation could *cure* certain cancers, but could also *cause* cancers. Advances in radiation delivery, understanding of radiation sensitivity of different cancer types, and combining radiation with computer technology have led to improved precision and improved cancer outcomes using radiation therapy. Despite these advances, there is still a small risk that modern radiation therapy may cause treatment-related cancers, termed **secondary malignancies**.

The objectives of radiation therapy in cancer patients can be curative or palliative. Curative radiation treatments take advantage of the fact that some, but not all, cancer types are more sensitive to radiation than their normal tissue counterparts. This discrepancy underlies the basis for radiation as a potent "organ-sparing" treatment for diseases such as head-and-neck cancers, bladder cancers, and gynecologic malignancies, among others. Some cancers, such as kidney cancer, are not very sensitive to radiation, and thus surgical resection remains the key modality for local control. The molecular basis for radiation sensitivity (or insensitivity) is not well delineated. For patients with metastatic (incurable) cancer, radiation can be used with palliative intent, to destroy cancer cells at a specific metastatic site that are causing symptoms, most commonly pain.

Modern radiation approaches can be largely categorized into three groups: **external beam radiation therapy** (EBRT), **brachytherapy** (insertion of a radioactive implant to deliver radiation from within an organ), and intravenous **radionuclides** (an internal radiation therapy in which radiopharmaceuticals, or radioactive drugs, migrate to cancer cells from the circulation). EBRT, as its name implies, uses radiation, typically X-rays (categorized as photons), electrons, or protons from external sources. The radiation beam is aimed at the targeted tumor, hoping to spare normal surrounding tissue.

During the early days of radiation, using X-rays without the benefit of computer-aided targeting, nearby healthy tissue also received a significant dose of radiation. Contemporary radiation therapy has increased precision because of **computed tomography** (CT) scans. CT scans offer advanced imaging of a cancer, are largely available in hospitals, and do not cost as much as other advanced imaging technologies. **Conformal radiation therapy** (CRT) uses CT scans to map the cancer in three dimensions; X-ray beams are delivered from multiple angles matching the shape of the tumor.

Even greater than the precision of CRT is that of **intensity-modulated radiation therapy** (IMRT), because IMRT allows for adjustment in the strength of the X-ray beams, which improves "sculpting" of radiation dose (Figure 2.5). The precision of IMRT makes it possible to deliver higher doses of radiation to the targeted cancerous tissue or organ with less collateral damage to normal surrounding structures. The use of IMRT as a cancer treatment has led to improved cure rates and fewer side effects. IMRT is now the standard form of EBRT in the United States for prostate cancer, breast cancer, lung cancer, and others.

Proton beam therapy is an alternative form of EBRT that uses protons as opposed to photons to deliver energy. Proton beam therapy has proven advantages for certain malignancies (for example, pediatric brain tumors and some malignancies of the eye), because of the precise energy delivery of proton beams that causes only limited damage to surrounding tissues; it has proven to be advantageous for tumors that are relatively superficial to the body surface. While it is attractive to conclude that the precision of proton beam delivery may be superior to conventional IMRT for the treatment of deeper structures like prostate cancer, it is important to note that a prospective, randomized trial comparing these techniques has not been published. Hence, as in the comparison of curative outcomes between robotic-assisted laparoscopic prostatectomy and traditional open prostatectomy for prostate cancer, the expensive therapy with theoretical advantages (proton beam) has not conclusively been proven to be an improvement over conventional IMRT.

Specialized EBRT techniques called **stereotactic radiosurgery** (SRS) have further refined the delivery of highly focused beams, resulting in high-energy radiation targeted to very small volumes. Despite its name, SRS does not involve actual surgery. SRS brands include CyberKnife and Gamma Knife, with treatment applications ranging from benign neurologic conditions (such as trigeminal neuralgia, a chronic pain condition affecting the fifth cranial nerve) to malignancies such as primary tumors (prostate, breast, lung, among others) and brain metastases.

Figure 2.5. Evolution of prostate cancer radiation treatment: intensity-modulated radiation therapy (IMRT). (A) 3-dimensional conformal radiation therapy (3D CRT). Radiation is delivered from multiple angles to target the prostate and a rim of normal tissue (target is outlined and labeled); however, the dose area receiving the highest dose of radiation (*solid shaded red area*) includes the rectum (*outlined*) and bladder (*outlined*). (B) IMRT. By comparison, the solid shaded red area is more limited, allowing for less "collateral damage" from radiation hitting the nearby structures. This technology improvement that can "sculpt" radiation is associated with decreased side effects without compromising cancer control. (From V. Ekambaram and R. Velayudham, *Int. J. Cancer Ther. Oncol.* 2(3):02032, 2014. https://creativecommons.org/licenses /by/3.0/)

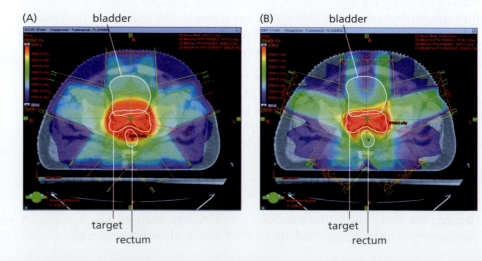

In brachytherapy, radioactive materials, such as titanium pellets impregnated with a radiation source, are inserted into the cancerous organ, delivering radiation internally. Brachytherapy is appropriate for select cancer patients with prostate, cervical, and endometrial cancers, sarcomas, and head-and-neck cancers, among others. For example, ^{125}I (iodine-125) seeds are used in prostate cancer, delivering a continuous rate of radiation to the normal and cancerous cells in the prostate (see Figure 6.12B). The half-life of iodine-125 is 60 days, meaning that the potency of radiation is diminished over 87% by 6 months and over 98% by 12 months. Because the radioactivity decays, the seeds become inert and are not removed.

Radionuclides are radioactive drugs that are administered intravenously and therefore travel in the circulation. Typically, these drugs migrate to a target site and, like brachytherapy seeds, deliver cytotoxic doses of radiation to nearby cancer cells as well as normal cells. Commonly used radionuclides include the β-emitting samarium (^{135}Sm) and strontium (^{89}Sr) isotopes, which are used for palliation of pain from bone metastases primarily in breast and prostate cancer. In contrast with this palliative (but not survival) benefit, the α-emitting radium isotope (^{223}Ra), which resides in the same column in the periodic table as calcium, homes to bone lesions that take up calcium, and is proven to prolong lives in men with bone-metastatic prostate cancer.

Medical cancer therapies

Medical cancer therapies are comprised of six general classes, found in Table 2.1. The term *chemotherapy* refers to treatment of a disease with a chemical substance, and thus the meaning of the term encompasses the use of any medication for any disease. In the context of cancer (and for the rest of this book), the term generally refers to conventional, or cytotoxic, chemotherapy. Other medical treatments include hormone therapy, therapeutic monoclonal antibodies, vaccines, cytokines, and targeted therapies. These classes are described in the chapters and cases that follow.

Cytotoxic chemotherapy drugs work by killing rapidly dividing cells. The identification of chemotherapy for cancer treatment dates back to World War II, when the U.S.S. John Harvey was bombed by German aircraft off the coast of Bari, Italy. The ship was carrying a secret cargo of 2000 bombs, which contained weaponized mustard gas, so named due to the odor. When the

Table 2.1. Categories of medical cancer therapies.

Category	General mechanism of action	Examples
Cytotoxic chemotherapy	Targets dividing cells and induces cell death	Cisplatin, cyclophosphamide
Hormone therapy	Diminishes hormone levels or hormonal signaling activity in hormone-sensitive cancers	Tamoxifen, aromatase inhibitors, GnRH agonists
Monoclonal antibodies	Antibody targets a specific protein and results in cell death, blockade of receptor–ligand interactions essential to the tumor, or other interference with other processes the tumor depends upon	Rituximab, trastuzumab
Vaccine therapy	Vaccine stimulates immune system to detect and remove cancer cells or prevents the induction of cancer by certain infectious agents	Sipuleucel-T, human papillomavirus quadrivalent vaccine
Cytokine therapy	Immunomodulatory activity, possible antiproliferative activity	Interferon, interleukin-2
Targeted therapy	Small molecules that inhibit a specific molecular target that is considered fundamental to cancer cell survival	Imatinib, erlotinib

bombs were exploded and mustard gas was released, sailors and civilians in the immediate vicinity suffered severe skin and lung injury, and many of these victims ultimately died. During preparation for World War II, researchers at Yale School of Medicine were recruited by the U.S. government to find an antidote to the poisonous mustard gas. Drs. Louis S. Goodman and Alfred Gilman evaluated the medical records of soldiers who had been exposed to mustard gas in World War I and discovered that many of them had unusually low numbers of lymphocytes, a subtype of white blood cells, as well as a decrease in the size of their lymph nodes and spleens. Because lymphocytes in lymph nodes and the spleen replicate faster than most cells in other organs, Goodman and Gilman hypothesized that a compound in the gas, nitrogen mustard, must attack rapidly dividing cells. Armed with the belief that their hypothesis could prove valid, they conducted human trials to study the potential of intravenously administering nitrogen mustard in patients with lymphoma, a cancer of lymphocytes. Regression of disease was documented. Nitrogen mustard, or mechlorethamine, became the first non-hormonal drug with significant antitumor activity. The success of mechlorethamine led to efforts to identify or create other chemical agents with antitumor activity.

Mechlorethamine falls under one of nine categories of chemotherapy drugs, called *alkylating agents*. The other categories are anthracyclines, anti-tumor antibiotics, antifolates, antimetabolites, platinum analogs, taxanes, topoisomerase inhibitors, and vinca alkaloids. A chemotherapy drug, or agent, is categorized according to its mechanism of action. (Table 2.2) In general, chemotherapy agents work by damaging, interfering with, or inhibiting replication of DNA or disrupting cell division to the point that cancer cells cannot survive. An in-depth discussion on the categories of chemotherapy agents is provided in Chapter 3, Cytotoxic Chemotherapy.

Table 2.2. Types of chemotherapy agents.

Class	Mechanism of action	Examples
Alkylating agents	Damage DNA by covalent binding of alkyl groups, preventing cell division and leading to cell death	Mechlorethamine Cyclophosphamide Dacarbazine
Anthracyclines	Interfere with DNA replication through a poorly defined mechanism, thought to occur via intercalation into DNA and displacement of DNA-binding proteins, or by causing DNA breaks	Daunorubicin Doxorubicin
Anti-tumor antibiotics	Inhibit DNA replication by causing single- and double-strand DNA breaks (bleomycin) or cross-linked DNA strands (mitomycin-C)	Bleomycin Mitomycin-C
Antifolates	Interfere with production of purine nucleotides for synthesis of DNA, preventing cell division (a type of antimetabolite)	Methotrexate Pemetrexed
Antimetabolites	Substitute components normally used for inclusion in RNA or DNA, interfering with gene transcription or chromosomal replication and causing damage during S phase; chemical structures are often similar to nucleosides	5-Fluorouracil Cytarabine Gemcitabine Hydroxyurea
Platinum analogs	Form covalent intra-strand DNA cross-links, preventing cell division and leading to cell death	Cisplatin Carboplatin
Taxanes	Stabilize microtubules, preventing completion of mitosis, leading to cell death	Paclitaxel Docetaxel
Topoisomerase inhibitors	Interfere with topoisomerases, which separate strands of DNA for copying during S phase	Irinotecan Etoposide
Vinca alkaloids	Unlike taxanes, these drugs interfere with microtubule assembly during M phase, preventing cells from dividing	Vinblastine Vincristine

Table 2.3. Common routes of chemotherapy administration.

Route	Notes	Examples
Intravenous	Administered into a vein (most chemotherapy drugs are delivered this way)	Platinum compounds Taxanes
Oral	Pills, tablets, capsules, and liquid	Capecitabine Temozolomide
Intrathecal	Used to treat cancers that have reached the central nervous system; delivers drugs to the cerebrospinal fluid (CSF) that surrounds the brain and spinal cord; delivered via lumbar puncture (spinal tap, a needle inserted between spine bones in the lower back) or a device (Ommaya reservoir, placed surgically between the skin and the scalp that delivers chemotherapy into a ventricle, a normal cavity of the brain filled with CSF)	Methotrexate Cytarabine
Intraperitoneal	Delivered via a catheter through the skin and abdominal wall into the abdominal cavity, to directly bathe cancer cells that have spread into the abdominal cavity; used most often for ovarian cancer	Cisplatin Paclitaxel
Intra-arterial	Delivered into artery supplying the tumor with blood via a thin catheter, often with image guidance; used most often for liver tumors (primary liver cancer or metastatic disease in the liver)	Doxorubicin Irinotecan
Subcutaneous	Injection into the space just under the skin, above the muscle	Cytarabine
Intramuscular	Injection through the skin into muscle	Methotrexate Bleomycin
Intravesicular	Delivered into the bladder via a Foley catheter; for bladder cancer	Mitomycin C Thiotepa
Topical	Cream applied to the skin; for skin cancers	5-Fluorouracil
Implantable	A drug-infused wafer placed in the space where a brain tumor was removed	Carmustine

Many chemotherapy agents are administered by intravenous infusion, allowing for rapid distribution via the circulation. However, chemotherapy can be administered through a variety of delivery routes (Table 2.3). Oral administration of chemotherapy can achieve high concentrations of the drug in the blood, a phenomenon called **bioavailability**. Because of anatomic barriers, some chemotherapy agents can reach high concentrations only if they are delivered directly to the cancer or the cancerous organ. For example, because of the blood–brain barrier, a highly selective filter that restricts permeability of substances into the brain, chemotherapy for cancers of the central nervous system can require intrathecal administration, wherein an agent is injected directly into the cerebrospinal fluid which surrounds the brain and the spinal cord.

The application of only one chemotherapy agent is referred to as single-agent chemotherapy. When more than one agent is administered, the treatment is referred to as **combination chemotherapy**. Combination chemotherapy was first described in 1965 as a breakthrough treatment for children with acute lymphoblastic leukemia (ALL; see Case 12-2).

Cancer cells generally divide faster than normal cells. Certain normal cells, such as those in hair follicles, skin, the gastrointestinal tract, and developing blood cells (in the bone marrow), also divide frequently, which means that chemotherapy will kill these rapidly dividing normal cells as readily as it destroys cancer cells. This accounts for the commonly experienced side effects associated with chemotherapy treatment (Table 2.4). Side effects are often temporary, but they can be permanent and even cause death. Organ dysfunction and decreases in healthy blood cells are examples of potentially dangerous side effects. Careful selection of the chemotherapy agent to be administered, with consideration of a patient's overall health and other

Table 2.4. Potential side effects of cytotoxic chemotherapy.

Hair loss
Skin sensitivity
Nausea and vomiting
Mouth sores
Decrease in white blood cells (leukopenia) with increased risk of infections; specifically a decrease in the neutrophil subset (neutropenia) that increases risk of serious bacterial infections
Decrease in red blood cells (anemia) with potential fatigue, shortness of breath, and lightheadedness
Decrease in platelets (thrombocytopenia) with increased bleeding or bruising
Neuropathy (nerve dysfunction), such as numbness and tingling in extremities and ringing in the ears
Infection
Diarrhea
Allergic reactions
Kidney damage
Heart damage
Liver damage
Infertility
Another cancer

Note: This is not an exhaustive list.

medical problems, is therefore imperative, as is follow-up care and monitoring. Side effects differ among chemotherapy agents, and thus side effects can be compounded with combination therapy as opposed to a single-agent treatment.

The routine use of multi-drug combinations for specific cancers is based on cumulative results from sequential clinical trials that confirm iterative improvements in patient outcomes without compromising patient safety. One might assume that all cancer cells are susceptible to any chemotherapy agent, but this has not proven to be true in experimental models or in patients. Further, one might think that combining chemotherapy agents with differing mechanisms of action would better kill cancer cells than would the application of only a single chemotherapy agent, but many clinical trials conducted on various malignancies have demonstrated that cancers do not uniformly respond to combinations of chemotherapy agents. In fact, combining chemotherapy agents may exacerbate toxicities without an improvement in disease response.

A "cycle" of chemotherapy defines a duration of time that a particular regimen is given at a specific dose and schedule. Cycles differ among diseases and regimens. Typically, a treatment regimen calls for more than one cycle of chemotherapy. Thus, the number of cycles determines the total duration of chemotherapy as well as the total dose of any given chemotherapy agent, which holds implications for chemotherapy dose-related side effects. In cases where the patient does not respond to an initial regimen (termed **first-line chemotherapy**), patients may receive **salvage chemotherapy**, which is second-line therapy and beyond, and typically composed of different chemotherapy agents than those used initially.

Genomics and oncology

Cancer is a disease of uncontrolled cell growth caused by the accumulating dysfunction of genes due to a combination of inherent defects and environmental insults. Recent technical advances have placed **genomics**, the use of DNA sequences and RNA transcripts to determine the structure and function of genomes, among the most powerful tools available for the study of cancer. Increasingly finding their way into the practice of oncology, genomic approaches based on sequencing and assays of gene expression promise to become clinical tools of unusual versatility for the practice of oncology. Genomics is already in wide use for the identification of cancer patient populations bearing genetic alterations that are treatable with molecularly targeted therapies. Advances in technologies for rapid and economical nucleotide sequencing and the development of analytical and interpretive bioinformatic algorithms greatly facilitate the use of genomics for identification of targetable cancer-supporting alterations. These oncogenic alterations can be point mutations, insertions, deletions, amplifications, chromosome alterations, or other types of genetic lesions. Specific examples of the clinical use of genomics include:

- Detection of chromosomal translocations and rearrangements—for example, the 9;22 translocation producing the *BCR–ABL* fusion, a signature chromosomal aberration of chronic myeloid leukemia and an indication for treatment with the protein tyrosine kinase inhibitor imatinib mesylate (see Case 4-1).

- Detection of gene amplification—for example, amplification of the *HER-2/neu* gene in breast cancers, an indicator for treatment with the monoclonal antibody trastuzumab (see Case 11-3).

- Identification of driver mutations—for example, the determination that a melanoma harbors activating mutations in the *BRAF* gene, indicating treatment with the kinase inhibitor vemurafenib (see Case 4-4).

- The determination that a patient is unlikely to benefit from a particular drug. For example, colorectal cancer patients bearing certain mutations in *KRAS* do not respond well to monoclonal antibodies that target the EGF receptor. Such a finding avoids the use of an expensive and likely futile treatment.

- The detection of genomic alterations that predict a more favorable prognosis. For example, glioblastoma patients with certain mutations in the gene encoding IDH-1, a Krebs cycle enzyme, have a more favorable prognosis than those without such mutations.

The examples cited above illustrate the role genomic data now plays in guiding cancer diagnosis, prognosis, and therapy. Although cytogenetic approaches, sometimes augmented by *in situ* hybridization with labeled probes, are still employed, microarrays and nucleic acid sequencing have become the key genomic tools of oncology. All of these approaches are examined in the sections that follow.

Cytogenetic approaches

Chromosomal aberrations are common in cancer and take many forms. These include internal rearrangements involving breakage and rejoining, **translocation** (a segment of one chromosome is broken off and joins to

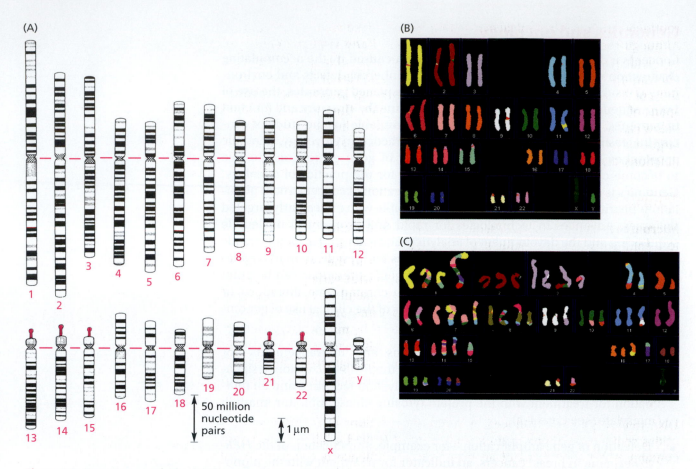

Figure 2.6. Karyotype analysis. (A) The normal human karyotype. The morphology and banding patterns of the 22 autosomes found in the cells of males and females and the sex chromosomes present in the cells of men (X and Y) or women (XX) are shown. Staining with appropriate dyes produces banding patterns distinctive for each type of chromosome. (B) Use of multiple fluorescent *in situ* hybridizations (mFISH) to distinguish each chromosome. Sets of fluorescent DNA probes can be constructed that hybridize across an entire chromosome, giving it a distinctive color. (C) Use of mFISH to detect chromosomal aberrations in a cancer cell. The karyotype of this pancreatic cancer cell shows many sorts of abnormalities including extra copies of some chromosomes, as well as breakage of pieces from one chromosome and their rejoining to another. (A, adapted from U. Francke, *Cytogenet. Cell Genet.* 31:24–32, 1981. B and C, courtesy of M. Grigorova, J.M. Staines and P.A.W. Edwards.)

another), deletion (a segment is lost), and **aneuploidy** (departures from the normal number of chromosomes). Many of these can be detected by analysis of a cell's **karyotype**, an image reflecting the number and structure of a cell's chromosomes. As shown in Figure 2.6A, conventional staining of the chromosomes reveals a pattern of banding that is distinctive for each of the 23 types of chromosome found in a human cell. However, the banding produced by conventional staining can detect only relatively large changes in chromosome structure and will not identify small regions of DNA gain or loss. Greater resolution and specificity can be gained by taking advantage of the ability of a sequence of DNA that is complementary to the sequence of a particular stretch of chromosomal DNA to hybridize to it. The direct hybridization of such complementary DNA probes to their chromosomal counterparts can be performed directly on the microscope slide (*in situ*) on which the chromosomes of the dividing cancer cell are displayed. Labeling of the DNA probe with a fluorescent dye allows the region of the chromosome to which it hybridizes to be easily detected. This method, dubbed **fluorescence *in situ* hybridization**, or FISH, is highly sensitive, and various modifications of it are routinely employed in clinical cytogenetic analysis (see Figures 2.6B and C). Although these cytogenetic approaches are useful and regularly employed in oncology, they are typically limited to the detection of large-scale genetic aberrations involving stretches of DNA of six megabases or more. However, most of the genetic changes of importance in oncology involve much shorter spans of DNA; many involve only a single base pair. The methods outlined below can detect changes in a single base pair and can be designed to detect large-scale genomic changes such as chromosome breaks, rearrangements, deletions, and amplifications.

Microarrays

Microarrays are collections of sequence-specific molecules that are attached to a solid substrate at specific sites. They can be used to determine a cancer's pattern of gene expression and then match it to the known profile of gene expression in a particular tissue. Along with a wide variety of diagnostic applications, in some cases microarrays might be useful for the detection of chromosomal defects, such as duplications, deletions, and rearrangements. Although some of these abnormalities can be seen by microscopy, microarrays provide better detection and finer resolution. Microarrays can be used to determine the tissue type in which a patient's cancer originated. This is a task that can be difficult to perform by histological examination but is readily done using microarrays.

DNA microarrays are produced by spotting collections of DNA oligonucleotides sequences (**oligos**) onto a solid support (called a "chip"). Each spot contains a few picomoles of an oligo whose sequence is complementary and specific for a particular gene or other segment of the genome. If DNA sequences that are complementary to the oligos arrayed on the chip are added, they will hybridize with those oligos on the chip. A variety of labeling strategies make it possible to both detect and quantify the hybridization of particular DNA sequences in a sample to complementary oligos attached to the solid support (Figure 2.7). Because a microarray can display over 10,000 distinct oligos, it can be used to study the expression or genetic variation of large numbers of genes simultaneously. The existence of very large libraries of oligos, some over 300,000, provides the ability to detect mutations in most regions of interest across the entire genome and allows rapid screening of samples for mutations known to be of interest for cancer diagnosis, prognosis, or the selection and monitoring of therapy.

Despite the greater precision and resolution of genomic approaches based on direct sequencing (discussed in the next section), microarrays offer many advantages, including being less expensive, less labor-intensive, faster, and easier to interpret. Additionally, the sample handling capacity of microarray approaches far exceeds that of sequencing. Large clinical trials or patient surveys can generate hundreds, even thousands, of samples that need to be analyzed over relatively short periods of time. Although such demands far exceed the capacity of sequencing approaches, they are readily accommodated by microarray-based genomic approaches. In addition, microarrays have been in clinical use for more than 15 years, and there is a background of familiarity with their strengths and limitations. The design of microarrays (for example, the sequences represented in the library of oligos used to construct the array) determines what genomic questions can be addressed and what data to expect. Consequently, microarrays are predictable. Paradoxically, although sequencing approaches have the potential to return novel and unexpected

(A) composition and layout of array

all of the oligonucleotides at a spot are identical to each other
and different from the oligonucleotides at another spot in the microarray

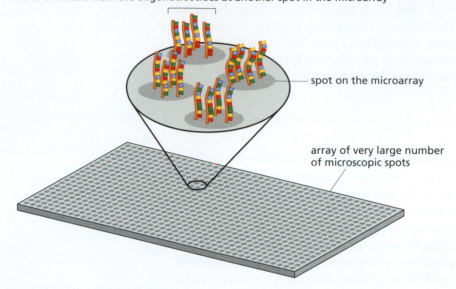

spot on the microarray

array of very large number
of microscopic spots

(B) microassays for mutated DNA

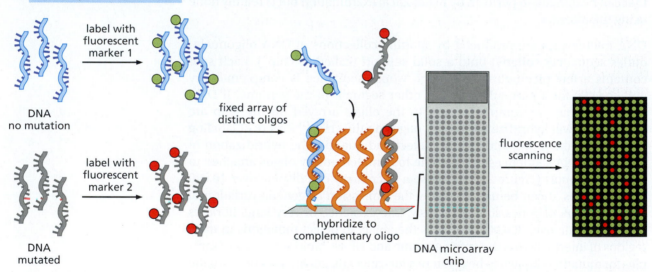

label with
fluorescent
marker 1

DNA
no mutation

fixed array of
distinct oligos

label with
fluorescent
marker 2

DNA
mutated

hybridize to
complementary oligo

DNA microarray
chip

fluorescence
scanning

Figure 2.7. Microarray analysis. (A) Composition and layout of an array. A solid surface is divided into a grid of many locations, and a microdrop containing a few picomoles of an oligonucleotide (oligo) of known sequence is spotted onto each known location. All of the oligos spotted at a given location are identical in sequence to each other, but different from the sequence of oligos spotted at all other locations on the microchip. In this way, very large arrays of distinct oligonucleotide sequences can be displayed on a chip the size of a microscope slide or smaller. (B) Microassays for mutated DNA. In this example, a microassay is made up from a library of oligos, some having the canonical sequence of the set of genes under examination and others bearing different sequences. The differences correspond to genetic alterations at various locations in selected segments of the genes under investigation. Addition of labeled unmutated DNA (green fluorescent marker in this example) and mutated DNA (red fluorescent marker) to the microarray results in fragments of these DNAs hybridizing with complementary oligos. The sequence of the oligos at locations fluorescing red allows one to detect and deduce the sequence of the mutated regions under examination.

results, this capacity to surprise is not welcome in clinical settings, where to be "actionable," laboratory tests must be capable of interpretation by an established and proven paradigm. Consequently, microarrays remain in wide use and will be only slowly displaced from some applications by sequencing technologies. It is likely that rather than an absolute displacement of one approach by the other, both technologies will continue to be deployed and microarrays will be used when the high resolution of sequencing is not required.

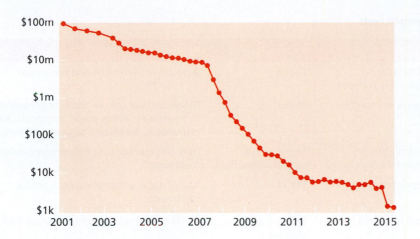

Figure 2.8. Falling costs of sequencing a human genome. The cost of sequencing a human genome fell steadily from 2001 until 2007, paralleling the drop in prices for computer chips. With the introduction of the set of technologies known as next generation sequencing (NGS), the decline became much more rapid and by 2015 approached $1000 under ideal conditions. (Data from the NHGRI Genome Sequencing Program, GSP.)

Sequencing

Sequencing is the definitive methodology of genomics, as it enables the determination of the precise sequence of nucleic acids for a given gene (or genome). For many years, DNA was sequenced by enzymatically mediated chain termination, an approach introduced by the British biochemist Frederick Sanger and his colleagues over 40 years ago and dubbed "Sanger sequencing." The first sequence of the human genome, a project that required more than 10 years and cost over $3 billion dollars, was performed by Sanger sequencing. Albeit the technical basis of a landmark achievement, time and cost made Sanger sequencing (first-generation sequencing) impractical for routine cancer diagnosis and the guidance of its therapy. The advent of new approaches to nucleic acid sequencing, collectively known as **next generation sequencing** (NGS), allows massive parallel sequencing of millions of fragments of DNA, making it possible to sequence an entire genome in one day at a cost that can be as low as around $1000 (Figure 2.8). The combination of speed and lowered costs afforded by NGS transformed nucleic acid sequencing from an expensive and labor-intensive research procedure into a tool that is increasingly used in the clinic. Sequencing has great potential for diagnostic applications, as an aid in the selection of cancer therapies, and as a biomarker for the effectiveness of some cancer therapies.

Next generation sequencing has made significant contributions to cancer biology, uncovering surprising levels of genetic diversity between and even within cancers. It has opened the possibility of classifying cancers according to genomic characteristics, enabling the clinical use of genomic features to guide the selection of therapy. Nevertheless, in spite of the power and promise of genomic technologies for the diagnosis of cancer, in almost all cases the diagnosis and staging of cancer still depends upon the microscopic examination of a biopsy or blood sample by a highly skilled pathologist. However, genomic markers are now included on the FDA's list of tests used for the diagnosis of cancer (Table 2.5), and the development of a variety of additional genomic-based tests can be expected. NGS-dependent genomic approaches are increasingly finding application in the identification of germ-line cancer risks and of somatically arising driver mutations in the genetic landscape of cancers. Among these somatic variants, one finds some that have prognostic value and others that can be used to guide therapy and monitor the course of drug therapies.

Identifying a patient's distinctive cancer genomic alterations often involves obtaining paired samples of normal and malignant tissue from the patient

Table 2.5. **Some FDA-approved cancer diagnostic tests.**

Tumor marker	Cancer type	Tissue analyzed	Application
ALK gene rearrangements	Non-small-cell lung cancer, anaplastic large cell lymphoma	Tumor	Determine treatment and prognosis
α fetoprotein (AFP)	Liver cancer, germ cell tumors	Blood	Diagnose liver cancer and follow response to treatment; assess stage, prognosis, and response to treatment of germ cell tumors
β-2-microglobulin	Multiple myeloma, chronic lymphocytic leukemia, lymphoma	Blood, urine, cerebrospinal fluid	Determine prognosis, follow response to treatment
β human chorionic gonadotropin (βhCG)	Choriocarcinoma, testicular cancer	Blood	Assess stage, prognosis, and response to treatment
BCR–ABL fusion gene	Chronic myeloid leukemia	Blood or bone marrow	Confirm diagnosis, monitor disease status
BRAF mutation V600E	Cutaneous melanoma, colorectal cancer	Tumor	Predict response to targeted therapies
CA15-3/CA27.29	Breast cancer	Blood	Assess effectiveness of treatment and recurrence
CA19-9	Pancreatic cancer, gall bladder cancer, bile duct cancer, gastric cancer	Blood	Assess effectiveness of treatment
CA-125	Ovarian cancer	Blood	Diagnose disease; assess response to treatment and recurrence
Calcitonin	Medullary thyroid cancer	Blood	Diagnose disease; assess response to treatment and recurrence
Carcinoembryonic antigen (CEA)	Colorectal cancer, breast cancer	Blood	Colorectal cancer metastasis; breast cancer recurrence and response to treatment
CD20	Non-Hodgkin lymphoma	Blood	Determine whether targeted therapy is appropriate
Chromogranin A	Neuroendocrine tumors	Blood	Diagnose disease; assess response to treatment and recurrence
Chromosomes 3, 7, and 9p21	Bladder cancer	Urine	Monitor tumor recurrence
Cologuard set	Colorectal cancer	Stool	Detection of colorectal cancer
Cytokeratin fragments 21-1	Lung cancer	Blood	Monitor tumor recurrence
EGFR mutation analysis	Non-small-cell lung cancer	Tumor	Determine treatment and prognosis
Estrogen receptor (ER)/ progesterone receptor (PR)	Breast cancer	Tumor	Determine whether treatment with hormonal therapy is appropriate
Fibrin/fibrinogen	Bladder cancer	Urine	Monitor progression and response to therapy
HE4	Ovarian cancer	Blood	Assess disease progression, monitor recurrence
HER2/neu	Breast cancer, gastric cancer, esophageal cancer	Tumor	Determine whether treatment with anti-HER2 therapy is appropriate
Immunoglobulins	Multiple myeloma, Waldenstrom's macroglobulinemia	Blood and urine	Diagnose disease, assess response to treatment, monitor recurrence
KIT	Gastrointestinal stromal tumor, mucosal melanoma	Tumor	Diagnose and determine treatment
KRAS mutation analysis	Colorectal cancer, non-small-cell lung cancer	Tumor	Determine whether targeted therapy is appropriate
Lactate dehydrogenase (LDH)	Germ cell tumors	Blood	Assess stage, prognosis, and response to treatment
Nuclear matrix protein 22	Bladder cancer	Urine	Monitor response to treatment
Prostate-specific antigen (PSA)	Prostate cancer	Blood	Diagnose disease, assess response to treatment, monitor recurrence

Tumor marker	Cancer type	Tissue analyzed	Application
Thyroglobulin	Thyroid cancer	Tumor and blood	Evaluate response to treatment, monitor recurrence
Urokinase plasminogen activator (uPa) and plasminogen activator inhibitor (PAI-1)	Breast cancer	Tumor	Determine aggressiveness of cancer, guide treatment
5-protein signature (OVA1)	Ovarian cancer	Blood	Preoperative assessment of pelvic mass for suspected ovarian cancer
21-gene signature (Oncotype DX)	Breast cancer	Tumor	Evaluate risk of recurrence
70-gene signature (MammaPrint)	Breast cancer	Tumor	Evaluate risk of recurrence
324 genes and selected gene rearrangements (FoundationOne CDx)	All solid cancers	Tumor	Identify patients who may benefit from one or another targeted therapy

Adapted from J.G. Vockley and J.E. Niederhuber, *BMJ* 350:h1832, 2015. With permission from BMJ Publishing Group Ltd.

(Figure 2.9). Fortunately, a sample of saliva provides an adequate source of normal tissue, making it unnecessary to obtain this essential reference tissue by invasive biopsy. Paired samples are essential to determine which genetic alterations are unique to the patient's cancer and which were present in all cells of the body, including cancer cells. The importance of paired-sample analysis was demonstrated in a comparative study showing that sequencing only cancer tissue yielded 31–65% false positives (genetic alterations that were present in, but not unique to, cancer tissue). In contrast, use of paired samples to correct for background genetic alterations present in normal cells allowed detection of most (95–99%) tumor-specific alterations with a specificity of greater than 99%. In definitive determinations of cancer-specific genetic alterations, paired tissue samples are subjected to one or more of the sequence-based analyses outlined below:

1. **Whole genome sequencing** (WGS). The genome is sequenced to provide a comprehensive comparison of genomic variation between normal and malignant tissue. WGS can identify point mutations; insertions and deletions (**indels**); re-arrangements and translocations (even when not detectable as chromosomal translocations/rearrangements by microscopy); and **copy number variations** (CNVs), the loss or gain in the number of copies of a gene. It can also detect the integration of viruses into the genome.

2. **Whole exome sequencing** (WES). The fraction of the genome (about 2%) that encodes proteins is referred to as the exome. The ability of WES to

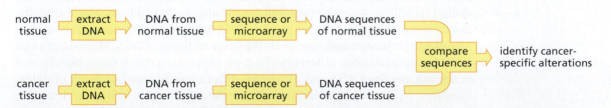

Figure 2.9. Paired-sample analysis for cancer-specific genetic alterations. Samples of DNA are extracted from cancer tissue and from normal tissue (an appropriately collected sample of saliva is often suitable) and subjected to genomic analysis using NGS or microarrays designed to detect particular types of genetic alterations. Comparison of the data allows detection of genetic alterations common to both normal and cancer tissue in addition to identifying alterations appearing only in the patient's cancer.

identify point mutations, insertions and deletions, and copy number variations is limited to their representation in the exome.

3. **RNA sequencing** (RNA-seq). Sequencing of the total RNA from a sample allows determination of the sequence of all of the RNA transcripts generated from the genome. It is performed by using reverse transcriptase to transcribe the collection of RNAs into corresponding **complementary DNAs** (cDNAs) and then using NGS to sequence the collection of cDNAs. RNA-seq detects mutations, changes in levels of gene expression, the occurrence of alternative or aberrant RNA splicing, and transcripts generated from gene fusions as a consequence of events such as chromosome translocations.

4. Identification of methylated cytosines. Chemical treatment of DNA with bisulfite converts nonmethylated cytosines to uracils, allowing one to discriminate them from methylated cytosines, an approach called **reduced representation bisulfite sequencing** (RRBS). Greater or lesser degrees of cytosine methylation in promoters can modulate the expression of genes, with greater degrees of promoter methylation being a modification that generally reduces the expression of the associated genes. Structural modifications of DNA, such as methylation, that do not involve changes in its base sequence are referred to as epigenetic rather than genetic changes. RRBS allows a comparison of patterns of **DNA methylation**-mediated epigenetic suppression of gene expression between normal and malignant cells. In addition to RRBS, other types of NGS-based epigenetic assays have been developed for specific research and diagnosic applications.

Some common genetic alterations and the genomic approaches suitable for their detection and characterization are summarized in Figure 2.10.

Each of these sequencing strategies has strengths and weaknesses. Whole genome sequencing is comprehensive and allows detection of genomic alterations throughout the genome. Unlike microarray approaches, where it is necessary to construct oligos representative of the alterations one desires to detect, WGS is unbiased and can discover and detect unanticipated alterations in cancer genomes. However, it is expensive and requires more time than other methods. Also, analysis of the large amount of data generated by WGS requires highly sophisticated bioinformatics. In contrast, by focusing the analysis on just the protein-encoding portion of the genome, WES can be conducted with considerable economies of time and labor, allowing higher throughput and lower costs. However, these advantages are gained at the expense of leaving more than 98% of the genome unexamined. An unbiased determination of differences in gene expression between cancer and normal cells can be obtained by RNA-seq, sequencing and comparing total RNA between cancer and normal cells. The isolation and processing of RNA and its rendering into cDNA for sequencing do require additional labor and skilled handling. Also, the possibilities for alternative splicing and transcription of noncoding DNA that encumber RNA-seq produce vast amounts of data resulting in bioinformatic demands that exceed those of WES. Microarrays can provide useful data on differences in gene expression much more rapidly and economically than RNA-seq. However, differences not within reach of determination by the oligos used to construct the microarrays will go undetected. In this regard, the unbiased approaches of WGS, WES, and, in the case of gene expression, RNA-seq are superior to more convenient but limited microarrays. Indeed these unbiased approaches enabled the discovery of significant and recurrent mutations in a wide variety of pathways, including those of cell division, regulation of chromatin structure, epigenetic phenomena,

Figure 2.10. Next generation sequencing (NGS) approaches for the analysis of genetic alterations. (A) A variety of genetic alterations found in genomic DNA. (B) Illustration of the effect of these alterations on the nature of the RNA transcript. (C) Listing of NGS approaches that can be used to detect and characterize the indicated genomic alteration. RNA-seq, RNA sequencing; RRBS, reduced representation bisulfite sequencing; SNP, single-nucleotide polymorphism; WES, whole exome sequencing; WGS, whole genome sequencing. (Adapted from S.S. Yadav et al., *Urol. Oncol.* 33:267.e1–e13, 2015. With permission from Elsevier.)

metabolism, DNA synthesis, and RNA processing, as well as several distinct ones involved in signal transduction.

NGS technologies

The decreasing costs of advanced sequencing technologies and development of interpretive bioinformatic algorithms have made it possible to sequence many thousands of cancer genomes. Large-scale cancer genome projects, such as **The Cancer Genome Atlas** (TCGA) and others, have revealed that different types of cancer vary greatly in the frequency of mutation. As shown in Figure 2.11, the highest numbers of mutation are found in cancers arising in tissues that have the highest exposure to environmental mutagens. Lung cancer (environmental mutagen: tobacco smoke) and melanoma (environmental mutagen: ultraviolet radiation in sunlight) are prominent in the category of highest mutation frequencies. In contrast, pediatric cancers, arising early in life and having a shorter span of time for accumulation of mutations, are at the lower end of the spectrum. The mutational landscape of the same type of cancer differs from patient to patient, sometimes showing similarities in driver mutations but, in some cases, displaying important differences. Surprisingly, even in the same patient, WGS has found differences in the mutational profiles of

Figure 2.11. Incidence of mutations in different types of cancer. The graph summarizes the range and median number of mutations found in a variety of types of cancer. Only nonsynonymous mutations, those capable of changing the amino acid sequences encoded by the affected gene, are tallied on the ordinate. EAC, esophageal adenocarcinoma; ESCC, esophageal squamous cell carcinoma; MSS, microsatellite stable; NSCLC, non-small-cell lung carcinoma; SCLC, small-cell lung carcinoma. (Adapted from B. Vogelstein et al., *Science* 339:1546–1558, 2013.)

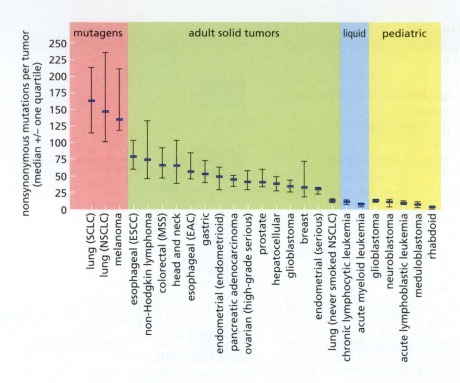

metastases and even in different parts of the same tumor mass. Though monoclonal in origin (that is, arising from a single cell), a cancer evolves over time to become a collection of different cancer cell populations that are to a greater or lesser degree genetically distinct from each other. This diversity has implications for treatment, since a targeted therapy that may be quite effective for combating the growth-promoting effects of one driver gene might be ineffective for a newly arising and different one.

Genomics

Diagnosis and staging are essential determinants of the therapy assigned to patients. Cancers are diagnosed and staged using conventional histologic examination of biopsy samples. However, there is increasing awareness that the use of genomics to identify the critical driver mutations supporting the growth and survival of a patient's cancer can be decisive in the selection of appropriate therapies. A large study examined the clinical outcomes in patients deemed eligible by genomic evaluation for treatment with an appropriate targeted therapy versus those not treated with the targeted agent, but who were retrospectively found to harbor the targeted mutation. The median overall survival in the group receiving the targeted therapy was 31.5 months compared with 9.6 months in the group that did not receive the drug. This study and others are making it apparent that for some cancers, when possible, patient workup should include both histological and genomic approaches.

Despite the extra time and expense, WGS and WES have an important advantage over tests that examine only a limited number of genes (targeted panels) or do not determine the complete sequence of the genes investigated (microarray approaches). Both WGS and WES make it is possible to examine variations across many biochemical and signaling pathways and identify genetic variations in pathways associated with cancer. This additional information can suggest new diagnostic approaches and provide insights for novel therapeutic approaches that simultaneously target multiple agents and pathways collaborating in the support of the cancer's growth and maintenance. WES is a less costly and more rapid alternative to WGS. It provides a way of comprehensively examining just the portion of the genome encoding exons. Similarly, RNA-seq can provide comprehensive information on genetic alterations of expressed genes as well as allowing qualitative and quantitative evaluations of gene expression, a considerable improvement over the capabilities of microarrays.

In those cases where it is possible to select a small subset of the exome known to be associated with particular cancers or groups of cancers, focusing these powerful sequencing technologies on a panel of genes comprising this subset has been quite useful. Use of appropriately selected gene panels allows exome sequencing to be carried out at great depth and much more quickly and economically than with WES or RNA-seq. The Solid Tumor Targeted Cancer Gene Panel by Next Generation Sequencing (CAPN) in use at the Mayo Clinic for solid tumors scans specific regions of a panel of 50 genes for genetic alterations relevant to the therapy of solid tumors. Some of the genes examined by CAPN are shown in Table 2.6. This panel detects alterations that are clinically actionable, helping to identify patients who might benefit from various targeted therapies. As the clinical cases in succeeding chapters will show, a growing number and variety of targeted therapies for the treatment of specific

Table 2.6. Representative members of the Mayo Clinic Solid Tumor Targeted Cancer Gene Panel.

Gene and encoded product	Function	Cancer association[a]
AKT1 Also known as protein kinase B (PKB)	A serine/threonine kinase associated with promotion of cell survival	Several including breast cancer, and lung cancer
APC Adenomatous polyposis coli	Regulation of β-catenin levels	Colorectal cancer
BRAF B-Raf proto-oncogene	A serine/threonine kinase of the MAP kinase pathway	Several including melanoma, lung cancer, and thyroid cancer
EGFR Epidermal growth factor (EGF) receptor	An EGF receptor with ligand-binding activated protein tyrosine kinase activity	Several including lung cancer, colorectal cancer, and cervical cancer
GNAQ Guanine nucleotide binding protein Q subunit alpha	Subunit of G-protein receptor that couples to receptor for phospholipase C	Spinal cord neoplasms
HRAS, KRAS, and NRAS Members of the RAS family of proto-oncogenes; each encodes a small G protein	Signal transducers from membrane receptors to intracellular signaling pathways such as the MAP kinase and PI3 kinase pathways	Many different types of cancer—mutations in RAS are among the most common in cancer.
IDH1 and IDH2 Isocitrate dehydrogenase 1 and 2	Isoforms of Krebs cycle enzymes that catalyze the oxidative decarboxylation of isocitrate	Gliomas and myeloid neoplasms
JAK2 and JAK3 Janus Kinase 1 and 2 are protein tyrosine kinases involved in cytokine receptor signaling	Nonreceptor protein tyrosine kinases that transduce signals from cytokine receptors via the JAK–STAT pathway	Stomach cancer and myeloproliferative disorders
MET Hepatocyte growth factor (HGF) receptor	Receptor for HGF with ligand binding activated protein tyrosine kinase activity	Several including hepatocellular carcinoma, lung cancer, and stomach cancer
TP53 Tumor protein p53	A 53 kDa protein that is a key tumor suppressor	A great many—TP53 is the most frequently altered gene in human cancers.

[a]In many cases there are associations with cancers in addition to those listed.

cancers are receiving FDA approval each year. Sequencing approaches such as CAPN are now being used to:

1. Scan multiple genes to identify patients with cancers that are likely to respond to a targeted therapy.

2. Detect mutations associated with response or resistance to particular therapies.

3. Identify mutations of prognostic value.

4. Aid in diagnosis.

Genomic assays of circulating tumor cells and tumor DNA

Cells and DNA of tumor origin are present in the bloodstream of cancer patients. However, isolation of **circulating tumor cells** (CTCs) or **circulating tumor DNA** (ctDNA) is challenging because they are present in small amounts. In 1 ml of blood, there are 10^6 white blood cells and 10^7 red blood cells but usually fewer than 10 CTCs, and of the 25 ng/ml of cell-free DNA found in blood, only 0.1% to 10% of circulating free DNA is ctDNA shed from the primary cancer and metastases. Despite the modest representation of CTCs and ctDNA in the bloodstream, recent technical advances enable their study. As illustrated in Figure 2.12, useful information on gene expression, the cancer exome, and even genomic rearrangements can be obtained from sequencing ctDNA or DNA extracted from CTCs. This approach also yields information on many types of cancer-specific genetic alterations, some of which can serve as markers for cancer detection that are specific and highly sensitive. Sequencing of ctDNA or DNA from CTCs provides a minimally invasive way of identifying driver genetic alterations—oncogenic point mutations, rearrangements, or amplifications—from a small sample, 5 to 10 ml, of patient

Figure 2.12. Genomic analysis of circulating tumor cells (CTCs) and circulating tumor DNA (ctDNA). Cancer cells and DNA can be obtained from samples of blood drawn from patients. The ctDNA can be analyzed by NGS approaches to determine the nature of genetic alterations present. NGS-derived sequence information and aspects of the cell biology of the CTCs can be derived from studies of cancer cells isolated from the bloodstream. In some instances, the CTCs can be cultured for further study.

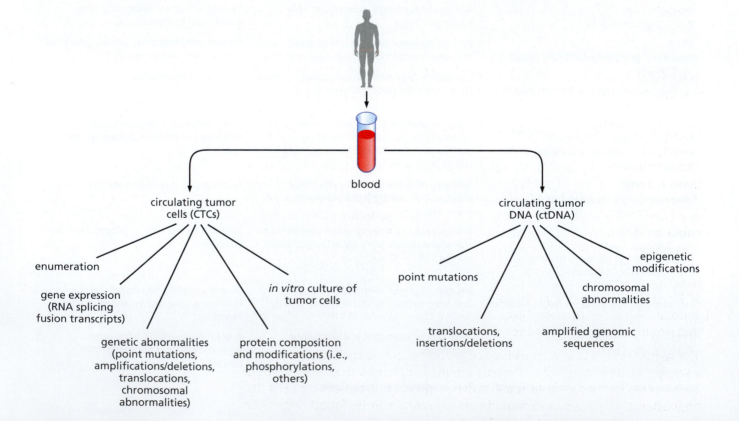

blood. Because it allows informative sampling of a patient's solid tumor by taking a blood sample instead of a surgically excised slice, this approach has been called **liquid biopsy**.

Pharmacogenomics and cancer therapy

Pharmacogenomics is the study of how the genome determines responses to drugs. Genetic variation affects individual responses, such as effective dose levels, metabolism, clearance, toxicities, and allergic reactions. An awareness of pharmacologically relevant genetic variation is an essential element of cancer drug design, development, and testing. For some cancer drugs, pharmacogenomic information is important for tailoring therapy to obtain optimum responses, and for other cancer drugs, such information explains how to avoid adverse responses to their administration. In the case of some cancer therapies, a combination of drugs is used. Some of these are anti-cancer drugs intended to kill or slow the growth of cancer cells, and other medications intended to impact normal tissue are given to manage side effects such as rashes, nausea, anemia, or infection. Genomically determined variations in drug metabolism by the members of the **cytochrome P450**-bearing group of enzymes, which play important roles in the metabolic alteration and detoxification of drugs, have important effects on the toxicity and dose levels of both anti-cancer drugs and drugs given to manage the side effects of treatment. In these conditions, the pharmacogenomics landscape can be quite complex because both somatic mutations in the tumor and host germ-line genetic variation affect the overall success of the treatment program. This can make it necessary to genomically evaluate both cancer tissue and normal tissue. Tests, some of which are FDA approved, are available for the evaluation of genomic variation, in particular genes associated with responses to a number of specific drugs used in the therapy and management of cancer. Use of WGS, WES, or RNA-seq and appropriately designed microarrays allows evaluation of a large and growing set of genes known to be of pharmacogenomic significance. Pharmacogenomic considerations are an essential part of drug therapy for cancer, and pharmacogenomic tests have a place in oncology alongside diagnostic and prognostic genomic tests.

Summary and future prospects

Genomic approaches have contributed greatly to studies of the biology of cancer. Aided by microarrays and especially the introduction of next generation sequencing technologies, a great deal has been learned. Application of these powerful genomic tools has enabled the discovery of new genes that drive cancer development, maintenance, and survival. Chromosomal aberrations detected by microscopy were the first class of genomic differences that could be definitively identified and associated with cancer. Because these aberrations—breaks, fusions, rearrangements, **hypodiploidy**, **hyperdiploidy**, and **microsatellite** instability—involve genomic changes, appropriate sequencing approaches and specifically designed microarrays can detect them at levels of resolution that far exceed those of microscopy. Consequently, genomic advances have enabled identification of chromosomal aberrations that would have otherwise gone undetected.

The sequencing of many thousands of genomes and exomes has uncovered a staggering level of genetic diversity among cancer genomes. Surprisingly, even within the same tumor mass, different populations of cancer cells distinguished by differences in mutational repertoire can be found. Ingenious mining of sequencing data has identified mutations in various steps of many cellular pathways, including those involved in signal transduction, cell cycle regulation, metabolism, chromatin remodeling, and others. Awareness of the steps in a pathway compromised by genetic alterations can direct or aid in the design and selection of therapies targeting the affected pathway.

Genomics is having a transformative effect on oncology. Cancer diagnosis, tumor classification, the choice of one therapy over another, and strategies for monitoring the course of treatment is increasingly impacted by genomics. For more than 100 years, microscopic examination of cytological and histological features of cells and tissues has been the basis for the diagnosis and classification of tumors. Now, there is growing appreciation of the usefulness of using both histological and genomic criteria for the precise classification of some tumors. The development and FDA approval of a DNA-based blood test for colorectal cancer encourages the expectation that similar tests will be devised for many other cancers. Studies by the Clinical Lung Cancer Genome Project (CLCGP), Network Genomic Medicine (NGM), and other institutions have demonstrated that genomic features are useful for classification and for the design of therapy programs. In fact, if a cancer of a given histology is found to be driven by an oncogene effectively treated by a targeted therapy for a histologically different cancer, the targeted agent may be employed in the different cancer based on the presence of a common genetic driver. As more oncogenic genetic alterations are successfully targeted with effective drugs, genomic criteria, not histological ones, increasingly will be the primary determinants of drug choices. Pharmacogenomic considerations, already in wide use in many areas of medicine, will steer drug choices toward those compatible with the patient's genetic constitution, increasing drug efficacy and reducing adverse events. This will have the economic benefit of avoiding the futile and often expensive use of drugs that the patient's genomic profile indicates will be ineffective.

Technological advances in DNA sequencing continue and will impact cancer genomics. Recently, the ability to sequence DNA from single cells made it possible to analyze the genome of CTCs or single cancer cells acquired directly from the tumor mass. Comparative studies of single cells sequentially obtained from the same cancer make it possible to monitor the evolution of its constituent cell populations and to conduct surveys to identify cells bearing genetic alterations that would make them resistant to specific chemotherapies. Single-cell analysis is especially well suited to studies of heterogeneity within tumors and for tracing the lineage of cells that migrate from the primary tumor and seed metastasis. Presently, the expense and technical challenges of single-cell sequencing prevent its general and widespread clinical use. However, sequencing technologies continue to evolve, and third generation sequencing, the probable successor to next generation sequencing (NGS), reads longer stretches of DNA and can make direct determinations of epigenetic modifications of DNA. Third generation sequencing includes **single-molecule sequencing** because it sequences an extensive stretch, thousands of bases, of a single molecule of DNA. Two important advantages of this approach are that it makes the bioinformatics of assembling the genome sequence simpler and requires smaller starting samples of DNA.

Take-home points

✓ Surgery, radiation therapy, and cytotoxic chemotherapy remain important pillars of cancer management.

✓ Integration of targeted therapies with standard treatments will likely be determined by sequential clinical trials to optimize efficacy.

✓ Genomics is the use of DNA sequences and RNA transcripts to determine the structure and function of genomes.

✓ Microarrays, next generation sequencing (NGS), and bioinformatics are the tools of genomics.

✓ Genomic approaches to the diagnosis, prognosis, and selection of targeted therapies are in wide and increasing clinical use.

✓ Although sequencing is the gold standard of genomics, microarrays are less expensive, less labor-intensive, and faster, and have far greater sample handling capacity than sequencing.

✓ Detection of cancer-specific genetic alterations often requires analysis of paired samples of a patient's normal and malignant tissue.

✓ Many sequencing options exist and include whole genome sequencing (WGS), whole exome sequencing (WES), RNA sequencing (RNA-seq), and reduced representation bisulfite sequencing (RRBS) for studies of DNA methylation.

✓ The decreasing costs and increasing speed of NGS has made it possible to sequence thousands of cancer genomes.

✓ Genomic studies of genetic alterations in cancer have shown that different types of cancer have different levels of mutation, mutational repertoires of the same type of cancer differ in different individuals, and cancer cell populations within the same cancer show variations in their mutational repertoire.

✓ Genomic analysis of circulating tumor DNA (ctDNA) or DNA from circulating tumor cells (CTCs) provide a "liquid biopsy" that can reveal the presence of cancer and shows promise as a means of evaluating the course of some therapies.

✓ Pharmacogenomics, the study of how the genome determines responses to drugs, is gaining wider use to guide the selection of drugs for cancer therapy and to reduce the incidence of adverse effects.

✓ Genomic analysis is finding a role as a complement to histology for tumor classification.

Chapter discussion questions

1. Whole genome sequencing provides the most comprehensive genomic information. Nevertheless, there are many instances when microarrays or limited sequencing of a selected panel of genes are the approaches chosen for clinical use. Why?

2. Microarrays can be designed to detect a variety of genetic alterations, including such chromosomal aberrations as the fusion of a piece of one chromosome with another. Assuming you know the exact sequence generated by a particular chromosome fusion, outline how you would generate a microarray that could detect it.

3. Suppose you have a sample of normal and cancerous tissue from a patient and want to determine the repertoire of tumor-specific genetic alterations present. Outline your approach, specifying the method you would use and why you would choose it over all of the others.

4. What might one be able to learn about a solid tumor from a sample of blood?

5. Genomic data can be a useful indicator for the selection of a drug in a patient. Can you outline situations where genomics provides contraindications for the use of a particular drug in a particular patient?

Selected references

Berger MF & Mardis ER (2018) The emerging clinical relevance of genomics in cancer medicine. *Nat. Rev. Clin. Oncol.* doi: 10.1038/s41571-018-0002-6, March 29.

Bunn PA Jr, Franklin W & Doebele RC (2013) The evolution of tumor classification: a role for genomics? *Cancer Cell* 24:693–694.

Chmielecki J & Meyerson M (2014) DNA sequencing of cancer: what have we learned? *Annu. Rev. Med.* 65:63–79.

Cohen JD, Li L, Wang Y et al. (2018) Detection and localization of surgically resectable cancers with a multi-analyte blood test. *Science* 359:926–930.

DeVita VT, Lawrence TS & Rosenberg SA (2015). DeVita, Hellman, and Rosenberg's Cancer: Principles & Practice of Oncology, 10th Ed. Philadelphia: Wolters Kluwer.

Ekambaram V & Velayduham R (2014) Analysis of low dose level volumes in intensity modulated radiotherapy and 3-D conformal radiotherapy. *Int. J. Cancer Ther. Oncol.* 2:02032.

Haber DA & Velculescu VE (2014) Blood-based analyses of cancer: circulating tumor cells and circulating tumor DNA. *Cancer Discov.* 4:650–661.

Jones S, Anagnostou V, Lytle K et al. (2015) Personalized genomic analyses for cancer mutation discovery and interpretation. *Sci. Transl. Med.* 7:283ra53.

Shen T, Pajaro-Van de Stadt SH, Yeat NC & Lin JC-H (2015) Clinical applications of next generation sequencing in cancer: from panel, to exomes, to genomes. *Front. Genet.* 6:215.

Vockley JG & Niederhuber JE (2015) Diagnosis and treatment of cancer using genomics. *BMJ* 350:h1832.

Vogelstein B, Papadopoulos N, Velculescu VE et al. (2013) Cancer genome landscapes. *Science* 339:1546–1558.

Wang L & Wheeler DA (2014) Genomic sequencing for cancer diagnosis and therapy. *Annu. Rev. Med.* 265:33–48.

Yadav SS, Li J, Lavery HJ et al. (2015) Next-generation sequencing technology in prostate cancer diagnosis, prognosis, and personalized treatment. *Urol. Oncol.* 33:267.e1–e13.

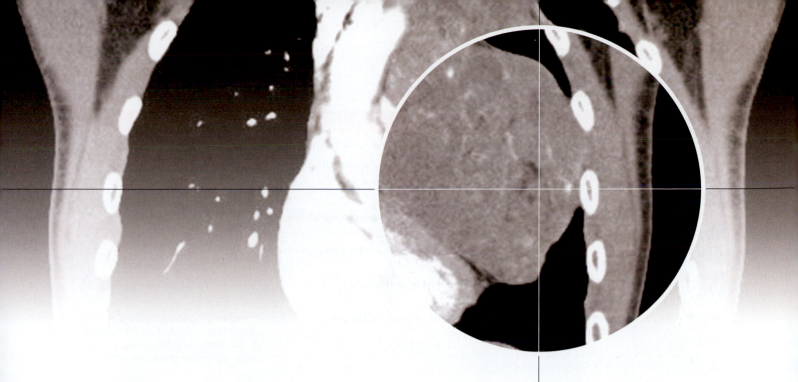

Chapter 3

Cytotoxic Chemotherapy

Introduction

Chemotherapy, in the broad sense of the term, refers to chemicals used to treat disease. Thus, almost any medication might be classified as chemotherapy. As it pertains to oncology, conventional chemotherapy is the application of **cytotoxic** drugs, substances that harm or destroy cells to control cancer (see Table 2.2). These cytotoxic chemotherapy agents kill rapidly growing cells, including cancer cells, but also rapidly dividing normal cells found in hair follicles, skin, gastrointestinal tract, and bone marrow (hereafter, "chemotherapy" will refer to these conventional cytotoxic chemotherapy agents). In contrast to the somewhat indiscriminate cell death produced by cytotoxic chemotherapy drugs, **molecularly targeted therapies** aim to selectively destroy cancer cells while leaving normal cells intact. Advances in cancer biology have provided a path toward personalized medicine using such targeted therapies. However, despite the promise of targeted therapies and the clear path they are providing toward the future of clinical oncology, chemotherapy continues to have an important role in the care and cure of cancer patients. Thus, it remains important for oncologists to thoroughly understand and deploy chemotherapy in the appropriate settings.

Chemotherapy has proven benefits in a variety of clinical settings. When given in the **perioperative** setting (either neoadjuvant or adjuvant, referring to before or after surgery, respectively) chemotherapy decreases recurrence rates and increases cure rates after surgery for many localized cancers including breast, colon, and bladder cancers. For advanced and metastatic cancers, chemotherapy may offer disease control and palliation (relief) of cancer-related symptoms, such as pain, and may prolong life. Chemotherapy may also be combined with targeted therapies for some cancers, as illustrated in

Table 3.1. Examples of combination chemotherapy regimens.

Regimen	Chemotherapy agents	Disease
ABVD	Doxorubicin (Adriamycin), bleomycin, vinblastine, dacarbazine	Hodgkin lymphoma
AC → T	Doxorubicin (Adriamycin), cyclophosphamide (Cytoxan), followed by paclitaxel (Taxol)	Breast cancer
BEP	Bleomycin, etoposide, cisplatin (Platinol)	Testicular cancer
R-CHOP	Rituximab, cyclophosphamide, hydroxydaunorubicin (doxorubicin), vincristine (Oncovin), prednisone	Non-Hodgkin lymphoma
FOLFOX, FOLFIRI	Leucovorin (folinic acid), 5-fluorouracil, oxaliplatin (or irinotecan for FOLFIRI)	Colon cancer
7 + 3	7 days of cytarabine (Ara-C) and 3 days of anthracycline (daunorubicin or idarubicin)	Acute myeloid leukemia
MOPP	Mechlorethamine (nitrogen mustard), vincristine (Oncovin), procarbazine, prednisone	Hodgkin lymphoma (historic regimen as the first combination introduced in the 1960s to achieve high success rates; no longer the standard)

chapters to follow. Last, chemotherapy has proven curative for patients with some advanced malignancies.

Chemotherapy agents employ an array of mechanisms to disrupt cell replication and thus tumor growth (see Table 2.2 for the classification and mechanisms of chemotherapy drugs). Cancers differ in their sensitivity to classes of chemotherapy agents. Susceptibility of a cancer type to a specific chemotherapy agent is often first identified in the laboratory, leading to early-phase clinical trials. The rational combination of chemotherapy agents, typically employing drugs using different mechanisms of action that might synergistically kill cancer cells, is then tested in clinical trials, with each successive trial aiming to optimize safety as well as disease control or cure rates by altering chemotherapy combinations, dosages, and/or schedules. Examples of chemotherapy combinations in routine clinical use can be found in Table 3.1.

In clinical oncology, it is instructive to understand the failures of cytotoxic chemotherapy. From the 1960s to 1980s, successive clinical trials using combination chemotherapy regimens produced remissions and cures in some leukemias and lymphomas, as well as testicular cancers, yielding optimism that chemotherapy would find the "Achilles' heel" in most cancers. This unfortunately proved overly optimistic, as the death rate from cancer has remained generally stable over the past 50 years in the United States, in contrast with the falling death rates from cardiovascular and cerebrovascular diseases (Figure 3.1).

The two cases that follow demonstrate extreme outcomes in conventional chemotherapy. Case 3-1 illustrates the curative potential of cytotoxic chemotherapy in metastatic testicular cancer. Case 3-2 describes pancreatic cancer, which is among the most lethal malignancies, and the failure of conventional chemotherapy to cure it.

Take-home points

✓ Cytotoxic chemotherapy remains an important pillar in management of many cancers, yielding improvement in cure rates when given before (as neoadjuvant chemotherapy) or after (as adjuvant chemotherapy) surgery for some malignancies.

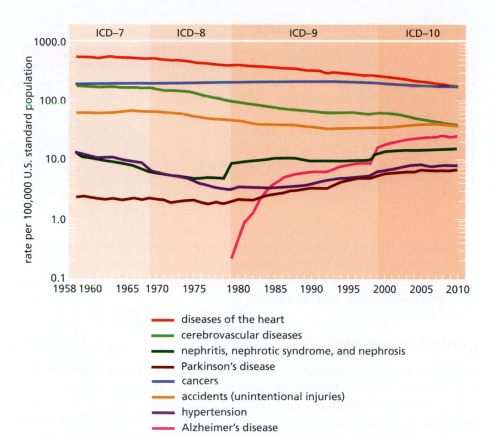

Figure 3.1. **Trends in death rates from leading causes of deaths in the United States, 1958 to 2010.** ICD is the *International Classification of Diseases*, which is periodically updated. (Adapted from S.L. Murphy, J.Q. Xu and K.D. Kochanek, *Natl. Vital Stat. Rep.* 61:1–117, 2013.)

— diseases of the heart
— cerebrovascular diseases
— nephritis, nephrotic syndrome, and nephrosis
— Parkinson's disease
— cancers
— accidents (unintentional injuries)
— hypertension
— Alzheimer's disease

✓ Chemotherapy has proven to be curative in a small number of advanced malignancies.

✓ Combinations of chemotherapy agents and molecularly targeted therapies may improve outcomes for some cancers.

✓ Recognition of the benefits and failures of conventional chemotherapy will remain important for training in oncology, appropriate management of patients, and insights into the biology of the disease.

Case 3-1	Testicular Cancer

Introduction

The treatment for testicular cancer represents one of the great triumphs of cytotoxic chemotherapy. Whereas 50 years ago, the diagnosis of testicular cancer carried a 90% chance of death, the prognosis for the disease today is essentially the opposite, due in large part to the efficacy of cytotoxic chemotherapy. Testicular cancers are also called **germ cell tumors** (GCTs). Germ cells are normal cells that reside in the gonads (testes and ovaries) and differentiate into the reproductive cells including sperm in men and eggs in women. Given their important role in reproduction, it may not be a surprise that tumors derived from germ cells are **pluripotent** (have the potential to develop into multiple types of cells). Some germ cell tumors, such as embryonal carcinoma or yolk sac tumor, have features that resemble early developmental stages. Perhaps the germ cell tumor that best exemplifies profound pluripotent capacity is the **teratoma**. Teratomas can contain

Topics bearing on this case

Trends in cancer mortality

Surgery, radiotherapy, and chemotherapy

Identifying useful targets of therapy

Table 3.2. Risk factors for testicular cancer.

Risk factor	Relative fold increase in risk
History of cryptorchidism (undescended testicle at birth)	10
Race: higher risk in Caucasian men; rare in African American men	—
Family history (first-degree relatives such as father, brother)	10
Personal history of testicular cancer: risk of second cancer in the other testicle	~100

components derived from one or more of the three germ layers that appear early in human development (endoderm, mesoderm, or ectoderm), with tissues contained within the tumor that can appear relatively normal, such as muscle cells, cartilage, hair, and bone.

Epidemiology and etiology

Testicular cancer is uncommon, with 9310 estimated new cases in the United States in 2018. Despite its rarity in general, testicular cancer is the most common malignancy in American men aged 20–35 years. A second, smaller peak in incidence of testicular cancer occurs in American men in their 50s. Fortunately, although the incidence has been relatively stable in the United States, the mortality rate has declined, with only 400 deaths estimated in 2018.

Risk factors for testicular cancer are summarized in Table 3.2. **Cryptorchidism**, an undescended testicle at birth, occurs in 2–5% of boys and raises their risk of developing testicular cancer 10-fold. Surgical correction of cryptorchidism (a procedure called orchiopexy) is recommended before puberty and often before the age of 2. A family history in a first-degree relative, such as a father or brother, will also raise the risk 10-fold, so patients' families should be made aware of the diagnosis when possible, with surveillance of sons of patients starting at an early age. A second testicular cancer arises in 2–3% of patients; hence, vigilance via testicular self-exams remains a lifelong task for men who have had the disease.

Diagnosis, workup, and staging

Men with testicular cancer most commonly present with a painless testicular mass. Some men present with pain in the scrotum, and testicular cancer is suspected if that pain is not traceable to a benign condition typical for the patient's age group, such as testicular torsion (a twist in the spermatic cord that leads to necrosis of the testis), **orchitis**, or **epididymitis** (with the latter two inflammatory diseases often associated with sexually transmitted bacteria). Some men have been diagnosed with testicular cancer as a result of infertility evaluation or due to **gynecomastia** (breast tissue enlargement) or **mastodynia** (breast pain). Symptoms of more advanced testicular cancer can include back pain, respiratory symptoms, and **hemoptysis** (coughing up blood). No standard recommendation exists for testicular cancer screening. One challenge for detection of testicular cancer is that young men infrequently see (or even have) primary care physicians, and consequently, they may ignore symptoms or delay reporting them.

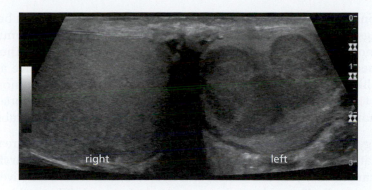

Figure 3.2. An ultrasound of a scrotum revealing left testicular cancer. Compared with the right testicle and its homogeneous appearance, the left testicle has nodular echotexture.

Evaluation of a testicular mass starts with an ultrasound of the scrotum (Figure 3.2). When a tumor is detected, prompt referral to a urologist (a surgeon who specializes in diseases of the urinary organs and male genital tract) is advised. Typically, urologists do not biopsy the tumor but recommend immediate resection of the diseased testicle, a procedure called radical inguinal orchiectomy ("radical" because the entire testicle and attached spermatic cord are removed; "inguinal" because the incision is made in the lower abdomen near the groin). Biopsy of a testicular tumor is not routinely performed, unlike many other malignancies in which a diagnostic biopsy occurs before a larger resection, because (1) the diagnostic ultrasound combined with clinical presentation make cancer the most likely diagnosis, and (2) sampling of the tumor increases the risk of spread. Fortunately, recovery from radical orchiectomy is typically very fast in this young male population.

(A)

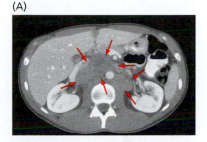

(B)

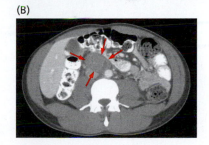

Figure 3.3. A CT scan of the abdomen and the pelvis from a patient with testicular cancer that has spread to retroperitoneal lymph nodes. Two different levels are shown in (A) and (B). Red arrows point to the enlarged lymph nodes.

At the time of diagnosis, the patient undergoes further evaluation to determine if the testicular cancer has spread elsewhere in the body, including computed tomography (CT) scans of abdomen and pelvis. Testicular cancer typically spreads to lymph nodes in the **retroperitoneum**, a space behind the peritoneal cavity that houses most of the digestive organs in the abdomen and pelvis (Figure 3.3). Blood tests are taken to check for **tumor markers**, which are proteins that may be expressed by testicular tumors, including lactate dehydrogenase (LDH), alpha-fetoprotein (AFP), and the beta subunit of human chorionic gonadotropin (β-hCG). More than half of patients with testicular cancer have elevated tumor markers. However, other conditions can cause elevated LDH, AFP, or β-hCG, so interpretation of the results must include clinical correlation, wherein the oncologist looks for other reasons the patient may have tested positive for these tumor markers (Table 3.3).

A CT scan of the chest is often performed when CT of the abdomen and pelvis or tumor markers are abnormal, or if the patient has respiratory symptoms. Brain imaging with CT scans or magnetic resonance imaging (MRI) is indicated for patients with especially elevated tumor markers or who exhibit neurological symptoms.

Table 3.3. Causes of LDH, AFP, and β-hCG elevation other than testicular cancer.

Tumor marker	May be elevated due to
LDH	Conditions associated with cell damage; myocardial infarction; lymphoma
AFP	Liver dysfunction, including hepatitis; hepatocellular carcinoma; gastrointestinal cancers
β-hCG	Pregnancy; hypogonadism; biliary cancer; pancreatic cancer; bladder cancer; gestational trophoblastic disease; laboratory artifact due to heterophile antibodies

Occasionally, tumors originating from the **mediastinum** (the area between the lungs including the heart, vessels, trachea, esophagus, thymus, and lymph nodes) are revealed to be germ cell tumors upon biopsy. These primary mediastinal germ cell tumors are malignancies that behave and are treated like testicular cancers, even in the absence of a primary tumor in the gonads (Figure 3.4). Such mediastinal tumors generally have a worse prognosis than tumors that originate in the testicle, but may still be curable.

Pathology of the primary tumor is important for prognosis and treatment. Testicular cancer subcategories include **seminoma**, **embryonal carcinoma**, **choriocarcinoma**, **yolk sac tumor**, and teratoma, which are all derived from pluripotent cells. Tumors may be comprised of several different histologies (and are then called "mixed germ cell tumors"). Cancers that are pure seminomas are distinguished from the other subcategories or mixed GCTs, which are collectively considered nonseminomas. Pure seminomas do not produce AFP, and are sensitive to radiation, unlike the other histologies.

Whereas many other cancers are divided among four stages, testicular cancer has three:

- Stage I: The disease is *confined* to the testis, spermatic cord, and scrotum.

- Stage II: The disease is *metastatic* but limited to retroperitoneal or pelvic lymph node groups (regional lymph nodes) and only mild elevation of the serum tumor markers.

- Stage III: The disease is *metastatic* and includes tumor markers that are highly elevated and associated with enlarged regional lymph nodes or distant nodal or visceral disease.

Tumor markers play an important role in staging. If the pathology and imaging indicate stage I disease but tumor markers remain elevated after orchiectomy, the disease is considered stage I-S due to the likely presence of microscopic metastases responsible for tumor marker production. Such patients are treated like patients with stage III metastatic disease.

Prognosis for metastatic disease and treatment recommendations for metastatic testicular cancer are largely driven by the International Germ Cell Consensus Classification (Table 3.4). Importantly, there is no poor-risk seminoma, speaking to the good overall prognosis for pure seminoma even if metastatic. Patients with good- and intermediate-risk metastatic seminoma have

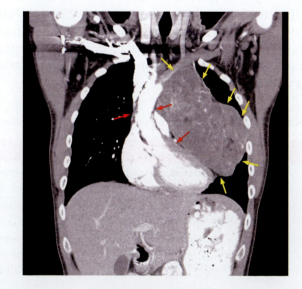

Figure 3.4. A CT scan of the chest from a patient with a mediastinal mass that was biopsy-proven to be embryonal carcinoma, a subtype of germ cell tumor. The very large mediastinal mass (outlined with yellow arrows) compressed against the heart and aorta (outlined with red arrows). This patient underwent chemotherapy and then resection of the residual mediastinal mass and remained clear of disease four years later.

Table 3.4. International germ cell consensus classification.

Prognostic category	Seminoma	Nonseminoma
Good risk	Any primary site *and*	Testis or retroperitoneal primary site *and*
	No nonpulmonary visceral metastasis *and*	No nonpulmonary visceral metastasis *and all of the following:*
	Normal AFP, any β-hCG, any LDH	AFP <1000 ng/ml β-hCG <5000 mIU/ml LDH <1.5 × upper limit of normal (ULN)
Intermediate risk	Any primary site *and*	Testis or retroperitoneal primary site *and*
	Nonpulmonary visceral metastasis *and*	No nonpulmonary visceral metastasis *and any of the following:*
	Normal AFP, any β-hCG, any LDH	AFP 1000–10,000 ng/ml β-hCG 5,000–50,000 mIU/ml LDH 1.5–10 × ULN
Poor risk	No seminoma patients classified as poor prognosis	Mediastinal primary site *and/or*
		Nonpulmonary visceral metastasis *and/or any of the following:*
		AFP >10,000 ng/ml β-hCG >50,000 mIU/ml LDH >10 × ULN

five-year overall survival rates of 86% and 72%, respectively. Patients with good-, intermediate-, and poor-risk metastatic nonseminoma have five-year overall survival rates of 92%, 80%, and 48%, respectively.

Management of stage I and II testicular cancer

Identifying stages I and II testicular cancer depends upon pathology (seminoma versus nonseminoma) and extent of lymph node burden. Decisions for treatment of a patient must balance the extent of disease, co-morbid medical conditions, commitment to close follow-up, and risk of side effects of therapy, including long-term risks of persistent treatment-related side effects in the young male patient with decades of life expectancy if cured of the testicular cancer.

Pure seminomas are sensitive to radiation and chemotherapy, unlike non-seminomatous histologies that are insensitive to radiation. For stage I pure seminoma, the prognosis is excellent, and <1% of patients will die from the disease. Management options of stage I pure seminoma include **active surveillance**, external beam radiation therapy to the para-aortic lymph nodes, and single-agent carboplatin chemotherapy. With relapse rates of approximately 20% (indicating 80% of patients are cured with orchiectomy), surveillance is a reasonable option for the patient who wishes to avoid overtreatment. Surveillance includes physical exams, blood tests for tumor markers, and periodic imaging studies, with the intervals between evaluations increasing especially beyond year three of follow-up. Radiation therapy was widely used and reduced relapse rates to 4–5%, but the risk of developing secondary malignancies decades after radiation, albeit rare at <1% of treated patients, has made this approach less popular. Single-agent carboplatin chemotherapy produces similar relapse rates as radiation therapy, but late toxicities from carboplatin remain unknown, as the largest studies of carboplatin adjuvant therapy were completed relatively recently and the young men in the studies are still in long-term follow-up.

Stage II pure seminoma patients may be treated with radiation therapy to the para-aortic lymph nodes and the iliac lymph nodes on the same side as the affected testicle. Patients with bulkier lymph nodes (>5 cm) are typically given similar treatment to that for stage III disease (see below).

Stage I nonseminoma generally has relapse risk of 30%. Stage I nonseminoma has historically been treated with **retroperitoneal lymph node dissection** (RPLND), an extensive operation that removes the lymph node basins behind the intestines that are the typical landing sites for metastatic testicular cancer. The RPLND can be both diagnostic and therapeutic. The absence of cancer in the resected lymph nodes would indicate a very low relapse rate, under 10%. Diagnosis of cancer in the nodes would lead to adjuvant chemotherapy, which can decrease relapse risk to 1%. The morbidity associated with RPLND has prompted oncologists to consider other management approaches to stage I nonseminoma testicular cancer, including active surveillance and one or two cycles of BEP (bleomycin, etoposide, and cisplatin) chemotherapy. The benefit of active surveillance is avoidance of overtreatment, as the corollary of a 30% chance of recurrence after orchiectomy is a 70% chance of cure without other therapy. Active surveillance may be best applied to those men who are committed to close follow-up. BEP chemotherapy can effectively reduce relapse risk to <3%. Given the toxicity of chemotherapy (discussed under stage III management, below) or RPLND, and the potential for overtreatment, the optimal management of stage I nonseminoma has not been well defined. Active surveillance is growing in popularity among academic medical centers.

Stage II nonseminoma is typically managed with RPLND or immediate chemotherapy, in a manner similar to stage III disease (below), depending on the lymph node metastatic disease burden. Post-RPLND, patients may require chemotherapy. Similarly, after "up-front" chemotherapy, patients with residual disease may require RPLND (described in more detail under stage III management).

For any treatment beyond orchiectomy, strong consideration is given to sperm banking, because of the 25% chance that normal fertility may not be possible after treatment. Radiation and chemotherapy may affect spermatogenesis, and RPLND may lead to retrograde ejaculation, in which semen is redirected to the bladder rather than ejaculated through the urethra.

Management of stage III testicular cancer

High tumor marker production after orchiectomy, and bulkier lymph nodes (>5 cm), typically warrant a stage III assignment. The standard initial management of stage III testicular cancer, irrespective of pure seminoma or nonseminoma histology, is combination chemotherapy. Successive clinical trials starting in the 1970s demonstrated high rates of response and ultimately cure for men in stage III whose disease would otherwise have claimed their lives in previous decades. The lessons from those multi-institutional trials have led to standardized therapies in contemporary practice in the United States. For men with good-risk metastatic disease (see Table 3.4), options include BEP for three cycles or EP (etoposide and cisplatin) for four cycles. For men with intermediate- or poor-risk metastatic disease, options are BEP or VIP (etoposide—also called

"VP-16," hence the "V" in the acronym—ifosfamide, and cisplatin), each for four cycles. For testicular cancer, one cycle of any of these regimens is three weeks (for more on chemotherapy cycles, see Chapter 2).

Chemotherapy selection takes into account risks of side effects and duration of treatment. As with most cytotoxic chemotherapy agents, the common side effects with the combinations BEP, EP, and VIP include nausea, vomiting, hair loss, decreases in blood cell counts (bone marrow toxicity raising risks of infection, fatigue, and bleeding from diminished populations of white blood cells, red blood cells, and platelets, respectively), **neuropathy** (dysfunction of peripheral nerves, often in the form of numbness or tingling), **tinnitus** (ringing in the ears), kidney dysfunction, and even death due to serious bacterial infections. Bleomycin carries the additional risk of interstitial pulmonary fibrosis, a condition whose mechanism is not well delineated but can be provoked even later in life upon exposure to high-fraction oxygen, such as the doses of oxygen given during anesthesia. Thus, for patients with co-morbid lung disease such as asthma or emphysema, a history of smoking, or lung metastases that may require further surgery, or for patients choosing to optimize lung capacity for other reasons (such as athletes preserving lung function), chemotherapy regimens that do not contain bleomycin may be considered. The trade-offs include additional chemotherapy treatment (choosing EP for four cycles over BEP for three cycles for good-risk disease) or increased bone marrow toxicity (choosing VIP over BEP, both for four cycles, for intermediate- and poor-risk disease).

It is expected that disease burden will regress and tumor markers (if initially elevated) will decline for most patients receiving chemotherapy. Following completion of chemotherapy, if tumor markers have resolved and imaging shows complete resolution of all metastatic disease, patients are placed on a surveillance program. If, however, tumor markers are normal, while imaging reveals residual masses, further treatment may be necessary. For a residual mass of pure seminoma that is <3 cm, observation is typically employed. If a residual mass is >3 cm and is identifiable on a subsequent PET scan, biopsy or resection is the next step, with salvage chemotherapy given for biopsy-confirmed residual cancer.

For nonseminomatous disease, residual masses can consist of active cancer in 10% of cases, fibrosis in 50%, and teratoma in 40%. Because imaging cannot distinguish among these possibilities, resection of all residual masses is considered, if surgically feasible. Resection is both diagnostic and therapeutic, in that teratomas, although slow-growing in general, can transform over time into carcinomas and sarcomas that are difficult to treat.

Relapses of testicular cancer are managed by second-line chemotherapy. Although second-line "salvage" chemotherapy has lower cure rates than first-line chemotherapy, the cure rate is still 25–50%. Select patients experiencing relapse can still achieve cures with high-dose chemotherapy, called **myeloablative chemotherapy**, which kills cells in the bone marrow. Immediately following myeloablative chemotherapy, patients are given **autologous stem cell transplantation**, wherein the patient's own stem cells are used to restore the bone marrow.

The case of Michael LaPorta, a healthy young man with a painless scrotal mass

Michael LaPorta was a 25-year-old postal worker in his usual state of good health when he noticed a painless swelling on the left side of his scrotum. His primary care physician promptly ordered a scrotal ultrasound that

Figure 3.5. Michael's CT scans demonstrating response of a para-aortic (retroperitoneal) lymph node to chemotherapy treatment. (A) Pre-treatment image: two separate lymph nodes are seen. (B) Post-treatment image shows regression of the lymph node indicated by the red arrow, but presence of a residual mass indicated by the yellow arrow.

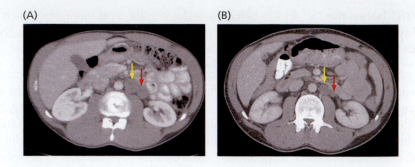

demonstrated a 5-cm left testicular mass. Michael was referred to urology and underwent a left radical inguinal orchiectomy that revealed a 5.5-cm mixed germ cell tumor comprised of yolk sac tumor (80%), seminoma (5%), and choriocarcinoma (15%) components. His tumor markers after surgery included LDH of 209 (upper limit of normal: 210 units/L), AFP of 5276 (upper limit of normal: 7.9 ng/ml), and β-hCG of 95 (upper limit of normal: <2.1 mIU/ml). CT scans revealed several enlarged retroperitoneal lymph nodes up to 3 cm but no other distant spread (Figure 3.5A).

Michael's medical history was notable only for his smoking half a pack of cigarettes daily for the previous six years. Neither he nor his parents could recall a history of cryptorchidism. There was no family history of testicular cancer. Michael was married without children. When he was questioned about symptoms, he described a two-month history of breast sensitivity. Gynecomastia was observed during his physical exam.

Soon after orchiectomy, Michael met with a medical oncologist. Chemotherapy was recommended for his metastatic, intermediate-risk nonseminoma. Sperm banking was performed. Due to his smoking history and difficulty committing to quitting cigarettes, VIP chemotherapy was recommended, to avoid the lung toxicity of bleomycin. He completed four cycles of VIP chemotherapy, which he tolerated with manageable nausea and expected fatigue from anemia. Follow-up after completion of chemotherapy showed that his tumor markers had completely normalized and his breast sensitivity and gynecomastia had resolved. However, post-chemotherapy CT scans revealed a residual 1.8-cm left para-aortic lymph node, but no other disease (see Figure 3.5B).

Due to residual disease, Michael underwent RPLND approximately four weeks after completion of chemotherapy. Teratoma was found in one lymph node that measured 1.7 cm; the other resected 19 lymph nodes were negative for metastatic disease. Thereafter he followed a close surveillance plan and is now six years free of disease. He enjoys an excellent quality of life, had no ejaculatory issues after RPLND, and was able to have two children naturally with his wife. His sons' pediatrician is well aware of Michael's cancer history and monitors Michael's sons with scrotal exams at least annually. Unfortunately, Michael continues to smoke cigarettes.

Targeted therapy and testicular cancer

With the high cure rates using combinations of chemotherapy and surgery, and occasionally stem cell transplant, as well as its being a relatively rare malignancy, the death rate from testicular cancer is low. Not surprisingly, other, more common and more fatal malignancies garner more attention and research into molecular insights for targeted therapy than does testicular cancer. Activation of KIT (also known as stem cell factor receptor) signaling, mutation of RAS, and altered DNA methylation have been described as potentially important molecular events for testicular cancer pathogenesis. To date, attempts to target KIT or VEGF signaling in patients whose testicular cancers have proven resistant to cisplatin-based chemotherapy have been disappointing, though it should be noted that such studies were very limited given the very low numbers of patients at risk.

Future prospects and challenges in metastatic testicular cancer

Management of testicular cancer represents perhaps the greatest accomplishment in the chemotherapy era. Further progress in treatment of metastatic testicular cancer and curing the small group of men at greatest risk for death will likely require multi-institutional studies involving cooperative groups and crossing national boundaries. Crossing "the last mile" in this disease may require combinations of chemotherapy, stem cell transplantation, and/or targeted therapies.

Take-home points

✓ Testicular cancer is rare but is the most common malignancy among men aged 20–35 years in the United States.

✓ Whereas death was formerly a common outcome in this disease, iterative improvements from clinical trials of combination chemotherapy regimens have produced high cure rates even in patients with metastatic disease.

✓ The current trend for stage I testicular cancer patients is management with active surveillance; other management approaches run the risk of overtreatment but may also be reasonable given the individual patient's scenario.

✓ Management of metastatic testicular cancer is relatively standardized in the United States, with treatment defined in part by the prognostic risk group (good-, intermediate-, and poor-risk categories).

✓ Residual and recurrent disease require specialized care but still may be treated with curative intent.

✓ Management decisions should take into account the age of the patient and the decades of life expectancy if cured of his testicular cancer, with respect to potential toxicities (short- and long-term) of therapy.

✓ Given the high cure rates, the development of combination chemotherapy for testicular cancer represents one of the greatest accomplishments of the chemotherapy era.

✓ There is currently no clear role for targeted therapies in this disease, given the cure rates afforded by cytotoxic chemotherapy.

Discussion questions

1. The trend in academic medical centers is to use active surveillance for stage I testicular cancer. Why might this approach be appealing? Why might active surveillance be challenging for this patient population?

2. Why is sperm banking important in this disease?

3. In what settings might you want to avoid bleomycin as a component of treatment for metastatic disease?

4. Why does gynecomastia occur in patients with elevated β-hCG?

Topics bearing on this case

Early clinical trials of toxicity and efficacy

Multi-drug treatment protocols

Hedgehog signaling and the desmoplastic stroma of pancreatic carcinomas

Case 3-2 Pancreatic Cancer

Introduction

Pancreatic cancer has a well-earned reputation as an aggressive, often fatal malignancy. In contrast with the successful implementation of cytotoxic chemotherapy to cure metastatic testicular cancer, the attempts to improve upon treatment for pancreatic cancer describe a humbling history dotted with few, modest successes. Unfortunately, only 20% of all cases are considered operable, and 20% of patients with operative cases survive 5 years, yielding only ~4% survival at 5 years among all patients with a new diagnosis of pancreatic cancer.

Epidemiology and etiology

Pancreatic cancer is common, with over 55,000 new cases diagnosed annually in the United States. Although it does not rank among the top 10 most common cancers in the United States in terms of incidence, pancreas cancer is the fourth leading cause of cancer death, with over 44,000 people dying from the disease every year. The incidence of pancreatic cancer increases with age, with slight male predominance.

Risk factors for pancreatic cancer, including inherited risks and modifiable behavior, are summarized in Table 3.5. Tobacco use is a major risk factor, with smokers having double the risk of nonsmokers in development of pancreatic cancer. Between 20 and 30% of pancreatic cancers are attributed to smoking. Obesity is associated with a 20% higher risk of development of pancreatic cancer.

Pancreatic malignancies are 90% **adenocarcinomas** (malignant tumors formed from gland-forming epithelial tissue) arising from the duct cells, of

Table 3.5. Risk factors for pancreatic cancer.

Cigarette smoking	
Diabetes mellitus	
Chronic pancreatitis	
Prior radiation to the pancreas (for other malignancies such as Hodgkin lymphoma or testicular cancer)	
Obesity	
Cirrhosis	
Hereditary syndromes:	BRCA2—associated with breast and ovarian cancers P16/CDKN2A—associated with familial melanoma PRSS1—associated with familial pancreatitis STK/LKB1—associated with Peutz–Jeghers syndrome MLH1/MSH2/MLH3/MSH6—associated with hereditary non-polyposis colorectal cancer (HNPCC) or Lynch syndrome VHL—associated with von Hippel–Lindau syndrome TP53—associated with Li–Fraumeni syndrome APC—associated with familial adenomatous polyposis (FAP)

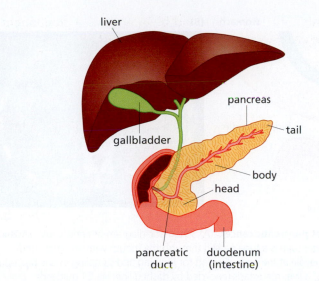

liver

pancreas

tail

gallbladder

body

head

pancreatic
duct

duodenum
(intestine)

Figure 3.6. Anatomy of the pancreas. The pancreas is situated in the upper abdomen. The pancreatic duct joins the common bile duct (draining the liver and gallbladder) and connects to the duodenum via the ampulla of Vater. Examination of this anatomy explains why a tumor at the head of the pancreas could obstruct the biliary flow, causing jaundice.

which two-thirds arise in the head and one-third in the tail of the pancreas (Figure 3.6). **Pancreatic neuroendocrine cell tumors** (PNETs, also called islet cell carcinomas) arise from cells within the pancreas that secrete hormones and comprise 5–10% of pancreatic malignancies. PNETs, with their very different biological origin from pancreatic adenocarcinoma, generally follow a less aggressive clinical course. The remainder of this case study will focus on pancreatic adenocarcinoma.

Diagnosis, workup, and staging

Patients with pancreatic cancer often present with nonspecific symptoms such as weight loss, fatigue, and progressive **anorexia** (loss of appetite). The classic presentation of painless **jaundice** (yellowing of the skin or whites of the eyes) is due to obstruction of the bile duct by a tumor of the pancreatic head, and the consequent inability for bilirubin to be excreted into the intestine as bile. Buildup of bilirubin leads to jaundice of the skin, pruritus (itching of the skin), darker urine, and paler stools. Increased bilirubin due to a pancreatic mass is generally distinguished from the many benign causes of jaundice that typically present with pain. Other symptoms of advanced disease include back pain, early **satiety**, nausea, vomiting, and **venous thromboembolic events** (blood clots) (Table 3.6).

Patients with multiple presenting symptoms (as enumerated in Table 3.6), a history of smoking, and/or a family history of pancreatic cancer may prompt suspicion of the diagnosis. Physical examination may reveal evidence of weight loss (such as temporal wasting, which is disproportionate thinning around the temples), jaundice, left supraclavicular (above the collarbone) lymph node enlargement, **ascites** (fluid buildup in the peritoneal cavity), enlarged liver or spleen, and **edema** (swelling) around the legs. Blood tests can be helpful but not necessarily diagnostic. Tumor markers CEA and CA19-9 may be elevated but are not specific; however, the latter is elevated in 70–90% of patients with advanced pancreatic cancer.

Abdominal symptoms often call for imaging tests. CT scans of the abdomen and pelvis may reveal a mass in the pancreas as well as metastasis to typical destinations, such as regional lymph nodes or the liver (Figure 3.7). A CT scan of the chest may reveal metastasis to lungs and the cervical/supraclavicular lymph nodes.

Table 3.6. Presenting symptoms of pancreatic cancer.

Painless jaundice, associated with pale stools, dark urine, pruritus
Weight loss
Fatigue
Anorexia
Abdominal pain
Back pain
Nausea or vomiting
Early satiety
Dyspnea
Abnormal blood sugar or glucose control
Venous thromboembolic event (blood clot)
Thrombophlebitis
Ascites

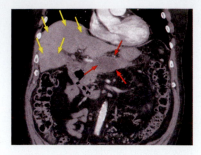

Figure 3.7. Pancreatic adenocarcinoma with metastasis to the liver. Abdominal CT scan demonstrating the pancreas primary tumor (*red arrows*) and numerous hypodense liver lesions, consistent with metastases (each yellow arrow indicates a separate metastatic nodule).

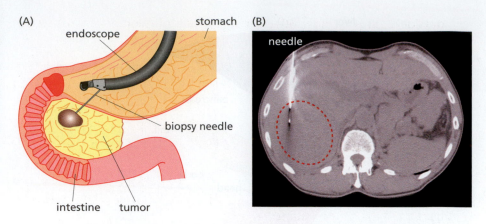

Figure 3.8. Evaluation of pancreatic cancer. (A) Schematic diagram of endoscopic ultrasound (EUS) with biopsy. The endoscope is a tube passed through the mouth with an ultrasound transducer and a biopsy needle at the end, allowing for imaging and sampling in one procedure. (B) Percutaneous biopsy of a liver metastasis (encircled by dashed line) via CT guidance.

The diagnosis is established by biopsy of the pancreatic lesion obtained by endoscopic ultrasound, typically performed by a gastroenterologist (Figure 3.8A), or by image-guided biopsy (percutaneous sampling) of a highly suspicious metastatic lesion, such as a liver lesion, if feasible, which allows for diagnosis of both primary malignancy and metastasis in the same procedure (Figure 3.8B).

To categorize the severity of each pancreatic cancer case, oncologists follow the TNM and staging system of the **American Joint Committee on Cancer** (AJCC) (Tables 3.7 and 3.8). The acronym "TNM" stands for the size of the primary tumor (T), the degree to which the cancer has spread to lymph nodes (N), and the presence of metastasis (M) or secondary tumors. TNM characteristics are combined to define stage 0, I, II, III, or IV:

- Stage 0 (Tis, N0, M0): The tumor is confined to the top layers of pancreatic duct cells (Tis), and it has not invaded deeper tissues (N0) or spread

Table 3.8. Pancreatic adenocarcinoma stage groupings.

Stage	T	N	M
0	Tis	N0	M0
IA	T1	N0	M0
IB	T2	N0	M0
IIA	T3	N0	M0
IIB	T1–3	N1	M0
III	T4	N0–1	M0
IV	T1–4	N0–1	M1

Table 3.7. American Joint Committee on Cancer (AJCC) TNM staging of pancreatic cancer.

Primary tumor (T)	Regional lymph nodes (N)
Tx: primary tumor cannot be assessed	Nx: regional nodes cannot be assessed
T0: no evidence of primary tumor	N0: no regional node metastasis
Tis: carcinoma *in situ*	N1: regional node metastasis
T1: limited to pancreas, ≤2 cm in greatest dimension	
T2: limited to pancreas, >2 cm	**Distant metastases (M)**
T3: extends beyond pancreas but without involvement of celiac axis or superior mesenteric artery	M0: no distant metastases
T4: involves celiac axis or superior mesenteric artery (unresectable primary tumor)	M1: distant metastases

outside of the pancreas (M0). Tumors at stage 0 are sometimes referred to as pancreatic carcinoma *in situ*.

- Stage IA (T1, N0, M0): The tumor is confined to the pancreas and is 2 cm across or smaller (T1), and it has not spread to nearby lymph nodes (N0) or distant sites (M0).

- Stage IB (T2, N0, M0): The tumor is confined to the pancreas and is larger than 2 cm across (T2), and it has not spread to nearby lymph nodes (N0) or distant sites (M0).

- Stage IIA (T3, N0, M0): The tumor is growing outside the pancreas but not into major blood vessels or nerves (T3), and it has not spread to nearby lymph nodes (N0) or distant sites (M0).

- Stage IIB (T1–3, N1, M0): The tumor is either confined to the pancreas or growing outside the pancreas but not into major blood vessels or nerves (T1–T3), and it has spread to nearby lymph nodes (N1) but not to distant sites (M0).

- Stage III (T4, any N, M0): The tumor is growing outside the pancreas into nearby major blood vessels or nerves (T4), it may or may not have spread to nearby lymph nodes (N), and it has not spread to distant sites (M0).

- Stage IV (any T, any N, M1): The cancer has spread to distant sites (M1).

Involvement of major arteries in the proximity of the pancreas (celiac axis, superior mesenteric artery) has a critical role in distinguishing stage III or IV disease from stage I or II, and determining whether resection is possible.

Management of localized pancreatic cancer

In the simplest terms, pancreatic cancer is often grouped into three categories: resectable, localized but unresectable, and metastatic. Imaging can help distinguish between the first two categories. Unresectable primary tumors have direct involvement of the local vascular structures including the celiac axis and superior mesenteric artery (Figure 3.9), and may even involve larger vessels such as the aorta and inferior vena cava. Encasement of the superior mesenteric vein may also render the disease unresectable.

Only 20% of pancreatic cancers are considered resectable. The approach to resection depends on the location of the primary tumor. For lesions in the head of the pancreas, **pancreaticoduodenectomy** (also known as the **Whipple procedure**, first performed by Dr. Allen Whipple at Columbia University in 1935) is the standard of care. The Whipple procedure removes the head of the pancreas, duodenum (first portion of the small intestine), gallbladder, part of the common bile duct, and lymph nodes. As expected for an operation removing this number of organs, the morbidity and mortality of the Whipple procedure are significant, and therefore this procedure is best performed at medical centers with surgical expertise performing a high volume of these operations. For tumors involving the body or tail of the pancreas, distal **pancreatectomy** and splenectomy are generally performed. Unfortunately, median survival for localized, resectable disease is only 18 months, with 20% of patients alive 5 years after surgery.

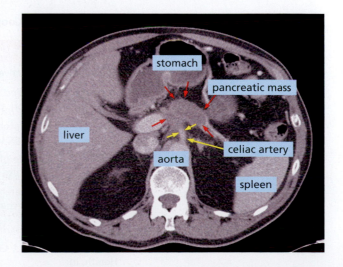

Figure 3.9. Unresectable pancreatic cancer. Abdominal CT scan showing a pancreatic tumor (*red arrows*) extending into and nearly obliterating the celiac artery (*yellow arrows*).

Neoadjuvant (preoperative) and adjuvant (postoperative) chemotherapy and/or radiation have been studied in many trials in the United States and Europe, to try to improve cure rates with surgery. Some trials have described a benefit to perioperative chemotherapy using gemcitabine or 5-fluorouracil (5-FU); however, clear evidence for a long-term survival advantage is lacking. Many patients receive chemotherapy or **chemoradiation** (combined chemotherapy and radiation) based on available data and given the dire overall outlook for this disease.

For localized but unresectable pancreatic cancer, chemotherapy or chemoradiation is applied. Chemoradiation with 5-FU or gemcitabine has demonstrated modest survival advantage over chemotherapy alone but has not led to increased cures. For example, chemoradiation with gemcitabine modestly improved survival compared with single-agent gemcitabine chemotherapy, from 9.2 months to 11.1 months. Combination chemotherapy regimens, such as 5-FU, leucovorin, irinotecan, and oxaliplatin (FOLFIRINOX), have demonstrated higher response rates in metastatic disease and have been shown to convert unresectable disease to resectable in a subset of patients, although long-term survival data to support this approach are still maturing.

Management of metastatic pancreatic cancer

Patients with metastatic pancreatic cancer have a **median overall survival** of 6 months. For these patients, chemotherapy treatment is given with palliative intent—that is, chemotherapy is administered to improve or maintain quality of life and to afford some degree of extended life. Single-agent chemotherapies have demonstrated low (5–20%) response rates. The standard management is single-agent gemcitabine, based largely on a trial showing that 24% of subjects receiving gemcitabine had a clinical benefit (improvement in pain, performance status, and weight) and a very modest but statistically significant improvement in median survival (5.7 versus 4.4 months) over 5-FU monotherapy. Numerous attempts to improve upon single-agent gemcitabine have added other chemotherapy agents, such as cisplatin, oxaliplatin, 5-FU, capecitabine, docetaxel, and irinotecan, with minimal survival benefit but significant increase in toxicity. In a clinical trial that combined gemcitabine and albumin-bound paclitaxel, overall survival was improved compared with gemcitabine alone (8.7 versus 6.6 months). FOLFIRINOX was examined in a randomized controlled trial compared with gemcitabine, and was associated

Table 3.9. Eastern Cooperative Oncology Group Performance Status (ECOG PS) scale.

Grade	Description
0	Fully active, able to carry on all pre-disease performance without restriction
1	Restricted in physically strenuous activity but ambulatory and able to carry out work of a light or sedentary nature, e.g., light housework, office work
2	Ambulatory and capable of all self-care but unable to carry out any work activities; up and about more than 50% of waking hours
3	Capable of only limited self-care; confined to bed or chair more than 50% of waking hours
4	Completely disabled; unable to carry on any self-care; totally confined to bed or chair
5	Dead

M. Oken et al., *Am. J. Clin. Oncol.* 5:649–655, 1982. With permission from Wolters Kluwer Health, Inc.

Table 3.10. Karnofsky Performance Status (KPS) scale.

General description	Score	Comment
Able to carry on normal activity and to work; no special care needed	100	Normal; no complaints; no evidence of disease
	90	Able to carry on normal activity; minor signs or symptoms of disease
	80	Normal activity with effort; some signs or symptoms of disease
Unable to work; able to live at home and care for most personal needs; varying amounts of assistance needed	70	Cares for self; unable to carry on normal activity or to do active work
	60	Requires occasional assistance, but is able to care for most of personal needs
	50	Requires considerable assistance and frequent medical care
Unable to care for self; requires equivalent of institutional or hospital care; disease may be progressing rapidly	40	Disabled; requires special care and assistance
	30	Severely disabled; hospital admission is indicated although death is not imminent
	20	Very sick; hospital admission necessary; active supportive treatment necessary
	10	Moribund; fatal processes progressing rapidly
	0	Dead

with a significantly improved response rate (32% versus 9%) and median overall survival (11.1 versus 6.8 months). FOLFIRINOX, however, is much more toxic than gemcitabine. Side effects with FOLFIRINOX include but are not limited to **neutropenia** (low neutrophil count), diarrhea, and neuropathy; therefore, FOLFIRINOX should be reserved for patients in reasonable physical condition, which is measured in oncology as **performance status** (PS). The Eastern Cooperative Oncology Group Performance Status (ECOG PS) and Karnofsky Performance Status (KPS) scores are often used to judge performance status (Tables 3.9 and 3.10). "Good" PS is generally considered ECOG PS <2 or KPS ≥80.

The case of Jeanne Brantley, a real estate broker with painless jaundice

Jeanne Brantley was 69 years old when her family noticed yellowing of her skin and eyes. She had quit smoking 10 years earlier and had diabetes that had become more difficult to control in the recent months. When she saw her primary care physician, weight loss of 10 lb (8% of her body weight) was noted compared with her visit the year before. Upon questioning, she described several months of "feeling full quicker" after meals. Whereas she normally maintained an energetic pace in her job as a real estate broker, she lately had noticed feeling more exhausted at the end of a busy day. The day before presentation, she noted darker urine and clay-colored stools.

On examination, Jeanne was noted to have more temporal wasting (thinning of the muscle mass near the temples seen with significant weight loss from cancer or nutritional deficiency) than might be expected in a patient with just 10 lb weight loss. Her skin was jaundiced, and her eyes were icteric (yellow where they should be white). She had fullness in the left supraclavicular lymph nodes, which were also firm to the touch. Her abdomen was slightly distended, but her liver was not enlarged and her spleen was not palpable.

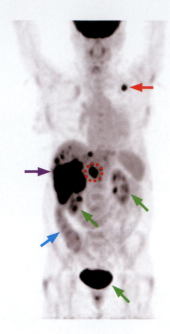

Figure 3.10. Jeanne's PET/CT scan demonstrating a pancreas tumor and metastases. The pancreatic tumor is encircled by a red dashed line. A metastatic left supraclavicular lymph node is indicated by the red arrow. A large mass made up of numerous confluent liver metastases is indicated by the purple arrow. Normal uptake of the fluorodeoxyglucose tracer can be seen in the kidneys (*two upper green arrows*), large intestine (*blue arrow*), and bladder (*lower green arrow*).

Jeanne's primary care physician ordered a CT scan of the chest and abdomen, which showed a 4-cm mass of the head of the pancreas as well as enlarged adjacent lymph nodes and multiple metastases in the liver up to 7 cm, and 2-cm lymph nodes in the left supraclavicular region (Figure 3.10). Jeanne was referred to a medical oncologist for the presumed diagnosis of metastatic pancreatic cancer.

Blood tests indicated that CA19-9 was within normal limits, but total bilirubin was elevated, as expected given her jaundice. She was referred to a gastroenterologist, who performed **endoscopic retrograde cholangiopancreatography** (ERCP), during which an external blockage of the common bile duct was seen and attributed to the tumor at the head of the pancreas (Figure 3.11). A brush biopsy of the bile duct was performed that confirmed pancreatic adenocarcinoma. During ERCP, a stent was placed to prop open the common bile duct, which eventually normalized the serum bilirubin and relieved Jeanne's jaundice and associated pruritus.

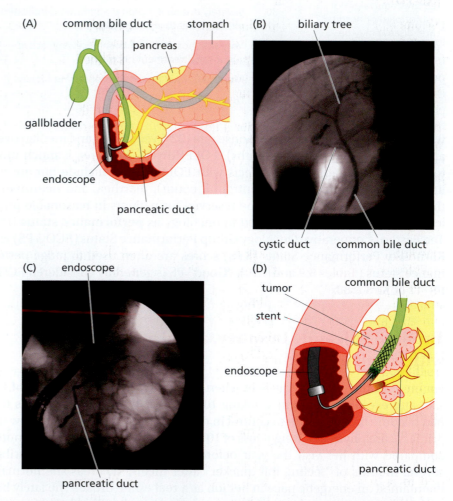

Figure 3.11. Endoscopic retrograde cholangiopancreatography (ERCP). (A) As in EUS, the endoscope for ERCP is passed through the mouth to the stomach and into the duodenum. Procedures can be performed using instruments extending from the end of the endoscope, including injection of dye into the common bile duct or the pancreatic duct. (B) Plain X-ray of an ERCP. Dye is injected into the common bile duct. The common bile duct branches into the cystic duct, which drains the gallbladder; and the biliary tree, which drains the liver. (C) Dye can also be injected to visualize the pancreatic duct. (D) During ERCP, a brush biopsy of the obstructed common bile duct can diagnose the cancer that is causing external compression of the duct. A stent can be placed to open the compressed common bile duct, relieving the obstruction caused by the cancer at the head of the pancreas. (A, Content supplied by the world's largest charitable funder of cancer research, © Cancer Research UK 2002. All rights reserved. B and C, courtesy of David Forcione, M.D.)

In the two weeks from presentation to final report of pathology from the biopsy, Jeanne unfortunately had continued to decline in her energy and lost another 5 lb. She felt unable to attend the open houses she had planned for the homes she had committed to sell. Given the diagnosis, Jeanne opted to retire from her work and focus her time on her health and family.

In discussion with her medical oncologist and her husband and three children, Jeanne agreed that her performance status had tumbled in the weeks since her initial visit to her primary care physician. Her performance status had gone from 0 to 1 to 2 (on the ECOG scale; see Table 3.9). She and her family understood that her disease was not curable and that it appeared to be rapidly progressive. There was no discussion about surgery or radiation; systemic therapy or best supportive care were discussed. Jeanne wanted "a try" at chemotherapy, and given her decline in performance status, single-agent gemcitabine was recommended, over the more toxic FOLFIRINOX regimen.

Despite eight weeks of gemcitabine chemotherapy, Jeanne's re-staging scans showed progression of disease in the liver, with larger lesions and new metastases. She had no pain but her performance status declined further to the point of being in bed or on the couch the majority of the day. It was clear that chemotherapy was not giving her much benefit and that visits to the clinic were taking away valuable time with friends and family. Jeanne opted to shift her focus to optimizing her quality of life rather than quantity of life and stopped chemotherapy. With the support of family, friends, and a home hospice service, Jeanne died at home two months later, just five months after her presentation with painless jaundice.

Targeted therapy and pancreatic cancer

Activation of signaling pathways involving KRAS, WNT, and growth factor receptors including EGFR, IGFR, and FGFR families, and also alterations in tumor suppressors such as p53, p16, and DPC4/SMAD4, have been demonstrated in pancreatic cancer. Abnormal KRAS is present in over 90% of cases and appears to be an early event, found in the low-grade precursor lesion **pancreatic intraepithelial neoplasia-1** (PanIN1). Overexpression of epidermal growth factor receptor (EGFR) is associated with worse prognosis in pancreatic cancer. Targeting of these altered pathways and angiogenesis has been attempted in numerous clinical trials.

To date, the only targeted therapy that has improved overall survival is erlotinib, an orally bioavailable tyrosine kinase inhibitor of EGFR (for more about targeted kinase inhibition, see Chapter 4). In a Phase III clinical trial of patients with advanced pancreatic cancer, 569 subjects were randomized to receive gemcitabine with either erlotinib or placebo. The patients who received erlotinib had improved median overall survival (6.2 versus 5.9 months; p = .038) compared with the placebo group, with generally manageable adverse events. The 1-year survival was improved from 17% to 23% (p = .023). Erlotinib was FDA approved on November 2, 2005, for treatment of locally advanced, unresectable, or metastatic pancreatic adenocarcinoma. Although erlotinib was hailed in 2007 as the "first to demonstrate statistically significant improved survival in advanced pancreatic cancer by adding any agent to gemcitabine," the modest improvement in survival (approximately 10 days) combined with additional cost and toxicity called into question whether this was truly an advance. Today, the gemcitabine and erlotinib combination remains an option for patients but is not widely used. In general, "fit" patients with good performance status seen at academic medical centers are steered toward novel clinical trials rather than the gemcitabine and erlotinib combination.

Future prospects and challenges in metastatic pancreatic adenocarcinoma

Many clinical trials have attempted to improve the outcome of pancreatic cancer for all of the clinically relevant stages (resectable, localized unresectable, and metastatic). While there have been improved response rates using contemporary combination chemotherapy regimens, it is not clear that these will change the long-term, dismal statistics on pancreatic cancer survival. Our best advance in molecular targeted therapy in this field has yielded an overall survival benefit of only a fraction of a month.

Clearly, our understanding of the biology of pancreatic cancer remains limited. Progress in the form of earlier diagnosis or conversion from unresectable to resectable disease might have significant impacts on overall survival. For metastatic pancreatic adenocarcinoma, an improved identification of the "driver" mutations and signaling pathways may yield better targets for therapy, although it seems that this highly aggressive malignancy will require an approach that targets multiple signaling pathways within the cancer cells and possibly other treatments that target the tumor microenvironment or immune system.

Take-home points

✓ Pancreatic adenocarcinoma is an aggressive malignancy with less than 5% long-term survival.

✓ Surgery remains the only curative modality, but even in a well-selected population, only 20% of surgical patients are cured.

✓ Tobacco use and obesity increase risk of the disease.

✓ Presentation is often insidious, with nonspecific symptoms. Painless jaundice is the classic presentation for a tumor at the head of the pancreas.

✓ Adjuvant chemotherapy and chemoradiation are often administered even though only modest survival advantages are seen in clinical trials.

✓ The median survival of patients with metastatic pancreatic cancer is 6 months.

✓ Cytotoxic chemotherapy options for metastatic disease include single-agent gemcitabine and combination chemotherapy of either gemcitabine together with albumin-bound paclitaxel, or FOLFIRINOX.

✓ The only FDA-approved targeted therapy for advanced pancreatic adenocarcinoma is erlotinib when combined with gemcitabine, but the survival benefit is extremely modest.

✓ Improved therapies based on molecular understanding of metastatic pancreatic cancer are sorely needed.

Discussion questions

1. Why is pancreatic cancer difficult to diagnose?

2. How does presentation differ when pancreatic adenocarcinoma arises from the tail rather than the head of the pancreas?

3. Why does pancreatic cancer typically present in advanced stages?

4. Does conversion from unresectable to resectable disease necessarily imply an increase in survival?

5. Preclinical data from cell lines suggested that EGFR might be a reasonable target in pancreatic cancer, but the clinical benefit was modest, at best. One possible explanation for the discrepancy is the high rate of KRAS mutation found in pancreatic cancers. How might this explain the lack of clinical benefit?

Chapter summary

Despite an expanding number of effective targeted therapies for many cancers (to be described in the remaining chapters), cytotoxic chemotherapy will maintain a key role as a curative treatment for at least some malignancies like testicular cancer and as an adjunct treatment in the perioperative setting for many other malignancies. The successful, curative application of cytotoxic chemotherapy in metastatic testicular cancer stands in stark contrast to the experience with pancreatic cancer, and underscores the many gaps in our understanding of cancer biology as it intersects with clinical oncology. The pancreatic cancer case also illustrates that statistically significant improvements in overall survival with targeted therapy may confer very modest benefits at best. However, the optimistic view is that such progress may represent the toehold that will lead to surmounting the barriers that maintain the stubbornly dismal prognosis for patients with pancreatic adenocarcinoma.

Chapter discussion questions

1. What factors might make some cancers more susceptible to cytotoxic chemotherapy than others?

2. Prevention or early detection would theoretically lead to an improvement in cancer statistics and outcomes. How would you design a screening program for a specific cancer?

3. Given the insidious nature of some cancers, would periodic whole-body CT scans be effective ways to screen for cancer? Why or why not?

Selected references

Introduction

Murphy SL, Xu J & Kochanek KD (2013) Deaths: final data for 2010. *Natl. Vital Stat. Rep.* 61:1–118.

Case 3-1

Einhorn LE, Williams SD, Chamness A et al. (2007) High-dose chemotherapy and stem-cell rescue for metastatic germ-cell tumors. *N. Engl. J. Med.* 357:340–348.

Hanna NH & Einhorn LH (2014) Testicular cancer—discoveries and updates. *N. Engl. J. Med.* 371:2005–2016.

International Germ Cell Cancer Collaborative Group (1997) International germ cell consensus classification: a prognostic factor-based staging system for metastatic germ cell tumors (GCT). *J. Clin. Oncol.* 15:594–603.

Oliver RT, Mead GM, Rustin GJ et al. (2011) Randomized trial of carboplatin versus radiotherapy for stage I seminoma: mature results on relapse and contralateral testis cancer rates in MRC TE19/EORTC 30982 study (ISRCTN27163214). *J. Clin. Oncol.* 29:957–962.

Siegel RL, Miller KD & Jemal A (2018) Cancer statistics, 2018. *CA Cancer J. Clin.* 68:7–30.

Case 3-2

Blazer M, Wu C, Goldberg RM et al. (2015) Neoadjuvant modified (m) FOLFIRINOX for locally advanced unresectable (LAPC) and borderline resectable (BRPC) adenocarcinoma of the pancreas. *Ann. Surg. Oncol.* 22:1153–1159.

Burris HA III, Moore MJ, Andersen J et al. (1997) Improvement in survival and clinical benefit with gemcitabine as first-line therapy for patients with advanced pancreas cancer: a randomized trial. *J. Clin. Oncol.* 15:2403–2413.

Conroy T, Desseigne F, Ychou M et al. (2011) FOLFIRINOX versus gemcitabine for metastatic pancreatic cancer. *N. Engl. J. Med.* 364:1817–1825.

Kanda M, Matthaei H, Wu J et al. (2012) Presence of somatic mutations in most early-stage pancreatic intraepithelial neoplasia. *Gastroenterology* 142:730–733.

Loehrer PJ, Feng Y, Cardenes H et al. (2011) Gemcitabine alone versus gemcitabine plus radiotherapy in patients with locally advanced pancreatic cancer: an Eastern Cooperative Oncology Group trial. *J. Clin. Oncol.* 29:4105–4112.

Moore MJ, Goldstein D, Hamm J et al. (2007) Erlotinib plus gemcitabine compared with gemcitabine alone in patients with advanced pancreatic cancer: a phase III trial of the National Cancer Institute of Canada Clinical Trials Group. *J. Clin. Oncol.* 25:1960–1966.

Morris JP IV, Wang SC & Hebrok M (2010) KRAS, Hedgehog, Wnt and the twisted developmental biology of pancreatic ductal adenocarcinoma. *Nat. Rev. Cancer* 10:683–695.

Siegel RL, Miller KD & Jemal A (2018) Cancer statistics, 2018. *CA Cancer J. Clin.* 68:7–30.

Von Hoff DD, Ervin T, Arena FP et al. (2013) Increased survival in pancreatic cancer with nab-paclitaxel plus gemcitabine. *N. Engl. J. Med.* 369:1691–1703.

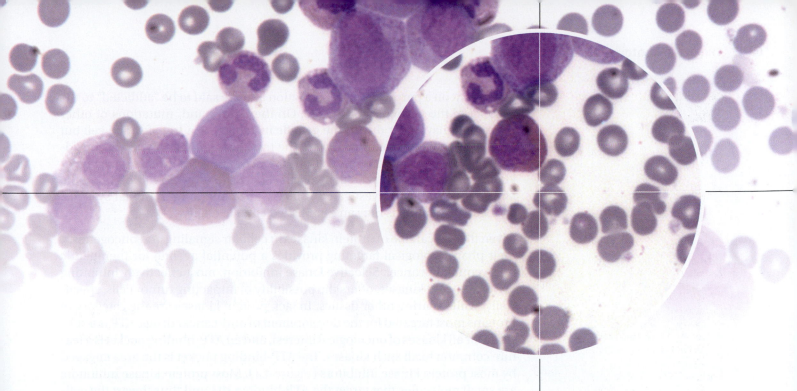

Chapter 4

Protein Kinase Inhibition

Introduction

Protein kinases are enzymes that catalyze **phosphorylation**, the transfer of phosphate from a donor such as adenosine triphosphate (ATP) to an amino acid residue of an acceptor protein, thereby altering the structure of the protein and affecting its function. The human genome encodes approximately 518 protein kinases and most fall into two major groups: serine/threonine kinases (385 members), which transfer phosphate from ATP to serine or threonine residues; and tyrosine kinases (90 members), which catalyze the ATP-dependent phosphorylation of tyrosine residues. Protein kinase-mediated phosphorylation can often act as an "on/off" switch for the target protein by altering its enzymatic activity, modifying its ability to associate with other proteins, changing its cellular compartmentation, or marking it for degradation. Consequently, almost every aspect of cell physiology—metabolism, proliferation, growth, movement, differentiation, and death—is significantly influenced by one or more protein kinases.

Given their pervasive importance, it is not surprising that disturbances in the function or levels of key protein kinases cause disease. More than 150 protein kinases have been implicated in disease, with over 120 different protein kinases connected to oncogenesis including tumor cell proliferation, survival, genomic instability, and tumor **angiogenesis**. Beginning with the discovery that the oncogenic agent of Rous sarcoma virus, Src, is a tyrosine kinase, the connection of protein kinases to cancer development remains a key focus of cancer research. Conceptually, the activating genetic mutations that convey a selective growth advantage are considered driver mutations, and these **oncogenes** are critical to the transformation of a normal cell to a cancer cell. Mutations in kinase-encoding genes are frequently the driver oncogenic alterations underlying many different cancers. Indeed, some cancers are overwhelmingly dependent on a particular driver mutation and are said to be "addicted" to the product of the particular oncogene. On the other hand, mutations of other genes that may occur due to the genetic instability of a transformed cell but are not essential for maintenance of the transformed phenotype are considered passenger mutations. Indications of **oncogene addiction** to a particular kinase are provided by sensitivity of cancer cells to targeted inhibition of the kinase *in vitro*. First investigated using *in vitro* and animal model systems, oncogene addiction is now an established clinical reality.

Given the critical role of protein kinases in cellular signaling and oncogenesis, their pharmacological targeting provides a potential avenue for therapeutic intervention in cancer. Selective kinase inhibition, most effectively inhibiting driver-mutated kinases, offers the possibility of impacting tumor cells, specifically, while sparing other tissues. In fact, protein kinases are the category of proteins most targeted for the development of anti-cancer drugs. ATP is a substrate for all kinases of oncological interest, and an ATP-binding pocket is a feature common to all such kinases. The ATP-binding pocket is the area engaged by most protein kinase inhibitors (Figure 4.1). Most protein kinase inhibitors are small molecules that target the ATP-binding site and thus thwart the activated protein kinase's key downstream pathways and consequent biologic effects. Alternatively, for protein kinases that depend upon ligand-induced activation of their kinase domains, such as **receptor tyrosine kinases** like the EGF receptor family, kinase activity can be inhibited by **monoclonal antibodies** that bind and sequester the ligand or prevent its binding by blockading the receptor. Such antibodies are in clinical use and will be discussed in Chapter 11. Therapies employing monoclonal antibodies or small-molecule inhibitors can be distinguished by their names, which generally end with "-mab" or "-inib," respectively (for example, rituximab or imatinib).

Development of inhibitors specific for a particular kinase is complicated by the existence of hundreds of kinases, all of which have some degree of similarity in their active sites due to the common evolutionary origin of their ATP-binding pockets (see Figure 4.1). Additionally, competitive inhibitors must be potent enough to outcompete the relatively high (~0.1 mM) intracellular concentration of ATP, the enzyme's normal substrate. Despite these challenges of specificity and potency, drug discovery programs have yielded a number of compounds that have obtained FDA approval (Table 4.1). Many of the tabulated compounds are the product of rational programs of drug design that created molecules that interact with both the ATP-binding site and hydrophobic areas near the ATP-binding pocket of the targeted kinase. Even though ATP, the natural substrate, is a formidable competitor for the ATP-binding site, its highly charged and strongly polar triphosphate tail has no affinity for the hydrophobic regions that line or lie quite close to the ATP-binding pocket of kinases. Most successful inhibitors have hydrophobic regions that interact strongly with these nearby hydrophobic pockets. Furthermore, since the activity of protein kinases can be affected by changes in protein conformation or interaction with other proteins, there are opportunities for inhibition that involve processes other than ATP binding.

It is fortunate that perfect specificity of protein kinase inhibitors has turned out to be unnecessary, since many inhibitors that inhibit more kinases than the one they were designed for fall within acceptable levels of toxicity. The ability of a drug to inhibit several kinases may be helpful in interfering with compensatory signaling pathways that could lead to drug resistance. A more obvious benefit of a less specific inhibitory profile is that the same drug can be useful as a therapy against a broader variety of cancers. This point is nicely illustrated by **imatinib mesylate** (Gleevec), the drug highlighted in Case 4-1, focusing on

(A)

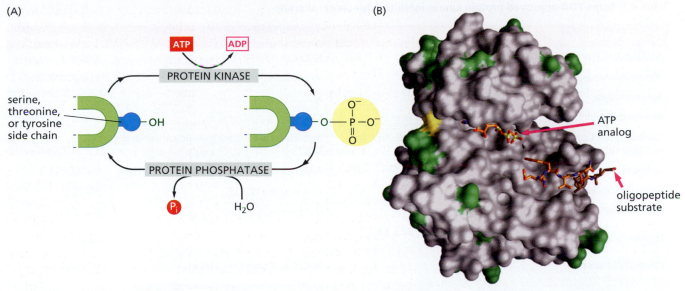

(B)

(C)

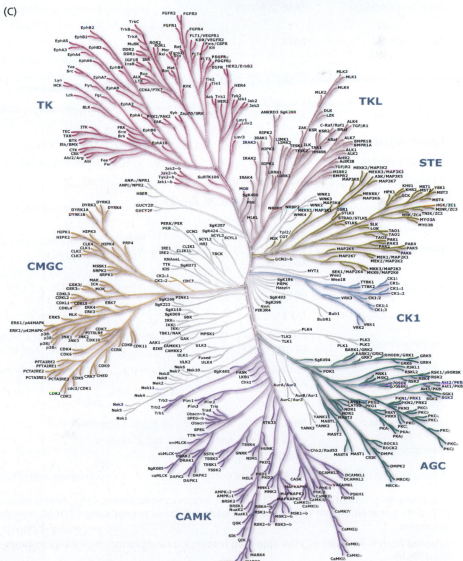

Figure 4.1. Protein kinases. (A) Amino acids of protein substrates are phosphorylated by protein kinases and dephosphorylated by the hydrolytic activity of protein phosphatases. (B) A schematic representation of a protein kinase highlighting the binding site of the substrate ATP (or ATP analog) and the polypeptide phosphate acceptor. (C) The human kinome. An evolutionary tree of protein kinases showing the major groups of protein kinases found in humans. The groupings are AGC, members of the protein kinase A, protein kinase C, and protein kinase G families; CAMK, calmodulin/calcium-regulated kinases; CMGC, set of CDK, MAPK, GSK-3, and CLK families; TK, tyrosine kinase; TKL, tyrosine kinase-like; STE, homologs of the yeast STE7, STE11, STE20 kinases; and CK1, casein-kinase-1. (B and C, from R.A. Weinberg, The Biology of Cancer, 2nd ed. New York: Garland Science, 2014. B, originally from S. Favelyukis et al., *Nat. Struct. Biol.* 8:1058–1063, 2001. Reprinted by permission from Macmillan Publishers Ltd. C, courtesy of Cell Signaling Technology, Inc.)

Table 4.1. Some FDA-approved protein kinase inhibitors for cancer therapy.

FDA name/trade name	Known protein kinase targets	Associated cancers	FDA approved
Afatinib/Gilotrif	EGFR	NSCLC	2/8/2013
Alectinib/Alecensa	ALK and RET	ALK-positive NSCLC	11/06/2017
Axitinib/Inlyta	VEGFR	RCC	2/5/2012
Bosutinib/Bosulif	BCR-ABL, Src, Lyn, and Hck	CML	1/8/2012
Cabozantinib/Cometriq or Cabometyx	RET, Met, VEGFR, Kit, TrkB, Flt3, Axl, and Tie2	metastatic medullary thyroid cancer and RCC	2/8/2012
Cobimetinib/Cotellic	MEK1/2	melanoma with BRAF V600E/K mutations	11/05/2015
Crizotinib/Xalkori	ALK and c-Met	ALK-positive NSCLC	3/6/2011
Dabrafenib/Tafinlar	B-RAF	melanoma	3/7/2013
Dasatinib/Sprycel	BCR-ABL, Src, Lck, Yes, Fyn, Kit, EphA2, and PDGFRβ	CML	10/28/2010
Erlotinib/Tarceva	EGFR	NSCLC and pancreas cancer	1/7/2004
Everolimus/Afinitor	FKBP12/mTOR	progressive neuroendocrine tumors of pancreatic origin, RCC, subependymal giant cell astrocytoma, and breast cancer	3/15/2009
Gefitinib/Iressa	EGFR	NSCLC	1/7/2003
Ibrutinib/Imbruvica	Bruton's tyrosine kinase (BTK)	CLL, mantle cell lymphoma, marginal zone lymphoma, and Waldenstrom macroblobulinemia	2/8/2013
Imatinib/Gleevec	BCR-ABL, Kit, and PDGFR	CML, ALL, aggressive systemic mastocytosis, and GIST	2/8/2003
Lapatinib/Tykerb	EGFR and ErbB2	breast cancer	2/8/2007
Neratinib/Nerlynx	HER2/ErbB2	HER2+ breast cancer	7/17/2017
Nilotinib/Tasigna	BCR-ABL and PDGFR	CML	2/8/2007
Pazopanib/Votrient	VEGFR, PDGFR, FGFR, Kit, Lck, Fms, and Itk	RCC and soft tissue sarcomas	3/9/2009
Ponatinib/Iclusig	BCR-ABL T315I, VEGFR, PDGFR, FGFR, EPH, Src family kinases, Kit, RET, Tie2, and Flt3	CML and Philadelphia chromosome positive ALL	1/7/2012
Regorafenib/Stivarga	VEGFR, BCR-ABL, B-RAF (V600E), Kit, PDGF, RET, FGFR, TIE2, and Eph2A	CRC	3/7/2012
Ruxolitinib/Jakafi	JAK1/2	myelofibrosis	1/6/2011
Sorafenib/Nexavar	C-RAF, B-RAF (V600E) Kit, Flt3, RET, VEGFR, and PDGFR	hepatocellular carcinoma, RCC, and dedifferentiated thyroid cancer	3/7/2005
Sunitinib/Sutent	PDGFR, VEGFR, Kit, Flt3, CSF1R, and RET	RCC, GIST, and pancreatic neuroendocrine tumors	3/6/2006
Temsirolimus/Torisel	FKBP12/mTOR	RCC	2/9/2013
Trametinib/Mekinist	MEK	melanoma	2/9/2013
Vandetanib/Caprelsa	EGFR, VEGFR, RET, BTK, Tie2, EphR, and Src family kinases	medullary thyroid cancer	1/6/2011
Vemurafenib/Zelboraf	A/B/C-RAF and B-RAF (V600E)	melanoma with BRAF (V600E)	8/17/2011

ALL, acute lymphoblastic leukemia; CLL, chronic lymphocytic leukemia; CML, chronic myelogenous leukemia; CRC, colorectal cancer; DTC, differentiated thyroid carcinoma, GIST, gastrointestinal stromal tumor; NSCLC, non-small-cell lung cancer; RCC, renal cell carcinoma.

Adapted from R. Roskoski Jr., FDA-approved protein kinase inhibitors, Blue Ridge Institute for Medical Research, http://www.brimr.org /PKI/PKIs.htm.

chronic myelogenous leukemia (CML), which follows. While moderately specific for the tyrosine kinase that drives CML, it also inhibits the activity of **Kit**, a receptor tyrosine kinase that drives a type of **gastrointestinal stromal tumor** (GIST), and the **platelet-derived growth factor receptor** (PDGFR), which drives certain **myeloproliferative neoplasms**. Imatinib's lack of complete specificity provides clinicians with a single drug that is effective for several diseases instead of only one.

Efforts to develop clinically effective protein kinase inhibitors have uncovered several key principles. First, in cancers that manifest oncogene addiction for the product of a protein kinase-encoding gene, inhibition of the kinase by an appropriate inhibitor will impair the growth and survival of the tumor. The cases in this chapter provide dramatic examples of the therapeutic efficacy of attacking the vulnerability of oncogene-addicted tumors. Second, since the targeted kinases play key roles in tumor proliferation or survival, treatment with an effective kinase inhibitor exerts strong selection for the expansion of resistant cell populations within the tumor, resulting in a loss of drug efficacy. Most resistance is caused by missense mutations in the sequence of the kinase, yielding changes in amino acid residues that have either of two effects on drug efficacy: some mutations block or hinder the inhibitor's access to vulnerable regions in or near the active site, whereas others result in conformational changes in the targeted kinase that make it less susceptible to the inhibitor. Efforts to overcome resistance to kinase inhibitors have involved the derivation of successive generations of inhibitors specifically tailored to overcome particular mutations known to result in escape from earlier-generation inhibitors.

Third, increasingly, genomic analysis is being used to profile the mutations in the gene encoding the targeted kinase, allowing the precise identification of escape mutations and guiding the switch to effective successor inhibitors. With the growing knowledge of genetic mutations in cancer, molecular profiling of a patient's disease at the time of diagnosis or relapse is being more broadly applied to identify targets for therapy, a feat that was not possible at the dawn of this millennium. Fourth, a particular kinase does not act alone but is part of a signaling pathway. This presents an attractive opportunity since downstream steps in the pathway provide targets of inhibition that may be attacked and thus prevent or slow tumor escape by the development of resistance at the initially targeted kinase step. Since kinase-dependent pathways typically impact yet other pathways important for tumor growth and maintenance, these pathways also provide a full menu of potential targets for inhibition.

The case studies presented in this chapter provide illustrations of the superior clinical efficacy of rationally designed protein kinase inhibitors, each designed to target and inhibit the dysregulated protein kinase driving the relevant oncogenic pathway. In each of these cases, inhibition of the driver protein kinase yields a strong clinical response with far less toxicity than conventional chemotherapy. Three of these cases (Cases 4-1, 4-2, and 4-3) describe landmark uses of **tyrosine kinase inhibitors** (TKIs). Case 4-4 describes a deployment of a combination of inhibitors: one, a **serine/threonine kinase inhibitor**; and the other, a **threonine/tyrosine kinase inhibitor**. All of the cases illustrate encouraging progress in the development of clinically effective protein kinase inhibitors.

Take-home points

✓ Phosphorylation mediated by protein kinases regulates almost every aspect of cell physiology including cell proliferation and survival.

✓ The human genome encodes 518 protein kinases, all of which have somewhat similar catalytic sites.

✓ Despite the similarities among protein kinases, more than two dozen kinase inhibitors having acceptable levels of specificity and potency have received FDA approval for clinical use.

✓ Mutations resulting in the dysregulation of protein kinases are drivers of many cancers.

✓ Inhibitors of protein kinases that selectively inhibit driver kinases impact the cancer while sparing normal tissues.

✓ Genomic profiling of a patient's cancer at diagnosis and during the course of disease is of increasing importance for the selection of appropriate kinase inhibitors for therapeutic intervention.

Topics bearing on this case

Chromosomal translocation

Imatinib and the story of CML

The Philadelphia Chromosome

Case 4-1   **Chronic Myelogenous Leukemia**

Introduction

Leukemia is a general term for cancers derived from white blood cells that predominantly circulate in the bloodstream and bone marrow. Among white blood cells, we have both **myeloid** and **lymphoid** cells, either of which may develop into leukemia (myelogenous or lymphocytic leukemia). Leukemias are typically also classified as chronic or acute, based on their clinical behavior and natural history. **Chronic myelogenous leukemia** (CML) is a **myeloproliferative** neoplasm characterized by uncontrolled production of mature **neutrophils**, and neutrophil precursors. Neutrophils are the most common type of white blood cell in humans, accounting for approximately 50–70% of healthy white blood cells in circulation, and are characterized by a multilobed nucleus and the presence of numerous granules in the cytoplasm. These cells normally function as "first responders" in the immune system, migrating to sites of infection and destroying bacteria and other pathogens via phagocytosis or release of antimicrobial toxins from their intracellular granules. Less abundant types of mature white blood cells include **lymphocytes**, **monocytes**, **eosinophils**, and **basophils**.

CML has three clinical phases: chronic phase, accelerated phase, and blast phase (also called blast crisis), with most patients presenting in chronic phase. The genetic hallmark of CML is an abnormal fusion of the *BCR* gene on chromosome 22 with the *ABL1* gene on chromosome 9, producing the t(9;22)(q34;q11) translocation, also known as the **Philadelphia chromosome**, so named due to its discovery by Drs. Nowell and Hungerford at the University of Pennsylvania in 1960 (Figure 4.2). The Philadelphia chromosome harbors the *BCR-ABL1*

Figure 4.2. Translocation involving chromosomes 9 and 22, resulting in the Philadelphia chromosome. (Adapted from R.A. Weinberg, The Biology of Cancer, 1st ed. New York: Garland Science, 2007. Original from A.S. Advani and A.M. Pendergast, *Leuk. Res.* 26:713–720, 2002.)

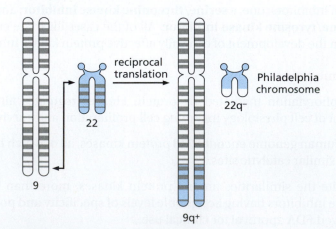

reciprocal translation

Philadelphia chromosome

22q⁻

22

9

9q⁺

fusion and was noted as an abnormally short chromosome 22; the corresponding chromosome 9 was found to be lengthened in a reciprocal translocation. This chromosomal translocation results in constitutive activity of the resultant fusion **BCR-ABL tyrosine kinase**, which enhances growth and proliferation of precursor myeloid cells via numerous downstream pathways including **RAS**, **RAF**, **JUN**, **MYC**, and **STAT** (see Figure 4.21 for an illustration of downstream signaling pathways).

Epidemiology

CML accounts for one to two new cases per year per 100,000 people and represents approximately 15% of newly diagnosed adult leukemias. Most cases are sporadic and the only known risk factor is exposure to very high doses of ionizing radiation (for example, occupational exposure, harm from nuclear fallout, or radiation therapy for other medical conditions). The median age of diagnosis is in the 50s, and there is a slight male predominance.

Diagnosis, workup, and staging

Close to half of patients are asymptomatic at diagnosis, with their disease detected incidentally as an elevated white blood cell count on a routine blood test. Symptomatic patients may experience fatigue, unintentional weight loss, symptoms of **splenomegaly** (splenic enlargement causing feelings of abdominal fullness, left upper quadrant abdominal discomfort, referred pain felt in the left shoulder, or early satiety, which describes fullness before finishing a normal-sized meal) (Figure 4.3), episodes of **gout** (an inflammatory disease caused by increased quantities of uric acid in the bloodstream), bleeding, and **thrombosis** (blood clots). Physical examination reveals splenomegaly in greater than half of patients at diagnosis, while **lymphadenopathy** (enlargement of lymph nodes) is usually absent. The cause for splenomegaly, which may be quite marked, is the diffuse infiltration of the spleen by excess myeloid cells of all stages of maturation.

Laboratory studies, including a complete blood count and examination of the peripheral blood smear, are notable for an increased white blood cell (WBC) count with a median count of approximately 100,000 cells/μl at diagnosis (normal range: 4500–11,000 cells/μl), predominantly composed of mature neutrophils.

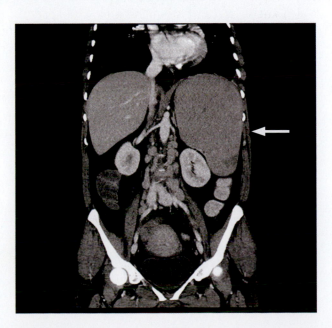

Figure 4.3. CT scan image of an enlarged spleen (arrow).

Figure 4.4. Normal maturation of neutrophils. Example of progression of myeloblast (A) to promyelocyte (B) to neutrophilic myelocyte (C) to neutrophilic metamyelocyte (D) to band (E) to segmented neutrophil (F). The other cells that lack nuclei are red blood cells.

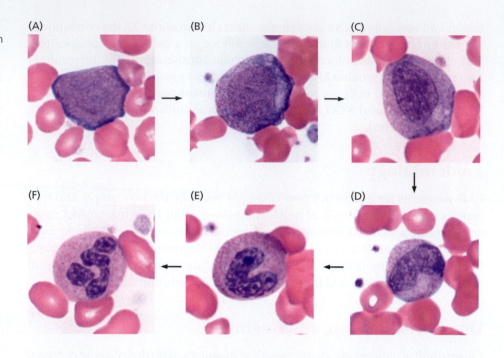

Increased immature white cell precursors are also typically seen including **myelocytes** and **band forms** (Figures 4.4 and 4.5). **Metamyelocytes** are typically not increased in proportion with other white blood cell precursors. Basophils are increased in approximately 90% of patients, and an increasing basophil count (≥20%) is characteristic of progression to accelerated-phase CML. **Myeloblasts** account for fewer than 5% of white blood cells with an increasing percentage marking the progression to accelerated phase (5–20%) and blast phase (>20%) disease. **Anemia** is present in approximately half of patients at diagnosis, and an elevated platelet count is observed in approximately one-third.

Because an increase in white blood cells may be a normal physiologic response to a broad range of infectious, inflammatory, or neoplastic conditions, diagnosis of CML requires confirmation of the t(9;22)(q34;q11) translocation. Prior to the availability of **cytogenetic analysis**, a diagnosis of CML was made based on suggestive clinical and pathologic features including an increase in mature

Figure 4.5. Peripheral blood smear from a patient with CML. This smear demonstrates mature neutrophils and occasional larger immature neutrophil precursors.

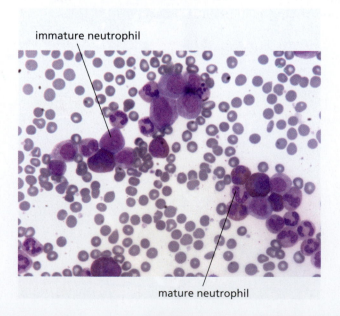

immature neutrophil

mature neutrophil

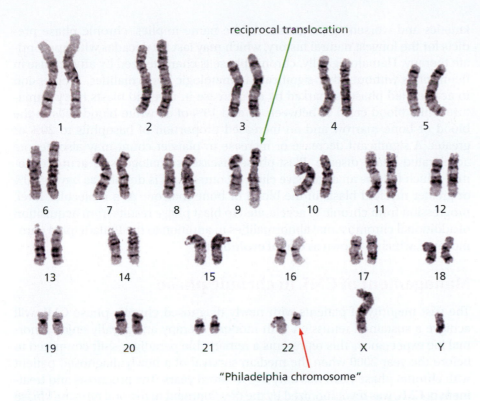

reciprocal translocation

1 2 3 4 5

6 7 8 9 10 11 12

13 14 15 16 17 18

19 20 21 22 X Y

"Philadelphia chromosome"

Figure 4.6. Visualization of the chromosomes in a karyotype. A karyotype is an arrangement of chromosome pairs at metaphase shown in descending order of size. This karyotype shows one normal copy each for chromosomes 9 and 22 with the other copy affected by a reciprocal translocation between the two chromosomes (*arrows*).

neutrophils, presence of neutrophil precursors, **basophilia**, splenomegaly, absence of an obvious infectious or inflammatory precipitant, and decreased activity of the enzyme leukocyte alkaline phosphatase, indicating the dysfunctional nature of the circulating neutrophils. Modern diagnosis, however, relies exclusively on identification of the culprit chromosomal abnormality or resultant gene product. Routine cytogenetics would typically show the characteristic translocation of chromosomes 9 and 22 in the karyotype of all tested metaphases (Figure 4.6). Fluorescence *in situ* hybridization (FISH) employs genomic probes for the *BCR* and *ABL* genes, and shows co-localization of the hybridization signals, which would be located on distinct chromosomes in the normal state (Figure 4.7). **Reverse-transcriptase PCR** (RT-PCR) is a highly sensitive test for detecting the *BCR-ABL* fusion gene product, and **quantitative PCR** (qPCR) also offers a quantitative tool for monitoring response to therapy by measuring transcript level to gauge depth of remission. These diagnostic modalities may be performed on either peripheral blood or bone marrow.

CML is classified at diagnosis as being in chronic phase, accelerated phase, or blast phase, with advancing phases of disease denoting more rapid disease

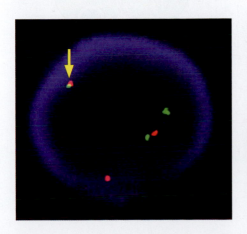
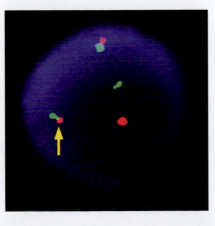

Figure 4.7. FISH analysis. *BCR* and *ABL* are found on distinct chromosomes. Using separate probes for *BCR* (*green*) and *ABL* (*red*) in a normal cell will show them to be distinct from one another, while cells with a fusion of *BCR* and *ABL* will have the *BCR* and *ABL* signals co-localized and forming a yellow color. These abnormal cells have one *BCR-ABL* fusion (*yellow arrow*), while the second chromosomes of the respective pairs are unaffected, with separable, normal green and red signals.

kinetics and worsening prognosis. As the name implies, chronic phase predicts for the longest natural history, which may last for decades with appropriate therapy. Hematologically, chronic phase is characterized by an increase in neutrophils without other significant hematologic abnormalities. Progression to accelerated phase is marked by an increase in myeloid blasts (very immature white blood cells) to between 10 and 19% of all white blood cells in the blood or bone marrow and an increased proportion of basophils to 20% or greater. A significant decrease or increase in platelet count may also denote accelerated-phase disease. Blast-phase disease is analogous to acute leukemia, which follows an aggressive clinical course, and is defined as having 20% or greater myeloid blasts in the blood or bone marrow. At a molecular level, progression from chronic to accelerated to blast phase results from acquisition of additional chromosomal abnormalities in addition to the Philadelphia chromosome, which is known as **clonal evolution**.

Management of CML in chronic phase

The vast majority of patients with newly diagnosed chronic-phase CML will achieve a sustained remission with modern therapy and ideally enjoy a normal life expectancy. This represents a remarkable paradigm shift compared to before the year 2000 when the median survival of a newly diagnosed patient with chronic-phase CML was only five to seven years. The prognosis and treatment of CML was revolutionized by the development of the oral tyrosine kinase inhibitor (TKI) imatinib mesylate, which binds to the ATP-binding domain of the BCR-ABL fusion protein, resulting in inhibition of the protein's oncogenic activity (Figure 4.8).

A Phase I study of imatinib in chronic-phase CML patients showed that nearly 100% of patients treated at therapeutic dose levels experienced a complete **hematologic remission** (normalization of all blood counts), and over half of patients experienced a **cytogenetic remission** (disappearance of the Philadelphia chromosome as measured by FISH), an uncommon achievement with the prior standard therapeutic combination of subcutaneous **cytarabine**, a standard chemotherapy drug (see Table 2.2), and interferon alpha.

A randomized trial then compared imatinib to cytarabine plus interferon, and found a complete hematologic remission rate of 95% in imatinib-treated

Figure 4.8. Imatinib blocks ATP binding to BCR-ABL and downstream phosphorylation of effector proteins.

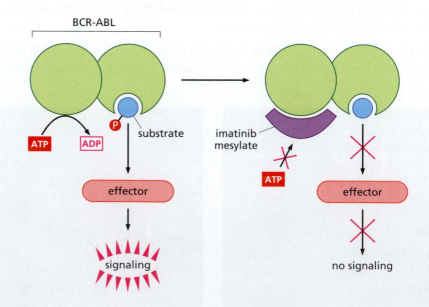

patients compared with 55% in patients treated with conventional therapy. Complete cytogenetic remissions were observed in 76% of patients treated with imatinib and only 15% in patients treated with cytarabine and interferon. Among patients achieving a complete cytogenetic remission on imatinib, approximately half achieved a **major molecular response**, defined as an equal to or greater than 3 log (1000-fold) reduction in *BCR-ABL* transcript by PCR. Interestingly, this increased to 80% by four years, demonstrating the possibility of ongoing improvement in response even years after initiation of therapy. After five years, only 7% of imatinib-treated patients had progressed to accelerated- or blast-phase CML, and 89% were alive. Risk of progression was lowest in patients who had a complete cytogenetic or major molecular response.

Adverse events on imatinib were common, but resulted in discontinuation of therapy in only 4% of patients. Common toxicities included peripheral edema (60%), muscle cramps (49%), diarrhea (45%), nausea (50%), musculoskeletal pain (47%), rash (40%), abdominal pain (37%), fatigue (39%), joint pain (31%), and headache (37%). By contrast, the combination of cytarabine and interferon was more toxic and less effective than imatinib, with only 3% of patients continuing to receive that therapy five years later due to both progressive disease and toxicities.

Imatinib became the first FDA-approved TKI for the treatment of cancer in 2001. Following the great success of imatinib, second-generation tyrosine kinase inhibitors were developed to try to improve efficacy with decreased toxicity. Two such inhibitors, **dasatinib** and **nilotinib**, were each compared head-to-head with imatinib for initial therapy of chronic-phase CML, and both demonstrated higher rates of complete cytogenetic and major molecular responses compared with imatinib, but without any difference in overall survival. All three tyrosine kinase inhibitors are currently FDA approved as initial therapy, and selection of agent for an individual patient requires consideration of both efficacy and unique toxicity profiles of the three agents. (Interested readers may read Robert A. Weinberg's *The Biology of Cancer* for additional background on the development of imatinib mesylate and perspective on resistance to TKIs and other drugs.)

The case of Jeffrey Paul, a man with a twisted ankle and an elevated white blood cell count

Jeffrey was a 53-year-old man who presented to the emergency department with left ankle pain after twisting his foot while lunging for a ball on the tennis court earlier in the day. He had no other complaints and reported being extremely healthy, though he admitted to not having seen his primary care physician for several years. Physical examination of the left ankle revealed swelling, tenderness, and some erythema (redness). An X-ray revealed no fractures; he was diagnosed with an ankle sprain. A routine complete blood count during his emergency department visit did show an elevated white blood cell count of 15,000 cells/μl (normal range, 4.5–11 $\times$ 10^3 cells/μl), which was attributed to the stress of his ankle injury. He was given ibuprofen and his ankle was wrapped in an elastic bandage. Follow-up with his primary care physician was recommended.

The ankle injury improved and Jeffrey did not seek further follow-up. One year later, he noted mild fatigue and the sensation of fullness on the left side of his abdomen. He also felt that he was getting full after eating less food than previously accustomed. He visited his primary care physician, who performed a physical examination and found Jeffrey to be well appearing overall but could feel the spleen to be enlarged on physical examination, palpable four fingerbreadths below the left ribcage (normally, the spleen cannot be palpated). The examination was otherwise unremarkable. A complete blood count showed a WBC of 37,000 cells/μl. The differential of types of white blood cells showed them to be 79% neutrophils, 1% blasts, 4% myelocytes, 3% lymphocytes, 1% monocytes, 1% eosinophils, and 12% basophils. The hemoglobin was 10.5 g/dl (normal range, 13.5–17.5 g/dl), and the platelet count was 325,000 cells/μl (normal range, 150,000–400,000 cells/μl). The primary care physician referred Jeffrey to a hematologist for evaluation of increased neutrophils with mild anemia (decreased red cells) and splenomegaly.

Based on the laboratory studies and physical examination findings, the hematologist made a presumptive diagnosis of CML. This was based on examination of the peripheral blood smear showing predominantly mature neutrophils along with immature myeloid forms (myelocytes) and increased numbers of basophils, slightly increased platelet count, and characteristic presence of splenomegaly on physical examination. To confirm the diagnosis, peripheral blood was sent for cytogenetic analysis, which showed a translocation of chromosomes 9 and 22, producing the t(9;22)(q34;q11.2) configuration, also known as the Philadelphia chromosome. FISH analysis confirmed the juxtaposition of the *BCR* gene from chromosome 22 adjacent to the *ABL* gene on chromosome 9. The diagnosis of CML in chronic phase was confirmed.

Jeffrey started imatinib mesylate, taking one pill by mouth daily. Two weeks after beginning therapy his blood counts already showed a response, with reduction in his WBC to 10,000 cells/μl and improvement in his hemoglobin to 11.5 g/dl. By three months after initiating therapy, his WBC was normal, with a normal **differential** (distribution of different types of white blood cells), and the anemia had completely resolved. He felt very well with no remaining abdominal discomfort or early satiety. Physical examination at that point demonstrated no remaining splenomegaly. Based on the return of all of his blood counts to normal, Jeffrey had a complete hematologic response to imatinib therapy. Cytogenetic evaluation at that time, however, showed a minor cytogenetic response with persistence of the t(9;22) chromosomal translocation in 40% of cells.

Jeffrey continued on imatinib therapy and had a sustained complete hematologic response and stable persistence of the t(9;22) transcript six months after initiation of therapy. Three months later, he was found to have an increasing WBC with an elevated proportion of mature neutrophils (80%). Basophils (normally less than 1%) represented 15% of the white blood cells, and occasional myelocytes were present (usually absent). There were no blasts seen. He was diagnosed as having progressive chronic-phase CML while on imatinib therapy. FISH confirmed presence of the t(9;22) transcript in 90% of cells. Mutational analysis was performed and showed the **T315I mutation**, a point mutation in BCR-ABL associated with imatinib resistance, present in the majority of tested cells.

Jeffrey's treatment was changed to ponatinib, an oral TKI with activity against the T315I mutation. He achieved a rapid complete hematologic response, and three months later, he had a complete cytogenetic response, with FISH detecting no cells harboring the t(9;22) translocation. qRT-PCR at that time showed a 1.5 log reduction in detectable *BCR-ABL* transcript. By six months after initiation of ponatinib, Jeffrey remained in hematologic response and the *BCR-ABL* clone was reduced greater than 3 logs. He now continues without detectable disease one year after initiation of second-line TKI therapy.

Management of accelerated- and blast-phase CML

Approximately 10 to 15% of patients with CML will present in accelerated- or blast-phase disease, and an additional 7% of patients will progress to accelerated- or blast-phase disease while receiving imatinib therapy (Table 4.2). A Phase II study of imatinib in accelerated phase found a complete hematologic response rate of only 34%, and 56% of patients had disease progression at one year. Higher response rates and improved response duration may be achieved with higher dosing than traditionally used for chronic-phase disease. Dasatinib and nilotinib have both been studied in patients with accelerated-phase CML that was resistant or intolerant to prior therapy, with complete hematologic responses seen in 45% and 26% of patients, respectively.

Blast-phase disease is defined as having ≥20% blasts in either the blood or bone marrow. These patients have poor responses to standard chemotherapy regimens employed for **acute myelogenous leukemia** (AML). Imatinib produces complete hematologic responses in only approximately 10% of patients, and only 7% of patients remain free from progression after three years. Dasatinib yields complete hematologic responses in about one-quarter of patients in blast crisis who had previously received imatinib, but very few responses are sustained. Similar results have been reported for the newer TKIs ponatinib and bosutinib.

Unlike TKI therapy in chronic phase, most responses in accelerated phase or blast crisis are not sustained, with the majority of patients ultimately progressing while on therapy. For this reason, eligible patients may undergo **allogeneic hematopoietic stem cell transplant** at the time of their best response to TKI therapy with the goal of achieving lifelong disease eradication. In an allogeneic stem cell transplant, patients receive donor stem cells from a healthy individual with the goal of having a new immune system recognize the patient's leukemia cells as "foreign" and thus leading to immune-mediated killing of the leukemia cells, known as a **graft-versus-leukemia** effect. This is in contrast to an autologous hematopoietic stem cell transplant in which patients are infused with their own previously collected stem cells following administration of high-dose chemotherapy with the goal of restoring healthy bone marrow function that would have been eradicated by the intensive chemotherapy.

Table 4.2. World Health Organization (WHO) criteria for CML phases.

Accelerated-phase CML	Blast-phase CML (blast crisis)
10 to 19% blasts in the peripheral blood or bone marrow	≥20% blasts in the peripheral blood or bone marrow
Peripheral blood basophils ≥20%	
Platelets <100,000/μl, unrelated to therapy	
Platelets >1,000,000/μl, unresponsive to therapy	
Progressive splenomegaly and increasing white cell count, unresponsive to therapy	
Cytogenetic evolution (defined as the development of chromosomal abnormalities in addition to the Philadelphia chromosome)	

Management of CML when initial TKI therapy fails

Among patients with CML of any phase whose disease progresses while receiving initial TKI therapy, mutational analysis of the BCR-ABL kinase can help guide selection of the next therapy, as the available TKIs do exhibit unique sensitivity patterns based on the types of mutations present. For example, the presence of certain mutations (Y253H, E255K/V, and F359V/C/I) predict improved sensitivity to dasatinib compared with other TKIs, while other mutations (T315A, V299L, and F317L/V/I/C) are more likely to be inhibited by nilotinib. The most difficult to treat mutation has been the T315I mutation, which predicts resistance to all three available TKIs used as initial therapy (imatinib, dasatinib, and nilotinib). The T315I mutation is caused by a single cytosine to thymine substitution at position 944 of the *ABL* gene, resulting in the amino acid threonine (T) being substituted by isoleucine (I). T315I causes steric hindrance to binding of most TKIs to the BCR-ABL fusion protein, thus allowing escape from imatinib inhibition and ongoing constitutive activity of the kinase. The novel TKI **ponatinib**, however, demonstrated encouraging clinical activity in a Phase II study of patients who were resistant to or intolerant of prior TKI therapy, including patients harboring the T315I mutation. These data garnered ponatinib FDA approval for second-line therapy of CML in 2012.

Future prospects and challenges in CML

CML remains the prototype for successful molecularly targeted therapy. Challenges in the modern era continue to be development of mutations in the BCR-ABL kinase that render the cells resistant to TKI therapy, as well as the intolerance of some patients to these generally well-tolerated oral agents. The availability of four different TKIs with activity in CML has helped overcome resistance and intolerance in many patients, but not all. Additional agents are currently in development to further target highly resistant mutations.

Take-home points

✓ Chronic myelogenous leukemia (CML) manifests with an increase in neutrophils, and may be asymptomatic or present with symptoms related to splenomegaly or fatigue.

✓ CML results from the t(9;22)(q34;q11) chromosomal translocation producing the BCR-ABL fusion protein, the defining cytogenetic marker of this disease.

✓ Molecular targeting of BCR-ABL with oral tyrosine kinase inhibitors revolutionized CML treatment and ushered in the modern era of molecularly targeted cancer therapy.

✓ Tyrosine kinase inhibitors available for initial therapy of CML include imatinib, dasatinib, and nilotinib.

✓ Patients with disease resistant to initial TKI therapy can have mutational analysis to determine if an alternative TKI may have increased activity against the specific mutation in their disease.

✓ Ponatinib is approved for patients whose disease is resistant to an initial TKI, and has activity against the most resistant mutation identified in CML, the T315I.

✓ Patients with accelerated- or blast-phase CML, as well as those with chronic-phase CML resistant to multiple TKIs, may be candidates for allogeneic hematopoietic stem cell transplantation.

Discussion questions

1. What is the underlying genetic defect that causes CML?

2. What are the clinical and laboratory features of CML?

3. How did the drug imatinib mesylate change the paradigm for treatment of CML?

4. How would you treat a CML patient whose disease has stopped responding to imatinib? Why do initially effective treatments eventually stop controlling the disease?

Case 4-2   Chronic Lymphocytic Leukemia

Topics bearing on this case

Somatic mutation

Various types of more common hematopoietic malignancies

Immunoglobulin light and heavy chains

Introduction

Chronic lymphocytic leukemia (CLL) is an **indolent** (slow growing) lymphoproliferative neoplasm characterized by progressive accumulation of functionally inactive malignant lymphocytes in the peripheral blood and in lymphoid organs. It is derived from mature B lymphocytes, which have multiple roles in fighting infection, including production of antibodies against pathogens. CLL most commonly involves the blood, bone marrow, and lymphoid organs, but may involve other sites as well. When CLL cells involve predominantly lymph nodes or other solid tissues (as opposed to primarily being found circulating in the blood), the disease is termed **small lymphocytic lymphoma** (SLL), but this term is considered synonymous with CLL rather than denoting a distinct disease. CLL is a clinically heterogeneous disease, with some patients having indolent disease that may never limit life expectancy whereas a small subset of patients will have more aggressive or treatment-resistant disease and a very guarded prognosis.

The clinical heterogeneity is reflected at the molecular level with a diverse range of genetic insults contributing to disease pathogenesis. CLL is characterized as having either mutated or unmutated **immunoglobulin** (Ig) **genes**. Following antigen stimulation many normal B lymphocytes undergo **somatic mutation** (as opposed to **germ-line mutation**) of their **immunoglobulin heavy-** and **light-chain genes** in order to improve their capacity to bind their target antigens. Approximately half of cases of CLL show evidence of mutated **Ig heavy-chain variable-region** (IGHV) genes, while the other half show unmutated IGHV, consistent with the latter subset arising from a less mature cell of origin. CLL patients with unmutated Ig variable-region genes have an inferior prognosis when compared with their mutated counterparts. Additional genetic events further contribute to pathogenesis and prognosis, including defects of tumor suppressors, such as **TP53** and **ATM**, activated oncogene products, such as BCL2, NOTCH1, and SF3B1, as well as defects in **microRNAs**. Mutations of the **B-cell receptor complex** (BCR) that enable antigen-independent signaling can further promote the proliferation and survival of the malignant cells.

Epidemiology

CLL accounts for two to six new cases per year per 100,000 people, or approximately 15,000 new cases annually in the United States. It is the most common leukemia, accounting for approximately 30% of all cases. The incidence varies

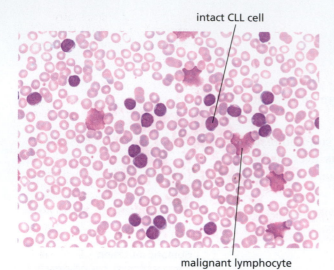

intact CLL cell

malignant lymphocyte

Figure 4.9. Peripheral blood smear of a patient with CLL. The smear is notable for an abundance of mature appearing lymphocytes and occasional "smudge cells," so called because these CLL cells become smudged during the slide preparation due to their increased fragility.

by race and geography, with a higher rate of CLL among Caucasians and only rare occurrence in patients of Asian descent. Most cases are sporadic, though there clearly is an inherited predisposition in a subset of patients. Approximately 15% of first-degree family members (parent, sibling, or child) of a CLL patient will also develop CLL or a related disorder, which constitutes an approximately sevenfold increased risk compared with the general population. CLL is a disease of the elderly, with a median age at diagnosis of 70 years. There is a slightly increased incidence among men compared with women.

Diagnosis, workup, and staging

Approximately one-third of CLL patients are asymptomatic at presentation, with the disease incidentally detected due to leukocytosis (high white blood cell count, or WBC count) observed on a routine complete blood count (CBC) (Figure 4.9). Symptomatic patients may present with painless lymphadenopathy (enlargement of lymph nodes), symptoms related to splenomegaly (spleen enlargement causing feelings of abdominal fullness, left upper quadrant discomfort, referred pain to the left shoulder, or early satiety), or fatigue. Patients may also present with symptoms related to **cytopenias** (low blood counts), including fatigue or shortness of breath due to anemia, bleeding or easy bruising due to thrombocytopenia (low platelets), or persistent or recurrent infections due to leukopenia (low white blood cells) or **hypogammaglobulinemia** (decreased antibodies). A minority of patients will present with **"B" symptoms**, which include fever, drenching night sweats, or unintentional weight loss. Physical examination may be completely normal, but common physical findings also include palpable lymphadenopathy or splenomegaly.

The CBC is notable for an increased WBC count in the majority of patients, with the differential of white blood cells showing the lymphocyte population to be increased and responsible for the overall leukocytosis. The malignant lymphocytes are small and mature and may not be easily distinguishable from normal lymphocytes on examination of a peripheral blood smear. The red blood cells and platelets are also evaluated on the CBC, as both can be decreased in advanced stages of CLL. Since many conditions other than CLL can cause lymphocytosis (Table 4.3), the diagnosis must be confirmed, most commonly with a technique called **flow cytometry**. In flow cytometry, the lymphocytes are assessed for **clonality** (derivation from the same cell) as well as for the pattern of surface proteins that typically characterizes CLL. Clonality

Table 4.3. Causes of mature lymphocytosis.

Leukemias and lymphomas (CLL, prolymphocytic leukemia, hairy cell leukemia, large granular lymphocyte leukemia, acute T-cell leukemia or lymphoma, mantle cell lymphoma, follicular lymphoma, Burkitt lymphoma, Sézary syndrome)
Infections (viruses, bacteria, protozoa, parasites, rickettsia)
Autoimmune diseases
Drug reactions
Serum sickness
Hyperthyroidism
Stress response (cigarette smoking, trauma)

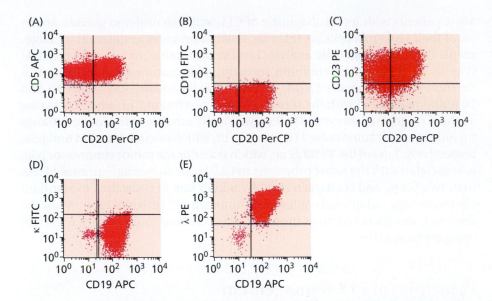

Figure 4.10. Flow cytometry plots confirming cell surface markers for CLL on the surface of lymphocytes. Each axis is for a different marker and measures strength of expression. The cross lines on these plots are quasi-translated axes: They divide what is considered positive and negative. (A) The lymphocytes are nearly all positive for CD5 and exhibit a range of CD20 expression with many of the cells only dimly expressing CD20. (B) The CD20-positive cells are uniformly negative for CD10, as is expected in CLL, though (C) they are positive for CD23. (D, E) Expression of either κ or λ light chain versus CD19, which is present on all CLL cells. These plots show the cells to be positive for CD19 as expected, and uniformly positive for the λ light chain and negative for the κ light chain. This test confirms clonality, as a normal lymphocyte population would have a κ:λ ratio of 1.5–2:1. This pattern of flow cytometric findings is consistent with a diagnosis of CLL. APC, allophycocyanin; FITC, fluorescein isothiocyanate; PE, phycoerythrin; PerCP, peridinin chlorophyll protein.

is confirmed by uniform expression of either **κ (kappa)** or **λ (lambda) light chains**. In a normal population of B lymphocytes, the ratio of cells displaying κ versus λ light chains is 1.5–2:1. Predominance of one light chain relative to the other, therefore, would suggest monoclonality. The typical pattern of proteins detected on the surface of CLL cells is **CD5**, **CD20** (dimly expressed), **CD19**, and **CD23** (Figure 4.10). If a diagnosis cannot be discerned from the peripheral blood, a bone marrow biopsy or lymph node biopsy may be required.

Diagnosis of CLL in the peripheral blood requires an absolute number of at least 5000 clonal lymphocytes/μl, as determined by flow cytometry and the complete blood count. Fewer than 5000 clonal cells/μl in the absence of lymphadenopathy or splenomegaly does not qualify for a malignant diagnosis and instead is called **monoclonal B-cell lymphocytosis** (MBL). This classification is based on the observation that approximately 5% of healthy blood donors will have a small population of these cells detected in their blood using flow cytometry, but at such a low number, they rarely behave in a malignant manner. The risk of a patient with MBL subsequently meeting criteria for CLL is only 15%, and fewer than 10% of patients with MBL will ever require therapy directed at CLL. MBL is therefore best considered a potentially premalignant condition that requires nothing more than routine surveillance. Patients with fewer than 5000 clonal cells/μl with presence of adenopathy, splenomegaly, or mass lesions related to the disease are technically classified as having small lymphocytic lymphoma (SLL), which is biologically identical to CLL and is managed accordingly.

CLL is staged with the Rai staging system, which employs only a CBC and physical examination to determine clinical stage. Stage 0 disease is characterized by circulating lymphocytosis in the absence of palpable lymphadenopathy or splenomegaly, or of anemia or thrombocytopenia on the CBC. Stage I disease is lymphocytosis plus adenopathy, stage II disease includes splenomegaly or **hepatomegaly** (liver enlargement), while stage III and stage IV disease are characterized by the presence of anemia and **thrombocytopenia**, respectively. Stage 0 disease is commonly referred to as low risk, stage I–II as intermediate risk, and stage III–IV as high risk. Average life expectancy for these groups had historically been estimated as approximately 13 years, 8 years, and 2 years for low-, intermediate-, and high-risk disease, respectively, though advances in therapy have substantially improved prognosis in the modern era.

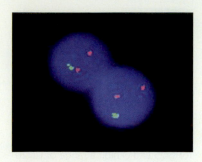

Figure 4.11. FISH for chromosome 17p and 11q deletions. FISH analysis performed on these two CLL cells, employing fluorescent probes evaluating both *TP53* on 17p13 (*red*) and *ATM* on 11q22 (*green*) shows two normal copies of *TP53*, but loss of one of the two copies of *ATM*.

Most patients with a new diagnosis of CLL will also undergo genetic analysis of their leukemia cells, as this can assist in prognostication and management. Targeted cytogenetic analysis is most commonly performed with FISH (Figure 4.11). The most common recurring chromosomal abnormality is a deletion of chromosome 13q14, which codes for microRNAs (miR) 15 and 16 and is associated with the most favorable survival in CLL when occurring as the sole genetic abnormality. Prognostically adverse cytogenetic findings include loss of chromosome 11q22 or 17p13, which encode for ATM and p53, respectively. Loss of the *TP53* gene, which encodes the tumor suppressor p53, is associated with the worst prognosis in CLL, with an average survival of less than two years, and is characterized by a high rate of resistance to standard chemotherapy. Additional copies of chromosome 12 (trisomy 12) may also be detected, though outcome in these patients is similar to those with a normal cytogenetic analysis.

Principles of CLL management

Numerous highly effective therapies are available to control CLL, though standard therapies cannot currently cure the disease. Given the usually indolent course of disease and myriad treatment options, the realistic goal of CLL management is to preserve a normal life expectancy and quality of life. Many patients will present with asymptomatic disease, prompting the question whether patients with newly diagnosed CLL that is asymptomatic and not interfering with healthy organ or bone marrow function actually benefit from early initiation of therapy. Several randomized controlled trials have addressed this question and found outcomes ranging from no benefit to immediate benefit by starting chemotherapy in asymptomatic patients compared with beginning chemotherapy at the time that symptoms or complications of CLL arise. In such studies, approximately one in five patients assigned to initial observation will never require chemotherapy for their disease, and among the majority who do ultimately require therapy, that treatment is often not required for five or more years from the time of diagnosis. That lengthy interval prior to therapy is a time when patients can avoid the toxicities of therapy when they otherwise feel well, and allows ongoing development of newer treatment strategies that may better target the disease and carry fewer toxicities than currently available therapies. As such, patients without indications for therapy at the time of CLL diagnosis are followed with clinical surveillance alone. Indications for beginning treatment of CLL include development of symptoms, rapidly progressing or bulky disease, impairment of organ function, decreased immune function producing frequent infections, or low blood counts related to bone marrow infiltration by CLL. A small proportion of patients with CLL will have their disease transform into a more aggressive lymphoma, called **Richter's transformation**, which is also an indication for immediate therapy.

Available treatments for CLL include single-agent **alkylating chemotherapy** (chlorambucil or **cyclophosphamide**), nucleoside analogs such as **fludarabine**, and other chemotherapy agents including **bendamustine**. Monoclonal antibodies have become a critical component of therapy in lymphoid malignancies as well, and act by binding to a specific target antigen on the surface of malignant cells, thereby targeting the body's immune response against those cells in a process known as **antibody-dependent cell-mediated cytotoxicity** (ADCC) (see Chapter 11). FCR—fludarabine, cyclophosphamide, and rituximab (a monoclonal antibody targeting the CD20 antigen on lymphocytes; see Case 11-1)—is the most effective combination chemotherapy to date, producing a median progression-free survival of approximately four years, and resulting in an improved overall survival when compared with patients treated with FC (fludarabine and cyclophosphamide). This regimen does carry toxicities including fatigue and bone marrow suppression, so use is limited to fit patients under the age of 70. Older patients may be treated with gentler but highly effective therapies combining less toxic chemotherapy drugs and monoclonal antibodies targeting B cells, with median progression-free survival of over two years.

The case of Patricia O'Halloran, a 68-year-old professor of German history with an elevated WBC count

Pat presented for her routine annual history and physical exam without any health complaints. She had a normal physical examination, though her CBC showed a mildly elevated white blood cell count of 18,000 cells/μl (normal range, 4.5–11 × 10^3 cells/μl). Her other blood counts including red blood cells and platelets were entirely normal. The laboratory test was repeated with a "differential" to show the different subpopulations of white blood cells, which revealed them to be 70% lymphocytes, 25% neutrophils, 3% monocytes, 1% eosinophils, and 1% basophils. In order to determine whether these lymphocytes were normal reactive cells or potentially malignant, the blood was sent for flow cytometry, which showed that 60% of her lymphocytes carried only the λ immunoglobulin light chain, dimly expressed the surface protein CD20, and strongly expressed the CD5 and CD23 proteins. A diagnosis of CLL was made. Cytogenetic analysis using FISH showed an isolated deletion of chromosome 13q14 in 28% of her lymphocytes. She was encouraged to hear that her stage was 0 on the basis of an isolated lymphocytosis and that the sole chromosomal abnormality of 13q deletion also predicted a favorable natural history and an excellent prognosis. Given the absence of indications for therapy, surveillance was recommended.

She was initially followed by her oncologist with physical examination and laboratory testing every three months. Over a three-year period, the total WBC was noted to only gently increase up to 23,000 cells/μl. She remained asymptomatic and without any low blood counts. Surveillance intervals were increased to every six months. She continued to do well with a slowly rising WBC for the next 17 years, by which time her WBC was 233,300 cells/μl, but still with a normal hematocrit (a measure of red blood cells) of 38% and platelet count of 208,000. At her visit with her oncologist, she described increasing fatigue to the point that it was becoming difficult to follow through with her daily activities. She was 83 years old at that time, but still teaching two courses each semester and had recently published a book on the cultural history of Bavaria. Due to progressive fatigue related to her CLL, she and her oncologist decided treatment would be needed. FISH analysis was sent prior to therapy and showed the known chromosome 13q14 deletion in 72% of cells, but now also with a 17p13 deletion in 9% of cells. She was treated with the anti-CD20 monoclonal antibody **obinutuzumab** plus the oral chemotherapy agent chlorambucil. She tolerated the treatment extremely well, and one week after beginning therapy, her WBC had plummeted to 3700 cells/μl. Her energy steadily improved, and at the end of six months of treatment, she felt well and was in a complete remission. She was again followed with observation.

Pat did very well for just greater than two years, at which time she noted swelling of lymph nodes in her neck and under her arms, as well as new fatigue. She brought this to the attention of her oncologist, who confirmed the presence of lymphadenopathy. He also noted an increased spleen size in the left side of her abdomen. A CBC showed her WBC newly elevated to 42,100 cells/μl with predominantly lymphocytes. She now had anemia with a hematocrit of 31% and a platelet count that was mildly decreased at 115,000. CT scans were obtained confirming diffuse lymphadenopathy and splenomegaly. In order to eliminate the possibility of a Richter's transformation, a large axillary lymph node was excised and showed only her known CLL. FISH analysis showed 13q14 deletion in 75% of cells but now with a 17p13 deletion in 68% of cells. She was prescribed the drug **ibrutinib** and instructed to take 420 mg by mouth once daily. She had mild diarrhea that quickly resolved, and within a few days of starting the drug she noted her lymph nodes to be rapidly decreasing in size. Her oncologist observed during that same interval that her WBC and lymphocyte count had rapidly increased in her CBC, with her WBC reaching 117,600 cells/μl. She felt better, so she continued on ibrutinib and over the subsequent four months her WBC gradually declined to normal at 8600 cells/μl. Her anemia and thrombocytopenia also improved to normal. Pat now continues on daily ibrutinib with a normal CBC and no adverse symptoms, lymphadenopathy, or cytopenias. She has been taking ibrutinib for two and a half years and her plan is to remain on this daily medication as long as she has no evidence of recurrent CLL.

Treatment of relapsed CLL

Patients with CLL that relapses after prior therapy may be followed with surveillance alone if their disease is not causing symptoms and other indications for the need of therapy are absent at the time of recurrence. Should the disease require therapy, considerations include the patient's age and fitness, medical co-morbidities, cytogenetic features of the CLL, and type of prior therapy. For patients who experienced a lengthy remission to their front-line therapy, the same regimen may be reused. The same range of chemotherapy agents and combinations available for front-line therapy may be used at relapse, as well as additional agents including the monoclonal antibody **ofatumumab** (anti-CD20), the inhibitor of PI3K delta idelalisib, the BCL2 inhibitor venetoclax, or high doses of steroids. Allogeneic stem cell transplantation may be considered for select younger patients with high-risk disease or disease that relapses quickly after initial therapy. Patients whose disease harbors the chromosome 17p13 deletion have very poor response rates and brief remissions to traditional chemotherapy drugs. Improved responses are seen with newer targeted therapies including ibrutinib, idelalisib, and venetoclax. Despite the potential for allogeneic stem cell transplantation to be curative in CLL, it is not more broadly applied due to the high risk of potentially fatal toxicities associated with this therapy, including infections related to chronic immunosuppression, and a complication in which the new transplanted immune system can attack the patient's native tissues including skin, gastrointestinal tract, liver, and other organs, called **graft-versus-host disease**.

Ibrutinib is a small-molecule covalent inhibitor of **Bruton's tyrosine kinase** (BTK) that has excellent activity in CLL, including patients who are resistant to chemotherapy or whose disease harbors the 17p13 deletion. BTK was named after Dr. Ogden Bruton, who initially described a disease of childhood known as **X-linked agammaglobulinemia**, which is a congenital immunodeficiency disease caused by a germ-line *BTK* mutation on the X chromosome, resulting

Figure 4.12. BCR signaling cascade and actions of ibrutinib. Activation of the B-cell receptor results in phosphorylation and activation of several critical proteins in the cascade, including BTK. This cascade results in activation of ERK and NF-κB, which promote cell survival and proliferation. Inhibition of BTK with ibrutinib successfully interrupts signaling and leads to down-regulation of pathways critical for lymphoma cell proliferation and survival, including ERK, NF-κB, JNK, MAPK, and NFAT.

Ibrutinib also results in decreased integrin $\alpha_4\beta_1$-mediated adhesion to fibronectin in the microenvironment and down-regulation of CD62L (L-selectin), which facilitates homing of lymphocytes to lymphoid tissues; as well as inhibiting chemotaxis toward CXCL12 and CXCL13 and release of CCL3 and CCL4. These multiple factors result in egress of CLL cells from their microenvironment in lymph nodes, bone marrow, and other lymphoid tissues, so that they can no longer receive pro-survival and pro-proliferative signaling from these tissues.

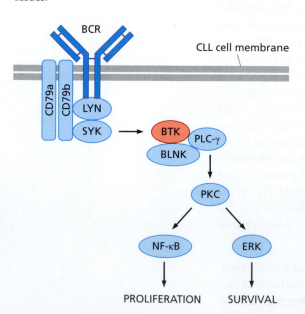

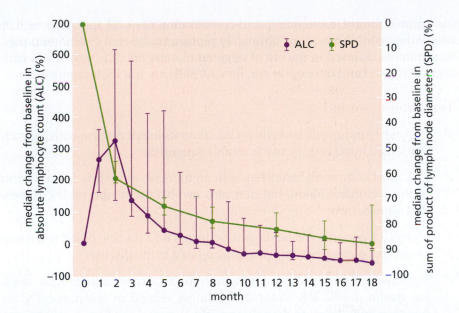

Figure 4.13. Ibrutinib effects on lymph node size and lymphocytosis. This graph demonstrates the effect of ibrutinib therapy over time on the absolute lymphocyte count (ALC) in the blood and the sum of the products of lymph node diameters (SPD). After the start of ibrutinib treatment, the size of lymph nodes rapidly declines and is associated with a transient spike in the number of circulating lymphocytes, which peaks one to two months into treatment and then gradually declines. (Adapted from J.C. Byrd et al., *N. Engl. J. Med.* 369:32–42, 2013. Reprinted with permission from Massachusetts Medical Society.)

in chronic infections related to absence of mature B lymphocytes and antibodies. In CLL, ibrutinib blocks the B-cell receptor (BCR) signaling cascade, which promotes survival of CLL cells by activating **ERK1/2** and **NF-κB**, among others (Figure 4.12). Ibrutinib also inhibits cellular adhesion, the binding of cells to substrate; and **chemotaxis**, the movement of cells toward chemicals. Inhibition of cell adhesion and chemotaxis is responsible for the surge of lymphocytes seen in the peripheral blood after initiation of ibrutinib as a result of the exodus of these cells from the confines of the lymph node and bone marrow compartments. A Phase I study in patients with relapsed or refractory CLL demonstrated an overall response rate of 69%. The drug is generally very well tolerated, with the most common toxicities being mild diarrhea, nausea, and fatigue. Decreased blood counts or infections related to ibrutinib were uncommon. The subsequent Phase II study found a response rate of 71% in relapsed or refractory patients, with 75% of patients remaining free from CLL progression two years after beginning therapy (Figure 4.13). Encouragingly, there was no difference in response rate in patients with chromosome 17p- deletion. Based on these encouraging data, ibrutinib was approved by the FDA in February 2014 for the treatment of relapsed or refractory CLL in the United States. A subsequent randomized clinical trial evaluated ibrutinib as initial treatment for CLL in adults 65 years or older, compared to the oral chemotherapy drug chlorambucil. Ibrutinib was superior with 85% of ibrutinib patients remaining free from progression three years later, compared to only 28% on chlorambucil. Ibrutinib is now a standard front-line treatment option for elderly CLL patients.

Future prospects and challenges in CLL

Ongoing studies are evaluating whether ibrutinib may be an appealing initial treatment for CLL in younger patients given the strong efficacy and excellent tolerability of this oral agent in elderly adults. Additional clinical trials are evaluating ibrutinib in combination with chemotherapy and antibody therapy. Novel therapies targeting distinct pathways are also showing promise in clinical trials for CLL, including inhibitors of PI3 kinase (idelalisib) (see Chapter 5) and the anti-apoptotic protein **BCL-2** (venetoclax). As the therapeutic landscape for CLL continues to evolve, a challenge will be determining the optimal timing, sequencing, and combination of novel agents, as well as where they combine with or ultimately replace traditional chemotherapies. Ibrutinib has ushered in the era of targeted therapy for CLL, which will continue to expand and offer optimism for CLL patients and their families.

Take-home points

✓ Chronic lymphocytic leukemia (CLL) is an indolent disease with a generally long natural history and favorable prognosis.

✓ CLL most commonly manifests with an increase in mature lymphocytes in the peripheral blood, but may also involve the lymph nodes, spleen, and other organs.

✓ Many CLL patients will be diagnosed based on a routine blood test result alone, and experience no symptoms related to the disease.

✓ For those patients experiencing symptoms, CLL may cause fatigue, swollen lymph nodes, left abdominal fullness related to splenomegaly, or, uncommonly, fever or drenching night sweats. Low blood counts may increase the risk of infections or easy bruising or bleeding.

✓ CLL is a biologically heterogeneous disease.

✓ CLL cases with an unmutated IGHV carry an inferior prognosis, as does CLL harboring deletions of either *TP53* on chromosome 17p13 or *ATM* on chromosome 11q22, which predict for relative resistance to chemotherapy.

✓ CLL cell survival relies on signaling through the B-cell receptor signaling cascade. Inhibition of BTK with ibrutinib shuts down this signaling pathway and produces sustained responses in the majority of CLL patients.

✓ Additional promising agents for CLL include those targeting BCL-2 or PI3 kinase.

Discussion questions

1. How does CLL usually present, and how is it diagnosed?

2. When is treatment required for a newly diagnosed patient with CLL?

3. How does ibrutinib work in the treatment of CLL?

Topics bearing on this case

Oncogene addiction

Targeted therapies

Mechanisms of drug resistance

Case 4-3   Lung Cancer

Introduction and epidemiology

Lung cancer is the third-most diagnosed cancer in the United States, after prostate and breast cancers. However, it is undisputedly the most common cause of cancer death, with an estimated 154,050 deaths among U.S. men and women in 2018. Worldwide, lung cancer due to prevalent tobacco use remains a significant cause of morbidity and mortality. Thus, there remain numerous opportunities for improvement in care of patients with advanced lung cancer.

Smoking causes lung cancer. Tobacco use is associated with 85 to 90% of lung cancer cases in the United States. Nonsmokers who live with smokers have a

20 to 30% increased risk of developing lung cancer from exposure to "second-hand" smoke. Other carcinogens associated with lung cancer development include radon and asbestos.

Diagnosis, workup, and staging

Symptoms of lung cancer include persistent cough, blood in sputum (hemoptysis), chest pain, and shortness of breath, among others (Table 4.4). Radiographic evaluation may lead to finding of a mass in the lungs with or without enlargement of regional lymph nodes. Radiologists occasionally describe lung masses as either **spiculated** ("spiked"; Figure 4.14) or rounded. While not a definitive description, spiculated masses raise suspicions of primary lung cancer. Masses in the lungs should also raise the possibility of cancer that has spread to the lungs from another location: metastatic cancer. Numerous malignancies can metastasize to the lungs, including cancers of the breast, gastrointestinal tract, head and neck, and kidney, among others. Presence of multiple lesions throughout the lungs should prompt consideration of metastatic disease.

Because of the high prevalence and mortality of lung cancer, screening for early-stage disease has been the subject of several large clinical trials. A study employing chest X-rays to screen for lung cancer in the general population (that is, among unselected patients) did not reveal a benefit in terms of increased overall survival. A study of chest CT scans in high-risk subjects, defined as ≥30 **pack-year** smokers (one pack-year = the number of cigarettes a one-pack-per-day smoker smokes in one year), revealed a 20% improvement in mortality compared with screening chest X-rays (National Lung Screening Trial, NLST). Adoption of these recent findings for screening in the general population was endorsed in 2012 by the American Society for Clinical Oncology (ASCO), the American College of Chest Physicians (ACCP), and the American Thoracic Society (ATS), but only in settings that can deliver appropriate care when cancers are detected. Whether broad uptake of this screening strategy will occur remains to be seen.

Nodules in the lungs are common sequelae of past infections and inflammatory processes. Distinguishing a malignant lesion from a benign nodule can be challenging. Thus, tissue diagnosis via biopsy remains key to the diagnostic workup. Tissue may be acquired via CT-guided **percutaneous biopsy** (that is, through the skin), during surgical evaluation (typically via **mediastinoscopy**, an invasive procedure involving camera-based evaluation of the chest cavity), or via **bronchoscopy** (evaluation of the airways using camera guidance) (Figure 4.15). Due to the common co-morbid problem of **chronic obstructive**

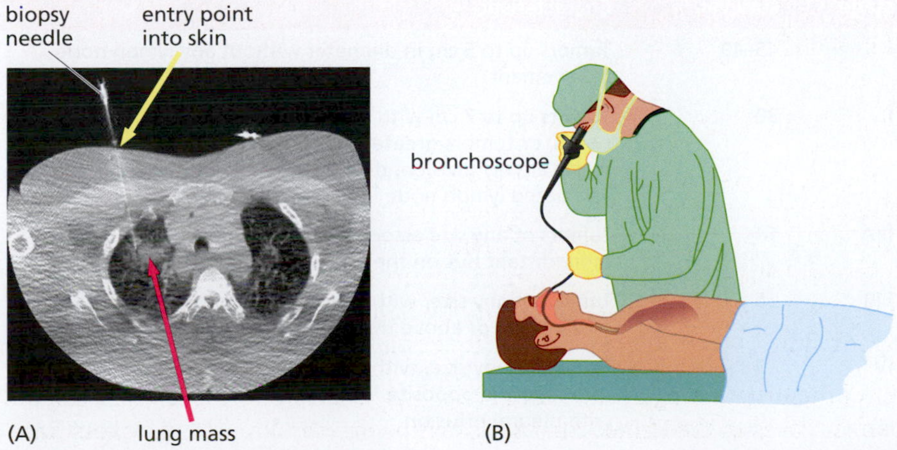

Table 4.4. Presenting symptoms of lung cancer.

Presenting symptom	%
Hemoptysis	17
Cough	17
Chest pain	15
Shortness of breath	12
None	12
Systemic symptoms	10
Pneumonia	8

Figure 4.14. Spiculated mass on chest CT.

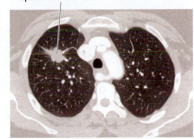

spiculated mass

biopsy needle   entry point into skin

bronchoscope

(A)   lung mass   (B)

Figure 4.15. Diagnostic sampling of lung cancer. (A) CT-guided biopsy of the lung mass in Figure 4.14. The interventional radiologist uses CT imaging to guide a needle toward the target lesion. (B) Schematic diagram of bronchoscopy.

pulmonary disease (COPD) in the smoking population, safe tissue acquisition is of paramount importance.

The tissue biopsy is evaluated by a pathologist to determine if cancerous cells are present, and if so, whether the cells appear to be of primary lung origin or have spread from another site (that is, are metastatic cancer to lung). Primary lung cancer is separated into two broad categories: **small-cell lung cancer** (SCLC) and **non-small-cell lung cancer** (NSCLC). SCLC is characterized by **neuroendocrine** features, and, like neuroendocrine cancers of other organs, is highly aggressive and often fatal. NSCLC is comprised of several different histologies, including adenocarcinoma, **squamous cell carcinoma**, **bronchioalveolar carcinoma**, and **large-cell carcinoma**. Lung cancers may also produce fluid in the space outside the lung called a **pleural effusion**. This fluid can also be sampled (by **thoracentesis**) for pathologic evaluation for malignant cells.

Chest CT evaluation is the standard imaging when a pulmonary mass has been identified. Due to the potential for lung cancer to spread to adrenal glands, liver, bones, lymph nodes, and brain, abdominal/pelvic CT scans and MRI of the brain are frequently obtained. The **fusion PET/CT** scan (merged CT and PET scans; Figure 4.16) is a relatively recent innovation and has become a routine tool in the staging of lung cancer (for more on imaging, see Chapter 1). **Fluorodeoxyglucose**-avid lymph nodes in the thorax connote high risk for locally invasive disease. Biopsy by a thoracic surgeon has the benefit of tissue sampling of the pulmonary lesion as well as sampling of regional lymph nodes.

SCLC is a highly aggressive lung cancer subtype accounting for 10 to 15% of all primary lung cancers. Radiation therapy combined with chemotherapy forms the basis of standard treatment for limited SCLC. The distinction between limited and extensive SCLC depends on whether all disease can be encompassed in one radiation field (limited) or not (extensive).

NSCLC can be categorized as localized (stages I–III) or metastatic (stage IV) (Table 4.5). For localized and locally advanced NSCLC, tumor size and involvement of lymph nodes determine stage (I–III).

Cancer stage is important for prognosis and treatment decision making. The five-year survival rates based on the National Cancer Institute's **Surveillance, Epidemiology, and End Results** (SEER) database are indicated in Table 4.5.

Figure 4.16. PET/CT fusion image of NSCLC in the right lung. The lung mass from the patient in Figures 4.14 and 4.15 (*red arrow*), with metastasis noted by abnormal uptake in the right hilar lymph node (*green arrow*).

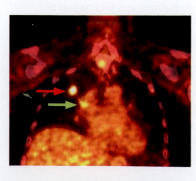

Table 4.5. Staging and survival of non-small-cell lung cancer.

NSCLC stage	Five-year survival, %	Characteristic
I	45–49	Tumors up to 5 cm in diameter without any lymph node involvement
II	30	Tumors up to 7 cm with involvement of nearby lymph nodes; or tumors greater than 7 cm or located near the major airway division, diaphragm, or chest wall, but without associated lymph node involvement
IIIA	14	Tumors of any size associated with lymph nodes that are more distant but on the same side of the chest
IIIB	5	Tumors of any size, with lymph nodes on the opposite side of the chest or above the collarbone
IV	1	Tumors of any size, with any lymph node involvement, and spread to the opposite lung or to distant organs; or tumors with pleural effusion

(Note that statistics are available for patients diagnosed between 1998 and 2000; see the American Cancer Society website.)

Management of localized lung cancer

SCLC. SCLC is a highly aggressive lung cancer subtype with early dissemination to other organs. Combination chemotherapy is used to kill cancer cells in both the primary tumor and microscopic deposits in other organs that may not be visible on imaging studies. Chemotherapy typically includes a combination of the cytotoxic agents **etoposide** and **cisplatin** (referred to as EP). Radiation therapy targeted to the primary lung tumor is combined with chemotherapy in limited SCLC (in which all disease can be encompassed in one radiation field). Because of the high propensity for SCLC to travel to the brain, prophylactic cranial irradiation (PCI) is often administered despite potential side effects, with Phase III clinical trial evidence that PCI improves local control in the brain as well as overall survival. Despite the localized nature of the disease, median overall survival for limited stage SCLC is only 14–20 months.

NSCLC. Surgery is typically the first step in management of localized NSCLC. For some patients who are not good surgical candidates because of other medical problems, radiation therapy may be used. For locally advanced disease (such as stage IIIA), chemotherapy is sometimes used to treat the primary tumor and lymph nodes, making surgical resection more feasible. Adjuvant chemotherapy after resection is employed in some early-stage patients with lymph node involvement.

Management of metastatic lung cancer

SCLC. Metastatic SCLC is treated with etoposide and cisplatin (EP) chemotherapy. The response rate is high, but unfortunately the durability of response is often short, leading to successive attempts at disease control with single-agent chemotherapy regimens, such as topotecan. Unfortunately, the success rate of disease control is low, with median overall survival of less than one year.

NSCLC. Metastatic NSCLC is similarly almost universally fatal. The mainstay of treatment has been combination chemotherapy, typically in "doublets" of chemotherapeutic agents, such as cisplatin combined with either pemetrexed or gemcitabine. More recently, the identification of molecular alterations that are "druggable targets" has paved the way for more elegant treatments that are now seeing positive results clinically. For patients with metastatic NSCLC without a targetable mutation, one first-line combination chemotherapy regimen includes cisplatin, pemetrexed, and bevacizumab, with median survival of just over one year.

Adjunctive treatments include palliative therapies for symptomatic metastatic lesions. For example, as with other malignancies, if pain correlates with a lesion in the bone, palliative radiation therapy may be employed for symptom control. Radiation cannot, however, be used for widespread metastatic disease due to toxicity of radiation therapy.

The importance of smoking cessation cannot be overstated. Even in the population that has metastatic cancer, clinical trials have exhibited greater response to therapy and overall survival in the subset of patients who quit smoking compared with those who did not. One study indicated that patients who quit smoking had double the survival rate of those who did not.

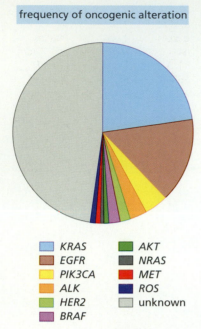

frequency of oncogenic alteration

KRAS	AKT
EGFR	NRAS
PIK3CA	MET
ALK	ROS
HER2	unknown
BRAF	

Figure 4.17. Genetic abnormalities in NSCLC. Pie chart depicting the approximate frequency of each oncogenic alteration. (Adapted from A.T. Shaw et al., *N. Engl. J. Med.* 365:158–167, 2011. Reprinted with permission from Massachusetts Medical Society.)

Genetic alterations and NSCLC

Genetic analysis has uncovered specific genetic alterations that appear to be driver mutations in NSCLC. Approximately half of lung adenocarcinomas exhibit mutations in known oncogenes, among which **epidermal growth factor receptor** (*EGFR*) and *KRAS* mutations are the most common (Figure 4.17). Testing for *EGFR* mutations is the most common in NSCLC because of the demonstrated clinical effectiveness of the EGFR inhibitors **erlotinib** and **gefitinib**. The following case will focus on a more recently described genetic alteration that activates the **ALK** protein.

The case of Alex Washington, a nonsmoker with a lingering cough

Alex Washington is a 51-year-old executive who presented to his primary care physician with six months of lingering cough. He attributed the cough initially to seasonal allergies, but when it persisted, he sought evaluation. His other medical history included hypertension and surgery for sports-related knee injuries. On examination, his chest was clear without wheezes. He was otherwise in excellent health, taking medications only for high blood pressure. He had never smoked cigarettes, other tobacco products, or illicit substances. A chest X-ray was obtained that revealed a mass in the left upper lobe of the lung.

He underwent standard further work-up for suspected lung cancer, including a PET/CT scan, which revealed a 5 × 3 × 3-cm lung mass, enlarged lymph nodes (**supraclavicular**, **mediastinal**, **paratracheal**, **subcarinal**), and a small liver mass. A biopsy of the dominant mass and surrounding lymph nodes was performed by a thoracic surgeon via mediastinoscopy. Pathology revealed adenocarcinoma, with metastasis to mediastinal lymph nodes.

Figure 4.18. Diagnosis of *EML4-ALK* positive NSCLC from a lung tumor biopsy. (A) Break-apart FISH assay; the green and red probes hybridize to the 5′ and 3′ regions of *ALK*, respectively; in cells with separation of the probes (*arrows*), a chromosomal rearrangement has occurred. (B) Hematoxylin and eosin staining reveals that the tumor is an adenocarcinoma. (C) DNA sequencing reveals the junction of *EML4-ALK*. (D) ALK protein (immunohistochemical stain) is expressed in a tumor (*brown*) but not in the normal lung epithelial cells. (From E.L. Kwak et al., *N. Engl. J. Med.* 363:1693–1703, 2010. Reprinted with permission from Massachusetts Medical Society.)

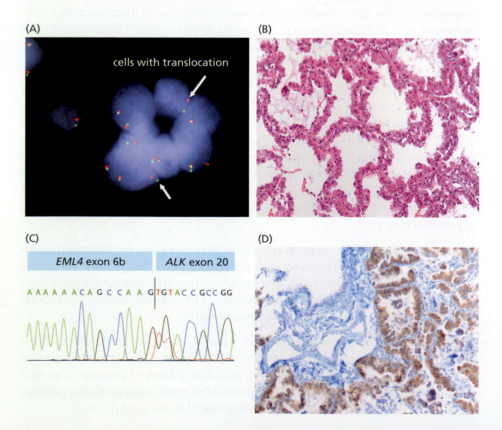

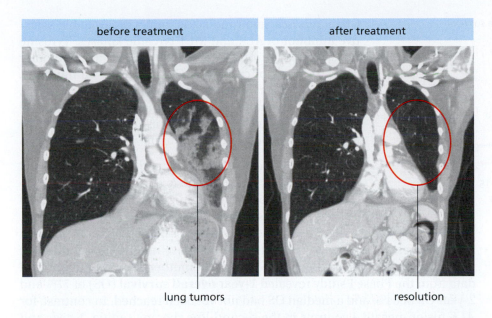

Figure 4.19. CT scans before and after treatment with crizotinib. The left lung masses have resolved with treatment. (From E.L. Kwak et al., *N. Engl. J. Med.* 363:1693–1703, 2010. Reprinted with permission from Massachusetts Medical Society.)

Alex received standard chemotherapy for metastatic NSCLC including carboplatin and paclitaxel, with initial decrease in size of the lymph nodes and lung mass and disappearance of the small liver lesion after six cycles of chemotherapy. Unfortunately, he developed pain with deep breathing, and imaging showed progressive disease. His primary cancer was then subjected to molecular testing. A fusion transcript of ***EML4*** and ***ALK*** (encoding anaplastic lymphoma kinase) was discovered (Figure 4.18). No other tested mutations were identified.

Alex enrolled on a clinical trial of **crizotinib**, an oral small-molecule inhibitor of ALK and **MET** tyrosine kinases. He had resolution of pain and cough within four weeks, and CT scans indicated resolution of all visible disease (Figure 4.19). His disease has been stable for four years and he enjoys an excellent quality of life. He notes that his initial prognosis at the time of diagnosis was just over one year.

EML4-ALK translocation-positive NSCLC

ALK gene rearrangements were first discovered in **anaplastic large-cell lymphoma**. In 2007, *ALK* rearrangements were first reported in NSCLC. The predominant *ALK* rearrangement involves a small inversion within chromosome 2, leading to fusion of a portion of the *EML4* gene with exons 20–29 of *ALK* (Figure 4.20). The resulting *EML4-ALK* fusion gene encodes a potent fusion kinase. This fusion event can be identified in tumors with a FISH assay using a **break-apart probe** (see Figure 4.18A).

ALK rearrangements are seen in 3–4% of NSCLC patients and are exclusively found in adenocarcinomas. *ALK* rearrangements are more common in patients who never smoked or have a history of light smoking (defined as less than or equal to a 10 pack-year history), seen in 15% of this population. Thus, more than 90% of patients with known *ALK* rearrangements have never smoked or were light smokers. These patients have a median age of 50, which is 10 to 15 years younger than the median for lung cancer patients without the *ALK* rearrangement.

The exact function of ALK is not well understood. There is little or no expression of ALK in adult tissues. ALK activates downstream signaling pathways that impact both cell proliferation and survival (Figure 4.21).

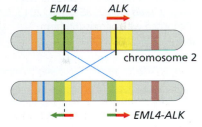

Figure 4.20. Inversion within chromosome 2p leading to *EML4-ALK* fusion.

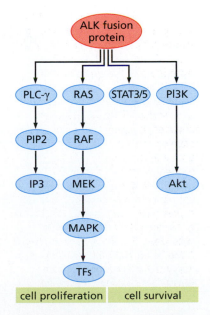

Figure 4.21. Signaling pathways downstream of activated ALK. Activated ALK can stimulate cell proliferation via PLC-γ and RAS pathways or cell survival via STAT and PI3 kinase signaling.

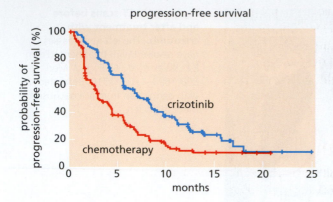

Figure 4.22. Crizotinib increased progression-free survival (PFS) in *ALK*-mutant NSCLC patients. Kaplan–Meier estimates of PFS in subjects randomized to crizotinib (*blue line*) or chemotherapy (*red line*). (From A.T. Shaw et al., *N. Engl. J. Med.* 368:2385–2394, 2013. Reprinted with permission from Massachusetts Medical Society.)

Crizotinib is a small-molecule, orally bioavailable selective inhibitor of ALK and MET tyrosine kinases. Tyrosine phosphorylation of activated ALK occurs at nanomolar concentrations of crizotinib. Crizotinib is well tolerated. Mild side effects include nausea, vomiting, diarrhea, peripheral edema, constipation, and visual changes. The lack of ALK protein in adult tissues may explain the low toxicity of this specific ALK inhibitor.

In a Phase I study including 82 patients with advanced *ALK*-rearranged NSCLC, the overall response rate to crizotinib was 57%, with an estimated probability of 6-month **progression-free survival** (PFS) of 72% (that is, 72% had no evidence that their cancer had progressed on scans at 6 months). These figures stand in stark contrast to the response rates of 10% and PFS of less than 3 months for standard single-agent chemotherapies. Overall survival data from the Phase I study revealed 1-year **overall survival** (OS) of 77% and 2-year OS of 64%, and a median OS had not yet been reached. In contrast, for *ALK* fusion-negative controls in the second-line therapy setting, 1-year and 2-year OS were 49% and 33%, respectively, with median OS of 11 months.

On the basis of these and other data, the FDA granted accelerated approval for crizotinib in August 2011. A Phase III study of crizotinib in 347 patients with *ALK*-positive lung cancer who had received one prior chemotherapy regimen was reported in 2013. Patients were randomly assigned to receive either crizotinib or intravenous chemotherapy (either pemetrexed or **docetaxel**). The primary endpoint was PFS. Crizotinib more than doubled the median PFS (7.7 versus 3.0 months, p < .001; Figure 4.22), and tripled the overall response rate (65% versus 20%, p < .001). Overall survival was not different between these groups, although the data were not mature and likely confounded by the high degree of protocol-allowed crossover to crizotinib in the subjects initially assigned to the chemotherapy group. One may infer that in these patients, all had *ALK*-mutant NSCLC and exposure to crizotinib was beneficial whether they were initially randomized to receive the drug or were allowed during the study to cross over to crizotinib upon disease progression while on chemotherapy. The FDA granted regular approval to crizotinib in November 2013.

Most patients develop resistance to targeted therapies. The development of resistance to ALK inhibitors follows the theme that targeted therapies, while effective, are not curative. Newer ALK inhibitors in development (for example, ceritinib) have been found to overcome some but not all acquired resistance mutations to crizotinib. Molecular analyses from patient samples have suggested mechanisms of resistance to crizotinib. The following changes were seen:

- mutation in the ALK kinase domain including G1269A, which confers resistance to crizotinib *in vitro*;

- *ALK* copy number gain;

- outgrowth of *EGFR* mutant NSCLC;

- *ROS1* mutation;

- *KRAS* mutation; or

- emergence of an *ALK* fusion-negative tumor compared with the baseline sample.

It is notable that medicine appears to be catching up with science. From the *ALK* chromosomal translocation discovery in NSCLC (2007) to a Phase I trial of a novel inhibitor (2010) to FDA approval (2011), just four years elapsed. Compare this timetable with the CML experience, with discovery of the Philadelphia chromosome in 1960 and FDA approval of imatinib in 2001 (see Case 4-1).

Future prospects and challenges in lung cancer

Management of lung cancer remains challenging. Even in patients with localized lung cancers, survival rates remain poor. Early detection, early intervention, and improvement in adjuvant therapies could improve survival. The true innovation in lung cancer therapy, however, has been the identification of mutated genes and the development of therapies that can target the mutations. Contemporary management of NSCLC frequently involves molecular profiling at the time of diagnosis, given the emerging understanding of the molecular underpinning of the disease. While it is tempting to speculate that the appropriately selected therapy could "cure" a patient's lung cancer, it is clear that lung cancer cells activate compensatory pathways that allow them to evolve resistance to the targeted therapy, highlighting the next great challenge in the era of targeted therapy: preventing and treating resistance to these novel agents.

Take-home points

✓ Smoking is a major risk factor for development of lung cancer.

✓ Histology of lung cancer still matters: small-cell lung cancer (SCLC) versus non-small-cell lung cancer (NSCLC) helps define the plan of care.

✓ Staging in lung cancer determines prognosis.

✓ Molecular testing of NSCLC has become standard in defining genetic alterations, which are seen in half of NSCLC patients.

✓ Targeted therapies are improving survival in selected patients, underscoring the need for molecular testing.

✓ *ALK* rearrangements define a subset of adenocarcinomas in patients who never smoked or were light smokers.

✓ The ALK inhibitor crizotinib has demonstrated potent activity and is FDA approved for *ALK*-mutant NSCLC.

✓ Emerging mechanisms of resistance may suggest combination therapies of targeted agents in the future or encourage development of other ALK inhibitors that are not susceptible to ALK kinase domain mutations.

Discussion questions

1. Does lung cancer only occur in smokers?

2. What genetic mutations are most common in lung cancer?

3. Is targeted therapy superior to cytotoxic chemotherapy for all patients?

4. What compensatory signaling pathways might explain resistance to the ALK inhibitor crizotinib?

Topics bearing on this case

Radiation as a risk factor

NRAS

B-RAF discoveries that have led to inroads into the melanoma problem

Case 4-4 | **Metastatic Melanoma**

Introduction and epidemiology

Melanoma is the fifth-most common cancer in the United States, with an estimated 91,270 new diagnoses and 9320 deaths in 2018. The incidence is on the rise, with diagnoses increasing in U.S. men more rapidly than any other cancer; for women, melanoma is second only to lung cancer with respect to rising incidence. The rising incidence may reflect an increase in ultraviolet (UV) radiation exposure related to indoor tanning use, as well as an increase in reporting, since many localized melanomas may be excised in the dermatologist's office and may not be logged with cancer registries. With a relatively young median age at diagnosis of 61, compared with the other common cancers, melanoma is second only to acute leukemia in adults in terms of lost years of potential life per melanoma-related death.

While melanoma is most common in fair-skinned individuals, who are prone to sunburns, the disease occurs in all ethnic groups and can arise in skin without substantial exposure to the sun. Numerous risk factors for development of melanoma have been described (Table 4.6).

Table 4.6. Risk factors for melanoma.

Family history of melanoma including inherited genetic mutations
Prior melanoma
Multiple clinically atypical moles or dysplastic nevi
Fair skin that sunburns easily
History of blistering sunburn
Red or blond hair

Melanoma arises from **melanocytes**. Melanocytes are cells found in the bottom layer of the epidermis of the skin that produce melanin, the pigment that is primarily responsible for skin color. Melanocytes are also found in the middle layer of the eye (**uvea**), inner ear, **meninges** (the connective tissue covering the central nervous system), heart, and bones. During embryonic development, melanocytes originate in the **neural crest** and then migrate to other organs. This chapter will focus on melanomas derived from the skin. Skin melanomas are typically considered to be initiated by DNA damage caused by exposure to UV radiation. Melanomas may arise from preexisting moles.

Other skin cancers, including squamous cell carcinoma (SCC) and **basal cell carcinoma** (BCC), are also caused by UV exposure but are rarely invasive beyond the skin or to other organs. Thus, the management of SCC and BCC is often limited to surgical removal.

Diagnosis, workup, and staging

When detected early, which is approximately 85% of the time, melanoma is very curable. The majority of cases are asymptomatic at presentation, other than the presence of a superficial skin lesion. Most skin lesions are not melanoma, but warning signs of suspicious skin lesions that should prompt consideration for melanoma are summarized as the "ABCDE" signs of melanoma (Figure 4.23):

- Asymmetry: moles are often symmetric but melanomas are not.

- Border: melanoma borders can be uneven.

- Color: melanomas may display different shades of the same color or different colors.

- Diameter: melanomas are usually larger than 6 mm (or 0.25 inch).

- Evolution: melanomas may change in size, shape, color, or elevation.

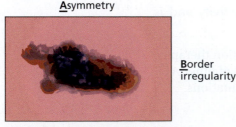

Asymmetry

Diameter >6 mm

Border irregularity

Color variegation

Figure 4.23. The ACBDE warning signs of melanoma. A, B, C, and D are depicted here, in a cutaneous melanoma. A, asymmetry; B, uneven borders; C, different colors of the pigmented lesions; D, larger diameter. E (not indicated in illustration), evolution over time.

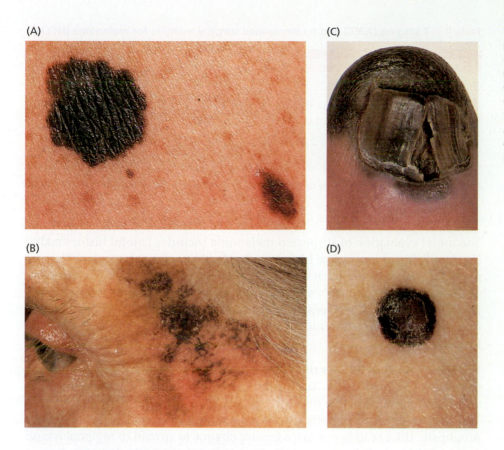

(A)

(B)

(C)

(D)

Figure 4.24. The four major subtypes of melanoma.
(A) Superficial spreading melanoma.
(B) Lentigo maligna melanoma.
(C) Acral lentiginous melanoma.
(D) Nodular melanoma. (From A.T. Skarin, Dana-Farber Cancer Institute Atlas of Diagnostic Oncology, 4th ed. Mosby Elsevier, 2010. With permission from Elsevier.)

There are four basic types of melanomas (Figures 4.24 and 4.25). The initial diagnosis is made by biopsy of a suspicious lesion.

Superficial spreading melanoma is the most common, accounting for 70% of cases and often seen in younger patients. As per its name, superficial spreading melanoma grows along the surface of the skin for a prolonged period before penetrating into deeper tissues. The initial appearance is typically flat or slightly raised, with irregular, asymmetric borders; and can arise from a preexisting mole. It is most commonly seen on the upper back among all patients, the trunk in male patients, and the legs of female patients.

Lentigo maligna melanoma similarly remains confined to the skin initially, with a flat or slightly elevated lesion that can have discoloration ranging from tan to dark brown. Lentigo maligna is seen in the elderly, especially in areas of skin with chronic sun exposure.

Acral lentiginous melanoma also spreads superficially in the early stages but appears to invade more quickly than superficial spreading and lentigo maligna melanomas. As the term "acral" pertains to extremities, this melanoma typically arises under the nails, on palms of the hands, or on soles of the feet. It is characterized by black or brown discoloration. Acral lentiginous melanoma is the most common melanoma among African-Americans and Asians and least common among white Americans.

Nodular melanoma, unlike the others, is typically invasive at the time of diagnosis, and as such is the most aggressive. The presentation is usually as a discolored bump, with colors including black, blue, gray, white, brown, red, tan, or skin-toned. Found in 10 to 15% of cases, nodular melanomas are most commonly located on the trunk, legs, and arms, or the scalp in men.

As with other malignancies, melanoma outcome depends on stage. Most patients (~85%) have localized disease, whereas 10% have regional disease (local lymph nodes) and 2 to 5% have distant metastatic disease.

frequency of melanomas

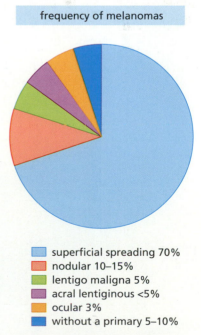

- superficial spreading 70%
- nodular 10–15%
- lentigo maligna 5%
- acral lentiginous <5%
- ocular 3%
- without a primary 5–10%

Figure 4.25. Frequency of the four types of cutaneous melanoma, ocular melanoma, and cases in which metastatic disease is found without a primary tumor.
(Reprinted with permission from the AIM at Melanoma Foundation, www.AIMatMelanoma.org.)

Table 4.7. T staging (AJCC) and recommended surgical margins for melanoma (NCCN).

T classification	Melanoma thickness (mm)	Recommended margin (cm)
Tis (*in situ*)	N/A	0.5–1
T1	≤1.0	1
T2	1.01–2.0	1–2
T3	2.01–4	2
T4	>4	2

AJCC, American Joint Committee on Cancer; NCCN, National Comprehensive Cancer Network

The initial evaluation of suspected melanoma includes careful history taking (for family history; personal history of dysplastic **nevi**, or atypical moles; and personal history of sunburn and UV exposure), complete skin examination, and evaluation for clinically apparent (palpable) lymph nodes. The suspected lesion is then surgically removed via an excisional biopsy. Where possible (depending on the anatomic location), a "margin" of 1–3 mm around the lesion is preferred.

Melanoma staging follows the TNM staging from the American Joint Committee on Cancer (AJCC). Tumor thickness, also called the **Breslow measurement**, describes the depth of penetration into the skin (Table 4.7). Thinner tumors are generally associated with lower-stage disease and better prognosis. Thicker tumors have a greater chance to spread to regional lymph nodes and beyond. Features of the tumor that describe higher-risk disease include the presence of ulceration (breakdown of the skin over the melanoma) and higher **mitotic rate** (number of cells in the process of mitosis per square millimeter).

Once diagnosed, the thickness and pathologic features of the melanoma may lead to additional surgery. Because of the possibility of superficial spread, a wide local excision is typically performed to ensure that the "margins"—the edges of the resection—do not contain cancer cells, because "positive margins" would signify that cancer cells were left behind. Recommended surgical margins differ based on thickness of the melanoma (see Table 4.7).

Lymph node involvement is the most important prognostic factor in melanoma. If a regional lymph node is palpable (generally >1 cm in diameter to be palpable), an excisional lymph node biopsy may be performed to evaluate for the presence of melanoma cells. A **sentinel lymph node** (SLN) **biopsy** is considered for patients whose melanoma has concerning features (for example, thicker tumors, presence of ulceration, higher mitotic rate). The SLN is the first lymph node that the melanoma would go to, if it has spread. To perform the SLN biopsy, the surgeon injects a radioactive substance and a blue dye into the area of the primary tumor, waits for approximately one hour, then evaluates for radioactivity in the lymph node areas that would be the typical lymphatic drainage for the tumor site (Figure 4.26). If a radioactive area is detected, an incision is performed to identify which lymph nodes are radioactive and blue. Those "sentinel" nodes are removed and evaluated under the microscope by the pathologists for the presence of cancer cells. The SLN technique has allowed for improved identification of nodal spread, as opposed to older techniques of complete **lymph node dissection**, which were not always necessary or informative, and which can cause side effects such as **lymphedema** (an abnormal collection of lymph in soft tissue, due to disrupted lymphatic drainage).

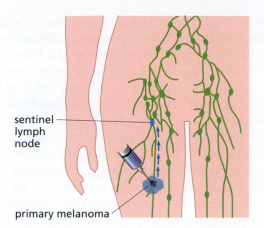

sentinel
lymph
node

primary melanoma

Figure 4.26. Sentinel lymph node (SLN) biopsy. Blue dye and a radioactive tracer are injected into the area of the primary tumor. Because many lymph nodes are in the normal drainage vicinity for lymphatic channels, the sentinel lymph node approach allows for the identification of the specific lymph node to which cells from the primary tumor might migrate. After allowing time for both the dye and the tracer to migrate along the lymphatic channels, the surgeon uses a Geiger counter or gamma probe to assess which nodes took up the radionuclide. Those sentinel lymph nodes with radionuclide are removed surgically and examined under the microscope for the presence of cancer cells.

When distant metastasis is suspected, imaging studies are performed, including CT scans of chest, abdomen, and pelvis, as well as brain imaging with an MRI. A PET/CT scan can be helpful to image areas of the body that are not typically seen on CT scans, such as the extremities.

A summary of TNM staging is found in Table 4.8. Stages I and II differ primarily by T stage, with "a" or "b" subclassifications distinguished by the absence or presence, respectively, of ulceration (for example, T2 tumors can be T2a or T2b depending on ulceration, with ulceration conferring a worse prognosis). Stage III melanoma involves regional lymph nodes (N1 involves 1 node; N2 involves 2–3 nodes; N3 involves ≥4 nodes). Stage IV melanoma includes distant metastasis, including distant lymph nodes or visceral sites such as lung, liver, or brain. The prognosis for T1 tumors is excellent, with >90% of patients alive five years after diagnosis. For thicker tumors without lymph node involvement, five-year survival ranges from 50 to 90%. For patients with stage III or IV disease, five year survival is 20–70% and <10%, respectively.

Management of localized melanoma

For localized melanoma, wide local excision with the appropriate surgical margin as described in Table 4.7 is the primary treatment. In certain circumstances, such as melanomas that are *in situ* (T stage Tis, defining cells that have not penetrated through the basement membrane of the epidermis), or patients whose disease cannot be resected due to co-morbid conditions or location of the primary disease, topical therapy such as imiquimod or radiation therapy may be employed.

For patients with lymph node metastasis determined by SLN biopsy, **completion lymph node dissection**, which is a more extensive removal of regional lymph nodes, can reveal additional nodal involvement in 15% of patients. While this may provide more understanding of the extent of the disease, the impact of a completion lymph node dissection on overall outcome is uncertain.

Table 4.8. AJCC TNM staging for melanoma.

Stage 0	Tis	N0	M0
Stage I	T1–T2a	N0	M0
Stage II	T2b–T4b	N0	M0
Stage III	any T	≥N1	M0
Stage IV	any T	any N	M1

Adjuvant therapy.   Patients with thicker melanomas or lymph node involvement are at high risk for recurrence and metastasis. Adjuvant therapies including **interferon** and radiation therapy have been evaluated. Interferon is a **cytokine**. Melanoma has long been known to be susceptible to the immune system, which has prompted the use of immunotherapies, vaccines, and cytokine therapies that provide "immunomodulatory" effects whose exact mechanisms are unknown but can cause melanoma regression (see Chapter 12, Case 12-3, for further description of immunologic strategies and treatment of melanoma). Adjuvant interferon at a variety of doses has been demonstrated to improve relapse-free survival but has not provided a statistically significant improvement in overall survival. Similarly, limited studies of adjuvant radiation have demonstrated a decrease in recurrence within the radiation field but no improvement in overall survival. Both interferon and radiation cause significant treatment-related morbidity. For these reasons, there was no standard adjuvant therapy for high-risk melanoma until October 2015, when the FDA approved adjuvant ipilimumab, a monoclonal antibody and immune checkpoint inhibitor (see Case 12-3 for further description of this approach).

Management of metastatic melanoma

The prognosis for patients with stage IV melanoma is generally poor. For a small, highly selected group of patients with disease isolated to just one distant site, termed "oligometastatic disease," surgery may offer long-term disease control. The best candidates typically have a long disease-free interval between diagnosis and recurrence, and a focus of disease that can be removed completely.

Chemotherapy for metastatic melanoma has had very limited success. Most standard cytotoxic agents have no activity. **Dacarbazine**, also known as **DTIC**, is an intravenous alkylating agent. Despite being considered the standard chemotherapy for metastatic melanoma, dacarbazine is associated with a low response rate of <20%, an overall survival of approximately six months, and common adverse events including pain, nausea, vomiting, and constipation. Attempts to combine chemotherapy agents have led to higher response rates but also higher toxicity and no improvement in overall survival.

Immunotherapy approaches are active in metastatic melanoma. Rare cases of spontaneous remissions have been described, presumably due to an antitumor immune response. Immune-based therapies including **interleukin-2** and immune checkpoint inhibitors such as **ipilimumab** and **pembrolizumab** are described further in Chapter 12. **Biochemotherapy** approaches, which combine chemotherapy and cytokines, have been attempted in Phase III clinical trials, with high overall response rates but also high toxicity, including high rates of admission to the intensive care unit, and ultimately no benefit in terms of overall survival.

Targeted therapies for known genetic alterations in melanoma include protein kinase inhibitors and represent an important recent development in the treatment of metastatic melanoma.

Genetic alterations and melanoma

Several important somatic mutations have been described in melanoma. Approximately half of melanoma tumors harbor an activating mutation of **BRAF**. **BRAF** belongs to the RAF family of serine/threonine kinases. Most *BRAF* mutations occur at codon 600 encoding valine (V600), with mutations to glutamic acid (V600E) or lysine (V600K). The activation of BRAF in melanoma was discovered in 2002. BRAF is an important kinase, integrating growth signals from RAS and PI3 kinase (PI3K) to phosphorylate and activate downstream MEK/MAPK signaling (Figure 4.27).

Mutations in **NRAS** have been described in 20% of metastatic melanomas. **NRAS** is a member of the RAS family of **GTPases**. Occupying a position just upstream of BRAF, activating mutations of NRAS are typically mutually exclusive with activating BRAF mutations. This mutual exclusivity is a common theme in other malignancies as well, probably because activating mutations that lead to malignant transformation are unlikely to require additional activating events within the same pathway.

Mutations of **GNAQ** have been described in half the cases of uveal (arising from the eye) melanoma. *GNAQ* encodes guanine nucleotide-binding protein G[q] subunit α. Activation of *GNAQ* leads to up-regulation of the MEK/MAPK pathway.

Mutations in c-KIT, a transmembrane receptor tyrosine kinase, have been described in 15 to 20% of acral and mucosal melanomas (a rare subtype of melanoma). While targeting of c-KIT is possible with tyrosine kinase inhibitors, as seen in other malignancies harboring c-KIT mutations, not all melanomas with c-KIT mutations respond to such therapies, indicating that the activating c-KIT mutation may not be one of the principal drivers in such cases.

The case of Tom Schaeffer, a retired machinist with an irregular brown skin lesion

Tom Schaeffer is a 63-year-old retired machinist who was first diagnosed with melanoma at the age of 59. A fair-skinned white man with blond hair, Tom first noted a lesion on his temple that was initially brown and roughly oval but over six months became more irregular in contour and border. His other medical history included early tobacco use and a heart attack at the age of 55 for which he had a coronary stent placed and received medications to minimize his risk of future events. He has experienced no further heart-related events. His primary care physician referred Tom to a dermatologist, who performed an excisional biopsy that demonstrated a lentigo maligna melanoma, 1.3 mm thick, without ulceration. Tom then underwent wide local excision with a skin graft, given the location of the lesion, and SLN biopsy. Biopsy margins were negative for tumor involvement. The SLN, on the left side of his neck near the jugular vein, was negative for spread of melanoma. Imaging was negative for metastatic disease. He was therefore staged as T2aN0M0 (see Table 4.8). At that time, there was no approved adjuvant therapy, so Tom was monitored with skin examinations and periodic imaging.

At the age of 62, three years after his initial diagnosis, Tom's dermatologist noted new subcutaneous nodules on his upper back and left deltoid muscle. Biopsy of two lesions from the upper back and deltoid revealed metastatic melanoma. CT scans of the neck, chest, abdomen, and pelvis revealed diffuse metastases in the lungs and scattered subcutaneous nodules of the chest, back, and trunk (Figure 4.28). An MRI of his brain did not reveal metastatic disease.

Molecular profiling of the biopsy specimens revealed the V600E mutation of *BRAF*. At the time of diagnosis of his recurrent, metastatic melanoma, Tom sought opinions at academic medical centers. He was considered a candidate

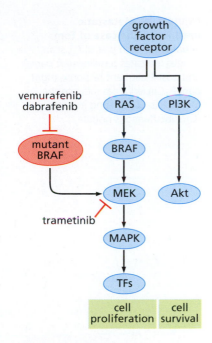

Figure 4.27. Signaling along the BRAF/MEK/MAPK pathway. Signaling to drive cell proliferation and survival frequently starts with engagement of growth factor receptors with specific ligands. The signal is transmitted via adaptor proteins to activate RAS, leading to activation of BRAF. BRAF then phosphorylates and activates MEK on serine and threonine residues. MEK activates MAPK by phosphorylation on tyrosine and threonine residues, ultimately leading to transcription factor activation in the nucleus to drive cell proliferation. An alternate pathway activated by receptors is via PI3 kinase to activate AKT and cell survival pathways. Mutant BRAF seen in melanoma can activate MEK without growth factor receptor signals. Inhibition via vemurafenib, dabrafenib, and trametinib occurs at critical steps of this signaling pathway.

Figure 4.28. Metastatic melanoma: the case of Tom Schaeffer. Each pair of CT scan images indicates baseline (*left panel*) and post-treatment response (*right panel*). Circled areas indicate the metastasis. (A, B) Lung metastases. (C) Subcutaneous nodule.

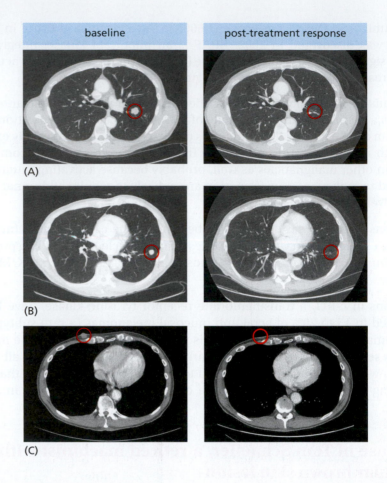

for interleukin-2 immunotherapy (albeit a marginal candidate given his history of heart disease), **vemurafenib** therapy (an inhibitor of activated BRAF), or a Phase II clinical trial combining the BRAF inhibitor **dabrafenib** with the MEK inhibitor **trametinib**. There was brief discussion of dacarbazine chemotherapy.

Tom opted for the clinical trial. The trial assigned him to receive 150 mg of dabrafenib and 2 mg of trametinib daily, both by mouth. By three months into the trial, the palpable subcutaneous nodules had regressed and were difficult to locate. The pulmonary nodules shrank greater than 50%, and no new nodules appeared (see Figure 4.28). His side effects from therapy included worsening of his hypertension (elevated blood pressure) that was readily managed by medication adjustment, joint aches that did not require medication, and rare occurrences of nausea and vomiting. His self-described quality of life was excellent. At nine months, his disease remains stable and he continues on the same doses of the trial therapies.

Kinase inhibition and melanoma

The discovery of activating *BRAF* mutations in 2002 led to the development of targeted inhibitors in melanoma. Much like tyrosine kinase inhibitors described earlier in this chapter, serine/threonine kinase inhibitors hold the promise of targeting the "oncogene addicted" cancer cell at its very core driver mutation. Sorafenib was the first known BRAF inhibitor but demonstrated little activity in patients with melanoma.

BRAF inhibition became a clinical reality with the development of the more potent and specific inhibitors vemurafenib (PLX4032) and

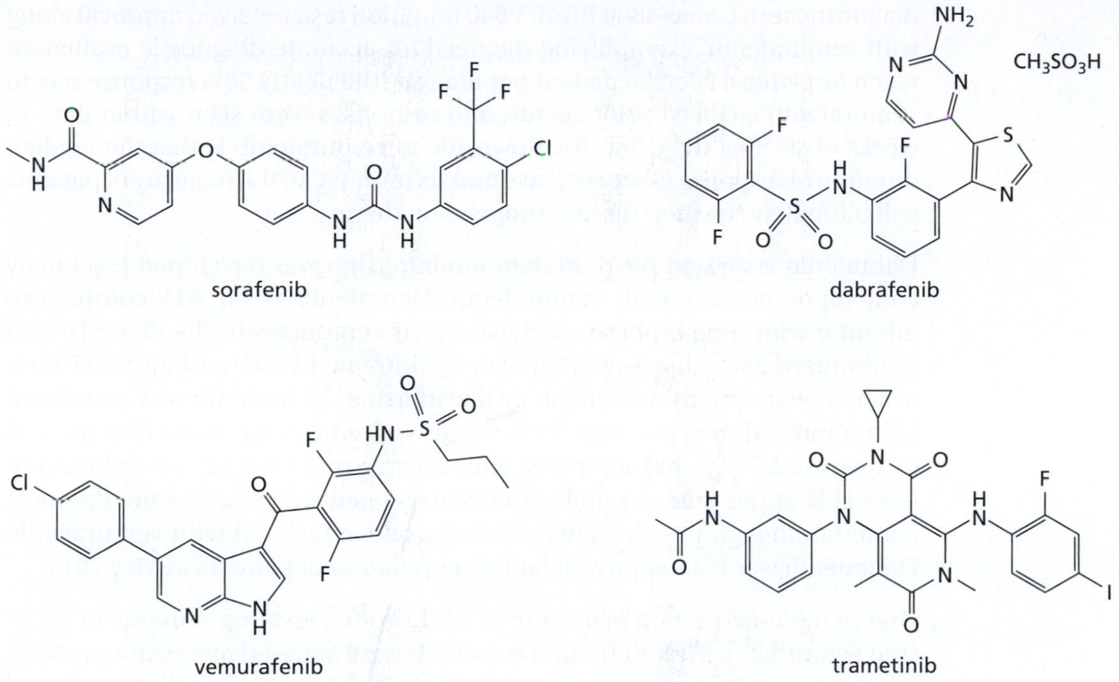

Figure 4.29. **Kinase inhibitors in melanoma.**

dabrafenib, which are inactive against wild-type (that is, nonmutated) BRAF (Figure 4.29). The Phase I study of vemurafenib established the maximal tolerated dose in subjects with metastatic melanoma, most of whom had progression despite chemotherapy. In subjects with the V600E *BRAF* mutation, 81% experienced complete or partial tumor response, an unprecedented response rate. This study *revolutionized* the approach to *BRAF*-mutant melanoma.

Further trials of vemurafenib confirmed the promising initial results. Both Phase II and Phase III trials enrolled subjects with previously untreated metastatic melanoma harboring V600 mutant *BRAF*. In the Phase II study of 132 subjects, the overall response rate was 53% (6% with complete response, 47% with partial response), median progression-free survival 6.8 months, and median overall survival 15.9 months, which compares extremely favorably with historical comparisons with dacarbazine chemotherapy. Vemurafenib was well tolerated in general, but 26% of patients developed a cutaneous squamous cell carcinoma (SCC) but not SCC of other organs. The mechanism of increased SCC is thought to involve activation of MAPK pathway signaling by BRAF inhibitors in cells that have wild-type BRAF; the cutaneous SCCs are readily managed by simple excision.

The Phase III trial randomized 675 subjects to vemurafenib versus dacarbazine chemotherapy. Vemurafenib was associated with improved response rate (48% versus 5%), progression-free survival (5.3 versus 1.6 months), and overall survival (84% versus 64% at six months). At the interim analysis, the data were so strongly in favor of vemurafenib that an independent data and safety monitoring board recommended crossover from dacarbazine to vemurafenib for patients in the control arm. The most common adverse events related to vemurafenib were joint aches, cutaneous SCC (12%), fatigue, and photosensitivity skin reactions (rash).

On the strength of these studies, vemurafenib was approved by the FDA for V600-mutant *BRAF* metastatic melanoma in August 2011. The companion diagnostic test, Cobas 4800 BRAF V600 mutation test, received approval along with vemurafenib, exemplifying the need for accurate diagnostic evaluation when targeting a specific patient population. The nearly 50% response rate to vemurafenib eclipsed prior agents, and responses were seen within days to weeks of starting the drug. The downside to vemurafenib is that the median duration of response is between five and six months, so the majority of patients will ultimately see their disease progress on therapy.

Dabrafenib is an oral BRAF-mutant inhibitor that was developed essentially contemporaneously with vemurafenib. Dabrafenib is an ATP-competitive inhibitor with similar pharmacodynamics as vemurafenib. The Phase III trial randomized 250 subjects with previously untreated V600-mutant *BRAF* metastatic melanoma to dabrafenib or dacarbazine. Dabrafenib was associated with improved response rate (50% versus 6%) and progression-free survival (5.1 versus 2.7 months) compared with dacarbazine. Overall survival was not evaluable at the time of publication. Skin-related side effects were the most common, though possibly lower than the rate associated with vemurafenib. Dabrafenib was FDA approved for this population of patients in May 2013.

Due to the downstream activation of MEK from activating mutants of BRAF (see Figure 4.27), MEK inhibition has also been evaluated in melanoma. MEK is a dual-specificity threonine/tyrosine kinase that phosphorylates and activates MAPKs (ERK1 and ERK2) at threonine 202 and tyrosine 204. In preclinical models of melanomas that develop resistance to BRAF inhibition, rapid recovery of MEK/MAPK signaling occurs, highlighting the importance of MEK signaling in melanoma. Additionally, the relatively short progression-free survival with BRAF inhibitors, despite their overall survival benefits, underscores that fact that acquired resistance to targeted therapies limits their utility as single agents.

Trametinib is an oral, small-molecule, allosteric, ATP-noncompetitive inhibitor of the two forms of MEK, MEK1 and MEK2. A Phase III study randomized 322 subjects with V600-mutant *BRAF* metastatic melanoma to trametinib or chemotherapy (either dacarbazine or paclitaxel). Trametinib was associated with improved progression-free survival (4.8 months versus 1.5 months), response rate (22% versus 8%), and overall survival (at six months, 81% versus 67%). Adverse events included rash, diarrhea, and fatigue, but no skin cancers were observed. On the basis of this study, trametinib was FDA approved in May 2013.

MEK and BRAF inhibition similarly improved progression-free and overall survival, but had similar disease progression within six months of initiation of therapy. Proposed mechanisms of resistance to the above therapies include new NRAS or MEK mutations, dimerization or variant splicing of V600-mutant BRAF, and MAPK-independent signaling activating cell proliferation pathways.

Given the rapid reactivation of MEK/MEPK signaling during acquired resistance to BRAF inhibition, the BRAF inhibitor dabrafenib was combined with the MEK inhibitor trametinib in a Phase I/II study, to attempt to delay resistance to BRAF inhibition. A total of 247 subjects with V600-mutant *BRAF* metastatic melanoma received varying doses of dabrafenib with trametinib. Full doses of these agents were tolerated well in combination (150 mg twice daily for dabrafenib, 2 mg once daily for trametinib). Median progression-free survival at these doses was 9.4 months, compared with 5.8 months with dabrafenib monotherapy in this study. The overall response rate was 76% with combined therapy, compared with 54% for dabrafenib monotherapy. The most frequent adverse events at the maximum doses were pyrexia (fever) and chills. Cutaneous SCCs occurred less frequently with the combination, although the difference was not statistically significant. On the basis of this study, the combination of dabrafenib and trametinib was granted accelerated approval by the FDA in May, 2013, for the treatment of patients with metastatic melanoma harboring V600-mutant *BRAF*, contingent upon successful completion of ongoing randomized controlled trials.

Future prospects and challenges in metastatic melanoma

The management of metastatic melanoma has undergone significant changes since 2011, when BRAF-targeted therapy and immune checkpoint modulation therapy (see Chapter 12) attained FDA approvals. While some durable responses to melanoma therapies exist, acquired resistance to the newer therapies highlight the cancer's ability to constantly evolve. Rational combinations of therapies may hold the key to improving disease response and overall survival. Exactly how targeted therapies and immune-based approaches will be used in sequence or in combination remains the topic of ongoing and future clinical trials.

Take-home points

✓ The incidence of melanoma is on the rise in the United States.

✓ Prevention of UV irradiation-related damage is paramount, especially in patients with elevated risk of developing melanoma.

✓ Early detection of thinner melanomas remains the best chance to cure the disease.

✓ Activated mutants of the serine/threonine kinase BRAF are found in nearly half of all melanoma cases.

✓ Mutant BRAF inhibition is associated with high response rates and improved survival, but the effect is short-lived (typically six months).

✓ Similarly, MEK inhibition improves disease control and survival, albeit briefly.

✓ Thus, similar to tyrosine kinase inhibition described in earlier cases, targeted therapies include serine/threonine and tyrosine/threonine dual-specificity kinase inhibitors.

✓ Accurate companion diagnostic tests for mutated proteins are critical for development of targeted therapies.

✓ The rational combination of BRAF and MEK inhibitors provides improved disease control; confirmatory trials are ongoing. That progression-free survival and response rates for a *combination* of targeted agents are better than single-agent strategies may represent a theme for future therapeutic developments.

✓ Combinations of targeted therapies that address mechanisms of acquired resistance to monotherapy may hold the key to better outcomes.

✓ For metastatic melanoma, with advances in targeted therapies and immune-mediated therapies (see Chapter 12), optimal sequencing or combination of effective therapies may further improve disease outcomes.

Discussion questions

1. Why is the incidence of melanoma on the rise?

2. Why might some BRAF inhibitors (vemurafenib, dabrafenib) work better than others (sorafenib) in melanoma?

3. Why don't some melanomas with c-KIT mutations respond to effective c-KIT inhibitors?

4. How might MAPK reactivation occur as a mechanism of acquired resistance to BRAF inhibitors in melanoma?

5. What might be the mechanism of resistance to combined BRAF and MEK inhibition?

  Chapter summary

Protein kinases are key contributors to the development and maintenance of many cancers. Appropriately targeted protein kinase inhibitors are making an increasing contribution to the treatment of cancer. Since 2001, more than two dozen protein kinase inhibitors have been approved for clinical use and hundreds more are currently in clinical trials. These cases illustrate the use of kinase inhibitors to treat a variety of cancers including leukemia, lymphoma, lung cancer, and melanoma. The oncogenes encoding the protein kinases driving the cancers discussed in three of these four cases arose in either of two ways. Chromosomal translocations create the novel oncogenic fusions that drive CML and an ALK-dependent subset of lung cancers, whereas point mutations generate the mutated *BRAF* driving a major subset of melanoma. On the other hand, CLL, whose primary drivers are mutated forms of other genes, also depends on the activity of normal unmutated BTK for survival, and inhibition of this kinase leads to the eventual death of these cells.

In each case, the use of an appropriate kinase inhibitor provided therapeutic benefit in situations where the standard of care was not of further benefit to patients. Unfortunately, kinase inhibitors are not a cure for most cancers, because effective inhibitors select for resistant cell populations, and eventually the drugs lose their efficacy. Encouragingly, the use of imatinib and its successors for the treatment of CML has shown that when a cancer continues to rely on pathways driven by mutated forms of the same oncogene, genomics can identify the mutation and even guide the switch from an ineffective inhibitor to an effective one (Figure 4.30). The melanoma experience has demonstrated that strategies targeting multiple steps in the same signaling cascade can be more effective than targeting a single kinase. However, if targeted inhibition leads to the selection of cell populations that have switched to alternative pathways, as revealed in some resistance mechanisms to ALK-targeted therapy in lung cancer, it is necessary to identify the pathway used by the new driver genes and, if possible, deploy a therapy effective for this pathway.

The ability to target key protein kinases depends on the identification of these kinases in patients and the availability of suitable and efficacious inhibitors for therapeutic use. Fortunately the number of FDA-approved protein kinase inhibitors for use in cancer therapy, now approaching three dozen, continues

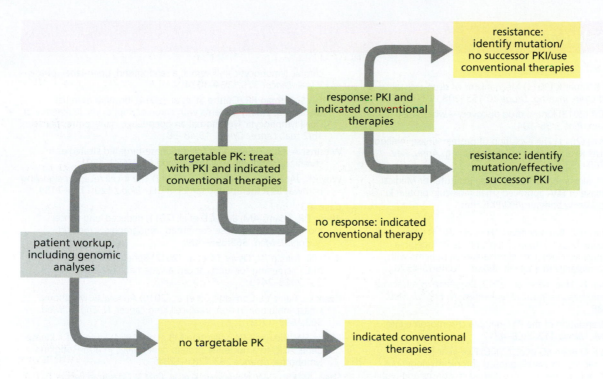

Figure 4.30. Decision tree for treatment of a targetable protein kinase.
The green boxes indicate protein kinase (PK) mutations that can be treated with an
available protein kinase inhibitor (PKI). Note that even effective PKIs are often used in
combination with conventional chemotherapies. The yellow boxes indicate an absence
of targetable PK mutations or the development of drug-resistant mutants for which there
is no available successor PKI. In the absence of a targetable mutation, or of a successor
inhibitor when resistance develops to the PKI initially deployed, indicated conventional
therapies are used.

to increase (see Table 4.1). However, given the variety and ubiquity of protein
kinases, and the additional diversity that can be generated by mutation, even
this already sizeable repertoire of kinase inhibitors will have to be much larger,
as more cancer-sponsoring kinases and their resistant mutants are identified.
At present, many kinase inhibitor candidates are in various stages of clinical
testing and it is reasonable to expect a continuing stream of these inhibitors
from the pipeline of development and testing.

Chapter discussion questions

1. What is the biological importance of protein kinases, and why are they of
   particular interest for understanding and treating cancer?

2. Outline the major challenges to the design and derivation of clinically useful protein kinase inhibitors.

3. Does a protein kinase inhibitor have to be absolutely specific to be a safe
   and effective drug?

4. Distinguish between a driver and a passenger gene mutation.

5. How might genomics be useful in guiding the use of protein kinase inhibitors and in detecting the selection of a population of cancer cells resistant
   to a particular inhibitor?

Selected references

Introduction

Barouch-Bentov R & Sauer K (2011) Mechanisms of drug-resistance in kinases. *Expert Opin. Investig. Drugs* 20:153–208.

Cohen P & Alessi DR (2013) Kinase drug discovery—what's next in the field? *ACS Chem. Biol.* 8:96–104.

Dar AC & Shokat KM (2011) The evolution of protein kinase inhibitors from antagonists to agonists of cellular signaling. *Annu. Rev. Biochem.* 80:769–795.

Roskoski R Jr (2017, November 14) Blue Ridge Institute for Medical Research database of FDA approved small molecule protein kinase inhibitors. http://www.brimr.org/PKI/PKIs.htm

Case 4-1

Cortes JE, Kantarjian HM, Brümmendorf TH et al. (2011) Safety and efficacy of bosutinib (SKI-606) in chronic phase Philadelphia chromosome-positive chronic myeloid leukemia patients with resistance or intolerance to imatinib. *Blood* 118:4567–4576.

Cortes JE, Kantarjian H, Shah NP et al. (2012) Ponatinib in refractory Philadelphia chromosome-positive leukemias. *N. Engl. J. Med.* 367:2075–2088.

Druker BJ (2008) Translation of the Philadelphia chromosome into therapy for CML. *Blood* 112:4808–4817.

Druker BJ, Guilhot F, O'Brien SG et al. (2006) Five-year follow-up of imatinib therapy for newly diagnosed chronic myelogenous leukemia in chronic-phase shows sustained responses and high overall survival. *N. Engl. J. Med.* 355:2408–2417.

Druker BJ, Tamura S, Buchdunger E et al. (1996) Effects of a selective inhibitor of the ABL tyrosine kinase on the growth of Bcr-Abl positive cells. *Nat. Med.* 2:561–566.

Jabbour E, Kantarjian H, O'Brien S et al. (2011) The achievement of an early complete cytogenetic response is a major determinant for outcome in patients with early chronic phase chronic myeloid leukemia treated with tyrosine kinase inhibitors. *Blood* 118:4541–4546.

Kantarjian HM, Shah NP, Hochhaus A et al. (2010) Dasatinib versus imatinib in newly diagnosed chronic-phase chronic myeloid leukemia. *N. Engl. J. Med.* 362:2260–2270.

Nowell PC & Hungerford DA (1960) A minute chromosome in human chronic granulocytic leukemia. *Science* 132:1497.

O'Brien SG, Guilhot F, Larson RA et al. (2003) Imatinib compared with interferon and low-dose cytarabine for newly diagnosed chronic-phase chronic myeloid leukemia. *N. Engl. J. Med.* 348:994–1004.

Saglio G, Kim D-W, Issaragrisil S et al. (2010) Nilotinib versus imatinib for newly diagnosed chronic myeloid leukemia. *N. Engl. J. Med.* 362:2251–2259.

Shah NP & Sawyers CL (2003) Mechanisms of resistance to STI571 in Philadelphia chromosome-associated leukemias. *Oncogene* 22:7389–7395.

Shtivelman E, Lifshitz B, Gale RP & Canaani E (1985) Fused transcript of abl and bcr genes in chronic myelogenous leukaemia. *Nature* 315:550–554.

Case 4-2

Byrd JC, Furman RR, Coutre SE et al. (2013) Targeting BTKwith ibrutinib in relapsed chronic lymphocytic leukemia. *N. Engl. J. Med.* 369:32–42.

CLL Trialists' Collaborative Group (1999) Chemotherapeutic options in chronic lymphocytic leukemia: a meta-analysis of the randomized trials. *J. Natl. Cancer Inst.* 91:861–868.

Goede V, Fischer K, Busch R et al. (2014) Obinutuzumab plus chlorambucil in patients with CLL and coexisting conditions. *N. Engl. J. Med.* 370:1101–1110.

Hallek M, Fischer K, Fingerle-Rowson G et al. (2010) Addition of rituximab to fludarabine and cyclophosphamide in patients with chronic lymphocytic leukaemia: a randomised, open-label, phase 3 trial. *Lancet* 376:1164–1174.

O'Brien S, Furman RR, Coutre SE et al. (2014) Ibrutinib as initial therapy for elderly patients with chronic lymphocytic leukaemia or small lymphocytic lymphoma: an open-label, multicentre, phase 1b/2 trial. *Lancet Oncol.* 15:48–58.

Wiestner A. (2012) Emerging role of kinase-targeted strategies in chronic lymphocytic leukemia. *Blood* 120:4684–4691.

Woyach, JA, Johnson AJ & Byrd JC (2012) The B-cell receptor signaling pathway as a therapeutic target in CLL. *Blood* 120:1175–1184.

Case 4-3

Aberle DR, Adams AM, Berg CD et al. (2011) Reduced lung-cancer mortality with low-dose computed tomographic screening. *N. Engl. J. Med.* 365:395–409.

Bach PB, Mirkin JN, Oliver TK et al. (2012) Benefits and harms of CT screening for lung cancer: a systematic review. *JAMA* 307:2418–2419.

Kwak EL, Bang YJ, Camidge DR et al. (2010) Anaplastic lymphoma kinase inhibition in non-small-cell lung cancer. *N. Engl. J. Med.* 363:1693–1703.

Morris SW, Kirstein MN, Valentine MB et al. (1994) Fusion of a kinase gene, *ALK*, to a nucleolar protein gene, *NPM*, in non-Hodgkin's lymphoma. *Science* 263:1281–1284.

Shaw AT, Kim D-W, Nakagawa K et al. (2013) Crizotinib versus chemotherapy in advanced *ALK*-positive lung cancer. *N. Engl. J. Med.* 368:2385–2394.

Shaw AT, Yeap BY, Mino-Kenudson M et al. (2009) Clinical features and outcome of patients with non-small-cell lung cancer who harbor *EML4-ALK*. *J. Clin. Oncol.* 27:4247–4253.

Soda M, Choi YL, Enomoto M et al. (2007) Identification of the transforming *EML4-ALK* fusion gene in non-small-cell lung cancer. *Nature* 448:561–566.

Case 4-4

Chapman PB, Hauschild A, Robert C et al. (2011) Improved survival with vemurafenib in melanoma with BRAF V600E mutation. *N. Engl. J. Med.* 364:2507–2516.

Davies H, Bignell GR, Cox C et al. (2002) Mutations of the BRAF gene in human cancer. *Nature* 417:949–954.

Flaherty KT, Infante JR, Daud A et al. (2012) Combined BRAF and MEK Inhibition in Melanoma with BRAF V600 Mutations. *N. Engl. J. Med.* 367:1694–1703.

Flaherty KT, Puzanov I, Kim KB et al. (2010) Inhibition of mutated, activated BRAF in metastatic melanoma. *N. Engl. J. Med.* 363:809–819.

Flaherty KT, Robert C, Hersey P et al. (2012) Improved survival with MEK inhibition in BRAF-mutated melanoma. *N. Engl. J. Med.* 367:107–114.

Hauschild A, Grob JJ, Demidov LV et al. (2012) Dabrafenib in BRAF-mutated metastatic melanoma: a multicentre, open-label, phase 3 randomized controlled trial. *Lancet* 380:358–365.

Saranga-Perry V, Ambe C, Zager JS et al. (2014) Recent development in the medical and surgical treatment of melanoma. *CA Cancer J. Clin.* 64:171–185.

Siegel RL, Miller KD & Jemal A (2018) Cancer statistics, 2018. *CA Cancer J. Clin.* 68:7–30.

Sosman JA, Kim KB, Schuchter L et al. (2012) Survival in BRAF V600-mutant advanced melanoma treated with vemurafenib. *N. Engl. J. Med.* 366:707–714.

Summary

Ferguson FM & Gray NS (2018) Kinase inhibitors: the road ahead. *Nat. Rev. Drug Discovery* 17:353–377.

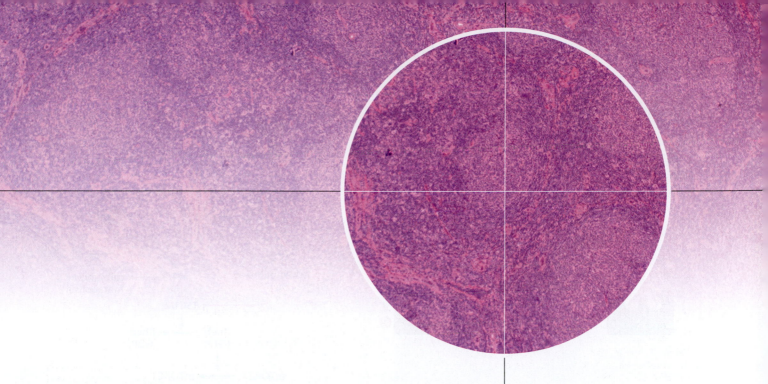

Chapter 5

Inhibition of the PI3-Kinase/ Akt/mTOR Pathway

Introduction

The **PI3K/Akt/mTOR** pathway plays a central role in regulating the normal function of cells and is one the most frequently dysregulated in human cancer. **Phosphatidylinositol 3-kinase** (PI3K), **Akt** (also called **protein kinase B**, PKB), and **mTOR** (mechanistic target of rapamycin, formerly known as mammalian target of rapamycin) are key nodes in a network that exerts a major influence on cell metabolism, proliferation, and survival. Dysregulation by mutation or other means that cause malfunctions of this pathway impact many of the hallmarks of cancer, including such cancer-supporting features of the tumor microenvironment as angiogenesis and inflammation.

The PI3K/Akt/mTOR pathway (Figure 5.1) is headed by PI3K, a lipid kinase that phosphorylates the membrane phospholipid phosphatidylinositol 4,5-bisphosphate (PIP2), converting it to phosphatidylinositol 3,4,5-triphosphate (PIP3). The membrane-bound PIP3 provides a docking site that recruits inactive Akt to the cell membrane, where its binding to PIP3 induces critical conformational changes that allow Akt to undergo a phosphorylation required for its conversion to an active serine/threonine protein kinase. Activated Akt mediates inhibition of apoptosis (a form of programmed cell death), promotion of anabolic processes (metabolic pathways involved in constructing complex biomolecules), and the activation of the pivotal **serine/threonine protein kinase**, mTOR, which plays a central role in deciding the cell's response to growth signals and nutrient availability. Not surprisingly, Akt activity is frequently increased in cancer. Although promising inhibitors of Akt have been developed, so far none have been approved for clinical use.

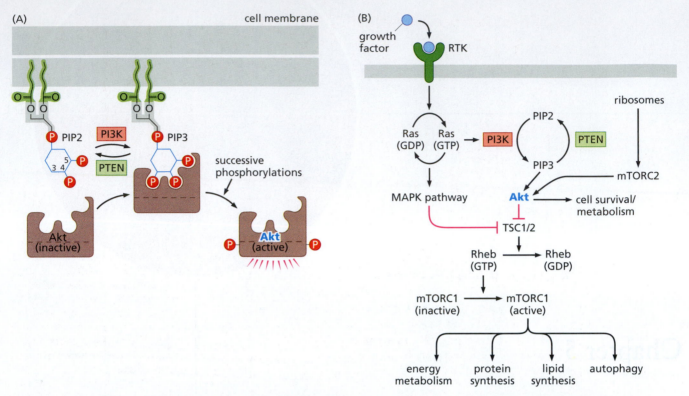

Figure 5.1. The PI3K/Akt/mTOR pathway. Three key kinases, PI3K, Akt, and mTOR—the core kinase of the two complexes mTORC1 and mTORC2—form a network that regulates many processes essential for cell maintenance, survival, growth, and proliferation. (A) PI3K enables activation of Akt. The conversion of membrane-resident 4,5-phosphatidylinositol bisphosphate (4,5-PIP2) by PI3K-mediated phosphorylation to 3,4,5-phosphatidylinositol triphosphate (3,4,5-PIP3) provides a docking site for the binding of inactive Akt. Conformational changes induced by its binding allow Akt to undergo the first of two phosphorylations required for its conversion to a fully active form. Note that the phosphatase PTEN destroys the ability of PIP3 to act as a docking site by removal of the phosphate added by PI3K. PTEN is a key tumor suppressor because it counteracts the tumorigenic effects of overly active PI3K. The anti-cancer effects of inhibitors of PI3K have the same effect as PTEN's action as a negative regulator. (B) A synopsis of the PI3K/Akt/mTOR pathway. This set of interlocking pathways regulates a broad spectrum of cellular activities. It can be initiated by growth factor-mediated activation of the small G protein Ras. Activation of mTORC1 is negatively regulated by TSC1/2, which inhibits conversion of the small G protein Rheb to the active GTP-bound state required for activation of mTORC1. The inactivation of TSC1/2 by Akt-mediated phosphorylation allows activation of mTORC1 by Rheb. Activated mTORC1 mediates protein and lipid synthesis, energy metabolism, and autophagy. The PI3K-mediated generation of PIP3 along with physical interaction with ribosomes is essential for the activation of mTORC2, the other complex of the mTOR pathway. Note that mTORC2 contributes an additional phosphorylation to Akt that results in its full activation. RTK, receptor tyrosine kinase; PTEN, phosphatase and tensin homolog; Rheb, Ras homolog enriched in brain; TSC1/2, tuberous sclerosis complex proteins 1 and 2. (Modified from M. Laplante and D.M. Sabatini, *Cell* 149:274–293, 2012.)

The activity of PI3K, which stands at the head of the PI3K/Akt/mTOR pathway, is negatively regulated by the tumor suppressor PTEN, a **phosphatase** that strips the activating phosphate from PIP3, converting it back to the inactive PIP2 and denying Akt the docking site necessary for its activation (see Figure 5.1). The negative regulation of the PI3K pathway by PTEN inhibits many pathways supporting cell growth, proliferation, and survival, and qualifies this phosphatase as a major tumor suppressor. Its importance is signified by the finding that it is lost or disabled in cancer cells at a frequency second only to the loss of p53 function, the most common genetic event in cancer.

The importance of the PI3K/Akt/mTOR network in cancer has made each of these kinases an attractive target for drug development. Insights from genomics, structural studies, and drug discovery programs led to the derivation of compounds targeting each of the nodes in this network. The search for PI3K inhibitors confronts a complex landscape involving eight different PI3Ks grouped into three classes, I, II, and III. A pan-PI3K inhibitor that inhibited all classes when used at doses high enough to have significant anti-tumor effects would impact a very broad spectrum of targets and have too high a level of toxicity to be well

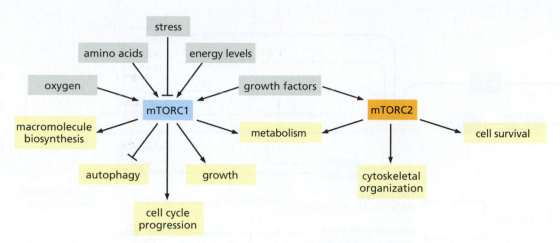

Figure 5.2. Inputs and outputs of mTORC1 and mTORC2. In addition to growth factors, other factors influence the protein kinase activity of the mTORC protein complexes. Although they both have the same catalytic unit—the serine/threonine protein kinase mTOR—as well as some subunits in common, mTORC1 and mTORC2 differ in some of the other proteins that comprise them. These differences are responsible for the distinctive input and output profiles outlined in the figure. (Modified from M. Laplante and D.M. Sabatini, *Cell* 149:274–293, 2012. With permission from Elsevier.)

tolerated. On the other hand, inhibition of only a single class of PI3Ks is more likely to come with acceptable levels of toxicity. Most efforts have focused on the class I enzymes, which appear to be the most important in cancer. The class I enzymes PI3Kα, PI3Kβ, PI3Kγ, and PI3Kδ are heterodimers of a 110-kilodalton catalytic subunit (p110) complexed with an 85-kilodalton regulatory subunit (p85). These **isoforms** are distinguished by their catalytic subunits (p110α, p110β, p110γ, and p110δ). Although these isoforms share structural homology, they do have differences that can be exploited to develop **isoform-specific inhibitors.** In addition to structural differences, some isoforms play greater roles in particular cell types or processes than others. This offers an opportunity for selective inhibition and more acceptable toxicity profiles. Consequently, there is active pursuit of isoform-selective drug discovery programs targeting particular PI3K isoforms, and the success of the PI3Kδ-selective inhibitor discussed in Case 5-2 validates the promise of the isoform-selective strategy.

The serine/threonine kinase mTOR is the core catalytic component of **mTORC1** and **mTORC2**, large multiprotein complexes that head interlocking regulatory signaling pathways. These complexes sense and use environmental cues to collaboratively determine the cell's transitions between **anabolic** (biosynthetic) and **catabolic** (degradative) pathways. The capacity to regulate when cells do or do not conduct processes that generate or consume large amounts of energy and nutrients is essential for growth and survival under conditions where nutrient availability is highly variable. As major sensors of nutritional status, mTOR pathways coordinate cellular responses to growth signals with nutrient availability, providing a beneficial and essential linkage of growth or stasis to nutritional status. As shown in Figure 5.2, mTORC1 and mTORC2 have different upstream inputs and downstream outputs. These pathways regulate such major processes as **protein synthesis**, cell growth, **cell survival**, and **autophagy** (controlled intracellular degradation of cellular organelles).

Dysregulation of mTOR-dependent processes can promote cancer in a number of ways. The mTORC1 complex supports mRNA translation, polypeptide synthesis/elongation, ribosome biosynthesis, synthesis of lipids for cell growth, and glycolytic and oxidative metabolism, and blocks the degradation of intracellular organelles by autophagy. The companion complex, mTORC2, is essential for the full activation of Akt, which can be a key element in fostering the survival of cancer cells since it protects against apoptosis. Furthermore, acting through Rho GTPases and PKC-α, mTORC2 affects the organization of the cytoskeleton, thereby influencing cell mobility and metastasis. Figure 5.3 presents an overview of the many ways mTORC1 and mTORC2 support cancer cells.

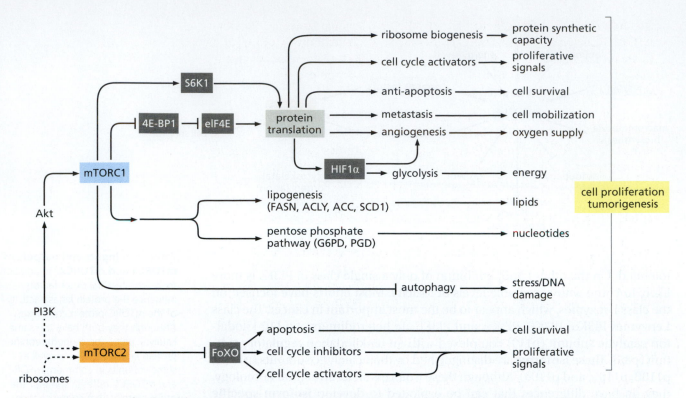

Figure 5.3. Support of cell proliferation and tumorigenesis by the PI3K/Akt/mTOR network. Activation of the PI3K pathway results in the activation of highly anabolic pathways under the control of mTORC1 and the inhibition of autophagy, which is catabolic. The activity of PI3K also activates the mTORC2 complex, which inhibits the activation of pathways that inhibit cell survival and proliferation. Inhibition of PI3K, Akt, mTORC1, or mTORC2 by protein kinase inhibitors targeting any one or a combination of these essential kinases has potential as an anti-cancer therapeutic approach. Akt also has mTOR complex-independent activities (*not shown*) that promote cell growth. 4E-BP1, 4E binding protein 1; SGK1, serum/glucocorticoid-regulated kinase 1; eIF4E, eukaryotic initiation factor 4E; SREBP-1, sterol regulatory element-binding protein 1; HIF1a, hypoxia-inducible factor a; FoXO, forkhead box protein class O. (Modified from M. Laplante and D.M. Sabatini, *Cell* 149:274–293, 2012. With permission from Elsevier.)

Recognition of the many roles played by mTOR pathways in cancer has stimulated interest in the development of drugs targeting the pathways. Efforts to target these pathways have pursued two major directions. One approach indirectly inhibits mTOR using the antibiotic **rapamycin** and its analogs, dubbed **rapalogs**, while the other strategy is direct inhibition by kinase inhibitors. Rapamycin, initially studied for its antifungal activity, was eventually found to suppress T cell–mediated immune responses and later was also found to have anti-tumor activity. A great deal of effort has been devoted to the synthesis of rapalogs, with the intent of discovering compounds with better efficacy than the parent compound, rapamycin. Rapalogs prevent substrates from gaining access to the active site of mTORC1, blocking the binding of substrates. Some rapalogs have shown promise in preclinical trials, and one, **temsirolimus**, was the first to receive FDA approval for cancer therapy (discussed in Case 5-1). However, when compounds belonging to this class are used as single agents, they are usually cytostatic, slowing or arresting the growth of the cancer, rather than being cytotoxic and producing tumor regression. Furthermore, rapalogs target mTORC1 but fail to inhibit mTORC2, an essential contributor to the complete activation of Akt, and a potential source of resistance to mTOR inhibitor therapy. The therapeutic efficiency of rapalogs can be compromised by the existence of negative-feedback loops in the mTOR pathway. Research revealed that when activated, the mTOR pathway mediates growth signaling but in some cells inhibits the expression of the platelet-derived growth factor receptor (PDGF-R), itself a receptor tyrosine kinase, or RTK, as well as **insulin-like growth factor-1** (IGF-1), the ligand for an RTK. Blockade of mTORC1 releases this inhibition and results in *increased* transcription of an RTK, such as PDGF-R, or an RTK ligand, such as IGF-1, producing the *paradoxical effect of greater signaling* from growth factors in the presence of rapamycin and rapalogs than in their absence.

Kinase inhibitors that inhibit both mTORC1 and mTORC2 have been developed, and these block the phosphorylation of all known targets of these complexes including the pathway leading to full activation to Akt. Studies both *in vitro* and *in vivo* have demonstrated that kinase inhibitors deliver greater inhibition of tumor cell proliferation and tumor growth than rapalogs. However, as mentioned above, the feedback processes negatively regulating signaling by some growth factor receptors are inhibited or lost upon inhibition of the mTORC1 kinase. Finally, mTOR and PI3K are members of the same kinase subfamily and have similar catalytic domains. Dual-specificity inhibitors have been developed that inhibit both of these kinases. These are highly potent and under evaluation in clinical trials. However, as expected, dual PI3K/mTOR inhibitors impact a broad spectrum of cellular processes, and management of their toxicity may pose problems.

Many approaches have been used to explore the utility of inhibiting components of the PI3K/Akt/mTOR signaling network. The spectrum includes pan-PI3K inhibition, p110 isoform-specific PI3K inhibitors, Akt inhibitors, and various mTOR inhibitors. Currently, there are more than 20 drug candidates in clinical development targeting various nodes of the network. It was expected that mutations conferring resistance to these targeted therapies would be common. Even a highly successful targeted therapy such as inhibition of BCR-ABL by imatinib (see Case 4-1) suffers from resistance arising from mutations that render the target insensitive to the inhibitor, and other examples of the acquisition of drug resistance mutations are well known. Unexpectedly, such resistance mutations have not been generally observed in the PI3K/Akt/mTOR network. Instead, resistance is mediated by loss or changes in feedback loops, the activation of compensatory pathways, and genomic alterations that activate processes downstream of the drug's site of action. These nonmutation-based problems will probably be best confronted by identifying mechanisms of resistance and, when possible, using this knowledge to devise combination therapies that effectively target those mechanisms.

Take-home points

✓ The PI3K/Akt/mTOR pathway is one of the most frequently dysregulated in human cancer.

✓ Drug discovery and development programs have derived inhibitors for each of the three nodal kinases PI3K, Akt, and mTOR.

✓ There are three classes and many isoforms of PI3K, a lipid kinase that stands at the head of the PI3K/Akt/mTOR pathway. Class I PI3 kinases are the most important in human cancer, and this class includes four isoforms of the p110 subunit: α, β, γ, and δ.

✓ The expression patterns of PI3K isoforms vary among cell types and tissues, with some expressing only one or a few and others displaying broader patterns of expression.

✓ The development of isoform-specific PI3K inhibitors has allowed targeting of processes dependent on the targeted isoform. More isoform-specific kinase inhibitors may lead to better-focused therapies with lower overall toxicity.

✓ mTOR, a serine/threonine kinase, is the catalytic component of mTORC1 and mTORC2, multiprotein complexes that head interlocking pathways that sense environmental cues, such as nutrient availability and growth signals, and determine the cell's transitions between anabolic and catabolic states.

✓ Akt is a serine/threonine kinase that is a key activator of the mTORC1 complex and, after full activation by mTORC2-mediated phosphorylation, plays essential roles in metabolism and cell survival.

✓ mTOR-dependent processes can support cancer in a number of ways.

✓ Two types of drugs targeting mTOR pathways have been developed: rapamycin or rapalogs that inhibit mTORC1-dependent processes, and kinase inhibitors that directly inhibit mTOR and target both mTORC1 and mTORC2.

<table>
<tr><td>

Topics bearing on this case

mTOR, a master regulator of cell physiology: an attractive target for anti-cancer therapy

von Hippel–Lindau (VHL) syndrome

hypoxia-inducible factor (HIF)

</td></tr>
</table>

Case 5-1 Renal Cell Carcinoma

Introduction

Renal cell carcinoma, or RCC, is just one of several types of cancers that can arise from the kidney; it derives from the cells lining the collecting tubules that are involved in the filtration process by which urine is produced as a waste product from blood. Cells from the more distal aspects of the urinary collecting system within the kidney—the collecting duct and the renal pelvis—give rise to different types of carcinomas with very different behaviors and treatments, and are not further discussed herein.

RCC is the eighth most common malignancy in the United States, with an estimated 65,340 new cases in 2018 and 14,970 deaths. With the ever-increasing access to and use of imaging studies such as CT or MRI, the majority of RCCs are now identified as "incidental findings" of a renal mass on imaging studies performed for other reasons. Accordingly, RCC incidence has been on the rise, and fortunately, if the disease is detected in early stages, the prognosis is excellent.

RCC contains several pathologic subtypes, including **clear cell**, papillary, chromophobe, and oncocytoma (oncocytic RCC; Table 5.1), which are distinguished from one another by unique biological, prognostic, and therapeutic differences. RCC is more common in men (male:female ratio, 1.6:1). Risk factors for developing RCC include hypertension, lifestyle factors such as smoking and obesity,

Table 5.1. RCC subtypes.

Histologic type	Frequency	Characteristics
Clear cell	60–70%	Cells have clear cytoplasm
Papillary	5–15%	Type I: cells have scant pale cytoplasm and low-grade nuclei Type II: cells have abundant eosinophilic (pink) cytoplasm and prominent nucleoli; aggressive subtype
Chromophobe	5–10%	Large cells with eosinophilic cytoplasm; indolent course
Oncocytic	5–10%	Benign tumors

Adapted from B.L. Rini, S.C. Campbell and B. Escudier, *Lancet* 373:1119–1132, 2009.

Table 5.2. Hereditary RCC syndromes.

Syndrome	Gene	RCC histology	Major clinical manifestations
von Hippel–Lindau	*VHL* tumor suppressor	Clear cell	Autosomal dominant (AD); retinal angioma; central nervous system hemangioblastomas; pheochromocytomas; pancreatic islet cell tumors; paragangliomas; cystadenoma of broad ligament or epididymis
Hereditary papillary renal carcinoma	*MET* proto-oncogene	Type I papillary	AD; multifocal bilateral renal cell tumors
Birt–Hogg–Dubé	*FLCN* tumor suppressor	Chromophobe, hybrid oncocytoma	AD; cutaneous fibrofolliculoma; pulmonary cysts; spontaneous pneumothorax
Hereditary leiomyomatosis and RCC	*FH* tumor suppressor	Type II papillary	AD; leiomyomas of skin and uterus; unilateral, solitary, and aggressive renal cell tumors
Succinate dehydrogenase-associated familial cancer	*SDHB* and *SDHD* subunits	Clear cell, chromophobe, papillary type II, renal oncocytoma	Head and neck paraganglioma; adrenal or extra-adrenal pheochromocytomas
Tuberous sclerosis complex	TSC1, TSC2	Clear cell	AD; facial angiofibromas; multifocal renal angiolipomas; neurologic disorders or seizures; lymphangiomyomatosis of the lungs

Adapted from B.L. Rini, S.C. Campbell and B. Escudier, *Lancet* 373:1119–1132, 2009.

and inherited predisposition such as **von Hippel–Lindau syndrome** and **Birt–Hogg–Dubé syndrome** (Table 5.2). The underlying genetic abnormalities for the hereditary syndromes that predispose patients to RCC have provided important insights into pathogenesis and management approaches.

Diagnosis, workup, and staging

With modern imaging, most renal masses are found incidentally, and patients are typically asymptomatic at the time of presentation. Classically, a triad of clinical features have been considered characteristic of RCC, including **hematuria** (blood in urine), abdominal pain, and a flank or abdominal mass; however, this triad is observed in approximately only 5% of cases at diagnosis. Additionally, symptomatic patients with RCC may present with anemia or constitutional symptoms such as weight loss.

Evaluation of suspected RCC should include a complete medical and family history, a thorough physical exam, and laboratory analyses including a complete blood count and blood chemistries including calcium, creatinine (a measure of kidney function), and liver function tests.

Evaluation of a suspected renal mass typically begins with a CT or MRI of the abdomen. An MRI can offer a more detailed evaluation of whether a tumor involves the renal vein or the inferior vena cava (Figure 5.4). RCC can spread to lung, lymph nodes, bone, liver, adrenal gland, pancreas, and brain. Because of the propensity for RCC to spread to the lungs, evaluation of the chest with an X-ray or CT scan is also performed, with the latter providing much more detail than the former. Bone scans and PET scans are not routinely performed for RCC as they are insensitive for detecting RCC metastasis. Brain imaging is typically performed only in patients with symptoms suspicious for involvement of the central nervous system, such as focal weakness or confusion.

The abdominal imaging studies above have high diagnostic accuracy for RCC. For patients undergoing surgery to remove the tumor, a needle biopsy is not always necessary, unlike many other malignancies in which a positive biopsy for cancer is necessary prior to the commitment for surgery. For some cases in which the imaging is not clearly pointing toward RCC (for instance, if

Figure 5.4. Imaging of a patient with a newly diagnosed 12-cm right kidney mass. (A) CT scan. (B) MRI scan. Both indicate coronal imaging (slicing through the body providing frontal imaging) from the same patient. These images do not demonstrate the maximum dimension of the lobular mass, but demonstrate internal necrosis (*yellow arrow*) and tumor extension into the right renal vein (tumor thrombus, *red arrows*), extending into the inferior vena cava (path of the inferior vena cava marked by the dotted pink line).

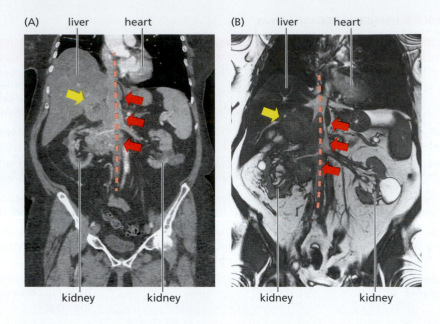

it might be suggesting either urothelial cancer arising from the renal pelvis or lymphoma), a CT- or ultrasound-guided needle biopsy can help to establish the diagnosis to guide appropriate management.

Staging of RCC employs the TNM system of the American Joint Committee on Cancer (AJCC; Table 5.3 and Figure 5.5). As with most cancers, survival correlates with stage at diagnosis, as indicated in the table.

Table 5.3. AJCC staging classification of RCC.

Primary tumor (T)	
Tx	Primary tumor cannot be assessed
T0	No evidence of primary tumor
T1	Tumor ≤7 cm and limited to kidney
T2	Tumor >7 cm and limited to kidney
T3	Tumor extends into major veins or surrounding tissues but not the adrenal gland and not beyond Gerota's fascia
T4	Tumor invades beyond Gerota's fascia

Regional lymph nodes (N)	
Nx	Nodes cannot be assessed
N0	No regional node metastasis
N1	Metastasis in regional node(s)

Distant metastasis (M)	
M0	No distant metastasis
M1	Distant metastasis

Stage grouping	T	N	M	5-year overall survival
Stage I	T1	N0	M0	96%
Stage II	T2	N0	M0	82%
Stage III	T1 or T2	N1	M0	64%
	T3	N0 or N1	M0	
Stage IV	T4	any N	M0	23%
	any T	any N	M1	

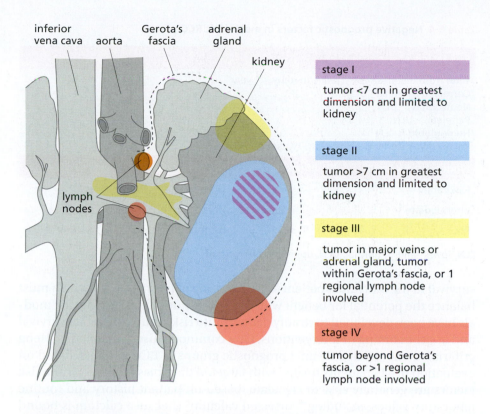

inferior vena cava aorta Gerota's fascia adrenal gland

kidney

lymph nodes

stage I

tumor <7 cm in greatest dimension and limited to kidney

stage II

tumor >7 cm in greatest dimension and limited to kidney

stage III

tumor in major veins or adrenal gland, tumor within Gerota's fascia, or 1 regional lymph node involved

stage IV

tumor beyond Gerota's fascia, or >1 regional lymph node involved

Figure 5.5. **Staging overview of renal cell carcinoma.** (From T. Cohen and F.J. McGovern, *N. Engl. J. Med.* 353:2477–2490, 2005. Reprinted with permission from Massachusetts Medical Society.)

Management of localized RCC

Fortunately, for many patients with RCC discovered incidentally, the cancer is detected at an early stage and can be surgically resected. The cure rates for stage I–II RCC are high at >90% (see Table 5.3). **Radical nephrectomy** (removal of the entire kidney) is the standard of care; it includes removal of the kidney, Gerota's fascia (a layer of connective tissue that encapsulates the kidney and adrenal glands; see Figure 5.5), **ureter**, and adrenal gland on the same side, as well as adjacent lymph nodes. A preferred alternative, where feasible, is partial nephrectomy, which removes the tumor and a rim of healthy tissue, allowing for "nephron-sparing," which maintains as much normal kidney tissue as possible to maintain renal function. For selected patients with small tumors (<4 cm) who may not tolerate surgery due to age or co-morbid medical issues, **percutaneous** (through the skin) **ablation** is an alternative to surgery with good outcomes and tolerability in reported case series. Using image guidance similar to performing a percutaneous biopsy, a needle is placed into the tumor, and the tumor is either frozen (**cryoablation**) or heated through an electrical current (**radio-frequency ablation**).

After surgical resection for apparently localized disease, approximately one-third of patients will eventually develop disease recurrence. Adjuvant therapies have been attempted to increase the cure rate after surgery but have not improved overall survival. Thus, there is no standard adjuvant treatment after partial or radical nephrectomy for stage I–III RCC.

Management of metastatic RCC

Metastatic RCC is considered incurable. Metastatic RCC has a variable natural history, with some cases exhibiting rapid spread and progression while others may have indolent (slow) growth, including rare cases of spontaneous tumor **regression**. This high degree of variability presents challenges to predicting

Table 5.4. Negative prognostic factors in metastatic RCC.

Risk factor			
Time from diagnosis to systemic therapy <1 year Corrected calcium > ULN Neutrophils > ULN Platelets > ULN Hemoglobin < LLN Karnofsky Performance Status <80% (see Table 3.10)			
Prognostic risk group	**No. of risk factors**	**Overall survival**	**2-year survival**
Favorable	0	Not reached	75%
Intermediate	1–2	27 months	53%
Poor	3–5	8.8 months	7%

LLN, lower limit of normal. ULN, upper limit of normal.

survival in the individual patient and the timing of interventions, which must balance the potential for benefit with the risk of side effects. Prognostic models have been developed to stratify patients into risk groups to predict survival and assist in timing of interventions. One commonly used model, the "Heng criteria," demonstrated distinct prognostic groups of RCC patients, based on evaluation of 645 patients treated with targeted therapies (Table 5.4). The risk factors are generally easy to calculate, based on patient history and routine laboratory values, including "corrected calcium" (because calcium is bound to serum albumin, and albumin levels vary between patients, the measured serum calcium is "corrected" for albumin via a simple formula), number of neutrophils (a subtype of white blood cells), number of platelets, and degree of anemia measured by hemoglobin. An increasing number of risk factors is associated with worse survival.

The occurrence of rare spontaneous regression hints at a role for the immune system in modulating RCC growth and progression. Immunotherapy with the cytokines interferon (IFN) and interleukin-2 (IL-2) has been extensively studied in metastatic RCC, and high-dose IL-2 was the only FDA-approved therapy for metastatic RCC until 2005. IL-2 is administered intravenously as a course of up to 14 doses over five days. The overall response rate was only 10–15%, with a small number (<5%) exhibiting a durable **complete response** (CR: complete disappearance of metastases on imaging). Importantly, some patients had durable CRs lasting longer than five years, suggesting that IL-2 may cure some patients. However, IL-2 is associated with significant toxicity and even 4% mortality, primarily due to **capillary leak syndrome**, in which IL-2 induces leakage of plasma and other blood components out of blood vessels and into other tissues, causing severe hypotension (low blood pressure) that can lead to organ failure and even death. These side effects improve rapidly after discontinuation of IL-2. Thus, IL-2 should be administered only in experienced centers with appropriate monitoring and supportive measures. Despite the high toxicity and low overall response rate, IL-2 therapy is still considered for appropriate patients with metastatic RCC, because it is the only approved therapy with proven durable CRs. Many patients, however, will not be sufficiently healthy to be considered candidates for IL-2 therapy. The role of immune checkpoint modulator therapy in RCC will likely change the landscape of RCC therapy, with FDA approval of the monoclonal antibody nivolumab for RCC in November 2015 (see Chapter 12 and Case 12-3 for more on this approach).

For most metastatic cancers, surgical removal of the primary tumor is not performed because it does not extend life and exposes the patient to the risk of an operation. For metastatic RCC, however, removal of the primary kidney tumor can improve overall survival due to spontaneous regression of metastatic lesions. Historically, under 1% of patients were reported to experience spontaneous regression of metastatic lesions after such **cytoreductive nephrectomy**. Two prospective Phase III randomized trials demonstrated a statistically significant increase in median survival in metastatic RCC patients treated with nephrectomy followed by IFN compared with patients who received IFN alone, although these studies were performed in an era when immunotherapies like IFN and IL-2 were the only available systemic treatments. In contemporary practice, cytoreductive nephrectomy is still performed for patients with good performance status. Whether nephrectomy improves survival in the current era of targeted therapies is unknown and is the topic of ongoing clinical trials.

In select cases, patients with an isolated metastatic RCC lesion (typically one lesion in one distant organ) might be candidates for local therapy such as surgical resection or percutaneous ablation of the metastasis. Although this approach to **oligometastatic** disease may not necessarily offer a cure, it may allow for a prolonged time off of systemic therapy, even lasting years in some patients. Favorable features for this approach include a solitary metastasis and a long interval between initial diagnosis and development of metastatic disease.

Because RCC is relatively resistant to conventional chemotherapy and radiation therapy, these treatments have only a limited role for this disease. Systemic therapies have changed dramatically since 2005, with the advent of targeted therapies.

Molecular alterations and RCC

The most common type of RCC, clear cell RCC, accounts for 60–70% of RCC cases and >90% of metastatic RCC. Von Hippel–Lindau (VHL) syndrome is an autosomal dominant inherited disease that predisposes patients to multiple cancers, including clear cell RCC, **pheochromocytomas** (arising from the adrenal medulla), and neuroendocrine tumors of the pancreas (see Table 5.2). Up to 45% of VHL patients will develop RCC. The *VHL* tumor suppressor gene, located on chromosome 3, is mutated in VHL disease, with most patients inheriting one mutated copy of the *VHL* gene and one normal copy. RCC in VHL patients develops as a result of alteration of the normal allele through deletion, **hypermethylation**, or mutational inactivation. This **loss of heterozygosity** (LOH) is observed in nearly 100% of hereditary cases and >75% of sporadic cases of clear cell RCC.

The VHL tumor suppressor normally inhibits cell growth and regulates the expression of several pro-angiogenic genes, which direct the formation of new blood vessels. Under normal oxygen conditions, VHL targets the **hypoxia-inducible factor** (HIF) family of transcription factors for ubiquitination and degradation. Inactivation of VHL through LOH leads to constitutive up-regulation of intracellular HIF and subsequent increased expression of downstream gene targets, including **vascular endothelial growth factor** (VEGF), **transforming growth factor-α** (TGF-α), **platelet-derived growth factor-β** (PDGF-β), and others (Figure 5.6).

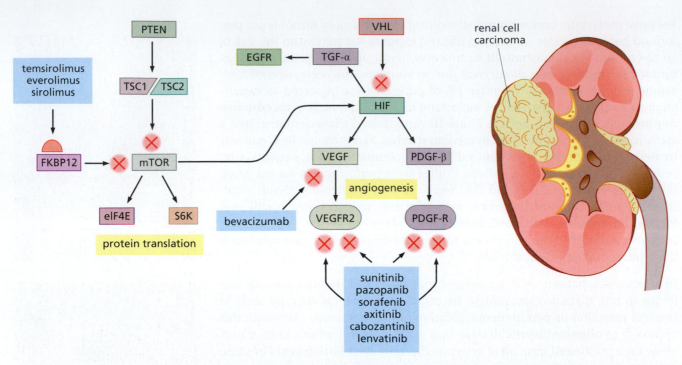

Figure 5.6. Important signaling pathways in RCC. Hypoxia-inducible factor (HIF) activity is modulated by the von Hippel–Lindau (VHL) protein and is up-regulated when VHL inactivation occurs either in spontaneous or inherited cases of RCC. HIF then activates downstream growth factor pathways, including angiogenesis via vascular endothelial growth factor (VEGF) and tumor cell growth via platelet-derived growth factor-β (PDGF-β) and transforming growth factor-α (TGF-α). The tyrosine kinase inhibitors (TKIs) sunitinib, pazopanib, sorafenib, axitinib, and cabozantinib and the monoclonal antibody bevacizumab inhibit the angiogenesis pathway. The mTOR pathway is also active in RCC and can interact with HIF activation to influence angiogenesis but can also mediate cellular growth through other pathways. The mTOR inhibitors temsirolimus and everolimus inhibit the binding of FKBP12 with mTOR. (From J. Brugarolas, *N. Engl. J. Med.* 356:185–187, 2007. Reprinted with permission from Massachusetts Medical Society.)

The case of Mary Carlton, a woman with an incidental kidney mass found during evaluation of abdominal pain

Mary was a 60-year-old female never-smoker with diabetes, hypertension, and coronary artery disease (a history of a heart attack at age 51), who presented to her primary care physician with one month of pain in her upper abdomen that led to diminished appetite and five-pound weight loss. She had no nausea or vomiting. She had no change in her bowel movements and no obvious blood in her stool. She had no urinary complaints and had not noticed blood in her urine. She had no known family history of cancer. On exam, mild tenderness was elicited on palpation of the upper abdomen, but no mass could be felt. Mary had been up to date with routine cancer screening, including colonoscopy and mammograms. Her blood tests were normal, including liver function tests and kidney function. An office urinalysis did not reveal any blood in the urine.

Mary had a trial on ranitidine, an antihistamine that diminishes acid production in the stomach, but felt no improvement after one month of consistent use. Her primary care physician opted to perform a CT scan to look for an explanation for her discomfort. There was no sign of gallbladder inflammation or any masses associated with her gastrointestinal tract. However, note was made of a 12-cm right kidney mass (Figure 5.7). The radiologist's interpretation of the CT scan was "likely consistent with renal cell carcinoma."

Mary had a CT scan of the chest that showed no evidence of other disease. She was referred to a urologist and had a radical right nephrectomy, because the size of the tumor precluded complete resection of the tumor via a partial nephrectomy. Pathology confirmed RCC, clear cell type, grade 3 (out of 4) with extension out of the kidney itself (extrarenal) but no involvement of the veins, and no involvement in any of the removed lymph nodes (T3 N0 M0, or stage III; see Table 5.3). She recovered well from surgery without any major complications, despite some concerns about her preexisting coronary artery disease. She noticed no change in the original abdominal pain symptoms. Although her stage III status placed Mary at high risk for eventual disease recurrence (36%; see Table 5.3), in the absence of any FDA-approved adjuvant treatments that could increase her chance at cure, she was placed on a surveillance plan.

Mary underwent CT scans every three months. At nine months after her nephrectomy, her chest CT scan revealed multiple round nodules in both left and right lungs (Figure 5.8). She had no symptoms at that time. She underwent a CT-guided percutaneous biopsy of a nodule in the right lower lung, in which a needle is advanced under CT guidance through the skin until the end reaches the lesion in question, allowing for sampling of the tissue without a major operation (see Figure 5.8, right panel). Unfortunately, the pathology confirmed metastatic RCC that appeared histologically identical to her original disease.

Mary otherwise felt completely well. Her laboratory tests were notable for normal calcium and albumin, and normal white blood cell count, but she had anemia (hemoglobin of 10.5 g/dl) and thrombocytosis (increased platelets of 510,000 per μl).

Mary discussed the options for treatment with her medical oncologist. Due to her other medical conditions, she was not considered a good candidate for IL-2 therapy. They discussed targeted therapies, including tyrosine kinase inhibitors such as **sunitinib** and **pazopanib**, and the mTOR inhibitor temsirolimus. With metastatic disease under a year from her original diagnosis, she met three risk factors for "poor-risk" metastatic RCC, and ultimately, temsirolimus was recommended. On therapy for three months, her lung nodules appeared stable to slightly smaller, but by six months, most lung nodules had significantly decreased in size (Figure 5.9). She continues on therapy at twelve months with her most notable side effect (change in her cholesterol and triglyceride profile) well managed by medication adjustment.

Targeted therapies and metastatic RCC

Targeted therapies for RCC generally inhibit either angiogenesis or the mTOR pathway. Angiogenesis, the development of new blood vessels, is a critical step to support growing tumors (see Case 10-1). Vascular endothelial growth factor (VEGF) is the ligand for VEGF receptors. A model for VEGF production in cancers is that cancer cell proliferation causes relative hypoxia (oxygen deficiency), leading to a need for angiogenesis to improve oxygen delivery to

(A)
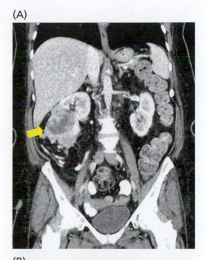

(B)
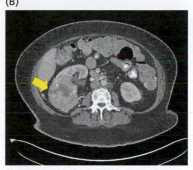

Figure 5.7. CT scans from Mary's case. (A) Coronal view (slicing through the body providing frontal imaging). (B) Axial view (horizontal slicing through the body, viewed looking up from the patient's feet as the patient is lying face-up on the scanning table; the patient's right side is on the viewer's left). Yellow arrows point to the renal mass in the lower pole of the right kidney. The left kidney is normal.

(A)
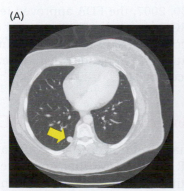

(B)
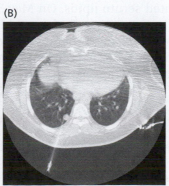

Figure 5.8. Mary's lung metastasis. (A) CT scan of the lungs, axial view, demonstrating a round nodule (*yellow arrow*) in the right lower lobe of the lung. Note that only one of her multiple lung nodules is demonstrated here because it is captured on this imaging "slice." (B) CT-guided percutaneous biopsy of the lung nodule. With the patient lying on her stomach, the nodule is localized by the CT scan and the needle is advanced slowly through the skin until it reaches the lesion. This image was rotated 180 degrees for the reader, to align the nodules in both A and B panels.

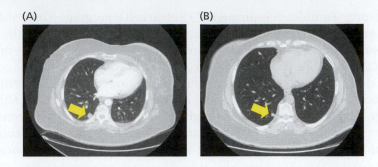

Figure 5.9. Mary's response to temsirolimus. (A) At three months of treatment, the right lung lesion was slightly smaller but the remaining lung metastases were essentially stable. (B) At six months of treatment, the right lung lesion and her other lung metastases were all decreased in size, but still visible.

the growing tumor. Hypoxia-inducible factor, HIF, mediates the production of VEGF. In RCC cells with inactivation of VHL, HIF turns on VEGF production even in the absence of hypoxia and thus plays a central role in driving growth and progression of RCC by proactive expansion of the blood vessel network supporting the growing tumor (see Figure 5.6). Effective therapies against VEGF-mediated angiogenesis have transformed care of metastatic RCC since the first approval of a targeted agent in this disease, **sorafenib**, on December 20, 2005, which led the way toward renewed interest, research, and development of new drugs for metastatic RCC.

FDA-approved therapies for metastatic RCC targeting angiogenesis include **bevacizumab**, a monoclonal antibody against VEGF (see Case 10-1), and the tyrosine kinase inhibitors sunitinib, pazopanib, sorafenib, and **axitinib**, which inhibit VEGF receptors among other targets (for more on TKI therapy, see Chapter 4). Notably, the clinical trials that led to approval of these therapies in RCC were primarily in patients with clear cell RCC, so the broader applicability of these drugs to other RCC histologies has not been proven.

The mTOR pathway is an important target in metastatic RCC. The mTOR pathway plays a central role in growth and survival (see Figure 5.3). In RCC, mTOR activity leads to HIF activation as well as cell proliferation. The upstream signals may include the GTPase activating proteins (GAPs) of the mTOR pathway **TSC1** and TSC2, which may underlie the pathogenesis of RCC development in the hereditary tuberous sclerosis complex syndrome (see Table 5.2).

The rapalog temsirolimus is an mTOR inhibitor that binds to the intracellular protein FKBP12, disrupting FKBP12 binding to mTOR and thereby decreasing mTOR signaling. Temsirolimus is administered as a weekly intravenous infusion. In a Phase III randomized controlled trial, 626 patients with previously untreated, poor-prognosis, metastatic RCC were randomized to receive temsirolimus, subcutaneous IFN, or both. This study included 20% of patients with non-clear–cell RCC histology. Temsirolimus significantly improved median overall survival compared with IFN (10.9 versus 7.3 months). Common side effects were generally manageable, including rash, peripheral edema (swelling in extremities), elevated blood sugars, and elevated serum lipids. On May 30, 2007, the FDA approved temsirolimus for the treatment of advanced RCC, the first mTOR-targeted agent to be approved for cancer therapy. A second mTOR inhibitor, **everolimus**, has also been FDA approved for metastatic RCC in patients with disease progression after therapy with a VEGF receptor TKI.

Future prospects and challenges in metastatic RCC

The management of metastatic RCC has been transformed by therapies targeting VEGF/angiogenesis and mTOR signaling. Mechanisms of resistance to targeted therapies are being identified with the hope of circumventing resistance or improving response to therapy. However, despite the considerable gains of having seven targeted therapies approved since 2005, none of these agents produce durable CRs like the rare, exceptional responses to IL-2, which comes at the cost of considerable toxicity. Novel immunotherapy approaches involving immune checkpoint inhibitors (see Case 12-3) in this historically immunotherapy-responsive disease have built upon the IL-2 experience and led to FDA approval, in November, 2015, of nivolumab for metastatic RCC after therapy with a VEGF receptor TKI. Whether immune checkpoint modulators produce durable CRs like IL-2 remains to be seen. The optimal approach may involve sequential or combination therapy with targeted agents and immunotherapy. Other areas of need include the development of adjuvant therapies that can increase the cure rate after surgical resection. As with other diseases that have multiple therapy options, identification of biomarkers that can guide the selection of optimal therapy remains an unmet medical need for RCC patients.

Take-home points

✓ Renal cell carcinoma (RCC) is a common cancer.

✓ In contemporary practice, most kidney tumors are diagnosed as incidental findings on imaging studies performed for other clinical indications.

✓ At the time of presentation, most patients do not have symptoms of RCC, such as the classic triad of hematuria, abdominal pain, and a flank or abdominal mass.

✓ RCC is associated with several hereditary syndromes.

✓ Prognosis of RCC is well correlated with staging.

✓ Localized RCC is often curable with surgery, or selective ablation in patients who may not be operative candidates.

✓ Approximately one-third of patients will have a recurrence.

✓ Surgery may still be of benefit for patients with metastatic RCC.

✓ Interleukin-2 therapy is highly toxic but is the only treatment that produces durable complete responses in metastatic RCC.

✓ Targets for therapy include angiogenesis, mediated by the VHL/HIF/VEGF pathway; the mTOR pathway; and immune checkpoints.

✓ There are nine FDA-approved targeted therapies for treatment of metastatic RCC, but none has consistently produced durable complete responses in patients.

✓ Immune checkpoint inhibitors represent the class of targeted therapy most recently approved in RCC.

✓ Future goals include optimal therapy selection for the individual patient as well as optimization of sequential or combinatorial approaches employing targeted therapies and immunotherapies.

Discussion questions

1. Kidney cancers are being diagnosed at earlier stages now compared with 20 years ago. Accordingly, cure rates have risen. Why?

2. What is the appropriate evaluation of a patient with a kidney tumor?

3. What is the evidence that kidney cancer may be responsive to immune manipulation?

4. What lessons from VHL have led to targeted therapy development?

5. What are the possible mechanisms of action of mTOR inhibitors in RCC?

Topics bearing on this case

Lymphomas

Expression of CD20 antigen by cells in the B-cell lineage

The PI3 kinase pathway

Case 5-2   Follicular Lymphoma

Introduction

Non-Hodgkin lymphomas (NHLs) are a heterogeneous collection of cancers derived from either B or T lymphocytes. There are currently over 60 recognized subtypes of NHL, each characterized by a unique biology, and often distinct natural history and therapy (Table 5.5). Understanding the discrete biology of a given histology is therefore critical in evaluating prognosis and treatment selection for a lymphoma patient. Approximately 90% of NHLs arise from B cells, with 10% arising from T cells. Among B-cell lymphomas, the clinical course can often be divided as diseases that follow either an indolent, aggressive, or highly aggressive natural history. Indolent lymphomas have an untreated natural history that may be measured in years or decades, while

Table 5.5. Subtypes of non-Hodgkin lymphomas.

Indolent non-Hodgkin lymphomas	
B-cell lymphomas	**T-cell lymphomas**
Follicular lymphoma (grades I–II)	T-cell large granular lymphocyte leukemia
Small lymphocytic lymphoma/chronic lymphocytic leukemia	Mycosis fungoides
Marginal zone lymphoma	
Lymphoplasmacytic lymphoma (Waldenström macroglobulinemia)	
Hairy cell leukemia	

Aggressive non-Hodgkin lymphomas	
B-cell lymphomas	**T-cell lymphomas**
Diffuse large B-cell lymphoma	Peripheral T-cell lymphoma, not otherwise specified
Follicular lymphoma (grade III)	Anaplastic large-cell lymphoma, ALK+ and ALK-
Mantle cell lymphoma	Angioimmunoblastic T-cell lymphoma
B-cell prolymphocytic leukemia	Extranodal NK T-cell lymphoma
High grade B-cell lymphoma	Enteropathy-associated T-cell lymphoma
Primary mediastinal (thymic) B-cell lymphoma	Subcutaneous panniculitis-like T-cell lymphoma
Plasmablastic lymphoma	Hepatosplenic T-cell lymphoma
Primary effusion lymphoma	T-cell prolymphocytic leukemia

Highly aggressive non-Hodgkin lymphomas	
B-cell lymphomas	**T-cell lymphomas**
Burkitt lymphoma	Precursor T-cell lymphoblastic lymphoma
Precursor B-cell lymphoblastic lymphoma	Adult T-cell leukemia/lymphoma

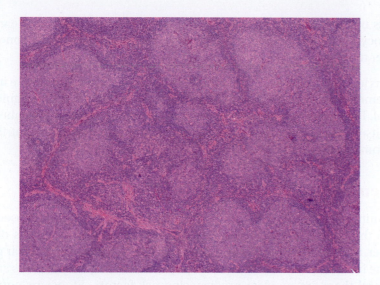

Figure 5.10. Low-power histologic view of follicular lymphoma (FL). Complete effacement of the normal lymph node architecture by back-to-back nodules of variably sized follicles (hematoxylin and eosin-stained) is shown.

aggressive and highly aggressive B-cell lymphomas would carry a life expectancy of weeks or even days without effective treatment. In general, indolent lymphomas are highly treatable but not considered curable with conventional therapy. Indolent lymphomas are treated as needed over the course of a patient's lifetime with the goal of preventing the disease from ever becoming life-threatening or from impairing quality of life. Among aggressive lymphomas, however, the goal of therapy is to cure the disease, because if not successfully eradicated, these diseases would be uniformly fatal.

Follicular lymphoma (FL) is the most common indolent lymphoma in the United States. The name is derived from the appearance of the lymph node structure under the microscope, which is characterized by effacement of the normal lymph node architecture by back-to-back nodules of malignant lymphocytes, known as follicles (Figure 5.10). The cells are mostly small in size, unlike cells in more aggressive lymphomas, and are derived from **germinal-center** B lymphocytes.

At a genetic level, FL is characterized by a chromosomal rearrangement of *BCL2*, a potent anti-apoptotic gene located on chromosome 18q21, which is translocated to the immunoglobulin heavy-chain promoter region on 14q32, resulting in constitutive expression of *BCL2* (Figure 5.11). Though identified in the vast majority of follicular lymphomas, this t(14;18) chromosomal translocation

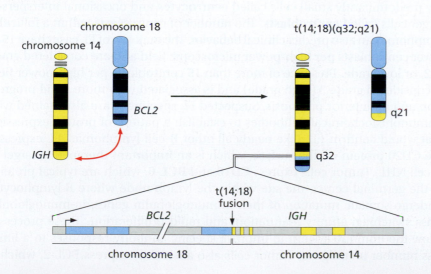

Figure 5.11. Chromosomal translocation in FL. The genetic sine qua non in FL is the translocation of *BCL2* on chromosome 18 with the *IGH* promoter region on chromosome 14, resulting in the t(14;18) and constitutive *BCL2* expression.

alone is not sufficient to produce lymphoma, but appears to occur early in a lymphocyte that subsequently develops secondary chromosomal abnormalities over time, ultimately leading to malignant transformation.

BCL-2 is located on the outer membrane of the **mitochondrion**, where it works as a member of a large family of pro- and anti-apoptotic proteins to regulate cell survival. Either a relative excess of anti-apoptotic BCL-2 family members, like BCL-2 itself, or a deficiency of pro-apoptotic family members may contribute to development of lymphoma.

Epidemiology

FL accounts for approximately 25% of the 74,000 cases of NHL diagnosed every year in the United States. Given the lengthy natural history of this disease, however, the **prevalence** (number of people alive living with FL) is far greater than the incidence (number of new cases). The vast majority of cases are sporadic, and there are no clear predisposing risk factors. FL occurs equally in men and women and the median age at diagnosis is in the mid-60s, but the disease can affect everyone from young adults to the elderly.

Diagnosis, workup, and staging

FL most commonly presents with painless lymph node enlargement. Lymph nodes may be noticed by patients themselves, or they may produce symptoms by pressing on a normal structure, producing discomfort or organ impairment, though this is uncommon. Lymph nodes in FL have been noted to wax and wane in size over time; that is, spontaneous reductions in lymph node size and number may be observed without specific intervention, implicating a role for immune surveillance in this particular disease. Occasionally, FL can present with fatigue related to anemia resulting from bone marrow infiltration by lymphoma. Systemic "B" symptoms, including unexplained fevers, drenching night sweats, and unintentional weight loss, are uncommonly seen in FL and are much more commonly present in aggressive subtypes of NHL. The presence of such systemic symptoms prompts suspicion of an aggressive transformation of follicular lymphoma into a higher-grade histology, which occurs in approximately 15% of FL patients during the course of their lifetime.

As noted, patients usually present with painless lymphadenopathy. Diagnosis is made by biopsy of an enlarged lymph node. The lymph node will show architectural disruption by frequent **neoplastic follicles**, and close inspection will identify predominantly small cells called **centrocytes** and occasional interspersed larger cells called **centroblasts**. The number of centroblasts within a follicular lymphoma can also predict clinical behavior. The majority of FL cases have 15 or fewer centroblasts per high-power microscopic field and are considered grade 1–2, or low grade. Presence of more than 15 centroblasts per high-power field is considered grade 3 (high grade) and is associated with more rapid progression and an inferior prognosis. Suspected FL specimens are also stained with immunohistochemical antibodies to establish a pattern of protein expression that would confirm FL. Like nearly all other B-cell lymphomas, FL expresses the CD20 protein on the surface, which is an important therapeutic target in B-cell NHL. Tumor cells express **CD10** and **BCL-6**, which are typical proteins of the germinal center, the site within the lymph node where B lymphocytes undergo somatic mutation of their immunoglobulin genes, immunoglobulin class switching, affinity maturation, and rapid proliferation. These processes allow the immune system to improve specific antibody responses to a limitless number of antigens. Tumor cells also aberrantly express BCL-2, which is

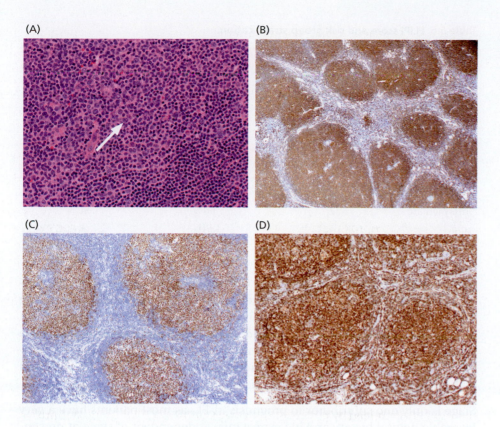

Figure 5.12. High-power histologic view and immunophenotype of FL. (A) High magnification of a single neoplastic follicle demonstrates a predominance of small, irregular centrocytes and rare large, oval centroblasts (*arrow*) comprising <15 cells per high-power field, consistent with low-grade FL (hematoxylin and eosin staining). Typical immunophenotype showing positivity of FL cells for the pan B-cell marker CD20 (B) and germinal-center marker BCL-6 (C), as well as aberrant expression of BCL-2 (D).

the anti-apoptotic gene product of the t(14;18) chromosomal translocation (Figure 5.12). Genetic analysis with either **karyotyping** or fluorescence *in situ* hybridization (FISH) for *BCL2* and *IGH* rearrangements can also identify the *BCL2* translocation at a molecular level, though this is generally not required to make the diagnosis.

FL is staged with the **Ann Arbor classification** (Figure 5.13). Stage I disease refers to involvement of a single lymph node region or isolated involvement of a single **extranodal** location. Stage II disease involves multiple lymph node regions confined to one side above or below the diaphragm. Stage III disease involves lymph node regions on both sides of the diaphragm, and stage IV disease is characterized by involvement of extranodal locations such as organs, bones, or bone marrow. These stages are often grouped as limited-stage (stages I and II) and advanced-stage (stages III and IV).

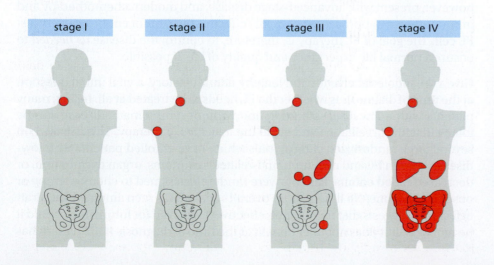

| stage I | stage II | stage III | stage IV |

Figure 5.13. Ann Arbor staging for lymphoma. Stage I disease involves a single lymph node region or isolated involvement of a single extranodal location. Stage II disease involves multiple lymph node regions on the same side of the diaphragm. Stage III disease involves lymph node regions on both sides of the diaphragm, and stage IV disease is characterized by involvement of extranodal locations, such as organs, bones, or bone marrow.

Table 5.6. FLIPI score and risk group stratification.

Risk factors
Age >60 years Ann Arbor stage III–IV Hemoglobin <12 g/dl Number of nodal regions >4 Elevated lactate dehydrogenase (LDH)

Risk groups	No. of risk factors	Estimated progression-free survival at 5 years	Estimated overall survival at 5 years
Low	0–1	80%	95%
Intermediate	2	60%	90%
High	3–5	40%	65%

Staging is performed using a full-body CT scan, often also including positron-emission tomography (PET), which is a functional imaging technique offering increased sensitivity over CT scans alone. Evaluation of the bone marrow with a bone marrow biopsy may be required. Laboratory tests at diagnosis include a complete blood count and measurement of chemistries including electrolytes and liver and renal function. **Lactate dehydrogenase** (LDH) level and **β-2-microglobulin** are also checked, as they carry prognostic value.

Stage is only one contributor to prognosis in FL, as most patients have a very favorable natural history and life expectancy, independent of stage at presentation. The majority (>85%) of FL patients will present with advanced-stage disease (stage III or IV). Despite this, the average life expectancy in FL patients is measured in decades, and many patients will have a life span that is not shortened by this disease. There is heterogeneity within FL as regards prognosis, with some patients following a much more aggressive course with shorter overall survival. The **FL international prognostic index** (FLIPI) identifies five adverse risk factors that can separate patients into low-, intermediate-, or high-risk groups, and may be useful in prognostication for a given patient (Table 5.6).

Principles of FL management

FL is generally considered to be highly treatable but not technically curable using conventional therapies. A potential exception to this rule is in rare cases of FL that present with a single localized region of involvement that can be treated with radiation, resulting in cure in over half of patients. The majority, however, present with advanced-stage disease, and modern chemotherapy and immunotherapy approaches are highly effective but cannot eradicate every last FL cell. The goal of FL therapy is, therefore, to control the disease *as needed* to ensure a normal life expectancy and quality of life, if possible.

Given the indolent behavior and lengthy natural history, a vital initial question at the time of diagnosis is whether the FL needs to be treated at all. In fact, many patients with a low disease burden and without symptomatic disease can be followed with surveillance and avoid the toxicities of therapy. This is based on several large randomized clinical trials which have enrolled patients with low-disease burden FL and no lymphoma-related symptoms, organ dysfunction, or decreased blood counts. Patients were randomly assigned to chemotherapy or observation alone. No difference in overall survival was seen among these strategies. This suggests that treatment is effective if reserved for future use, should it be required, but does not offer benefit at the time of diagnosis if the patient has non-bulky asymptomatic disease. These studies also demonstrate that nearly 20% of FL patients will never require therapy at all, and that among the majority of patients who do ultimately need treatment, the median time until requiring that therapy is four to five years from diagnosis. For these reasons, patients with low disease–burden FL and no adverse symptoms, organ impairment, or cytopenias are followed with surveillance, and treated only at such time as indications for therapy arise.

For patients who do require therapy, treatment options include radiation, anti-CD20 monoclonal antibody therapy, or anti-CD20 monoclonal antibody therapy combined with chemotherapy. Radiation therapy is employed for patients with localized (stage I) disease with the goal of cure. Lower-dose radiation may also be used in patients with advanced-stage FL for selective treatment of a discrete symptomatic site of disease. Most patients with advanced-stage disease in need of therapy, however, will receive systemic therapy. **Rituximab** and obinutuzumab are anti-CD20 monoclonal antibodies used in combination with chemotherapy for FL and other B-cell NHLs (see Case 11-1). When used alone in low disease–burden FL patients, rituximab will induce responses in the majority of patients, though only a minority will achieve a complete remission. The average duration of remission is between two and three years. Chemotherapy added to rituximab increases the overall and complete response rates, and now produces remissions that last on average between three and five years. Several clinical trials have demonstrated that the addition of rituximab to chemotherapy improves overall survival compared with that same chemotherapy alone, making rituximab the only medication to show an overall survival benefit in Phase II trials in this disease. Options for chemotherapy regimens combined with rituximab (R) include R-bendamustine, R-CVP (cyclophosphamide, vincristine, prednisone), or R-CHOP (cyclophosphamide, doxorubicin, vincristine, prednisone). No clinical trial has demonstrated an overall survival benefit favoring one regimen over another, so the choice of regimen is guided by weighing the risk and benefit ratio of the respective options. R-bendamustine is the most widely used chemoimmunotherapy for FL based on a randomized trial showing improved complete response rate and progression-free survival over R-CHOP. Toxicity was also generally improved favoring the R-bendamustine regimen, leading to this as a widely used standard of care. R-CHOP remains the standard of care for most aggressive lymphomas (see Case 11-1) and continues to be preferred for high-grade FL (grade 3) as well as FL that has undergone transformation into a more aggressive NHL. Among low-grade FL patients who respond to initial therapy, another randomized clinical trial compared ongoing maintenance infusion of rituximab administered once every two months for up to two years versus observation alone, and found that maintenance rituximab significantly improved progression-free survival, though overall survival was identical and excellent in both groups. Not surprisingly, there were increased side effects in the patients receiving ongoing rituximab therapy, though most of these were mild. Based on these data, maintenance rituximab is an option for patients following initial therapy. More recently, a randomized clinical trial in advanced-stage FL compared the next generation anti-CD20 monoclonal antibody obinutuzumab plus chemotherapy versus rituximab plus chemotherapy, each arm followed by maintenance with the respective monoclonal antibody. That trial demonstrated an improved progression-free survival favoring obinutuzumab, but without a difference in overall survival. There are increased side effects with obinutuzumab, including increased rates of infusion reactions and neutropenia, so the risk:benefit ratio must be considered on a patient-by-patient basis.

The case of Stephen Kim, a community organizer with kidney stones

Stephen was a healthy man of 56 years when he presented to his local emergency room with acute-onset right flank pain. The pain was sharp and excruciating, and made it difficult to even lie still. Physical examination was notable for exquisite tenderness on palpation of the right flank, and it was difficult for the patient to participate in the rest of the examination due to discomfort. He was treated with pain medication with some relief, and a urine analysis showed no evidence of infection but did show a small amount of occult blood. An abdominal CT scan was performed confirming the clinical suspicion of a stone, which measured 10 mm × 5 mm and was lodged in the distal right ureter and caused a mild amount of **hydronephrosis** (swelling of the kidney due to backing up of urine). Surprisingly, however, the CT scan also demonstrated multiple enlarged lymph nodes involving the **mesenteric** (referring to the region in the abdomen that provides a blood supply and lymphatic drainage for the intestines) and **retroperitoneal** (referring to the area behind the cavity housing the abdominal organs) nodal regions. None of the nodes were large, and all measured 20 mm or less. The lymph nodes were not compressing or invading any organs or structures. For the kidney stone, Stephen's pain was treated with ketorolac, a nonsteroidal anti-inflammatory drug, as well as morphine. Kidney function on laboratory studies was normal, as was a complete blood count. He was started on intravenous fluid with the goal of "flushing out" the kidney stone with increased urine output. Once the pain was better controlled, his physician also performed a more detailed lymph node exam and identified multiple prominent lymph nodes, including 15-mm lymph nodes in the neck, a 2.5-cm lymph node under the right arm, and multiple **inguinal** (near the groin) nodes all measuring less than 15 mm. Stephen was unaware of any lymph node enlargement and had felt entirely well prior to the onset of his renal colic (sharp pain from the lower back radiating forward toward the groin). After three hours of intravenous hydration, Stephen passed the kidney stone and was discharged from the emergency department, with instructions to follow up with his primary care physician regarding the asymptomatic lymphadenopathy.

He was referred to a surgeon who performed an excisional biopsy of the enlarged axillary lymph node. That biopsy showed follicular lymphoma, grade 1–2. Ann Arbor stage was III, and his FLIPI score was 1, with advanced stage being his only risk factor. He met with a medical oncologist, and since his disease burden was low and he was entirely asymptomatic without any impairment of his blood counts or organs by lymphoma, they elected to follow with surveillance. Six months later, a repeat CT scan showed that all nodes had spontaneously regressed in size, by approximately 25%, and with no new sites of disease. He continued to be followed with surveillance, with imaging showing some waxing and waning behavior of his lymph nodes. Five years after diagnosis at age 61, however, his lymphadenopathy had slowly progressed to the point where he had developed uncomfortable lymph nodes under his arms and in his neck, where they were visible to coworkers, prompting him to wear turtlenecks. CT scans showed multiple lymph nodes in his thorax and abdomen measuring 3–5 cm. LDH was now minimally elevated. At this time, a repeat biopsy was performed to ensure the disease had not transformed into aggressive lymphoma, and confirmed ongoing FL, grade 1–2. Lymphoma therapy was now required.

He was treated with the combination of rituximab and bendamustine, which resulted in rapid reduction in all lymph nodes and return of his normal neck contours. A CT scan performed after three cycles of R-bendamustine therapy confirmed a uniformly excellent response, and at the completion of six total cycles of treatment he was in complete remission. After discussing risks and benefits of maintenance rituximab at that time, he preferred to follow with observation and allow complete recovery of his immune system. He recovered quickly from treatment and was followed with surveillance. He did well for six years, at which time he noted recurrent lymphadenopathy. At this time, a CT scan showed extensive lymphadenopathy as well as hydronephrosis on the left side, which, unlike the swelling at his initial presentation, was caused not by a kidney stone but by a large lymph node compressing the ureter. Kidney function was mildly impaired as a result. A core needle biopsy of that lymph node again showed FL grade 1–2. This time he was treated with R-CHOP, to which he responded well. After completing six cycles of this **chemoimmunotherapy**, he was placed on maintenance rituximab to optimize the length of that remission. Unfortunately, his disease progressed one year later. At that time, his oncologist placed him on **idelalisib**, a PI3K inhibitor, at 150 mg twice daily by mouth. Encouragingly, the lymph nodes all regressed again. His disease remains controlled on idelalisib one year later, now at age 69, and his plan is to continue this medication until such time as the disease progresses again and additional treatment is required.

PI3-kinase inhibition in follicular lymphoma

Of the four isoforms of class I PI3Ks, PI3Kα and PI3Kβ isoforms are broadly expressed, with the PI3Kα isoform playing a role in angiogenesis, which has garnered attention in carcinomas, where angiogenesis remains a valuable therapeutic target. The PI3Kγ and PI3Kδ isoforms, however, are more prominently expressed in white blood cells, with the PI3Kδ isoform predominantly expressed in B lymphocytes, where it plays a key role in B-cell development, signaling, and survival. PI3Kδ has therefore emerged as a rational therapeutic target in B-cell lymphomas (Figure 5.14).

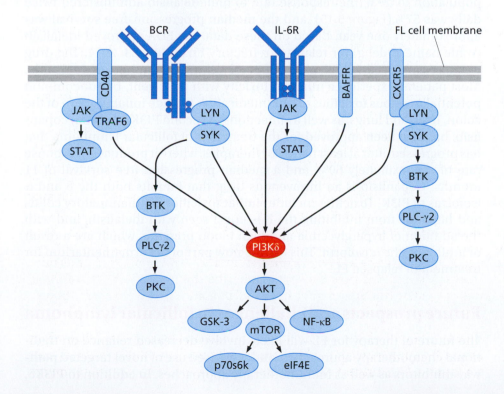

Figure 5.14. PI3Kδ pathway in lymphoma. PI3Kδ is activated by multiple signals, including the B-cell receptor and engagement of CD40, BAFF-R, and CXCR5, by infiltrating cells in the tumor microenvironment. Activation of PI3Kδ leads to activation of multiple downstream signaling molecules, resulting in cellular proliferation, survival, and trafficking.

Figure 5.15. Waterfall plot of response to idelalisib in refractory indolent lymphomas. The best response with respect to tumor size during idelalisib treatment is shown. Each column represents a single study patient's percent change in the sum of perpendicular dimensions (SPD) of tumors. Ninety percent had improvements in lymphadenopathy. The dashed line shows the percentage change that represents the criterion for formal lymphadenopathy response (greater than 50% reduction in tumor size). Included subtypes are FL, follicular lymphoma; LPL/WM, lymphoplasmacytic lymphoma/Waldenström macroglobulinemia; MZL, marginal zone lymphoma; SLL, small lymphocytic lymphoma. (From A.K. Gopal et al., *N. Engl. J. Med.* 370:1008–1018, 2014. Reprinted with permission from Massachusetts Medical Society.)

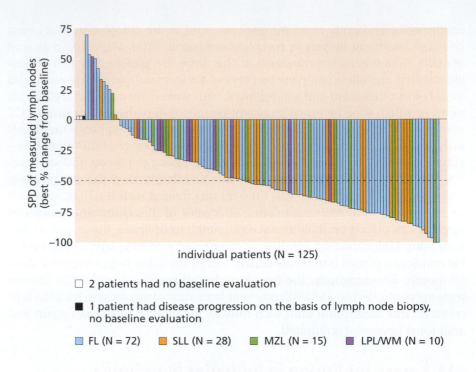

Idelalisib is a potent, highly selective inhibitor of PI3Kδ. *In vitro* studies showed the drug to be effective at decreasing the downstream activity of the PI3K pathway in lymphoma cells, including decreased activation of AKT and mTOR. This led to a Phase I clinical trial, which showed remarkable clinical activity of idelalisib in numerous B-cell malignancies, including indolent B-cell lymphomas and chronic lymphocytic leukemia. Additional clinical trials were then conducted to confirm efficacy. A Phase II clinical trial was conducted in 125 patients with indolent B-cell lymphoma, all of whom had lymphoma that was refractory to both rituximab and traditional alkylating chemotherapy. The majority of patients (58%) had FL. In this very difficult population to treat, the response rate to oral idelalisib administered twice daily was 57% (Figure 5.15), and the median progression-free survival was approximately one year. Based on these data, the FDA approved idelalisib (trade name Zydelig) for relapsed/refractory FL on July 23, 2014. The drug was also approved for relapsed/refractory chronic lymphocytic leukemia. Most patients experience minimal toxicity with this agent, but uncommon potentially serious toxicities on this medication include inflammation of the colon, liver, and lungs, as well as infections. A second PI3K inhibitor, copanlisib, has also been approved for the treatment of follicular lymphoma that has progressed after at least two prior therapies, where it produces a response rate of approximately 60%, and a median progression-free survival of 11 months. Copanlisib is an intravenous drug that inhibits both the δ and α isoforms of PI3K. Toxicities include similar toxicities of inflammatory colitis and hepatitis from inhibiting the δ isoform seen with idelalisib, and with the addition of hyperglycemia and high blood pressure, which are a result of inhibiting the α isoform. This drug is now part of the armamentarium for treatment of relapsed FL.

Future prospects and challenges in follicular lymphoma

The future of therapy for FL will likely involve decreased reliance on traditional chemotherapy approaches, and increased use of novel targeted pathway inhibitors as well as immunotherapy approaches. In addition to PI3Kδ, additional appealing targets in patients with FL include the B-cell receptor signaling cascade, including Bruton's tyrosine kinase (BTK; see Case 4-2); the *BCL2* proto-oncogene, which is constitutively expressed in nearly all FLs; and EZH2, a histone methylase, which is mutated in approximately 28% of FLs. Targeted inhibitors of all of these three targets are under active investigation for treatment of this disease. Additionally, immunotherapy strategies have demonstrated promise in early clinical trials in FL. The combination of rituximab with the immunomodulating agent lenalidomide, which increases T-cell and NK-cell activity and improves antibody-mediated lymphoma cell killing *in vitro*, has shown excellent clinical activity in previously treated and previously untreated disease. The combination of rituximab and lenalidomide is being compared with standard chemoimmunotherapy in a large randomized trial seeking to unseat traditional chemotherapy as the initial standard of care in this disease. Novel antibodies and antibody drug conjugates (see Chapter 11) are also under active development and may further add to the treatment armamentarium against FL.

Take-home points

✓ Non-Hodgkin lymphomas (NHLs) are a heterogeneous collection of more than 60 discrete diseases, which constitute a broad range of biologies, natural histories, and treatment approaches.

✓ NHLs are often clinically categorized based on their natural history as indolent, aggressive, or highly aggressive diseases.

✓ FL is the prototype for indolent B-cell lymphomas, which are typically not curable but highly treatable with an overall favorable prognosis.

✓ FL may initially be observed without therapy in low-disease burden, asymptomatic patients.

✓ For high-disease burden symptomatic FL, chemotherapy plus the anti-CD20 monoclonal antibody is the current standard of care.

✓ FL resistant to both rituximab and chemotherapy constitutes a high-risk disease.

✓ Therapeutic targets in treating FL include PI3Kδ, BCL-2, BTK, and EZH2.

✓ The PI3Kδ inhibitors idelalisib and copanlisib are FDA approved for relapsed FL based on excellent clinical activity in patients with highly refractory FL.

Discussion questions

1. What is included under the heading of non-Hodgkin lymphoma (NHL)?

2. How are lymphomas staged, and what goes into determining prognosis in FL other than stage?

3. Describe the initial approach to management when FL is diagnosed.

4. How do idelalisib and copanlisib work in FL?

5. What additional novel treatment strategies are emerging in FL?

Chapter summary

The PI3K/Akt/mTOR pathway orchestrates a complex signaling network that is essential to the life of normal cells. Because this pathway supports many of the hallmarks of cancer and is highly active in a wide variety of malignancies, intense effort has been focused on targeting and controlling the key kinases of the network: PI3K, Akt, and the mTOR complexes mTORC1 and mTORC2. Basic biochemical and structural studies and extensive programs in drug discovery and development have led to the generation of many candidate drugs. Two of these, temsirolimus and idelalisib, are highlighted in the case studies presented in this chapter. The effectiveness of idelalisib for the treatment of follicular lymphoma provides an important insight. Since it was anticipated that the development of inhibitors against PI3K would be greatly complicated by the existence of four distinct isoforms of this enzyme, the derivation of a drug that is selective for one of them, PI3Kδ, is an important milestone. It demonstrates that differential involvement of some PI3K isoforms on cell processes in specific tissues can be exploited, allowing concentration of inhibition on the relevant isoform and avoiding the toxicity of broad inhibition of PI3K. In the case of idelalisib, a PI3K isoform important for B-cell proliferation and survival is targeted to treat a susceptible type of B-cell lymphoma.

An overview of the PI3K/Akt/mTOR pathway, indicating points where targeted therapies have been directed, is shown in Figure 5.16. The figure also reminds us that these pathways do not act in isolation but are most often triggered by inputs from receptor tyrosine kinases, notably growth factor receptors but also including others such as various immunoreceptors. This overview also highlights the "crosstalk" of this network with other major signaling pathways such as the Ras-MAP kinase pathways that can activate this network at the level of PI3K or bypass this kinase at the head of the pathway and activate segments of the pathway below it. This network of interacting pathways offers many opportunities for therapeutic intervention,

Figure 5.16. Targeting the PI3K/Akt/mTOR pathway. The key nodes of this pathway, PI3K (a heterodimer of regulatory p85 and catalytic p110 subunits), Akt, and the two complexes mTORC1 and mTORC2, are the targets of the variety of protein kinase inhibitors shown in the diagram. The Ras pathway interacts with the PI3K/Akt/mTOR pathway at two levels. In addition to triggering the MAP kinase pathway, activated Ras can directly activate PI3K. ERK can contribute to the activation of mTORC1, promoting protein synthesis and cell growth. mTORC1 also negatively regulates growth signals by activation of S6 kinase (S6K) and down-regulation of transcription of PDGF-R and IGF-1. Thus, inactivation of mTORC1 with a rapalog can paradoxically cause enhanced RTK pathway signaling via increased transcription of RTKs or their ligands—a potential mechanism for resistance to mTOR inhibitor therapy and possibly an explanation for the cytostatic nature of rapalog therapy. p85, 85-kDa regulatory subunit of PI3K; p110, 110-kDa catalytic subunit of PI3K.

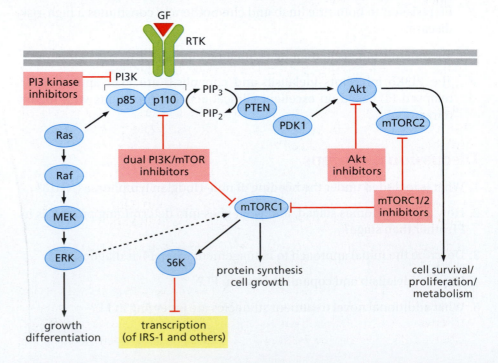

and a broad variety of inhibitors have been designed for its inhibition. Consideration is also being given to combinations of inhibitors of the PI3K pathway with inhibitors of steps in the Ras-MAPK pathway, since cross-talk from this pathway can influence the effectiveness of drugs inhibiting PI3K. For example, patients with high levels of MEK or ERK might be less sensitive to drugs targeting PI3K since these kinases can activate mTORC1 by a route that does not involve PI3K. In such a situation, combination therapies involving inhibitors of MEK or ERK may improve the efficacy of some PI3K/Akt/mTOR pathway inhibitors in two ways. First, they might synergize with them, and second, they could block escape from PI3K inhibitors by MAP kinase-mediated bypass of PI3K. Much of the success of drugs targeting the PI3K/Akt/mTOR pathway will depend upon selection of appropriate patient populations and gaining an understanding of which drugs or immunotherapies will make the most efficacious inhibitors of this important and central signaling network.

Take-home points

✓ Discovery and development programs have generated a large number of candidate drugs for the targeting of each of the three kinases of the PI3K/Akt/mTOR pathway.

✓ Idelalisib and copanlisib are inhibitors of an isoform of PI3K that plays a major role in the B-cell and other immunocyte lineages and have been approved for treatment of follicular lymphoma.

✓ Temsirolimus and everolimus are synthetic analogs of rapamycin and have been approved for use in the treatment of several cancers including renal cell carcinoma, breast cancer, and pancreatic neuroendocrine tumors.

✓ There are a wide and growing variety of inhibitors of one more enzymatic nodes of the PI3K/Akt/mTOR pathway, and their successful use will likely depend upon selection of the appropriate inhibitor and its use in the appropriate patient population.

Chapter discussion questions

1. What is the role of PI3K in the PI3K/Akt/mTOR pathway, and how is this role regulated by PTEN?

2. PI3K is an umbrella term that includes many classes and isoforms. Why isn't major emphasis placed on discovery and development of a potent inhibitor that would inhibit all PI3Ks?

3. Both mTORC1 and mTORC2 include mTOR as their catalytic component yet have some biological activities that are different and distinctive. What activities are distinctive to mTORC1 and which to mTORC2? What is the structural basis for the difference in the activities of these two complexes?

4. Since rapamycin and rapalogs are effective inhibitors of mTORC1, why is there an effort to develop direct inhibitors of mTOR?

5. Anti-cancer drugs targeting particular enzymes often select for cells bearing mutations in the gene encoding the targeted protein that make the enzyme insensitive to the inhibitor. Why is selection of such mutations uncommon when drugs targeting the PI3K/mTOR pathway are employed?

6. Explain why a safe and efficacious PI3K inhibitor might not produce a good clinical response in a patient with elevated ERK activity.

7. The range of inhibitors targeting the PI3K/Akt/mTOR pathway makes it possible to inhibit all three nodal kinases. Discuss the advantages and disadvantages of such inhibitors.

Selected references

Introduction

Fruman DA & Rommel C (2014) PI3K and cancer: lessons, challenges and opportunities. *Nat. Rev. Drug Discov.* 13:140–156.

Hirsch E, Ciraolo E, Franco I et al. (2014) PI3K in cancer-stroma interactions: bad in seed and ugly in soil. *Oncogene* 33:3083–3090.

Saxton RA & Sabatini DM (2017) mTOR in growth, metabolism and disease. *Cell* 168:960–976.

Song MS, Salmena L & Pandolfi PP (2012) The functions and regulation of the PTEN tumour suppressor. *Nat. Rev. Mol. Cell Biol.* 13:283–296.

Case 5-1

Cohen T & McGovern FJ (2005) Renal-cell carcinoma. *N. Engl. J. Med.* 353:2477–2490.

Flanigan RC, Salmon SE, Blumenstein BA et al. (2001) Nephrectomy followed by interferon alfa-2b compared with interferon alfa-2b alone for metastatic renal-cell cancer. *N. Engl. J. Med.* 345: 1655–1659.

Fyfe G, Fisher RI, Rosenberg SA et al. (1995) Results of treatment of 255 patients with metastatic renal cell carcinoma who received high-dose recombinant interleuken-2 therapy. *J. Clin. Oncol.* 13:688–696.

Heng DYC, Wanling X, Regan MM et al. (2009) Prognostic risk factors for overall survival in patients with metastatic renal cell carcinoma treated with vascular endothelial growth factor-targeted agents: results from a large, multicenter study. *J. Clin. Oncol.* 27:5794–5799.

Hudes G, Carducci M, Tomczak P et al. (2007) Temsirolimus, interferon alfa, or both for advanced renal-cell carcinoma. *N. Engl. J. Med.* 356:2271–2281.

Marcus SG, Choyke PL, Reiter R et al. (1993) Regression of metastatic renal cell carcinoma after cytoreductive nephrectomy. *J. Urol.* 150:463–466.

McDermott DF, Regan MM, Clark JI et al. (2005) Randomized phase III trial of high-dose interleukin-2 versus subcutaneous interleukin-2 and interferon in patients with metastatic renal cell carcinoma. *J. Clin. Oncol.* 23:133–141.

Mickisch GH, Garin A, van Poppel H et al. (2001) Radical nephrectomy plus interferon-alfa-based immunotherapy compared with interferon alfa alone in metastatic renal-cell carcinoma: a randomised trial. *Lancet* 358:966–970.

Rini BI, Campbell SC & Escudier B (2009) Renal cell carcinoma. *Lancet* 373:1119–1132.

Siegel RL, Miller KD & Jemal A (2018) Cancer statistics, 2018. *CA Cancer J. Clin.* 68:7–30.

Weidemann A & Johnson RS (2008) Biology of HIF-1α. *Cell Death Differ.* 15:621–627.

Yagoda A, Petrylak D & Thompson S (1993) Cytotoxic chemotherapy for advanced renal cell carcinoma. *Urol. Clin. North Am.* 20:303–321.

Case 5-2

Ardeshna KM, Smith P, Norton A et al. (2003) Long-term effect of a watch and wait policy versus immediate systemic treatment for asymptomatic advanced-stage non-Hodgkin lymphoma: a randomised controlled trial. *Lancet* 362:516–522.

Gisselbrecht C, Schmitz N, Mounier N et al. (2012) Rituximab maintenance therapy after autologous stem-cell transplantation in patients with relapsed CD20+ diffuse large B-cell lymphoma: final analysis of the collaborative trial in relapsed aggressive lymphoma. *J. Clin. Oncol.* 30:4462–4469.

Gopal AK, Kahl BS, de Vos S et al. (2014) PI3Kδ inhibition by idelalisib in patients with relapsed indolent lymphoma. *N. Engl. J. Med.* 370:1008–1018.

Marcus R, Imrie K, Belch A et al. (2005) CVP chemotherapy plus rituximab compared with CVP as first-line treatment for advanced follicular lymphoma. *Blood* 105:1417–1423.

Nooka AK, Nabhan C, Zhou X et al. (2013) Examination of the follicular lymphoma international prognostic index (FLIPI) in the National LymphoCare study (NLCS): a prospective US patient cohort treated predominantly in community practices. *Ann. Oncol.* 24:441–448.

Summary

Dienstmann R, Rodon J, Serra V et al. (2014) Picking the point of inhibition: a comparative review of PI3K/AKT/mTOR pathway inhibitors. *Mol. Cancer Ther.* 13:1021–1031.

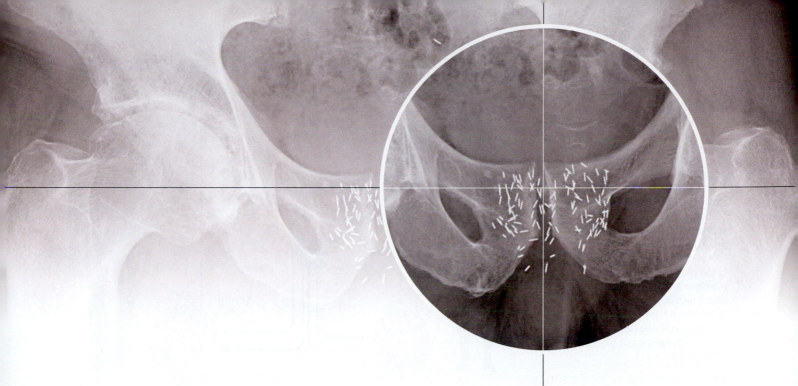

Chapter 6

Hormone Therapy

Introduction

Reproductive tissues such as breast, endometrium, ovaries, and prostate develop and grow in response to **hormones**, signaling chemicals that are produced in **endocrine** glands. Produced in the **hypothalamus**, gonadotropin-releasing hormone (GnRH) is secreted into the hypophyseal portal system of blood vessels in the brain, which allows for rapid transport to the anterior **pituitary gland**. In response to GnRH binding to GnRH receptors on the cell surface, gonadotrope cells of the pituitary gland produce and release the gonadotropins, luteinizing hormone (LH), and follicle-stimulating hormone (FSH) into the bloodstream. LH and FSH, in turn, stimulate the gonads (testes or ovaries) to produce sex steroids (Figure 6.1). **Steroidogenesis**, mediated by cellular pathways in the adrenal glands (organs that are located above the kidneys) and the gonads (testes or ovaries), is the production of steroid hormones, which include the sex steroid groups progestogens (progesterone, 17-hydroxyprogesterone, and others), **androgens** (notably, testosterone and androstenedione), and estrogens (estrone and estradiol) (Figure 6.2). Rising levels of sex steroids in the serum, in a process known as **negative feedback**, cause the hypothalamus to stop producing and releasing GnRH, thereby causing the pituitary gland to stop producing LH and FSH.

The cases in this chapter focus on cancers of tissues and organs that are stimulated by the sex steroids—the breast and the prostate. Breast development in women occurs during puberty, and throughout the many menstrual cycles of a woman's life, the cyclic secretion of estrogen and progesterone, stimulated by LH and FSH release, leads to cell proliferation in and subsequent enlargement of breast ducts and lobules. Similarly, the endometrium undergoes enormous changes during menstrual cycles, with estrogen stimulating

**Figure 6.1. The hypothalamic/
pituitary/gonadal axis.**
Gonadotropin-releasing hormone
(GnRH) secreted in a pulsatile fashion
by the hypothalamus stimulates
release of the gonadotropins
luteinizing hormone (LH) and
follicle-stimulating hormone (FSH).
In men, LH stimulates testosterone
production from Leydig cells in the
testes, and FSH stimulates sperm
production. In women, LH and
FSH stimulate follicle maturation
in the ovary and production of
estrogen and progesterone. Elevated
blood testosterone and estrogen
levels "feed back" and shut down
GnRH and LH/FSH production in
the hypothalamus and pituitary,
respectively.

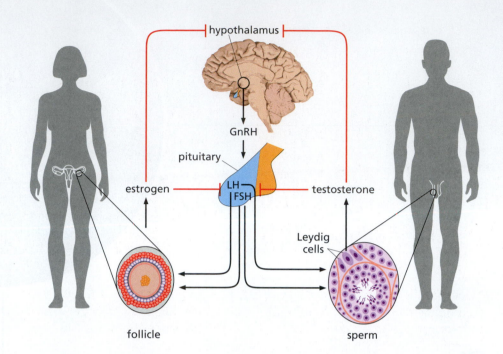

proliferative changes of the uterine lining to create a thick, blood vessel-rich
glandular tissue capable of supporting a growing fetus. In men, the prostate
gland—which plays a critical reproductive role by creating the slightly alkaline seminal fluid that protects and conveys sperm for travel along the mildly
acidic vaginal tract—does not undergo cyclical changes but is nevertheless
acted upon by sex steroids. The normal prostate is subject to growth over time,
and it is thought that all elderly men will eventually experience **benign prostatic hypertrophy** (BPH) as a result.

In this context, it is not surprising that tissues that coordinate cell proliferation and death in response to hormonal stimuli would be prone to neoplastic
transformation. Indeed, breast cancer and prostate cancer are the most common malignancies in women and men, respectively, in the United States, and
both rank as the second-leading cause of cancer death after lung cancer for
their respective genders. Fortunately, hormonal therapy can be employed to
exploit this underlying disease biology.

**Figure 6.2. Steroidogenesis
pathways.** The enclosed yellow
area contains the pathways shared
by the adrenal glands and gonads.
Pathways producing primarily
mineralocorticoids, glucocorticoids,
and sex steroids occur in the adrenal
gland (in enclosed blue areas).
Enzymes are labeled in red. 11β,
11β-hydroxylase; 17,20, C-17,20-
lyase (CYP17); 17α, 17α-hydroxylase
(also CYP17; a single CYP17 enzyme
possesses both C-17,20-lyase and
17α-hydroxylase activities); 17βR,
17β-reductase; 18, aldosterone
synthase; 21, 21-hydroxylase; 3β,
3β-hydroxysteroid dehydrogenase;
5αR, 5α-reductase; A, aromatase.

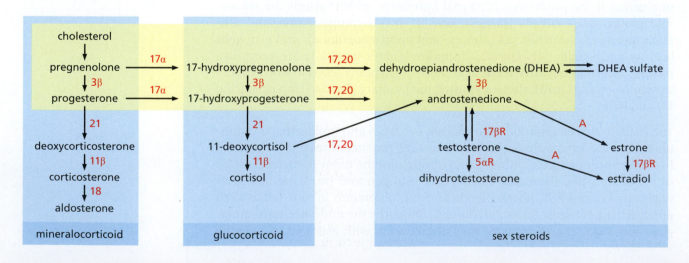

Table 6.1. Strategies to disrupt hormonal pathways.

Strategy	Examples	Mechanism
GnRH inhibition	Leuprolide, degarelix	Disruption of the GnRH signal from hypothalamus to pituitary; blocks the release of LH and downstream production of steroid hormones from gonadal sources
Steroidogenesis inhibition	Abiraterone	Selective inhibition of enzymes in the steroidogenesis pathway from cholesterol to androgens; effective for adrenal and gonadal sources of androgens, as well as androgen production in prostate cancer cells
Inhibition of peripheral conversion of steroids	Anastrazole, exemestane, letrozole	Aromatase inhibitors block the conversion of androgens into estrogens in peripheral (e.g., adipose) tissues
Steroid hormone receptor antagonist	Enzalutamide, tamoxifen	Blocks binding of steroid hormone to the ligand-binding domain of the receptor, thereby preventing nuclear translocation, recruitment of co-factors, binding to DNA, and driving of transcription

There are four general types of hormonal therapy to treat cancers whose growth is stimulated by sex steroids: inhibition of gonadotropin release, inhibition of steroidogenesis, inhibition of conversion of steroids in peripheral tissues, and direct inhibition of steroid hormone receptors (Table 6.1; Figure 6.3).

Hormonal therapies serve as the foundation of treatment for hormone-sensitive malignancies, but are generally not considered curative. They may be used as a single agent for disease control, or may be combined with surgery, radiation, or additional medical therapies. When used as a single agent, duration of response is highly variable, ranging from short-lived to sustained remissions lasting many years.

Inhibition of gonadotropin release

The release of the gonadotropins LH and FSH from the pituitary gland can be blocked by GnRH **agonists**, chemicals that bind to GnRH receptors to stimulate signaling; or by GnRH **antagonists**, chemicals that interfere with GnRH receptor signaling (see Figure 6.3A). GnRH agonists cause tonic (continual) stimulation of the GnRH receptors on the gonadotrope cells of the anterior pituitary gland, as opposed to the normally pulsatile (rhythmic) GnRH signal. Tonic GnRH stimulation initially causes LH and FSH release (and a surge in gonadal production of sex steroids) followed by down-regulation of GnRH receptors on the gonadotrope cells, shutting down LH and FSH release.

Examples of GnRH agonists include leuprolide and goserelin. GnRH agonists are FDA approved to treat breast cancer by stopping LH release and subsequent ovarian production of estrogen, which can fuel the growth of breast cancer cells that express the estrogen receptor (ER). Similarly, GnRH agonists are used in prostate cancer to stop LH release and testicular production of testosterone, which stimulates growth of prostate cancer cells. The GnRH antagonist degarelix is also used in prostate cancer treatment and essentially causes the same phenomenon of shutdown of LH production, but without the initial stimulation of LH release seen with GnRH agonists. Down-regulation of testosterone production in prostate cancer by inhibiting LH release is considered a form of "medical castration," with similar—but reversible—side effects of surgical castration (removal of both testes). Potential side effects from medical castration are numerous but largely reversible upon stopping the GnRH agonist or antagonist; they include hot flashes (similar to the hot flashes experienced by postmenopausal women), weight gain, loss of lean muscle mass, loss of bone density with increased potential for fractures, loss of erections,

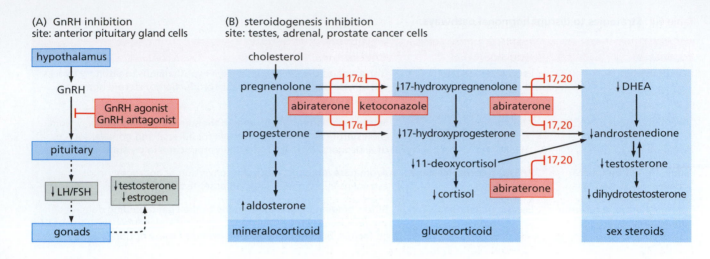

(A) GnRH inhibition
site: anterior pituitary gland cells

(B) steroidogenesis inhibition
site: testes, adrenal, prostate cancer cells

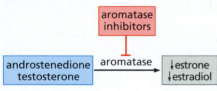

(C) inhibition of steroid conversion
site: muscle, fat cells

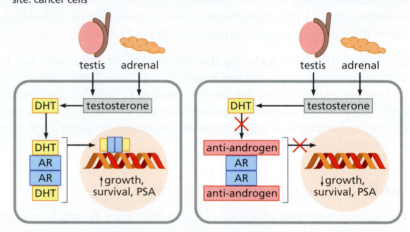

(D) steroid hormone receptor antagonists
site: cancer cells

Figure 6.3. Four general strategies to disrupt hormone signals. (A) Inhibition of gonadotropin release. GnRH agonists initially stimulate LH and FSH production before tonic activation of the GnRH receptors on the pituitary causes down-regulation of the GnRH receptors, eventually shutting down LH and FSH production, and consequently shutting down sex steroid hormone production by the gonads. GnRH antagonists have similar effects but without the initial LH/FSH simulation. (B) Steroidogenesis inhibition. In this simplified version of Figure 6.2, abiraterone and ketoconazole inhibit CYP17 pathways, decreasing production of all products of the glucocorticoid and sex steroid pathways, while producing increased levels of the mineralocorticoid aldosterone. This strategy is used in prostate cancer treatment. (C) Inhibition of conversion of steroids in peripheral tissues. In postmenopausal women, estrogen is largely derived from the androgens testosterone and androstenedione made in the adrenal glands and converted in peripheral tissues such as fat and muscle into the estrogens estradiol and estrone, respectively. The enzyme aromatase catalyzes this conversion. Aromatase inhibitors (AIs) effectively block production of estrogens. This strategy is used in ER (estrogen receptor)-positive breast cancer in postmenopausal women. (D) Direct inhibition of steroid hormone receptors. Testosterone is produced by testes and adrenal glands, and is converted into the more potent dihydrotestosterone (DHT) by 5-α-reductase. (*Left panel*) Testosterone or DHT binds to AR (androgen receptor) in the cytoplasm of prostate cells, leading to translocation into the nucleus, where the androgen-bound AR recruits co-factors, binds DNA, and stimulates transcription leading to prostate-specific antigen (PSA) production and cell proliferation. (*Right panel*) Anti-androgens block all of these activities. ER antagonists perform similar activities in breast cancer cells.

loss of libido, and potential development of diabetes mellitus and cardiovascular complications.

Steroidogenesis inhibition

The disruption of steroid hormone production is an effective treatment approach for prostate cancer (see Figure 6.3B). Steroidogenesis is the production of steroid hormones from the cholesterol precursor (see Figure 6.2 for steroidogenesis pathways leading to production of testosterone as well as the mineralocorticoid aldosterone, which is involved in blood pressure control, and the glucocorticoid cortisol, which mediates inflammatory and stress responses). In males, steroidogenesis normally occurs in cells of the adrenal glands and testes, and has been demonstrated to occur in prostate cancer cells themselves. Steroidogenesis inhibitors block individual steps in the enzymatic conversion of cholesterol to androgens.

Ketoconazole, an antifungal drug, was found in the 1980s to cause gyneco-mastia (breast enlargement) in some men being treated with high doses for a fungal infection. The discovery was a clue that this drug could be useful in treating hormone-sensitive cancers. Eventually, this property of ketoconazole was attributed to its ability to interfere with CYP17, a cytochrome P450 enzyme localized to the endoplasmic reticulum that is critical for adrenal and gonadal steroidogenesis. Due to potentially severe side effects including hypertension and hypokalemia (low serum potassium) seen at higher doses than are required for its antifungal effect, ketoconazole currently has only a limited role as a treatment for prostate cancer. A safer steroidogenesis inhibitor, **abiraterone**, more potently and selectively inhibits CYP17 and is an effective prostate cancer treatment (see Case 6-2 for more on abiraterone). Because of the potential to interfere with production of steroid hormones other than sex hormones in the adrenal glands, care must be taken when evaluating steroidogenesis inhibitors for potential disruption of normal mineralocorticoid- or glucocorticoid-dependent processes (see Figure 6.2).

Inhibition of conversion of steroids in peripheral tissues

The disruption of peripheral conversion of steroids is an effective treatment approach for breast cancer (see Figure 6.3C). In pre-menopausal women, estrogen is primarily made by the ovaries. In postmenopausal women, estrogen is largely derived from the androgens testosterone and androstenedione made in the adrenal glands and subsequently converted in peripheral tissues such as fat and muscle into the estrogens estradiol and estrone, respectively. The enzyme aromatase, a member of the cytochrome P450 superfamily, catalyzes this conversion. Postmenopause, circulating estrogen levels are approximately 20% of those in pre-menopausal women.

Aromatase can be blocked by **aromatase inhibitors** (AIs) such as exemestane, anastrazole, and letrozole. AIs thus block the critical signal driving ER-mediated breast cancer cell growth, as described in Case 6-1. AIs cannot effectively block ovarian production of estrogen, and therefore are not used in pre-menopausal patients. In postmenopausal women whose ovaries are no longer producing estrogen, AIs can effectively diminish estrogen levels, which are primarily produced in peripheral tissues.

Steroid hormone receptor antagonists

Unlike the above categories that decrease hormone production, steroid hormone receptor antagonists directly inhibit the hormone signal at the level of the cancer cell. These drugs are referred to as "anti-androgens" or "anti-estrogens" because they can block the engagement of the androgen receptor (AR) or ER, respectively, by the normal hormones. AR and ER are both located in the cytoplasm. Ideally, a steroid hormone receptor antagonist blocks the interaction of the hormone with the ligand-binding domain of its receptor, as well as the downstream activities of the activated AR or ER, such as translocation from the cytoplasm to the nucleus, recruitment of co-factors, binding to the specific DNA response elements, and driving a transcriptional program leading to cell proliferation (see Figure 6.3D).

The anti-androgen **enzalutamide** is described in Case 6-2. Other AR antagonists include apalutamide, bicalutamide, flutamide, and nilutamide. ER antagonists in active clinical use for breast cancer treatment include tamoxifen and fulvestrant.

The cases presented in this chapter provide examples of all four types of hormonal therapy. The breast cancer case (6-1) illustrates the role of aromatase inhibitors in blocking peripheral conversion of sex steroids. The prostate cancer case (6-2) illustrates the use of the anti-androgen to directly inhibit AR-driven cell growth. Additionally, the prostate cancer case describes the use of GnRH agonists to disrupt the hypothalamic/pituitary contribution to testosterone production, as well as the role for the steroidogenesis inhibitor abiraterone in metastatic castration-resistant disease.

Take-home points

✓ Breast cancer and prostate cancer, two hormonally sensitive cancers, are the most common cancers among women and men, respectively, in the United States.

✓ Because of the central role of estrogen and testosterone in driving cell proliferation in breast and prostate cancers, respectively, many patients respond to hormone therapies.

✓ Strategies to disrupt the hormone pathways that contribute to and support breast and prostate cancers include inhibition of hypothalamic/pituitary signals, inhibition of steroidogenesis, inhibition of peripheral conversion of steroids, and direct blockade at the level of the steroid hormone receptor.

Topics bearing on this case

Human epidermal growth factor receptor 2 (HER2/Neu)

Selective estrogen receptor modulator (SERM)

Estrogen and breast cancer

Hereditary breast cancer syndromes

| Case 6-1 | **Breast Cancer** |

Introduction

Breast cancer is the most common malignancy for women in the United States, with an estimated 266,120 new cases in 2018. One in eight women in the Western world will develop breast cancer. Additionally, 2550 new cases of male breast cancer will be diagnosed in 2018 in the United States. The majority of breast cancers are curable if caught early. However, breast cancer remains the second-leading cause of cancer death in U.S. women after lung cancer, with 40,920 deaths in 2018. Among ethnic groups, African-American women have lower rates of breast cancer incidence but higher rates of more aggressive disease, leading to higher breast cancer death rates compared with white women. Thus, there remains considerable room for improvement in our approaches to targeted screening for and treatment of breast cancer.

Etiology

Major risk factors for breast cancer development are summarized in Table 6.2. Approximately 5–10% of breast cancers are hereditary, with most due to mutations in genes *BRCA1*, *BRCA2*, and *PALB2*, and

Table 6.2. Major risk factors for breast cancer.

Risk factor	Relative (fold) risk of breast cancer compared with general public
Increasing age	18 (age 70 vs. age 30)
Genetic mutation	200 (for age <40) 15 (for women in their 60s)
Atypical hyperplasia	5
Prior chest wall radiation	5
Increased breast density	4
Strong family history	3–4

rare cases involving mutations in genes *PTEN*, *p53*, *CDH1*, and *STK11* (see Table 9.1 for more details about hereditary breast cancer syndromes). Another 20% are associated with familial predisposition, likely related to a combination of factors including low-**penetrance** genes (referring to a low proportion of individuals with inherited genes associated with a particular trait who actually develop the disease) and environmental factors.

In the absence of family history or a known genetic mutation that predisposes to breast cancer formation, the highest risk factors for breast cancer development are advanced age, incidental breast irradiation at a young age (for example, to treat Hodgkin lymphoma), presence of atypical hyperplasia in the breast, and having dense breast tissue after menopause. Prolonged endogenous estrogen exposure—due to early menarche (initiation of menstrual cycles), late menopause, nulliparity (never bearing children), or age greater than 30 years at birth of first children—is associated with higher risk of developing breast cancer. Use of hormone replacement therapy after menopause, obesity, and alcohol consumption increase a woman's lifetime risk of breast cancer only slightly. Oral contraceptive use is not clearly associated with breast cancer risk. Breastfeeding may confer a protective effect from breast cancer risk.

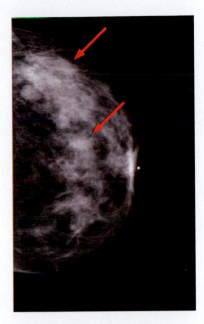

Figure 6.4. Breast cancer on mammogram. Multiple nodules can be seen within one breast, as indicated by the red arrows.

Diagnosis, workup, and staging

As described in Chapter 1, screening for cancer is considered effective when lives are saved. Breast cancer screening has been proven to save lives. In countries where breast mammography is widely available, screening programs detect many early-stage cancers. Mammograms are X-ray images of the breast, which can be captured on film or stored as computer files (digital). With image enlargement and enhancement capabilities, digital mammography holds some advantages over film mammography, and is the gold standard for screening. Breast MRI may detect additional disease in a subgroup of high-risk women and is recommended for those who underwent chest irradiation in adolescence or have a known genetic mutation predisposing them to breast cancer, such as mutations in *BRCA1* or *BRCA2* (see Case 9-1). Screening programs differ worldwide, but most recommend a bilateral mammogram for all women of average risk at least once every two years between ages 50 and 70; the benefits of screening before age 50 are controversial. Figure 6.4 illustrates an abnormal mammogram.

In most of the world, there is no screening for breast cancer, and consequently, in many countries, the disease remains undiagnosed or patients present with symptoms of advanced local or distant disease (Figure 6.5). Unfortunately, access to palliative care is also limited in these same countries. Therefore, global efforts are best focused on education about early detection and encouraging women to seek medical attention if they feel a breast mass.

Figure 6.5. Locally advanced breast cancer. Woman presenting with a large breast mass causing distortion of the breast, "*peau d'orange*" skin changes (*yellow arrow*), and nipple retraction, as well as cervical lymphadenopathy (enlarged lymph node; see shadow marked by the red arrow). (Courtesy of Michaela Higgins, M.D.)

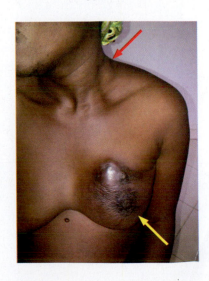

Some women present with a lump in their breast or **axilla** (armpit). This may cause nipple retraction or skin tethering or discoloration. Pain and ulceration are usually associated with advanced disease that has been present for many months or even years. A woman presenting with a palpable breast lump should undergo "triple assessment," which refers to clinical examination, mammography, and core needle biopsy for pathological diagnosis. Breast cancer has a predilection for metastasizing to the ipsilateral (same side) axillary lymph nodes and later via **hematogenous** (carried by the bloodstream) spread to bone, liver, lung, and brain. For women who present on a mammogram or clinical exam with a suspicious abnormality that is less than 2 cm in size, with no other signs or symptoms, a chest X-ray and liver function tests are the only staging examinations recommended, as the likelihood of detecting occult metastases in such cases is very low. For those presenting with a larger lump or symptoms, staging examinations should include CT scans of chest and abdomen and a nuclear bone scan. Brain scans are not routinely performed

Table 6.3. American Joint Committee on Cancer (AJCC) staging classification for breast cancer.

Primary tumor (T)				
Tx	Primary tumor cannot be assessed			
T0	No evidence of primary tumor			
Tis	Carcinoma *in situ*			
T1	Tumor $\leq$2 cm and limited to breast			
T2	Tumor $>$2 cm but $\leq$5 cm			
T3	Tumor $>$5 cm			
T4	Tumor of any size growing into the chest wall or skin			
Regional lymph nodes (N)				
Nx	Nodes cannot be assessed			
N0	No regional node metastasis			
N1	Spread to 1–3 axillary lymph nodes and/or spread to internal mammary lymph nodes on sentinel lymph node biopsy			
N2	Spread to 4–9 axillary lymph nodes, or spread to and enlargement of internal mammary lymph nodes			
N3	Spread to $>$10 axillary lymph nodes, or spread to lymph nodes under or above the clavicle			
Distant metastasis (M)				
Mx	Distant spread cannot be assessed			
M0	No distant metastasis			
M1	Distant metastasis			
Stage grouping	**T**	**N**	**M**	**Comment**
Stage 0	Tis	N0	M0	This is ductal carcinoma *in situ* (DCIS)
Stage I	T1	N0	M0	
	T0 or T1	N1mi	M0	N1mi = microscopic spread (0.2–2 mm size of cancer growth within a lymph node)
Stage II	T0 or T1	N1	M0	
	T2	N0 or N1	M0	
	T3	N0	M0	
Stage III	T0 or T1 or T2	N2	M0	
	T3	N1 or N2	M0	
	T4	N0 or N1 or N2	M0	
	any T	N3	M0	
Stage IV	any T	any N	M1	

in the absence of neurologic symptoms such as new onset of weakness or seizures. Table 6.3 describes the staging classification for breast cancer.

Breast cancer subtypes

Breast cancer is a heterogeneous disease. Routine morphological examination can distinguish the most common pathological subtypes of ductal and lobular breast cancers although other types exist. The majority (50–70%) express estrogen receptor α (ERα) and/or progesterone receptor (PR) and are therefore amenable to hormonal therapies. Approximately 25–30% overexpress HER2 (human epidermal growth factor receptor 2, also known as Neu or ErbB2), a receptor tyrosine kinase of the epidermal growth factor receptor (EGF-R) family that may be targeted by either of the monoclonal antibodies trastuzumab (Herceptin) or pertuzumab (Perjeta), the small-molecular tyrosine kinase inhibitor lapatinib (Tykerb), or the antibody–drug conjugate trastuzumab–emtansine (Kadcyla; see Case 11-3). A minority of breast cancers do not express ER, PR, or HER2 and are called "triple negative" disease.

Gene expression arrays and RNA sequencing experiments have been used to classify breast cancer into molecular categories. One classification separates breast cancer into five distinct "signatures" with varying prognoses (Table 6.4). The five signatures are luminal A, luminal B, triple negative/basal-like, HER2-positive, and normal breast-like types. These subtypes can be approximated using routine immunohistochemical stains. The majority of luminal A-type breast tumors are ER/PR positive and have a low proliferative index (measured by staining for Ki-67, a nuclear protein associated with cell proliferation).

Table 6.4. Characteristics of the five major molecular subtypes of breast cancer.

Subtype	Molecular characteristics	Prevalence	Clinical features
Luminal A	ER$^+$ and/or PR$^-$ HER2$^-$ Low Ki-67	30–50%	Generally slow growth Responds to hormone therapy Associated with best prognosis overall
Luminal B	ER$^+$ and/or PR$^+$ HER2$^+$ or HER2$^-$ with high Ki-67	10–25%	Responds to hormone therapy Greater response to chemotherapy compared with luminal A
Triple negative/ basal-like	ER$^-$ PR$^-$ HER2$^-$	15–20%	Particularly common in African-American women No response to hormone or HER2-directed therapies Sensitive to platinum-based chemotherapy and PARP inhibitors (see Case 9-1) Most aggressive natural history
HER2 type	ER$^-$ PR$^-$ HER2$^+$	5–15%	Responds to trastuzumab and other HER2-directed therapy Responds to anthracycline-based chemotherapy Generally poor prognosis
Normal breast-like	Similar to normal breast tissue	5–10%	Significance is undefined; some studies indicate better prognosis

Luminal B-type breast cancers are also ER positive but have a higher Ki-67, and some may overexpress HER2. The HER2 subtype overexpresses HER2, and the basal-type subgroup are negative for both ER/PR and HER2. The luminal cancers portend the best prognosis (with luminal A better than luminal B), and the basal subgroup is associated with the most aggressive natural history.

Management of localized breast cancer

The majority of women with early-stage breast cancer can be cured with local therapy, usually a lumpectomy or mastectomy (see Figure 2.3), which may be followed with adjuvant radiation, hormone therapy, and/or chemotherapy. Randomized trials comparing mastectomy versus a combination of breast-conserving surgery (lumpectomy) with radiation have been conducted. On the whole, they showed a statistically significant reduction of the risk of local relapse with mastectomy, but no significant influence on overall survival and disease-free survival. A mastectomy typically includes removal of axillary lymph nodes, which are the most common sites for metastasis. In some centers, surgical expertise allows for nipple-sparing breast mastectomies, where the natural nipple is left intact (if the tumor is sufficiently distant from the nipple/areolar complex). There are many reconstructive options available to women that offer good cosmetic outcomes, including the use of either breast implants or the patient's own natural tissues, for example, a flap of skin, fat, or muscle from another area of the body to approximate the shape of the removed breast. Common types of flaps mobilize muscle from the upper back (a latissimus dorsi flap) or lower abdomen (a transverse rectus abdominis myocutaneous, or TRAM, flap).

At the time of primary breast-conserving surgery, blue-colored or radioactive dye is routinely injected into the affected breast, and the lymph nodes in the **ipsilateral** (same side) axilla are then examined to identify any lymph nodes to which the dye has traveled. The first lymph node (or nodes) to which the dye drains, designated the sentinel lymph node, is then removed and sent for immediate intraoperative pathological assessment, known as a sentinel lymph node biopsy. If cancer is identified in that node, the surgeon then proceeds to complete dissection (removal) of the axillary nodes, as there is a significant risk of further disease in the axilla. If, on the other hand, the sentinel lymph node does not contain evidence of cancer, the patient is spared the morbidity of further axillary surgery, and her risk of lymphedema (fluid retention and tissue swelling due to disrupted lymphatic drainage) is far less. (See Case 4-4 and Figure 4.26 for an example of sentinel lymph node testing in melanoma surgery.)

High-risk breast cancer features include ER/PR-negative disease, positive lymph nodes, and larger tumors. Adjuvant chemotherapy is offered to patients with high-risk early breast cancer, even if the cancer is ER-positive, to reduce the risk of recurrence. The magnitude of benefit is typically greater for younger compared with older women. The Early Breast Cancer Trialists' Collaborative Group overview concluded that chemotherapy can reduce the risk of relapse and death by 37% and 30%, respectively, in women under 50 with positive lymph nodes, and by 19% and 12%, respectively, in women between 50 and 69 years of age. The TAILORx trial demonstrated in 2018 that additional information from molecular testing can inform which patients at low/intermediate risk of recurrence might forgo adjuvant chemotherapy. Two or even three drug combinations are used in the adjuvant setting, typically including an anthracycline such as doxorubicin (Adriamycin), an alkylating agent such as cyclophosphamide (Cytoxan), and a taxane such as paclitaxel (Taxol). For breast cancer that overexpresses HER2, a HER2-blocking drug, such as the monoclonal antibody trastuzumab, is also given as adjuvant therapy (see Case 11-3 for more on HER2-positive breast cancer). Adjuvant chemotherapy causes side effects, including hair loss, nausea, vomiting, fatigue, mucositis (inflammation of mucous membranes lining the digestive tract), diarrhea, and vulnerability to serious infections due to diminished white blood cells.

Breast radiation is recommended for all women who undergo breast-conserving surgery with the exception of some women over 70 years of age with low-risk ER-positive breast cancer who take anti-estrogen therapy. Axillary and supraclavicular lymph nodes are routinely irradiated if axillary spread is found on sentinel lymph node biopsy, as these patients are at particularly high risk of relapse in those regions. Radiation is typically performed after completion of adjuvant chemotherapy.

Cancers that express ER and/or PR are sensitive to steroid hormones, and even exquisitely low doses of circulating estrogens (as found in postmenopausal women, for example) are sufficient to drive mitosis and growth of breast cancer cells. Thus, adjuvant hormone therapy strategies that disrupt this pathway are highly effective and associated with decreased recurrence rates by 40% and risk of death by 35% for most women with localized disease. The typical adjuvant hormone therapy treatment begins after adjuvant chemotherapy (if given), may start concurrently with adjuvant radiation (if given), and lasts for at least five years.

For pre-menopausal women with intact ovaries, estrogen is produced in a cyclical manner and the only FDA-approved therapies are tamoxifen, which is a **selective estrogen receptor modulator** (SERM) that blocks the activity of the ligand estrogen at the level of the ER, and a combination of a GnRH agonist such as leuprolide or goserelin—which stops ovarian production of estrogen—with tamoxifen. Tamoxifen is an oral medication taken once daily, whereas GnRH agonists are delivered by injection. Young women who have had their ovaries removed or irradiated can receive the same treatments prescribed to postmenopausal women.

After menopause, the ovaries no longer produce estrogens. For postmenopausal women, estrogens are produced from the peripheral conversion of androgens via the enzyme aromatase in the adrenal gland and in fat and muscle cells. Aromatase inhibitors (AIs) block this estrogen production and have been used routinely as adjuvant treatment in postmenopausal breast cancer patients. AIs are oral medications that are taken once daily.

The most recently approved hormone therapy for breast cancer is the **selective estrogen receptor down-regulator** (SERD) fulvestrant, which is given as an intramuscular injection monthly.

Any therapy that inhibits estrogen will mimic and/or exacerbate menopause-type symptoms, such as hot flashes, vaginal dryness and irritation, loss of bone mineral density, mood swings, and weight gain. The younger the woman, the more pronounced these symptoms can be; however, these treatments are generally among the most well-tolerated cancer therapies. As with any oral medication, it is vital to encourage patients to take the medication as prescribed, which is often necessary for years.

Management of metastatic breast cancer

Metastatic breast cancer is not curable. The goal of treatment is to palliate (lessen) symptoms and control the disease, balanced against treatment-related toxicities. The median survival for metastatic breast cancer is 30 months and has been increasing with the advent of new therapies. The survival rate varies among breast cancer subtypes. Prognostic indices have been derived based on clinical factors, such as history of adjuvant chemotherapy, distant lymph node metastasis, liver metastasis, lactate dehydrogenase (LDH) level and disease-free interval. Increased risk factors were associated with worse survival; Yamamoto et al. (1998) described median survival of 45.5, 24.6, and 10.6 months, respectively, among low-, intermediate-, and high-risk groups.

For women with ER/PR-positive breast cancers, the standard initial therapy is hormone therapy. Like adjuvant hormone therapy for postmenopausal women, this treatment involves AIs. For pre-menopausal women, the management includes the SERM tamoxifen, with or without ovarian suppression with a GnRH agonist. Response rates of 40–70% to hormone therapy are typical and may last more than 12 months in patients without prior exposure to adjuvant hormone therapy. Progression of breast cancer after an initial response with one hormonal manipulation may predict an inability to respond to other hormone treatments. Ultimately, patients whose disease becomes resistant to hormone therapy are then treated with cytotoxic chemotherapy.

Many chemotherapies work to some degree in the treatment of breast cancer; therefore, breast cancer patients can be treated with, and often respond to, multiple lines of chemotherapy agents. Unlike the use of combinations of agents in the adjuvant setting, women with metastatic breast cancer are typically treated with sequential single-agent regimens to minimize toxicity because the intention of treatment is palliative. Chemotherapy agents with proven activity in metastatic breast cancer include antimetabolites (capecitabine, gemcitabine), anthracyclines (doxorubicin), taxanes (paclitaxel, docetaxel) and other microtubule inhibitors (eribulin, ixabepilone), and vinca alkaloids (vinorelbine).

For breast cancer that overexpresses HER2, a HER2-blocking drug such as trastuzumab or lapatinib would typically be added to hormone therapy or chemotherapy to synergistically enhance the response (see Case 11-3).

The case of Brenda Nathan, a paramedic with a breast mass

Brenda found a lump in her left breast while in the shower at the age of 45. She was otherwise healthy and unaware of any family history of breast cancer. Given the size of her tumor at 4 cm in diameter, she was fortunate that initial staging studies proved negative for distant spread. She opted for mastectomy rather than lumpectomy. At the time of mastectomy and axillary lymph node dissection, she was found to have involvement of multiple axillary lymph nodes. Her breast tumor was ER/PR positive and HER2 negative. She was diagnosed with stage III breast cancer and offered adjuvant radiation, chemotherapy, and hormone therapy to reduce her risk of recurrence. Brenda tolerated combination chemotherapy and radiotherapy extremely well and commenced tamoxifen daily on completion of chemotherapy. Tamoxifen was prescribed because Brenda was pre-menopausal at the time of her breast cancer diagnosis. The plan was for five years of tamoxifen therapy.

Brenda continued to work. Several years later, she began to experience occasional bone aches and pains, but she attributed them to her job as a paramedic that required strenuous lifting and transporting of patients. She mentioned the bone discomfort to her oncologist. A bone scan unfortunately displayed multiple abnormal areas on her ribs, vertebrae, and pelvis (Figure 6.6). Further CT scans were negative for other abnormalities.

Figure 6.6. Brenda's bone scan. Multiple abnormal areas are seen on her ribs (*red arrow*), vertebrae (*blue arrows*), and pelvis (*green arrow*). Increased signal seen in common areas of degenerative changes (arthritis) is considered normal (black arrows pointing to knees).

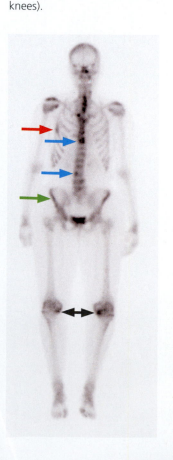

The GnRH agonist leuprolide was added to Brenda's regimen of tamoxifen. The leuprolide induced a "chemical menopause" because her ovaries no longer produced estrogens. Brenda's disease remained stable for a period of nine months before progressing further in several bone areas. She received palliative radiation to painful areas of metastases in her lumbar vertebrae and right hip. At that point, her hormone therapy was switched to the aromatase inhibitor letrozole while the leuprolide was continued. After 18 months of hormonal manipulations, her cancer continued to spread in the bone and now appeared in her liver. Brenda continued to work as a paramedic; however, she could no longer ride in the ambulance due to reduced mobility and was confined to office work. She was treated with several lines of chemotherapies until her performance status declined, at which point hospice care was recommended. Brenda died from her disease at the age of 53.

Hormone therapy and metastatic breast cancer

Targeted therapy in metastatic breast cancer has for decades started with hormone manipulation, for those cancers that express ER and/or PR, similar to the hormonal manipulation of the androgen receptor in prostate cancer to be described in Case 6-2. In the late 1990s, three AIs were FDA approved for the treatment of metastatic ER/PR-positive breast cancer: the steroidal AI exemestane (Aromasin), and the nonsteroidal AIs letrozole (Femara) and anastrazole (Arimidex). Because AIs only inhibit peripheral conversion of androgens into estrogens, they are not used in pre-menopausal women, who produce abundant gonadal estrogens.

AIs have higher response rates and longer times to progression compared with tamoxifen, but no difference in overall survival. Evaluation of potential toxicities is therefore important in choosing an AI versus tamoxifen. Both AIs and tamoxifen cause similar levels of hot flashes, bone or joint aches, and fatigue. Because tamoxifen has partial agonist effects on ER in the uterus, there is a very low—but not zero—risk of developing endometrial cancer due to tamoxifen. Tamoxifen is also associated with a low risk for developing venous thromboembolic events (blood clots). AIs do not have agonist effects on ER and thus are not associated with endometrial cancer, which is a recognized risk of tamoxifen, nor are they associated with venous thromboembolic events. However, AIs cause more loss of bone mineral density and are associated with more fractures compared with tamoxifen, whose partial ER agonist activity in the bone is protective. On balance, most postmenopausal women with metastatic breast cancer therefore receive an AI.

Future prospects and challenges in metastatic breast cancer

Unfortunately, nearly all metastatic breast cancers eventually become resistant to strategies that disrupt estrogen pathways. The vast majority of these cancers still express ER, but it is likely that alternative oncogenic signaling pathways are activated that bypass the ER signaling pathway. This is termed "acquired endocrine resistance," and there are many putative causes, including up-regulation of the HER2 pathway, activation of the PI3K pathway (*PIK3CA* mutations are found among 25% of breast cancers overall, and the percentage is highest in ER-positive tumors; see Chapter 5 for more on PI3K inhibition), and point mutations in *ESR1*, the gene encoding ER. This is an area of very active research, and most approaches being studied are combinations of standard hormone therapy and novel targeted agents. The first

FDA-approved (in 2012) combination for the treatment of metastatic breast cancer that has progressed during treatment with an AI combines the steroidal AI exemestane with the mTOR inhibitor everolimus (see Chapter 5), which improved progression-free survival significantly when compared with exemestane alone.

Further combinations are currently in clinical trials, such as fulvestrant and PI3K inhibitors and AIs in combination with selective inhibitors of the cyclin-dependent kinases CDK4 and CDK6 (such as palbociclib). The combination of palbociclib and letrozole was FDA approved in February, 2015, for ER-positive, HER2-negative metastatic breast cancer.

Take-home points

✓ Breast cancer is the most common non-skin cancer among females.

✓ Most breast cancers are readily detected by mammography.

✓ Where mammography is available, the majority of breast cancers are diagnosed at an early stage.

✓ Early breast cancer is highly curable and treated with multimodality therapy, including surgery, radiation, chemotherapy, and hormonal manipulations.

✓ Breast cancer is increasingly recognized as comprising several different subtypes with unique biologies, natural histories, and treatment approaches.

✓ The majority of breast cancers express estrogen receptor (ER) and progesterone receptor (PR). These cancers respond to hormonal manipulations and generally have a good prognosis.

✓ "Triple negative" breast cancer is ER/PR/HER2 negative, more common in young women and mutant *BRCA1* carriers. It typically displays aggressive behavior.

✓ HER2-overexpressing breast cancer accounts for about 15% of breast cancers and is associated with a relatively aggressive course. However, therapies targeting the HER2 receptor are highly effective and have improved prognosis dramatically for these patients (see Case 11-3).

✓ The development of resistance to hormone therapy is a common challenge in the breast cancer clinic.

✓ Mechanisms of resistance to hormone therapy include activation of the PI3K/mTOR pathway, mutations of ER, and development of HER2 overexpression.

✓ Novel approaches to overcome resistance to hormone therapy include the combination therapies of anti-estrogens with the mTOR inhibitor everolimus, which was FDA approved in 2012; of the aromatase inhibitors combined with the CDK4/6 inhibitors palbociclib, ribociclib, or abemaciclib, which were FDA approved in 2015–2018; and of anti-estrogens and PI3K inhibitors, which are currently in clinical trials.

Discussion questions

1. What are the most common risk factors for breast cancer development?

2. Where does breast cancer most commonly spread?

3. How should a woman presenting with a breast mass and bone pain be evaluated?

4. How do aromatase inhibitors work? Who should receive an AI?

5. How do some breast cancers develop resistance to endocrine therapies?

Case 6-2   Prostate Cancer

Topics bearing on this case

Cancer screening

Androgen hormones

Chromosomal translocations, including *TMPRSS2-ERG*

Introduction

Prostate cancer is the most common cancer among men in the United States, with 164,690 new diagnoses estimated for 2018. The incidence of prostate cancer reached a peak in the early 1990s due to the introduction of the prostate-specific antigen (PSA) blood test for prostate cancer screening (Figure 6.7). The best explanation for this peak is that more cases were *detected* due to widespread PSA testing, as opposed to a suddenly higher national disease burden, because despite the rise in cases, the death rate from prostate cancer has not followed a similar pattern. In fact, the death rate has fallen gradually, suggesting that prostate cancer is either highly treatable, or perhaps readily diagnosed but clinically inconsequential for many patients (cases may represent an *overdiagnosis* of an indolent condition). Nonetheless, prostate cancer is the second-leading cause of cancer death after lung cancer in U.S. men, with 29,430 deaths estimated for 2018.

Prostate cancer affects one in seven American men. The average age at diagnosis in the United States is 69 years. Among ethnic groups, African-American men have higher rates of prostate cancer development and are 2.5 times more likely than Caucasian men to die from the disease. Table 6.5 lists the risk factors for development of prostate cancer.

Prostate cancer is an adenocarcinoma, and 80% of cancers originate from the posterior "peripheral zone" of the gland. Approximately half of all prostate cancers harbor a somatic translocation fusing the androgen-responsive gene *TMPRSS2* with a gene encoding an ETS-family transcription factor, which could be ERG, ETV1, ETV4, or ETV6. These translocations, which were discovered in 2005, together make up the most common chromosomal translocation event among all human carcinomas. In the prostate gland, they have been found in precancerous lesions known as prostatic intraepithelial neoplasia (PIN). Cells in prostate glands with multifocal cancers have been demonstrated to harbor different variations of these translocations. Although these *TMPRSS* translocations are common in prostate cancers, their presence has not been associated definitively with better or worse outcomes.

Figure 6.7. Trends in incidence rates for selected cancers in men, United States, 1975 to 2012. Note that the first PSA test became commercially available in 1986, and the increase in prostate cancer incidence largely reflected increased diagnosis due to screening. (Modified from R.L. Siegel, K.D. Miller, and A. Jemal, *CA Cancer J. Clin.* 66:7–30, 2016. With permission from John Wiley & Sons.)

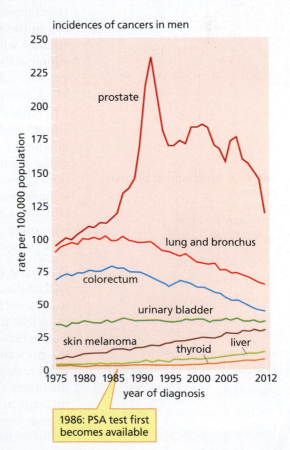

Table 6.5. Risk factors for prostate cancer.

Age >50
Race (higher risk in African-American men)
Family history (first-degree relatives such as father, brother, son)
BRCA2 gene mutation

Figure 6.8. Anatomy of the prostate and palpation by digital rectal exam. The prostate is situated in the pelvis between the bladder and rectum. The urethra drains urine from the bladder and passes through the prostate, which explains why urinary symptoms are associated with both benign and malignant prostate disease.

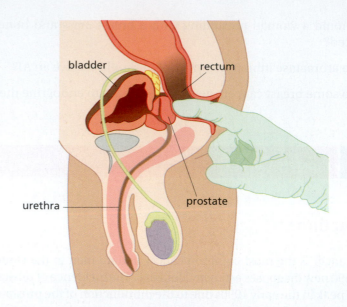

Table 6.6. Symptoms of prostate cancer.

Frequent urination
Difficulty starting urinary stream
Urgency to urinate
Weak urinary stream
Erectile dysfunction
Pain during urination or ejaculation
Blood in urine or semen
Bone pain in pelvis, back (spine)
Weight loss
Incontinence of urine or stool
Weakness or paralysis of legs

Other common genetic alterations seen in prostate cancer include amplification of the androgen receptor (*AR*) gene and somatic (in cells other than germ cells) mutations of *AR* associated with resistance to anti-androgens, a group of drugs that block androgen binding to AR. While many such alterations have been described in advanced prostate cancer, amplification and mutation have also been described as potential early events in prostate cancer pathogenesis.

Diagnosis, workup, and staging

Most prostate cancers in the United States are detected by screening and are thus found in men without symptoms. The prostate is located deep in the pelvis, under the bladder, and is palpable via a digital rectal exam (DRE). The urethra, which conveys urine from the bladder to the penis, runs through the center of the prostate (Figure 6.8). Nearly all men will have enlargement of the prostate gland with age, a process called benign prostatic hypertrophy (BPH). Many symptoms that are associated with prostate cancer (Table 6.6)—such as frequent urination, urgency to urinate, difficulty initiating urination, and slower stream—occur due to obstruction of the urinary stream, which can also result from benign causes such as BPH and prostatitis (inflammation of the prostate). Erectile dysfunction is also common with age and therefore does not necessarily portend a diagnosis of prostate cancer. The symptoms of bone pain, weight loss, incontinence, and weakness of legs are associated with metastatic prostate cancer, and are nonspecific because they can be seen with other malignancies. Fortunately, with prostate cancer screening, fewer men present with symptoms of metastatic prostate cancer.

Screening is typically performed by primary care physicians and includes a DRE and a PSA blood test. PSA is a highly controversial biomarker as a screening test for prostate cancer. Because PSA is produced by normal prostate cells as well as prostate cancer cells, elevation in the serum PSA level often results from a benign condition, such as BPH or prostatitis, or from manipulation of the prostate, which can occur with ejaculation, palpation, or insertion of a Foley catheter to drain the bladder. Traditionally, 4 ng/ml has been considered the upper limit of normal for PSA. Frequently, a PSA result greater than 4 ng/ml leads to a referral to a urologist for a biopsy. In the United States, approximately one million biopsies are performed annually, with 25–30% diagnostic for prostate cancer. Prostate cancer screening has led to the detection of earlier-stage disease, with the majority of patients diagnosed with low-grade disease that is generally slow to grow and spread.

As described in Chapter 1, screening is considered effective when lives are saved. Screening for prostate cancer is highly controversial. Several key features about prostate cancer have fueled the debate about prostate cancer screening, and warrant description. First, many men are subjected to PSA screening, estimated at 30 million per year in the United States. Second, with one million biopsies performed per year and a 70–75% negative biopsy rate, many men are thus being subjected to an invasive procedure but most will not be diagnosed with cancer. Third, it is clear that most prostate cancers grow slowly—so slowly that even if a screening test revealed prostate cancer, it is unlikely that the patient's life span would be affected. In fact, an autopsy study of 152 young men who died of trauma found incidental prostate cancer in 27% and 34% of men in their 30s and 40s, respectively. It is unlikely that many of those cancers would have become clinically significant, even if diagnosed decades later as a result of PSA screening. Fourth, as a disease that is generally diagnosed in older men, those patients may have accumulated other medical conditions (co-morbidities) that will likely impact life span more than an indolent prostate cancer. Said differently, most men diagnosed with prostate cancer will die of a disease other than prostate cancer. Fifth, when told of the diagnosis of prostate cancer, many patients undergo treatment even if clinical and biological indicators point to an indolent form of the disease, and treatment is associated with significant potential side effects, such as impotence and erectile dysfunction, which can negatively impact quality of life. Finally, the two largest randomized controlled trials, one in the United States and the other in Europe, had conflicting conclusions about the benefit of prostate cancer screening. Whereas the U.S. study indicated no survival benefit with screening, the European study results indicated that there was a survival benefit, but that 1000 men would need to be screened to save one life. Considering the low return in terms of numbers of men screened versus lives saved, prostate cancer screening could be viewed as a high-cost proposition.

Without clear evidence of a benefit to prostate cancer screening, the U.S. Preventive Services Task Force recommended against PSA screening in 2012. Despite this recommendation, it is clear that as the second-leading cause of cancer death in the United States, the diagnosis of prostate cancer, appropriate management decision making, and improved treatments remain critically important.

Prostate cancer is diagnosed by the use of a transrectal ultrasound-guided needle biopsy, an outpatient procedure performed with local anesthesia (Figure 6.9). A patient typically referred for biopsy has an elevated PSA or a

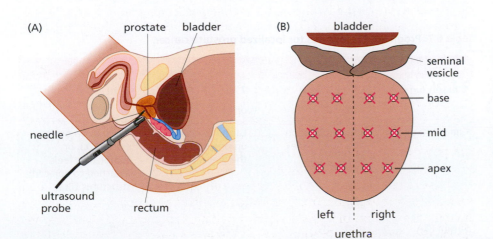

Figure 6.9. Prostate biopsy.
(A) Schematic of transrectal ultrasound-guided biopsy. The probe inserted through the rectum can measure the prostate size and guide the needle biopsies. (B) Schematic diagram of a 12-core biopsy. The prostate is divided visually into six sextants (base, mid, and apex on left and right sides). Two core samplings are taken from each sextant (each core marked with a red X).

nodule palpated on DRE by the primary care physician. The standard biopsy includes two samples (core biopsies) from each of six regions of the prostate. Unlike other malignancies that are diagnosed when a radiographic, visible, or palpable lesion can be detected and specifically sampled, the standard prostate biopsy is a blind sampling of all areas of the prostate. Because 60–80% of prostate cancers are multifocal, there is a risk of undersampling of the prostate, that is, missing cancerous areas of the gland.

When prostate adenocarcinoma is diagnosed on biopsy, the grade is defined by the pathologist using the **Gleason score**. Designed by Dr. Donald Gleason, the scoring system standardized grading based on the architecture of the prostate cancer cells and surrounding stroma. Two cancerous areas describing the majority and the minority of the disease are defined in each core biopsy and labeled with a number between 1 and 5, with higher numbers indicating more aggressive, less-differentiated cells. In contemporary practice, the Gleason grade ranges between 3 and 5; grades 1 and 2 are no longer scored as cancers. The majority and minority grades are then listed as the first and second addends, respectively, and summed as the Gleason score (range: 6–10). Aggressiveness of the cancer as defined by grade includes low-risk Gleason score 6, intermediate-risk Gleason score 7, and the high-risk cancers Gleason score 8–10. The low-grade, Gleason score 6 prostate cancers are generally slow to grow and spread and constitute ~70% of diagnoses. Intermediate- and high-risk disease comprise approximately 15% each of the balance of prostate cancers.

Patients with localized cancers are grouped into prognostic categories: low-, intermediate-, and high-risk disease (Table 6.7). For prostate cancer, the "risk" describes the chance that the disease might recur despite effective primary therapy, with chances of recurrence for low-, intermediate-, and high-risk disease of approximately 10%, 20%, and 40%, respectively. Low-risk cancers have all three of the following characteristics: Gleason score of 6, a small nodule on DRE, and PSA under 10. Intermediate risk cancers have Gleason score of 7, a larger nodule on DRE, or PSA between 10 and 20. High-risk cancers are defined as having Gleason score 8–10 or bilateral nodules on DRE or PSA > 20.

Intermediate- and high-risk cancers warrant staging evaluation, whereas low-risk disease, with its characteristic indolent growth pattern, is highly unlikely to demonstrate metastasis. Prostate cancer characteristically metastasizes to bone and to lymph nodes in the pelvis and retroperitoneum, a space behind the peritoneal cavity that houses most of the digestive organs in the abdomen and pelvis. Like many other cancers that spread to bone, prostate cancer targets bones where there is active **hematopoiesis** (formation of blood) and

Table 6.7. Prognostic categories for localized prostate cancer.

Characteristic	Low-risk	Intermediate-risk	High-risk
Gleason sum	6	7	8–10
	and	*or*	*or*
Clinical stage	Small nodule on one side of the gland (T2a)	Larger nodule on one side of the gland (T2b)	Nodules on both sides (T2c), or tumor that feels attached to adjacent structures (T3)
	and	*or*	*or*
Serum PSA	<10	10–20	>20

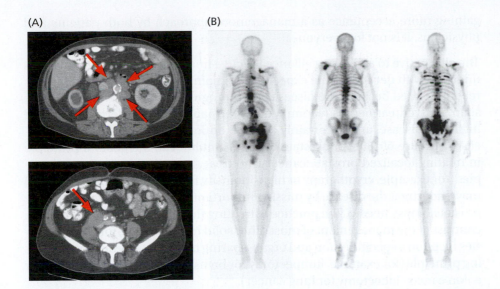

Figure 6.10. Metastatic prostate cancer. (A) Two CT scan images from the same patient demonstrating mildly enlarged retroperitoneal lymph nodes (*red arrows*). (B) Bone scans from three individual patients demonstrating abnormal uptake in metastatic lesions (*dark spots*).

a rich supply of growth factors, such as the pelvis, spine, sternum, ribs, and proximal femur and humerus. Unlike cancers that cause **lytic** bone metastases that carve out gaps in the normally calcified cortical bone, prostate cancer causes predominantly **blastic** metastases, characterized by the appearance of newly deposited bone that is structurally disorganized and significantly weaker than normal bone. The typical staging studies include an abdominal and pelvic CT scan (for lymph node involvement) and a bone scan (Figure 6.10); see Chapter 1 for more on these scans. Generally, little benefit is added by subjecting the patient to more extensive imaging of the chest or the brain, or more expensive imaging studies such as PET scans.

Unlike other diseases, which are categorized by TNM staging (such as Table 6.3 for breast cancer), prostate cancer is most often categorized as metastatic or localized, with localized cancers subdivided among low-, intermediate-, and high-risk disease. With recent improvements in therapy, the average life span of patients with metastatic disease continues to lengthen and frequently surpasses four years in contemporary practice.

Management of localized prostate cancer

Due to the indolent nature of most diagnosed prostate cancers, with 70% being low-risk disease, monitoring the most favorable disease on a program of active surveillance offers the advantages of avoiding both overtreatment and treatment-related side effects. The most favorable disease has low-risk features (see Table 6.7) and very low-volume disease (up to three of the 12 core samples may be involved). By monitoring the disease with serial DRE exams, PSA tests, and a scheduled program of repeat prostate biopsies, an active surveillance program provides the element of time, allowing the patient and his team to judge whether and how quickly the disease may be progressing. For older patients, this may be preferable to the risks of treatment, especially if the disease shows very little change over time. The primary disadvantage of active surveillance is the lack of guarantee that the disease is fully understood, with the possibility that higher-grade or higher-volume cancer has eluded the prostate biopsy needle. A second disadvantage is the potential anxiety that a patient will naturally experience about having an untreated cancer, and the need for serial invasive diagnostic procedures. While active surveillance is

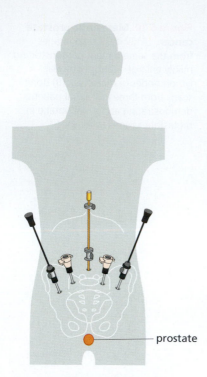

— prostate

Figure 6.11. Laparoscopic prostatectomy. Unlike an open prostatectomy, using an incision from the umbilicus to the suprapubic bone, the laparoscopic approach uses four to five small incisions. A camera allows visualization of the internal organs, and other instruments are inserted via the other ports, with the urologist holding the instruments and guiding their action. The robotic-assisted laparoscopic approach uses a robot to hold the instruments while the surgeon manipulates the robotic arms from a nearby console.

gaining more acceptance as a management approach by both patients and physicians, it is not for everyone.

The alternative to active surveillance is active treatment. For younger patients under 60 with decades of life expectancy remaining, active treatment is often recommended because the disease could progress over time and impact life span. Active treatment is recommended for patients with intermediate- and high-risk disease, and is certainly a viable option for men with low-risk disease. Because 60–80% of prostate cancer is multifocal, the general principle in treating localized prostate cancer is to treat the entire gland. Focal therapies (for example, cryotherapy or high-intensity focused ultrasound, or HIFU) may undertreat the disease by missing important tumors that have eluded the prostate biopsy needle. The practice of treating the entire gland stands in stark contrast to the management of most other solid malignancies, for which biopsies focus on a specific lesion and organ-sparing treatment has become a guiding principle (for example, lumpectomy for breast cancer, hemi-colectomy for colon cancer, lobectomy for lung cancer).

Active treatment options for localized disease are divided between surgery and radiation. These approaches have never been compared in a randomized controlled trial, but contemporary series studies indicate similar disease control rates. Thus, the treatment choice involves the patient's candidacy for either option based on co-morbid medical conditions, and if given a choice, what his personal preference may be, based on potential side effects. This is a highly nuanced discussion that is personalized for the individual patient.

Surgical management is via a radical prostatectomy, performed by a urologist. Due to the prostate's envelopment of the urethra (see Figure 6.9), when the prostate is removed, the urethra is reattached to the bladder. Urinary control then becomes dependent on the muscles of the pelvic floor; thus, a small percentage of men will have trouble with urinary incontinence after recovery from surgery. The nerves that control erections run alongside the prostate. Whenever feasible, a "nerve-sparing" approach is followed, to allow recovery of erectile capability after the initial trauma of surgery. In recent decades, advances in surgical management have introduced less-invasive operations including **laparoscopic** prostatectomy, wherein a camera and instruments are used to perform the operation via "portholes" rather than a larger open incision, and robotic-assisted laparoscopic prostatectomy, using a robot to hold and manipulate the instruments (Figure 6.11). Ultimately, the techniques all result in the same operation, with similar results, so patients are advised to seek expert urologic care, not necessarily a specific type of operation.

Radiation options include brachytherapy and external-beam radiation therapy (EBRT) (Figure 6.12). Brachytherapy involves insertion of radiation-emitting seeds, generally impregnated with iodine-125, into the prostate gland, causing

Figure 6.12. Radiation therapy for prostate cancer. (A) Schematic diagram of external-beam radiation therapy of the prostate. The radiation delivery apparatus is on a gantry, a device that can rotate around the patient lying on the table. This approach allows lower-energy radiation beams from multiple angles to converge on the prostate, limiting radiation exposure and damage to nearby structures. (B) X-ray of the pelvis demonstrating brachytherapy seeds in the prostate (*red arrow*). The entire prostate is treated with this distribution of seeds.

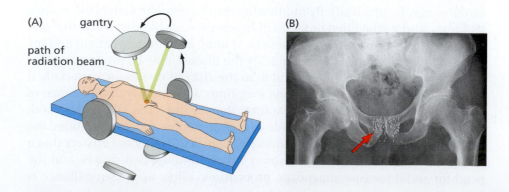

(A)   gantry

path of
radiation beam

(B)

death of surrounding cells. Seeds are not removed. Side effects include potential acute worsening of existing voiding symptoms, and therefore brachytherapy is not often recommended for patients with significant lower urinary tract issues. EBRT generally uses X-rays delivered from multiple external angles, to decrease collateral damage to normal structures while delivering high doses of radiation where the beams overlap, specifically, at the prostate gland. In recent decades, just as surgical techniques have improved, technology advances have led to an increase in EBRT options including intensity-modulated radiation therapy (IMRT) and proton beam therapy. Both brachytherapy and EBRT may cause bladder and bowel urgency and bleeding in the short and long term, respectively, owing to the "margin" of overlap needed to ensure complete treatment of the prostate gland. As after surgery, erections are affected by radiation treatment, although the time course leading to erectile dysfunction is more gradual with radiation.

Hormone therapy has been shown in randomized controlled trials to improve cancer control and cure rates when incorporated with EBRT for high-risk and some intermediate-risk prostate cancers. There is no proven role for hormone therapy with surgery. Hormone therapy is also called "medical castration" and **androgen deprivation therapy** (ADT) because the treatment shuts down testosterone production by the testes, much like surgical castration (bilateral orchiectomy, or removal of both testes)—but reversibly. Potential side effects from ADT are numerous but largely reversible once the course of treatment is complete, including hot flashes (similar to hot flashes experienced by postmenopausal women), weight gain, loss of lean muscle mass, loss of bone density with increased potential for fractures, loss of erections, loss of libido, and potential development of diabetes mellitus and cardiovascular complications. ADT may extend over a period as long as three years for patients with high-risk localized disease.

Management of metastatic prostate cancer

The standard initial management of metastatic prostate cancer is ADT. The hormonal responsiveness of prostate cancer was discovered in 1941, and earned the Nobel Prize in Physiology or Medicine in 1966 for Dr. Charles Huggins. He found that androgens stimulate prostate cancer growth and that surgical castration can inhibit prostate cancer dissemination. We now know that the AR mediates the stimulation of prostate cancer cell growth by the androgens testosterone and dihydrotestosterone. Thus, inhibiting the AR pathway represents the oldest and most established form of targeted therapy for prostate cancer. Castration therapy can be accomplished by surgical castration (bilateral orchiectomy) or medical castration using GnRH agonists or antagonists (see Figure 6.3A and the introduction to this chapter for a description of the mechanism of action). Unlike the use of ADT for a finite time period for high-risk, localized prostate cancer, ADT for metastatic prostate cancer should be lifelong, and hence, irreversible surgical castration is a reasonable option. Given the choice of surgical versus medical castration, most American men choose medical castration, while surgical castration remains the standard of care in developing countries with limited health care resources.

ADT is not curative therapy. Nearly all patients respond to ADT with a decrease in PSA and improvement in symptoms or radiographic lesions, but the durability of response is highly variable. For most patients, ADT offers several years of disease control before the development of metastatic castration-resistant prostate cancer (mCRPC). ADT is prescribed for life, because recovery of testosterone would stimulate regrowth of dormant prostate cancer cells.

Until 2010, the only FDA-approved treatment for mCRPC was docetaxel chemotherapy. Docetaxel is a taxane, which causes cell death via microtubule stabilization, preventing disassembly of the microtubules in cells and inhibiting cell division (see Chapter 2 regarding conventional chemotherapy). Two randomized controlled trials with results published simultaneously led to FDA approval of docetaxel in 2004. Docetaxel chemotherapy was associated with 30–50% of patients having an improvement in PSA (defined as a decline of >50% of serum PSA) and a 2–3 month improvement in overall survival. Numerous trials adding other agents to docetaxel have failed to improve overall survival compared with standard docetaxel treatment.

The case of Adam Presley, a retired high school principal with hip pain

Adam is a 67-year-old retired high school principal with high blood pressure and high cholesterol that are well managed with medication, diet, and exercise. At the age of 65, he had noted increasing left hip discomfort over several months that led to more difficulty walking the golf course. His primary care physician ordered X-rays of the left hip, which demonstrated abnormal areas of the pelvis and left femur that were considered indeterminate. Adam then had an MRI of the pelvis, which revealed a blastic lesion in the left femur as well as similar lesions in the right side of the pelvis. A bone scan demonstrated multiple abnormal lesions in his left femur, pelvis, spine, and ribs (Figure 6.13). CT imaging also confirmed retroperitoneal lymph node enlargement (Figure 6.14A).

Because enlarged retroperitoneal lymph nodes and blastic bone metastases are characteristic of metastatic prostate cancer, Adam's primary care physician performed a DRE, which revealed a nodular, hard prostate gland. A serum PSA test was performed, which had been in the normal range (<4 ng/ml) approximately three years earlier. The PSA was measured at 97 ng/ml.

Figure 6.13. Adam's imaging demonstrating metastatic prostate cancer. (A) MRI of the pelvis demonstrating a bone lesion in the left femur (*red arrow*). (B) Bone scan showing lesions in the pelvis, spine, ribs, and left femur (*arrow*, corresponding to the lesion on the MRI).

Given his disease features, Adam was presumed to have metastatic prostate cancer. He was started on ADT, using leuprolide, a GnRH agonist. His testosterone reached castration levels within four weeks. His PSA fell to under 1 ng/ml within six months of starting ADT. His hip discomfort resolved, but his golf game did not appreciably improve because he had weight gain around his abdomen and decreased strength while on ADT, despite a conscious effort to get more exercise. He lost his libido, which made the impotence from ADT less of a concern than he had anticipated. By staying on calcium and vitamin D, as well as an active walking program, he was able to maintain his bone density unchanged despite ADT.

Two years after starting ADT, Adam's PSA started to rise from its very low level under 1 ng/ml to 3.2 ng/ml. He had no new symptoms. With confirmation of the PSA rise over two weeks to 3.9 ng/ml and scans indicating growing lymph nodes but stable bone disease, he was considered to have mCRPC.

Given progression of his prostate cancer, Adam discussed treatment options with his medical oncologist. Although docetaxel chemotherapy was an option, they agreed that hormonal manipulation was more appealing given the recent FDA approval of novel hormone agents and Adam's lack of symptoms attributable to progressive prostate cancer. Adam

(A)

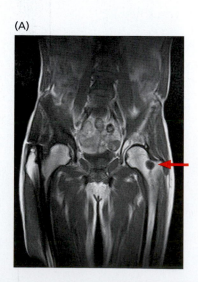

(B)

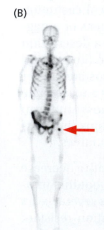

(A)   (B)

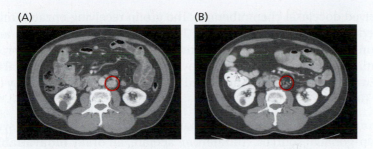

Figure 6.14. Adam's improvement on enzalutamide. (A) CT scan of abdomen and pelvis prior to treatment demonstrating an enlarged retroperitoneal lymph node (*red circle*). (B) CT scan after three months of treatment revealed improvement in the size of the lymph node.

started the oral anti-androgen enzalutamide, at 160 mg daily. While he had slightly more fatigue, he was still able to perform all of his activities of daily living. At three months of treatment, his PSA fell and stabilized under 2 ng/ml. Imaging of his lymph nodes indicated a decrease in the size of the nodes, consistent with the PSA decline and an overall improvement in prostate cancer disease control (see Figure 6.14). At two years on therapy, Adam shows no evidence of disease progression and enjoys an excellent quality of life. His golf game has never been better.

AR targeting and prostate cancer: beyond ADT

Targeted therapy for prostate cancer has been successfully applied for decades in the form of ADT. However, prostate cancer cells eventually develop resistance to ADT, and this development of mCRPC is the key reason that prostate cancer remains the second leading cause of cancer death in American men. The mechanisms of resistance include amplification of the *AR* gene, activating mutations of *AR*, expression of aberrantly spliced AR mRNA variants that are insensitive to anti-androgen therapy, and **up-regulation** of androgen biosynthesis from non-gonadal sources, including the adrenal glands and the cancer cells themselves. It has become clear that strategies that can block AR pathway activity beyond castration can improve overall survival in mCRPC. The term for the disease has evolved from "hormone-refractory prostate cancer" to "castration-resistant prostate cancer," to reflect the fact that the disease may be resistant to castration alone but not be refractory to further hormonal manipulation.

Strategies targeting the AR have led to three new FDA-approved drugs since 2011. The first approach uses the anti-androgen enzalutamide to block residual androgens from stimulating prostate cancer growth. Anti-androgens competitively bind to the ligand-binding domain of the AR, thus inhibiting translocation of the AR to the nucleus, where the AR recruits other co-factors and binds the DNA to stimulate transcription (see Figure 6.3D). Anti-androgens are not new in concept, as numerous early-generation anti-androgens were developed and used for prostate cancer, including bicalutamide, nilutamide, and flutamide. The early anti-androgens were only weak antagonists of AR and, in some cases, could have partial agonistic effects after prolonged use. Enzalutamide lacks the partial agonist activity and is a more potent inhibitor of AR activity compared with its predecessors.

Enzalutamide has been proven to improve overall survival in men with mCRPC in two randomized controlled trials. The first study tested enzalutamide versus placebo in 1199 subjects with mCRPC who had previously received docetaxel chemotherapy. Enzalutamide prolonged overall survival from 13.6 to 18.4 months, leading to FDA approval on August 31, 2012. The second study tested enzalutamide versus placebo in 1717 subjects with mCRPC who had not yet received docetaxel chemotherapy. The group receiving enzalutamide had improved progression-free survival at 12 months (65% versus 14%) and 29% reduction in the risk of death. Enzalutamide was also associated with prolonged time until the initiation of chemotherapy and time until the first skeletal-related event. Enzalutamide was FDA approved for use prior to chemotherapy on September 10, 2014.

The second strategy targeting the AR pathway diminishes androgen synthesis from non-gonadal sources. Abiraterone acetate, an orally **bioavailable prodrug** of abiraterone, is a selective inhibitor of CYP17, an enzyme that plays a critical role in testosterone biosynthesis from cholesterol (see Figure 6.2 in the introduction of this chapter). Abiraterone can thus block androgen synthesis in adrenal glands, testes, and prostate cancer cells. In clinical trials, abiraterone acetate has been administered to patients with mCRPC and demonstrated significant antitumor activity. Because abiraterone interferes with adrenal production of cortisol (see Figure 6.3B), there is compensatory increase in pituitary gland production of adrenocorticotropic hormone (ACTH), and consequent increase in the steroid products upstream of the CYP17 blockade, which produces an excess of mineralocorticoids (specifically, aldosterone). The mineralocorticoid excess explains most of the significant adverse events related to abiraterone, including fluid retention, hypertension, and hypokalemia (low serum potassium). Co-administration of low-dose prednisone (5 mg given twice daily), which exhibits strong glucocorticoid and modest mineralocorticoid activities, can diminish the ACTH up-regulation and abrogate most side effects related to mineralocorticoid excess.

Like enzalutamide, abiraterone has been proven in two randomized controlled trials to improve overall survival in men with mCRPC. The first study tested abiraterone versus placebo in 1195 subjects with mCRPC who had previously received docetaxel chemotherapy. Abiraterone prolonged overall survival from 10.9 to 14.8 months, along with improvements in progression-free survival and PSA response, leading to FDA approval on April 28, 2011. The second study tested abiraterone versus placebo in 1088 subjects with mCRPC who had not yet received docetaxel chemotherapy. Abiraterone was associated with improved progression-free survival (16.5 versus 8.3 months) and overall survival (34.7 versus 30.3 months). Abiraterone was also associated with prolonged time until the initiation of chemotherapy, decreased opiate use for cancer-related pain, and improved performance status. Abiraterone was FDA approved for use prior to chemotherapy on December 10, 2012.

Enzalutamide and abiraterone are generally well tolerated. Enzalutamide is associated with fatigue in some men, and increased incidence of seizures in patients with a known seizure disorder. Abiraterone therapy can cause issues with diabetes control due to the required co-administration of prednisone. The optimal sequencing between enzalutamide and abiraterone is unknown. Faced with choices, medical oncologists and patients will often decide between these drugs based on co-morbid medical issues.

Resistance eventually develops to these AR pathway agents. Splice variants of AR, including AR-V7, which lacks the ligand-binding domain of AR, have been described in prostate cancer cells of patients who develop resistance to enzalutamide. Activating mutations of AR or up-regulation of other compensatory pathways may someday be countered by the development of newer therapies or combinations of treatments that can circumvent development of such avenues of resistance.

Table 6.8. FDA-approved therapies for mCRPC since 2010.

Drug	Description	FDA approval	
		Indication for use	Date
Sipuleucel-T	Autologous cellular immunotherapy	Before or after docetaxel	4/29/2010
Cabazitaxel	Taxane chemotherapy	After docetaxel	6/17/2010
Abiraterone	Steroidogenesis inhibitor	After docetaxel	4/28/2011
		Before docetaxel	12/10/2012
Enzalutamide	Anti-androgen	After docetaxel	8/31/2012
		Before docetaxel	9/10/2014
Radium-223	Radiopharmaceutical	Before or after docetaxel	5/15/2013

Future prospects and challenges in metastatic prostate cancer

The advent of enzalutamide, abiraterone, and three other FDA-approved therapies for mCRPC since 2010 have ushered in a new era for prostate cancer therapy and promise for prostate cancer patients (Table 6.8). Sipuleucel-T is an **autologous** cellular immunotherapy that attempts to induce an effective immune response against the patient's prostate cancer and has demonstrated modest efficacy. Radium-223 dichloride is a **radiopharmaceutical** that employs radium to home in on areas of blastic bone metastases where its high-energy, short-range alpha particles kill nearby prostate cancer cells. Cabazitaxel is a taxane chemotherapy that may be less sensitive to cellular efflux via the multidrug resistance (MDR) P-glycoprotein pump compared with docetaxel. The optimal sequencing and potential combination of therapies may hold the key to improving disease response and overall survival. Whether other processes, pathways, or proteins will prove to be viable targets for prostate cancer therapy remains to be seen. For now, the road to improve overall survival in mCRPC is littered by previous unsuccessful therapies targeting endothelin-1, VEGF, VEGF-R, and MET, among many others.

Take-home points

✓ Prostate cancer is the most common cancer among men in the United States.

✓ Prostate cancer is the second-leading cause of cancer death among U.S. men.

✓ The use of serum PSA for prostate cancer screening is controversial.

✓ The majority of diagnosed prostate cancers are low grade.

✓ Appropriate management of localized prostate cancer could include active surveillance or active treatments including radical prostatectomy or radiation options including brachytherapy (seeds) or external-beam radiation therapy.

✓ Hormone therapy, or androgen deprivation therapy (ADT), has numerous side effects but a proven role for high-risk and some intermediate-risk localized prostate cancers.

✓ Prostate cancer generally metastasizes to lymph nodes and bones.

✓ ADT remains the standard initial therapy for metastatic disease.

✓ Metastatic cancer that is resistant to ADT is considered to be metastatic castration-resistant prostate cancer (mCRPC).

✓ The androgen receptor (AR) remains a critical target, even in mCRPC.

✓ Mechanisms of resistance to ADT include *AR* amplification, activating AR mutations, and up-regulation of non-gonadal production of androgens.

✓ Anti-androgens and steroidogenesis inhibition represent viable approaches to improve overall survival for men with mCRPC.

✓ Although optimal combinations and sequencing of the five therapies for mCRPC approved by the FDA since 2010 remain unclear, the successful establishment of these therapies has led to increased drug research and development for the management of mCRPC.

✓ Other molecular targets for prostate cancer beyond the AR pathway that can be inhibited and affect overall survival have not been established.

Discussion questions

1. Why did prostate cancer incidence rise in the 1990s without much change in the death rate?

2. What features would be ideal in a blood biomarker to make it an accurate screening tool for cancer? In what ways does PSA fall short?

3. How does medical castration work?

4. What are the mechanisms leading to resistance to ADT?

5. How might prostate cancers develop resistance to the newer AR pathway agents enzalutamide and abiraterone?

Chapter summary

Sex hormones are key drivers of cell proliferation in some hormonally sensitive reproductive tissues and therefore also of cancers of these tissues. It is thus not surprising that hormone manipulation therapy strategies provided the first examples of targeted therapies against specific growth pathways in cancer cells.

The success of hormonal manipulation in metastatic breast and prostate cancer has led to earlier implementation of these strategies in localized disease. There is an established role for tamoxifen (a selective ER modulator) and the aromatase inhibitors in the adjuvant therapy of ER-positive localized breast cancers. Likewise, androgen deprivation therapy (ADT) has an established role when incorporated with external-beam radiation therapy for men with high-risk and selected intermediate-risk localized prostate cancer. Whether the incorporation of the newer AR pathway agents may cause "enhanced" androgen suppression and improve clinical outcomes, such as lower recurrence rates, remains to be seen but is the subject of active clinical research. Certainly, improvements in therapy of localized disease that decrease metastatic cases would be an ideal use of these targeted approaches.

Chapter discussion questions

1. How do menstrual cycles, age at menarche, age at menopause, pregnancy, and age at first pregnancy influence breast cancer development?

2. Describe the potential opportunities for inhibition of steroid hormone pathways.

3. Name the hormone therapies currently available for breast and prostate cancers, and describe how resistance to each hormone therapy develops.

4. What toxicities are of particular consequence with steroidogenesis inhibitors that are not seen with the other hormone therapy strategies?

Selected references

Introduction

Siegel RL, Miller KD & Jemal A (2018) Cancer statistics, 2018. *CA Cancer J. Clin.* 68:epub ahead of print.

Case 6-1

Baselga J, Campone M, Piccart M et al. (2012) Everolimus in postmenopausal hormone-receptor-positive advanced breast cancer. *N. Engl. J. Med.* 366:520–529.

Early Breast Cancer Trialists' Collaborative Group (2005) Effects of chemotherapy and hormonal therapy for early breast cancer on recurrence and 15-year survival: an overview of the randomized trials. *Lancet* 365:1687–1717.

Fisher B, Anderson S, Redmond CK et al. (1995) Reanalysis and results after 12 years of follow-up in a randomized clinical trial comparing total mastectomy with lumpectomy with or without irradiation in the treatment of breast cancer. *N. Engl. J. Med.* 333:1456–1461.

Perou CM, Sørlie T, Eisen MB et al. (2000) Molecular portraits of human breast tumours. *Nature* 406:747–752.

Sørlie T, Perou CM, Tibshirani R et al. (2001) Gene expression patterns of breast carcinomas distinguish tumor subclasses with clinical implications. *Proc. Natl. Acad. Sci. USA* 98:10869–10874.

Sparano JA, Gray RJ, Makower DF et al. (2018) Adjuvant chemotherapy guided by a 21-gene expression assay in breast cancer. *N. Engl. J. Med.* 379:111–121.

Yamamoto N, Watanabe T, Katsumata N et al. (1998) Construction and validation of a practical prognostic index for patients with metastatic breast cancer. *J. Clin. Oncol.* 16:2401–2408.

Case 6-2

Antonarakis ES, Changxue L, Wang H et al. (2014) AR-V7 and resistance to enzalutamide and abiraterone in prostate cancer. *N. Engl. J. Med.* 371:1028–1038.

Beer TM, Armstrong AJ, Rathkopf DE et al. (2014) Enzalutamide in metastatic prostate cancer before chemotherapy. *N. Engl. J. Med.* 371:424–433.

The Cancer Genome Atlas Network (2015) The molecular taxonomy of primary prostate cancer. *Cell* 163:1011–1025.

de Bono JS, Logothetis CJ, Molina A et al. (2011) Abiraterone and increased survival in metastatic prostate cancer. *N. Engl. J. Med.* 364:1995–2005.

Huggins C, Stevens RE & Hodges CV (1941) Studies on prostatic cancer: II. The effects of castration on advanced carcinoma of the prostate gland. *Arch. Surg.* 43:209–223.

Ryan CJ, Smith MR, de Bono JS et al. (2013) Abiraterone in metastatic prostate cancer without previous chemotherapy. *N. Engl. J. Med.* 368:138–148.

Sakr WA, Haas GP, Cassin BF et al. (1993) The frequency of carcinoma and intraepithelial neoplasia of the prostate in young male patients. *J. Urol.* 150:379–385.

Scher HI, Fizazi K, Saad F et al. (2012) Increased survival with enzalutamide in prostate cancer after chemotherapy. *N. Engl. J. Med.* 367:1187–1197.

Tomlins SA, Rhodes DR, Perner S et al. (2005) Recurrent fusion of TMPRSS2 and ETS transcription factor genes in prostate cancer. *Science* 310:644–648.

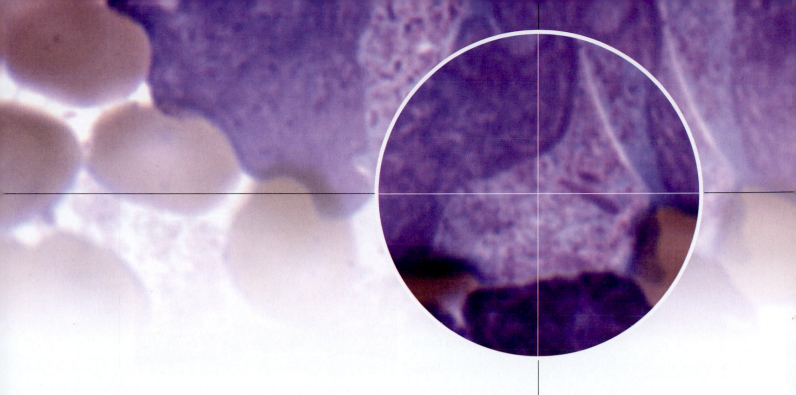

Chapter 7

Differentiation Therapy

Introduction

The normal development of tissues generally progresses from a pluripotent stem cell giving rise to daughter cells that acquire traits of maturation and eventually populate an organ with post-mitotic, mature, differentiated cells. Consequently, the bulk of tissue in a healthy organ is usually comprised of **differentiated** cells, which are often cells that no longer divide or divide only to maintain tissue turnover or repair, are programmed to express specific proteins that follow distinct cellular pathways for normal tissue formation and function, and are interconnected to allow coordinated function within an organ (Figure 7.1). There is evidence that some cancer cells share many features with stem cells—the undifferentiated cells whose progeny always include at least one identical stem cell (a process providing for self-renewal) and may produce a more differentiated cell as well. These cells are often called **cancer stem cells**. In fact, malignant cell **transformation** is often described as a **dedifferentiation** process, in which the cancer cells acquire traits of more primitive precursor cells (Figure 7.2 gives an example of pathologic grading of prostate cancer based on degree of dedifferentiation). Conceptually, then, cancer cells have reverted from a fully mature state or have failed to completely differentiate into fully mature cells and, thus, are arrested in a premature state of cell development. They have been described as having morphologic features of normal cellular precursors, expressing proteins, such as cell surface markers, that suggest a block (whether partial or complete) in the cell maturation process. Is it possible to overcome the differentiation blockade in cancer cells as a therapeutic strategy? The answer is yes—to an extent.

The concept of **differentiation therapy** is to compel cancer cells to differentiate. Therefore, the rational and deliberate application of the treatment requires (1) a detailed understanding of the key obstruction in the cell differentiation of the particular cancer, (2) discovery of inducible targets within the

Figure 7.1. Schematic diagram of normal differentiation of skin. The basal layer (stratum basale) is the source of skin progenitor cells, giving rise to maturing keratinocytes.

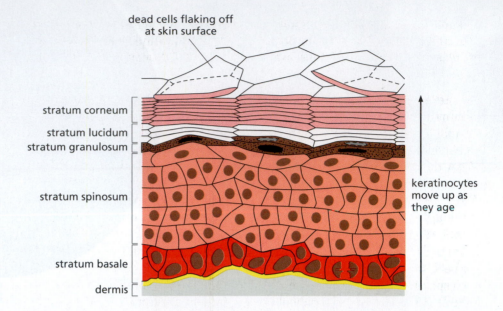

cells of the cancer that can be manipulated to overcome the key obstacle, and (3) an effective drug that can reach the target in cancer cells to induce differentiation with limited toxicity to other cells.

Although differentiation therapy to treat established solid cancers has not yet been completely realized, lessons can be gleaned from its use in the treatment of precancerous lesions. An example is in the treatment of **oral leukoplakia** (Figure 7.3), a clinical description for white lesions on the lining of the mouth (oral mucosa) that are thought to be premalignant lesions that can develop into invasive carcinomas; the reported rate of malignant progression from oral

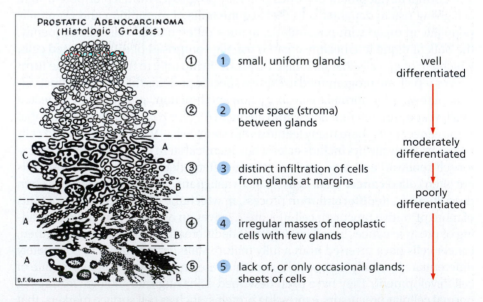

Figure 7.2. Gleason classification categorizes prostate cancer by the apparent degree of differentiation (or dedifferentiation) of the cells. The prostate gland is itself a collection of many microscopic glands consisting of specialized, or differentiated, cells. The Gleason grades 1–5 describe the degree to which cells within a tissue sample retain their glandular differentiation, with 5 representing the fullest dedifferentiation. Subpatterns A, B, or C for certain grades were described by Gleason. Gleason grading remains the key determinant of prognosis for prostate cancer patients today, and has not been unseated by molecular or genetic characterization of a patient's tumor. (From P.A. Humphrey, *Mod. Pathol.* 17:282–306, 2004. Reprinted by permission from Macmillan Publishers Ltd.)

leukoplakia to invasive carcinoma is up to 43%. Prevention of oral carcinomas could obviate the need for potentially disfiguring surgical procedures and decrease the mortality of a disease that causes 10,030 deaths annually in the United States.

For prevention of malignant transformation of premalignant oral lesions, the minimum requirements for differentiation therapy include high clinical efficacy, tolerability, and prolonged durability of response. Prevention strategies for oral leukoplakia employ **retinoids**, such as 13-*cis*-retinoic acid, a metabolite of vitamin A, or of beta-carotene, a precursor of vitamin A. Retinoids are associated with a 27 to 57% resolution of oral leukoplakia and a 45 to 90% partial response. However, continued therapy is critical, as 50% of cases recur upon discontinuation of therapy. Additionally, the toxicity of retinoid treatment, including skin rash, nosebleeds, **conjunctivitis**, oral **mucositis** (mouth sores), and **teratogenicity** (birth defects) limits the tolerability of even short-term therapy. Beta-carotene has been associated with a 52% response rate (4% complete resolution, 48% partial response) and 40% disease stability. Responses to six months of therapy were durable at one year in many patients, and treatment was not associated with significant toxicity.

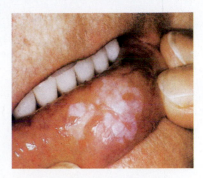

Figure 7.3. Oral leukoplakia. Leukoplakia presents as a white lesion on the inner lip of a patient. (From A.T. Skarin, Dana-Farber Cancer Institute Atlas of Diagnostic Oncology, 4th ed. Mosby Elsevier, 2010. With permission from Elsevier.)

Case 7-1	Acute Promyelocytic Leukemia

Introduction

Acute promyelocytic leukemia (APL) is a form of acute myeloid leukemia (AML), a malignancy of immature myeloid cells, which are a subtype of white blood cells. The proliferation of aberrant cell precursors in bone marrow, as is seen in AML, leads to insufficient production of healthy circulating cells in the blood. AML is markedly heterogeneous, with diverse morphologic, molecular, and chromosomal subtypes (Table 7.1; Figure 7.4). (In Table 7.1, APL is classified as AML-M3.) The molecular and chromosomal characteristics of AML often impact the risk and prognosis of disease, with some patients having relatively good-risk disease and others having poor-risk features and outcomes

Topics bearing on this case

Retinoic acid

Controls of differentiation

Retinoic acid receptor

Table 7.1. French, American, and British (FAB) classification of acute myeloid leukemia (AML) subtypes.

Classification	Affected cell type	Description
M0	Myeloblastic	Minimally differentiated AML
M1	Myeloblastic, with minimal maturation	Myeloblasts dominant leukemic cell in the marrow
M2	Myeloblastic, with maturation	Many myeloblasts present, but some cells developing toward fully formed blood cells
M3	Promyelocytic (acute promyelocytic leukemia)	Translocation between chromosomes 15 and 17
M4	Myelomonocytic (acute myelomonocytic leukemia)	Often associated with an inversion of chromosome 16
M5	Monocytic (acute monocytic leukemia)	Features of developing monocytes
M6	Erythroleukemic (acute erythroid leukemia)	Features of developing red cells
M7	Megakaryocytic (acute megakaryocytic leukemia)	Features of developing platelets

Figure 7.4. Karyotypic analysis revealing translocation of chromosomes 15 and 17 in a patient with APL.

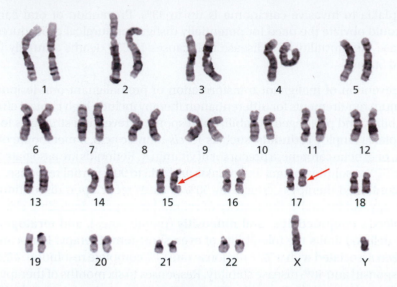

(Table 7.2). However, all cases of AML are highly aggressive and, without immediate diagnosis and successful treatment, are uniformly fatal.

APL is characterized by a translocation between the long arms of chromosomes 15 and 17 [t(15;17)]. This highly specific genetic alteration in APL was first described in 1977 and occurs in 98% of APL cases. The translocation involves the *PML* (promyelocytic leukemia) and the *RARA* (retinoid acid receptor α) genes, resulting in the formation of the abnormal fusion protein **PML–RARA**. (Other less common alterations have been detected in APL, usually involving alternative translocations of chromosome 17.) The aberrant PML–RARα gene product triggers a block in the normal differentiation of myeloid precursors and causes the subsequent accumulation of abnormal promyelocytes in the bone marrow. Based on this disease biology, a successful treatment strategy would be to restore successful cellular differentiation and maturation, thus causing the leukemic population of cells blocked at the promyelocyte stage of development to further differentiate into normal and fully functional mature leukocytes. Such a strategy is behind differentiation therapy.

RARα is a nuclear protein that normally heterodimerizes with retinoid X receptors (RXRs), a family of retinoid receptors that act as co-factors for other nuclear receptors, including vitamin D_3 and thyroid hormone receptors. The RARα–RXR complex binds to retinoic acid (RA) response elements (RAREs) of gene promoters, resulting in gene transcription. RA and other ligands for RAR are known as retinoids and are derived from vitamin A (retinol).

Table 7.2. Prognostic factors in acute myelogenous leukemia (AML).

Adverse prognostic features	Favorable prognostic features
Age over 65	Presence of an *NPM1* mutation (without concurrent *FLT3-ITD* mutation)
Presence of an *FLT3-ITD* mutation	Presence of a *CEBPA* mutation
Poor-risk cytogenetics [e.g., –7, –7q, –5, or complex (>3 chromosomal abnormalities)]	"Good" risk cytogenetics [e.g., t(15;17), inv(16), t(16;16), t(8;21)]
AML arising from preceding myelodysplasia or chronic myeloproliferative states	
Therapy-related disease (AML secondary to prior chemotherapy or radiation exposure)	
Presence of granulocytic sarcoma (extramedullary disease)	
Acute bilineal or biphenotypic leukemia	

(A)

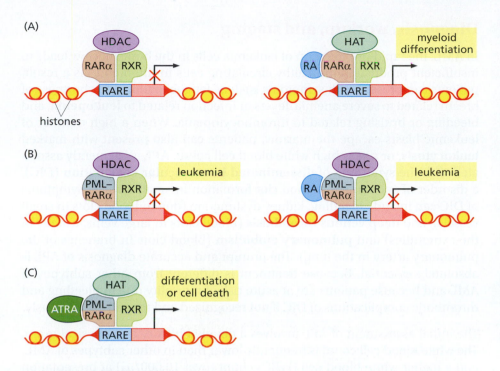

Figure 7.5. Mechanism of transformation by PML–RARα.
(A) Normally, in the absence of ligand (RA or ATRA), the RARα–RXR heterodimer binds to the RA response element (RARE) but transcription is inactivated by the binding of histone deacetylase (HDAC). Physiological doses of RA release HDAC from the RAR–RXRα complex and allow for the histone acetyltransferase (HAT) complex of transcriptional co-activators to bind, in turn allowing expression of target genes and myeloid differentiation. (B) In APL, PML–RARα binds RXR and HDAC but will not release HDAC at physiologic doses of RA, resulting in a block in transcription of RA-dependent differentiation genes and thus giving way to leukemia. (C) In APL, pharmacologic dosing of ATRA can induce HDAC dissociation and HAT/co-activator binding to trigger either differentiation or death of the leukemic blasts. (Modified from P. Salomoni and P.P. Pandolfi, *Nat. Med.* 6:742–744, 2000.)

In the absence of ligand, the RARα–RXR dimer binds to a histone deacetylase complex (HDAC) that epigenetically silences gene transcription programs. Binding of a ligand such as RA dissociates the co-repressor protein from RARα–RXR, allowing for recruitment of a co-activator protein complex that includes histone acetyltransferase (HAT), which can thus promote gene transcription (Figure 7.5A). The genes activated by retinoids are involved in proliferation arrest, cell cycle exit, prevention of apoptosis, and myeloid cell differentiation. In the nonmalignant state, RA promotes the differentiation of immature myeloid cells into neutrophils, and blocking RA *in vitro* slows down the differentiation of bone marrow stem cells.

The PML–RARα fusion protein includes the RARα ligand-binding domain and DNA-binding domain. The fusion of PML–RARα apparently enhances the recruitment of HDAC to the PML–RARα–RXR heterodimer. Target gene silencing cannot be relieved by physiologic concentrations of RA (see Figure 7.5B); however, the use of supraphysiologic levels of RA analogs induces dissociation of the HDAC silencing complex and can activate differentiation programs and result in killing of leukemia cells.

Epidemiology and etiology

APL comprises approximately 10% of AML in adults. Although the precise incidence of APL is unknown, it is certainly rare. The number of newly diagnosed cases per year in the United States is estimated to be 1000. Incidence is impacted by age, being very uncommon in children less than 10 years old, increasing steadily during the teen years, reaching a plateau during early adulthood, and then holding constant until decreasing after age 60. This is in contrast to other subtypes of AML, which are markedly more common among the elderly. With respect to the incidence of APL among ethnic groups, contradictory data regarding a presumed higher incidence among those from Central and South America and parts of Europe have been reported. No clear environmental, behavioral, or inherited genetic precipitants have been identified.

Diagnosis, workup, and staging

In AML, the rapid proliferation of leukemia cells in the bone marrow leads to insufficient production of healthy circulating cells in the blood. As a result, patients can present with symptoms such as fatigue, pallor, and shortness of breath related to severe anemia, fevers or infection related to **leukopenia**, and bleeding or bruising related to thrombocytopenia. When a high number of leukemic blasts escape the marrow, patients can also present with marked **leukocytosis**, or a very high white blood cell count. APL is frequently associated with the syndrome of **disseminated intravascular coagulation** (DIC), a disorder of spontaneous blood clot formation in small vessels. Symptoms of DIC can include bleeding, kidney dysfunction (due to blood clots in small vessels), and **deep venous thrombosis** (blood clots in large veins, usually in the extremities) and **pulmonary embolism** (blood clots in branches of the pulmonary artery in the lung). The prompt and accurate diagnosis of APL is absolutely essential, because treatment is different from other subtypes of AML and because patients are at acute risk of mortality due to bleeding and thrombotic complications of DIC if not recognized and treated expeditiously.

The initial assessment of APL involves a detailed history and physical exam. The white blood cell count is frequently lower than in other subtypes of AML, and a higher white blood cell (WBC) count (over $10,000/\mu l$) at presentation predicts for higher-risk disease. A suppressed platelet count below $40,000/\mu l$ at presentation has also been associated with an inferior prognosis in some studies (Table 7.3). Because of the frequent association of APL with DIC, initial laboratory evaluation should always include a complete blood count, **coagulation** tests (prothrombin time, or PT; and activated partial thromboplastin time, or aPTT—which are often prolonged due to consumption of clotting factors with DIC), and levels of **fibrinogen** (an essential protein for clotting, often low in patients with DIC due to consumption) and **D-dimer** (a protein fragment from a degraded blood clot that is often markedly elevated in DIC patients).

The diagnosis of AML is established upon finding leukemia cells in a bone marrow biopsy (see Table 7.1 for distinguishing characteristics among the AML subtypes). The characteristic morphology in APL includes the presence of abundant **promyelocytes** with intense azurophilic (blue) granules and **Auer rods**, which are needle-shaped cytoplasmic inclusion bodies first described in 1906 (Figure 7.6).

However, morphologic diagnosis can be challenging, especially with the less common microgranular variant, in which the granules are often significantly smaller and more subtle and, therefore, difficult to visualize. Nevertheless, the detection of the t(15;17) chromosomal translocation by cytogenetic evaluation, FISH (fluorescence *in situ* hybridization), or PCR (polymerase chain reaction) techniques establishes the diagnosis in almost all cases. Although karyotypic analysis can detect the classic chromosomal translocation, molecular diagnostics can confirm the presence of the *PML–RARA* gene product.

Table 7.3. Prognostication of APL, based on peripheral leukocyte and platelet count.

	Low risk	Intermediate risk	High risk
White blood cells	<10,000/μl	≤10,000/μl	>10,000/μl
Platelets	>40,000/μl	≤40,000/μl	

Adapted from M.A. Sanz et al., *Blood* 115:5137–5146, 2010.

In those patients who are found to be affected by DIC, prompt initiation of therapy is essential, as it helps to limit potentially life-threatening coagulopathy (blood clot formation). Indeed, once the diagnosis of APL is suspected based on clinical, laboratory, or morphological criteria, differentiation therapy should be urgently initiated. In cases of DIC with bleeding, it may also be necessary to support the patient with transfusions of blood products, including **cryoprecipitate** (which replaces fibrinogen), **plasma** (which replaces blood clotting factors), **platelets**, and red blood cells.

Initial treatment of APL

Data suggest that up to 17% of APL patients die before receiving therapy for their leukemia, most likely related to sequelae of DIC. Upon suspicion of APL, prompt initiation of differentiation therapy with **all-*trans* retinoic acid** (ATRA) is imperative. ATRA was first used to treat patients in Shanghai, China, in 1988, and with evidence of complete response, clinical trials expanded to France and then the rest of the world. Supraphysiologic levels of ATRA can induce dissociation of HDAC from the PML–RARα/RXR complex, allowing for HAT complexes containing transcriptional co-activators to bind, eventually leading to either differentiation of the leukemic cells or cell death (see Figure 7.5C).

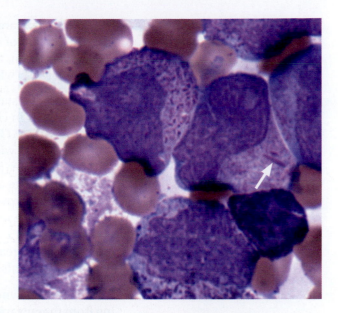

Figure 7.6. Abnormal promyelocytes of acute promyelocytic leukemia (APL), with prominent azurophilic granules seen within the cytoplasm. Also seen are Auer rods (*arrow*), needle-shaped inclusion bodies containing clumps of azurophilic granules.

Cytotoxic chemotherapy should not be started on presentation, because it can further worsen the coagulopathy. Instead, chemotherapy usually starts several days following initiation of ATRA. For those patients with high-risk disease (WBC over 10,000/μl), concurrent treatment with ATRA and hydroxyurea (a cytotoxic chemotherapy agent that inhibits DNA synthesis) can be used to limit the adverse clinical effects of **differentiation syndrome**, which is the rapid release of inflammatory cytokines from the malignant promyelocytes often triggered by ATRA. The initial phase of cytotoxic chemotherapy is known as **induction chemotherapy**, because the goal of this phase of treatment is to induce **remission**. This may then be followed with chemotherapy with the goal of further increasing the depth of remission, known as **consolidation therapy**, and then additional chemotherapy to maintain that remission, known as **maintenance therapy**.

Induction chemotherapy commonly includes an anthracycline chemotherapy agent, such as daunorubicin, as well as the drug cytarabine. ATRA is continued concurrently. Consolidation following induction chemotherapy often consists of a combination of ATRA with the chemotherapy agents anthracycline and **arsenic trioxide** (ATO), although, as before, consolidation approaches can vary. The role of prolonged maintenance therapy (with ATRA alone or in combination chemotherapy) is controversial.

ATO is an inorganic compound that was found to have potent antileukemic activity *in vitro* via inhibiting proliferation and directly inducing apoptosis. ATO appears to act by binding to the PML moiety of the PML–RARα protein, resulting in **polyubiquitination** and degradation of the **oncoprotein**. Clinical activity of ATO was first observed in China, and an initial report from Japan demonstrated complete remissions in 11 of 14 APL patients who had been refractory to ATRA.

The incorporation of ATO into therapeutic regimens for APL has led to recent improvements in outcomes. ATO is the most active single agent against APL and is thought to act by direct degradation of the PML–RARα protein, allowing transcription of target genes and normal differentiation. ATO is often included in the treatment of relapsed APL, in which case it is very effective. To achieve a second remission, a second regimen of ATO followed by autologous stem cell transplant (SCT) will cure approximately 60% of relapsed patients.

Initial studies established that elderly patients can be treated with ATRA alone or ATRA in combination with ATO, with impressive clinical outcomes. More recently, ATRA in combination with ATO was studied as initial therapy for patients with low-risk APL. This combination was compared with a traditional ATRA plus chemotherapy regimen, and ATRA/ATO produced an event-free rate and overall survival rate of 97.1% and 98.7%, respectively, significant improvements over the traditional ATRA-plus-chemotherapy approach. These results have led to a shift in the treatment of patients with non-high-risk APL. But the ATRA and ATO combination is not free of toxicity, including an increased frequency of differentiation syndrome and liver toxicity, which warrants careful monitoring of patients on this therapy.

Differentiation syndrome continues to be a unique clinical challenge resulting from therapy with ATRA and ATO and is related to the infiltration of organs with differentiated granulocytes once the differentiation arrest is lifted on the accumulated neoplastic promyelocytes. Clinically, this typically appears within the first few days of therapy. Manifestations include a rapid rise in the WBC count, fever, weight gain, and shortness of breath, with or without **pulmonary infiltrates** found on a chest X-ray. Differentiation syndrome can be treated with initiation of dexamethasone, a type of steroid. Severe cases may require temporary discontinuation of ATRA or initiation of cytotoxic chemotherapy with hydroxyurea.

A less common complication of ATRA therapy is **pseudotumor cerebri**, characterized by increased intracranial pressure, headaches, nausea, and visual changes. This complication is more common in children and is treated by discontinuation of ATRA and addition of **diuretics**, such as mannitol. Pseudotumor cerebri can, at times, be difficult to distinguish from severe headaches, a known and common toxicity associated with ATRA.

APL has the most favorable prognosis of all the acute leukemias in adults. More than 90% of patients will achieve complete molecular remission after induction and consolidation therapy, and these favorable outcomes are now being seen with non-chemotherapy-containing regimens as well. As with remission-inducing chemotherapy used for other types of AML, a midtreatment bone marrow biopsy, typically performed 10–14 days after start of treatment, is often performed to determine the need for further therapy. With APL, however, a midtreatment biopsy will often be difficult to interpret due to the process of differentiation. Therefore, unlike other AML subtypes, the bone marrow should be assessed 4–5 weeks post-chemotherapy, at the time of blood count recovery. The majority of patients will have molecular evidence of disease after induction therapy, but after consolidation therapy, molecular testing for *PML–RARA* by PCR should be negative.

The case of Mark Gable, a young man with bleeding gums following dental cleaning

Mark was a 20-year-old man who had been in his usual state of good health until approximately 10 days prior to presentation. At that time, he started to feel fatigued. His parents noted that he was asleep for most of the day. Soon thereafter, Mark noticed that his bedsheets were wet from sweating in the middle of the night. On the day prior to presentation, he had gone to the dentist for a regularly scheduled visit to have his teeth cleaned. This visit was uneventful, but soon thereafter Mark's gums were oozing blood. He went to bed, and awoke the next morning with significant bleeding from his mouth. He then presented to the hospital for evaluation.

On presentation to the emergency room, Mark was found to be weak, with low-grade fever, and with bleeding from his gums. His blood counts were markedly abnormal, with a WBC of 34,000/μl (normal is 4000–11,000/μl), hemoglobin of 8.5 g/dl (normal, 12–16 g/dl), and a platelet count of 12,000/μl (normal, 150,000–400,000/μl). Given these abnormal values, the hematology–oncology team was urgently consulted, and they noted a circulating population of large, abnormal-appearing cells with azurophilic granules and **bilobed nuclei** on the peripheral blood smear. Mark was also markedly **coagulopathic**, with an elevated prothrombin time (PT) and a very low fibrinogen level. These laboratory abnormalities suggested DIC. Blood product support with plasma, cryoprecipitate, and platelets was immediately administered. A bone marrow biopsy was performed, and samples were sent for hematopathologic and cytogenetic evaluation. Given the range of symptoms, presence of DIC, and appearance of the blood smear, APL was suspected and he was urgently initiated on ATRA.

Mark was admitted to the hospital, and he was closely monitored for DIC. Slowly, his fibrinogen and coagulation levels stabilized, and the oozing of blood from his mouth ceased. After 48 hours, a translocation of chromosomes 15 and 17 was detected on molecular cytogenetic analysis, and the diagnosis of APL was confirmed. Induction chemotherapy was initiated on day 4. It was noted that his WBC was steadily climbing to 40,000/μl and then to 55,000/μl on ATRA, and over the course of the next week, he developed worsening **dyspnea** (shortness of breath), requiring oxygen. This was consistent with differentiation syndrome, which complicates the initial treatment in approximately one-quarter of APL patients. Mark was treated with intravenous dexamethasone (a steroid), which led to an improvement in his hypoxia (insufficient oxygenation) and respiratory symptoms. He went on to tolerate induction chemotherapy with the chemotherapy drugs daunorubicin and cytarabine without other complications.

Approximately 30 days later, a bone marrow biopsy was performed that revealed a morphologic remission; however, *PML–RARA* was still detectable by PCR. Subsequently, Mark went on to receive four cycles of consolidation chemotherapy, after which a repeat bone marrow biopsy no longer revealed the *PML–RARA* PCR product, suggestive of a molecular remission. He was placed on prolonged maintenance therapy with ATRA, which he tolerated well, apart from intermittent mild headaches. He remains in remission and free of disease three years later.

Management of relapsed APL

Fortunately, the rate of APL relapse with modern conventional therapy is very low, but it commonly occurs in those with high-risk APL. Conversion from PCR-negative to PCR-positive for the *PML–RARA* transcript is a harbinger of **relapse**, for which therapy is again indicated. The most commonly used and effective approach for relapsed APL is ATO, typically without concurrent use of ATRA. Therapy for relapsed disease, although not as successful as initial treatment, does lead to disease remission in the majority of patients, and over 80% of patients treated for APL will be long-term survivors.

Discussion questions

1. What is the underlying chromosomal abnormality that defines APL?

2. What are the most common complications of treatment with ATRA?

3. What are acceptable therapeutic paradigms for the initial management of APL?

4. Why might differentiation therapy as a treatment strategy be difficult to generalize to other cancers?

Chapter summary

The success of differentiation therapy as a treatment for acute promyelocytic leukemia (APL) is a compelling argument for further exploitation of differentiation therapy across the spectrum of clinical oncology. However, we have not seen success using differentiation therapy in other cancers. Why, then, has this approach not found more widespread application? The uniform differentiation response of APL cells, probably including the cancer stem cell population, leads to a high rate of complete molecular remission and cure. By contrast, for most solid tumors, it is thought that therapies that destroy the bulk of the tumor cells but fail to eliminate cancer stem cell populations eventually lead to relapse due to a replenishing of cancer cells from this less-differentiated reservoir of cancer cells. Thus, the heterogeneity of cells found in most human cancers may thwart a uniform response to differentiation therapy.

Unfortunately, our vast fund of knowledge defining important molecular pathways leading to tumorigenesis has led to few insights into effective differentiation therapy across the oncologic spectrum; nevertheless, differentiation therapy remains a treatment strategy worthy of research. A key feature of differentiation therapy for APL was the identification of the specific block in maturation of cells within that particular cancer and the availability of a drug that can induce the cancer cells to overcome the differentiation obstacle. The application of differentiation therapy to other malignancies would require (1) a detailed understanding of the key obstruction in differentiation, (2) an element that can be recruited to overcome the obstacle (in the case of APL, that is restoration of wild-type RARα–RXR activity by ATRA to induce normal transcription of downstream genes), and (3) an effective drug that can induce differentiation with limited toxicity to other tissues. In optimistic terms, APL represents a perfect confluence of these elements. Less optimistically, this approach may not be broadly applicable to other diseases.

Chapter discussion questions

1. What are the key features for successful differentiation therapy?

2. Why is the success of differentiation therapy limited at present to APL?

3. Might differentiation therapy be a useful strategy for the treatment of cancers other than APL—that is, could it be used for premalignant lesions to prevent the development of invasive solid cancer and the associated morbidity and mortality? What might be the limitations of this approach?

Selected references

Introduction

Garewal HS, Katz RV, Meyskens F et al. (1999) β-carotene produces sustained remissions in patients with oral leukoplakia: results of a multicenter prospective trial. *Arch. Otolaryngol. Head Neck Surg.* 125:1305–1310.

Garewal HS, Meyskens FL Jr, Killen D et al. (1990) Response of oral leukoplakia to beta-carotene. *J. Clin. Oncol.* 8:1715–1720.

Gorksy M & Epstein JB. (2002) The effect of retinoids on premalignant oral lesions: focus on topical therapy. *Cancer* 95:1258–1264.

Humphrey PA. (2004) Gleason grading and prognostic factors in carcinoma of the prostate. *Mod. Pathol.* 17:282–306.

Siegel RL, Miller KD & Jemal A (2018) Cancer statistics, 2018. *CA Cancer J. Clin.* 68:7–30.

Case 7-1

Altucci L, Leibowitz MD, Ogilvie et al. (2007) RAR and RXR modulation in cancer and metabolic disease. *Nat. Rev. Drug Discov.* 6:793–810.

Avvisati G, Lo-Coco F, Paoloni FP et al. (2011) AIDA 0493 protocol for newly diagnosed acute promyelocytic leukemia: very long-term results and role of maintenance. *Blood* 117:4716–4725.

Lo-Coco F, Avvisant G, Vignetti M et al. (2013) Retinoic acid and arsenic trioxide for acute promyelocytic leukemia. *N. Engl. J. Med.* 369:111–121.

Lowenberg B, Downing JR, Burnett A et al. (1999) Acute myeloid leukemia. *N. Engl. J. Med.* 341:1051–1062.

Park JH, Qiao B, Panageas KS et al. (2011) Early death rate in acute promyelocytic leukemia remains high despite all-trans retinoic acid. *Blood* 118:1248–1254.

Powell BL, Moser BK, Stock W et al. (2011) Adding mercaptopurine and methotrexate to alternate week ATRA maintenance therapy does not improve the outcome for adults with acute promyelocytic leukemia (APL) in first remission: results from North American leukemia intergroup trial C9710. ASH Annual Meeting Abstract. *Blood* 118:258.

Raelson JV, Nervi C, Rosenauer A et al. (1996) The PMR/RARα oncoprotein is a direct molecular target of retinoic acid in acute promyelocytic leukemia cells. *Blood* 88:2826–2832.

Rowley JD, Golomb HM & Dougherty C (1977) 15/17 translocation, a consistent chromosomal change in acute promyelocytic leukaemia. *Lancet* 1:549–550.

Sanz MA, Martin G, González M et al. (2004) Risk-adapted treatment of acute promyelocytic leukemia with all-trans-retinoic acid and anthracycline monotherapy: a multicenter study by the PETHEMA group. *Blood* 103:1237–1243.

Sanz MA, Tallman MS & Lo-Coco F (2005) Tricks of the trade for the appropriate management of newly diagnosed acute promyelocytic leukemia. *Blood* 105:3019–3025.

Tallman MS & Altman JK. (2009) How I treat acute promyelocytic leukemia. *Blood* 114:5126–5135.

Tallman MS, Nabhan C, Feusner JH & Rowe JM (2002) Acute promyelocytic leukemia: evolving therapeutic strategies. *Blood* 99:759–767.

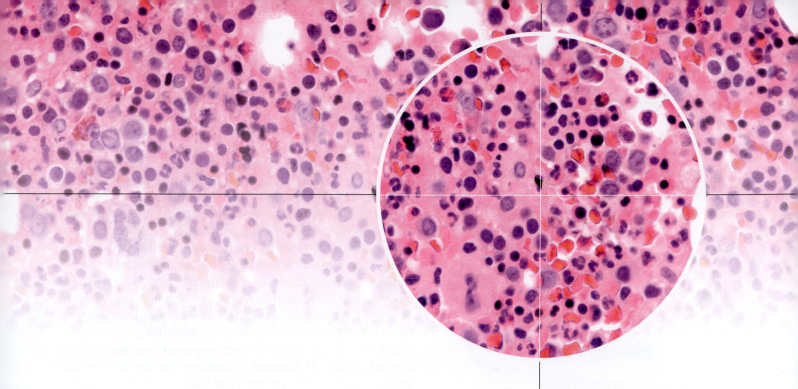

Chapter 8

Epigenetic Therapy

Introduction

Gene expression is regulated by multiple mechanisms. Among these, transcription has been extensively studied and involves a complex interplay of transcription factors, RNA polymerases, a vast repertoire of transcriptional co-activators and co-repressors, and key regulatory DNA elements, such as promoters and enhancers. However, studies of cancers demonstrate that rates of transcription frequently do not correlate with levels of protein product, suggesting an even more complex dynamic than the mRNA transcription that controls protein expression. Protein expression levels are regulated by mRNA transcription, mRNA degradation, mRNA translation, and protein degradation. Mutations in the DNA sequence and conformational alterations in the DNA "architecture" can impact mRNA transcription. **Epigenetics** refers to the study of changes in gene expression that are driven by external modifications rather than alteration of the DNA sequence. Broadly, there are four classes of epigenetic modifications that impact gene transcription, including **histone** modification, DNA **methylation**, microRNA (miRNA) targeting, and **long noncoding RNA** (lncRNA) activity (Figure 8.1). The study of epigenetics has revealed specific alterations conferring a selective growth advantage that occur in cancer cells but not in normal cells.

Histone modification

DNA is packaged into **chromatin**, a complex of DNA, RNA, and proteins. DNA is wound around **nucleosomes**, which are octamers (complexes of eight proteins) composed of two copies each of four histone proteins (H2A, H2B, H3, and H4; Figure 8.2). A fifth histone, H1, may be associated with the nucleosome octamer. The electrostatic force between the positive charges of lysine

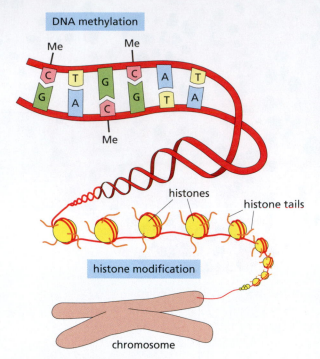

DNA methylation

histone modification

chromosome

Figure 8.1. Epigenetic targets for therapy. DNA methylation and histone modification offer two potential targets for therapy. Note that these modifications do not alter the DNA sequence but impact gene expression. In DNA methylation, methyl groups added to cytosines near gene promoters can repress gene transcription. Histone modifications alter the binding of DNA to histone proteins, permitting or preventing access of transcriptional machinery to DNA. (From I. Qiu, *Nature* 441:143–145, 2006. Reprinted by permission from Macmillan Publishers Ltd.)

residues from the histones and the negative charges from the phosphate backbone of DNA binds these constituents into compact units of chromatin. Sixty histone-modifying enzymes have been discovered and more may emerge with further research. Histone modifications can include **acetylation**, methylation, and **ubiquitination** of lysine residues, and phosphorylation of serine residues. Histone modifications may alter the electrostatic force and thus the adherence of the histone core to DNA, allowing or preventing access to DNA by components of the transcription machinery. Histone acetylation is generally associated with active gene *expression*, whereas histone methylation is generally associated with gene *repression*, although methylation of certain lysine residues correlates with gene activation. Histone ubiquitination (addition of **ubiquitin** groups to proteins) targets the histone for **proteasomal** degradation (the normal process of disposal of proteins that can also be targeted for cancer therapy; see Case 9-2).

Degrees of histone acetylation on specific lysine residues is a result of a balance between **histone deacetylases** (HDACs) and **histone acetyltransferases** (HATs; Figure 8.3A). Acetylation of histones neutralizes the positive charge of the lysine residues, decreasing the electrostatic force between the histone octamer and the DNA, thereby opening the chromatin structure and allowing exposure of transcription factors and RNA polymerase to DNA binding sites that were previously inaccessible. Cancers in general exhibit aberrant histone modification compared with normal cells, with some cancers demonstrating increased HDAC over HAT activity.

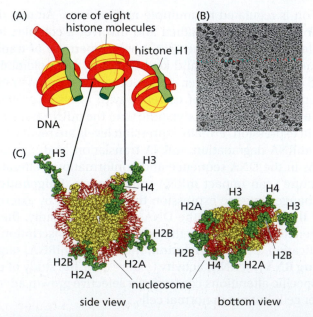

Figure 8.2. Organization of chromatin. (A) DNA is wrapped around nucleosomes, comprised of the octamer of histone proteins, whose positively charged lysine residues exert electrostatic attraction forces with the negatively charged phosphate backbone of DNA, compacting the chromatin. The histone H1 also interacts with the DNA and octamer complex. Chromatin can thus take on a "beads on a string" appearance seen in the micrograph (B). (C) The DNA double helix (*red*) wraps around the four histone molecules H2A, H2B, H3, and H4. (From R.A. Weinberg, The Biology of Cancer, 2nd ed. New York: Garland Science, 2014. Upper schematic adapted from W.K. Purves et al. (1998) Life: The Science of Biology, 5th ed. Sinauer. Micrograph from F. Thoma, T. Koller, and A. Klug, *J. Cell Biol.* 83:403–427, 1979. With permission from Rockefeller University Press.)

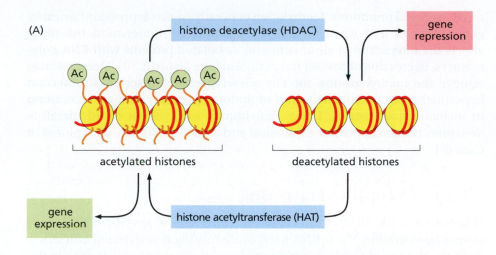

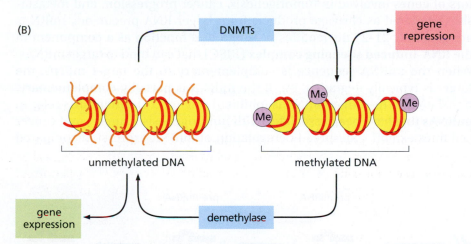

Figure 8.3. Impact of epigenetic modifications. (A) Acetylation of lysine residues on histones neutralizes their electrostatic attraction to DNA, allowing access of transcriptional machinery and expression of genes. Acetylation is mediated by histone acetyltransferase (HAT), while histone deacetylase (HDAC) removes acetyl groups, compacting the chromatin and repressing gene expression. (B) Methylation of certain cytosines in CpG dinucleotides found near gene promoters causes gene repression. Methylation is mediated by DNA methyltransferases (DNMTs). Demethylases remove the methyl groups, thereby allowing gene transcription. (From H.-W. Yao and J. Li, *J. Pharm. Exp. Ther.* 352:2–13, 2015. With permission from the American Society of Pharmacology and Experimental Therapeutics.)

HDAC inhibitors promote acetylation of histones, allowing for transcription of genes that might normally be repressed, including genes controlling differentiation, apoptosis, and cell cycle arrest. In this way, HDAC inhibitors could be effective cancer therapies, as described in Case 8-2. HDAC inhibitors may impact other cellular functions that lead to apoptosis, for example, by increasing the production of **reactive oxygen species** leading to cysteine-aspartic protease (**caspase**) activation; by acetylation of nonhistone proteins, such as p53 and DNA repair proteins; and by induction of apoptosis-related proteins, such as tumor necrosis factor-related apoptosis-inducing ligand (TRAIL) and Fas.

DNA methylation

DNA methylation is a covalent modification of DNA. DNA methyltransferases (DNMTs) attach methyl groups to the cytosine bases of CpG sequences (dinucleotides of cytosine followed by guanine) that frequently occur near transcriptional promoters. Methylation generally causes repression of nearby genes through a mechanism that is incompletely understood but likely affects the interaction of chromatin and associated proteins with RNA polymerases, decreasing access for transcription (see Figure 8.3B). Demethylases remove the methyl groups, thereby allowing gene transcription. Aberrant hypermethylation and inactivation of tumor suppressor genes are common in myeloid malignancies. Drugs that impact methylation could therefore *de-repress* tumor suppressor expression and lead to cell death, as explored in Case 8-1.

MicroRNA (miRNA) targeting

Discovered in the 1990s, miRNAs affect translation of specific mRNA transcripts by degrading the mRNA itself or disrupting translational efficiency, or both. The list of miRNAs continues to grow, some of which are regulators of genes involved in tumorigenesis, cancer progression, and metastasis. Generated as cleavage products from larger RNA precursors, miRNAs are generally 21–25 nucleotides in length and function as a component of the **RNA-induced silencing complex** (RISC) that can bind to target mRNAs. When the miRNA sequence is complementary to the target mRNA, the target is typically degraded, but if the miRNA is partially complementary, translation of the target mRNA is inhibited (Figure 8.4). **OncomiRs** refer to miRNAs that have been associated with tumorigenesis. Cancers and cancer cell lines exhibit widespread deregulation of miRNA expression compared

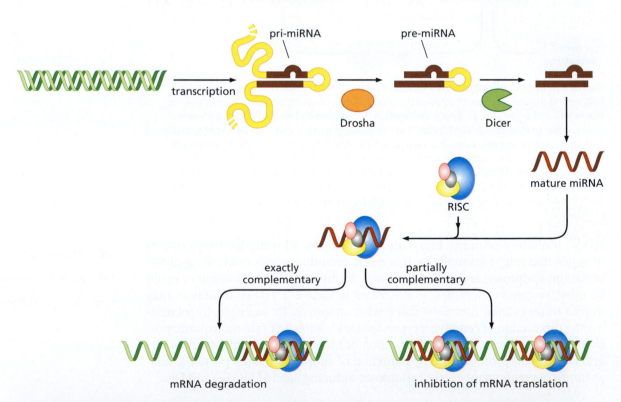

Figure 8.4. MicroRNAs and gene regulation. MicroRNAs are cleaved from the precursor pri-miRNA transcript with an enzyme complex containing the protein Drosha. The resulting product, the pre-miRNA, folds on itself into a double-strand RNA hairpin via self-complementary nucleotide sequences. A second processing step involving the Dicer enzyme generates the mature miRNA. The miRNA binds to the multiprotein complex termed RISC (RNA-induced silencing complex), which can associate with cytoplasmic mRNAs. If the miRNA is fully complementary to a sequence on the mRNA, the mRNA target is degraded. If the miRNA has only partial sequence complementarity, translation of the mRNA is inhibited. (From R.A. Weinberg, The Biology of Cancer, 2nd ed. New York: Garland Science, 2014.)

with normal cells. The precise roles of miRNAs in malignant transformation and the use of miRNAs as biomarkers that predict aggressiveness of disease are areas of active exploration. Targeting miRNAs is a rational but as yet unrealized therapeutic avenue.

Long noncoding RNAs (lncRNAs)

Distinct from the short miRNAs, the lncRNAs are typically >200 nucleotides in length. The lncRNAs are found in both the nucleus and the cytoplasm, but have no protein-coding sequences. The function of lncRNAs is not well understood but they may be associated with proteins that regulate transcription. The ongoing discovery of lncRNAs and association of some lncRNAs with metastasis in human cancers indicate that our understanding of the complexity of gene regulation remains limited. To date, given the unclear function of lncRNAs, targeting them as an epigenetic therapy for cancer is quite premature.

Take-home points

✓ Epigenetic modifications impact gene expression without altering the DNA sequence of the gene.

✓ Epigenetic modifications include histone modification, DNA methylation, microRNA (miRNA) targeting, and long noncoding RNA (lncRNA) activity.

✓ Cancers often exhibit transcriptional silencing of tumor suppressors and other genes that could lead to cell death.

✓ Targeting epigenetic activities may be an effective therapy for cancer patients.

✓ Targeting miRNAs is not yet a clinical reality.

✓ Further study of lncRNAs may yield a very different understanding of regulation of gene expression and the roles of the extensive regions of the genome that do not code proteins directly.

Case 8-1	Myelodysplastic Syndromes

Topics bearing on this case

DNA methylation

Cell dysplasia

Immunomodulatory agents

Introduction and epidemiology

Myelodysplastic syndromes (MDSs) are clonal disorders impacting myeloid **progenitor cells** in the bone marrow compartment, leading to aberrant hematopoietic (blood cell) differentiation and altered function of mature blood cells. Patients often present with various degrees of cytopenias (reduction in number of blood cells), including anemia (reduced number of red blood cells), thrombocytopenia (of platelets), and neutropenia (of neutrophils, a type of white blood cell), with the attendant clinical consequences **hypoxia** (oxygen deficiency), bleeding, and infection, respectively. Myelodysplastic syndromes are heterogeneous disorders with highly variable clinical courses, depending on underlying molecular and chromosomal lesions that drive their genesis and evolution. The majority of cases are pre-leukemic and will progress to acute myeloid leukemia (AML) over time, although the rapidity of that progression will depend on underlying

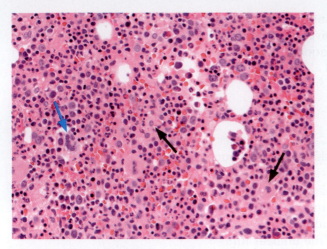

Figure 8.5. **Bone marrow core biopsy in a case of myelodysplastic syndrome.** The marrow is hypercellular for the patient's age, with increased numbers of marrow elements (hematoxylin and eosin stain). Abnormal maturation is evident in the megakaryocytic lineage (which produces platelets), as many megakaryocytes contain simplified, monolobated nuclei (*black arrows*), in contrast to more normal-appearing megakaryocytes, which contain complex, hyperlobated (multilobed) nuclei (*blue arrow*).

morphologic, cytogenetic, and molecular features of the MDS type. Most patients will die from MDS, either as a result of progression to AML or from the clinical impact of bone marrow failure.

Unlike cancers that are characterized by specific, recurrent genetic mutations, MDS is characterized by frequent epigenetic abnormalities, suggesting an important role for dysregulation of gene expression by epigenetic alteration in the pathogenesis of MDS. Aberrant DNA methylation is the dominant epigenetic finding in MDS and is associated with poor prognosis. The epigenetic silencing of genes involved in the regulation of cell division, DNA repair, and apoptosis, among others, may play key roles in the progression from MDS to leukemia.

The exact incidence of MDS is challenging to estimate, given that many patients with lower-risk disease and mild cytopenias may not be diagnosed. Nevertheless, the incidence of MDS increases with age and is significantly higher in older populations, estimated at 20 per 100,000. Age is indeed the dominant risk factor for disease. However, other risks associated with the development of MDS are exposure to radiation and alkylating agents. An association with certain agricultural solvents and chemicals and tobacco has also been reported.

Diagnosis and workup

Bone marrow evaluation is essential for the diagnosis of MDS. The bone marrow compartment in patients with MDS is usually **hypercellular** (contains an excess number of cells), but uncommonly, the bone marrow may be hypocellular depending on the type and stage of disease. Hematopathologists assess the bone marrow for evidence of **dysplasia** (abnormal appearance under the microscope, but not cancerous) in one or more cell lineages, as well as for the presence of myeloblasts (blood stem cells that develop into white blood cells). A typical bone marrow biopsy in MDS is shown in Figure 8.5. A clue to the presence of MDS is the presence of unusual appearing bilobed hypogranular neutrophils in the peripheral blood, known as pseudo-Pelger–Huët cells (Figure 8.6).

Various classifications of MDS have been developed over the past three decades. The French American British (FAB) classification organized MDS into five subtypes, namely, refractory anemia (RA), refractory anemia with ringed

Figure 8.6. **Differences between normal and dysplastic neutrophils.** (A) Normal neutrophil, characterized by segmented or multilobed nuclei and light pink cytoplasmic granules. Normal red blood cells surround the neutrophil. (B) In contrast, dysplastic neutrophils in patients with myelodysplastic syndrome (also known as pseudo-Pelger–Huët cells) are characterized by bilobed (*red arrow*) or unilobed (*green and blue arrows*) nuclei, highly condensed chromatin, and, in some cases, loss of normal cytoplasmic granulation (best depicted with the blue arrow). Note that all three of these features can be seen in a blood smear from the same patient.

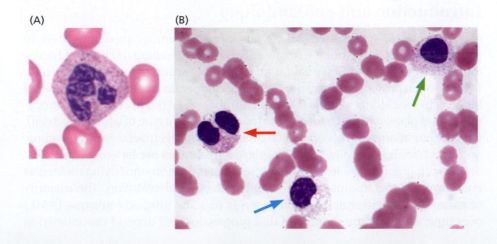

(A)   (B)

sideroblasts (RARS; atypical precursors to red blood cells with iron granules), refractory anemia with excess blasts (RAEB), refractory anemia with excess blasts in transformation (RAEB-T), and chronic myelomonocytic leukemia (CMML). This classification system was predominantly based on the presence of ringed sideroblasts on microscopy and the extent of involvement of the marrow and peripheral blood with immature myeloblasts, ranging up to 30%. Marrow blast involvement of greater than 30% would suggest progression to acute myeloid leukemia (AML). Since the FAB system was developed, the classification of MDS has further evolved, with the more recent World Health Organization (WHO) guidelines establishing two RAEB categories (based on marrow myeloblast percentage), decreasing the marrow threshold of AML to 20% myeloblasts, assessing the extent of marrow dysplasia across multiple cell lineages, and removing CMML from the classification system (Table 8.1). Additionally, the WHO classification specifies an uncommon category of MDS that exhibits a clonal cytogenetic abnormality, deletion of chromosome 5q (del5q).

Table 8.1. World Health Organization (WHO) classification of myelodysplastic syndromes.

Disease	Blood findings	Bone marrow findings
Refractory cytopenia with unilineage dysplasia [refractory anemia (RA), refractory neutropenia (RN), or refractory thrombocytopenia (RT)]	Unicytopenia or bicytopenia <1% blasts	Unilineage dysplasia in ≥10% of the cells in one myeloid lineage <5% blasts <15% of erythroid precursors are ringed sideroblasts
Refractory anemia with ringed sideroblasts (RARS)	Anemia <1% blasts	≥15% of erythroid precursors are ringed sideroblasts Erythroid dysplasia only <5% blasts
Refractory cytopenia with multilineage dysplasia (RCMD)	Cytopenia(s) <1% blasts No Auer rods $<1 \times 10^9$/l monocytes	Dysplasia in ≥10% of the cells in 2 myeloid lineages <5% blasts No Auer rods
Refractory anemia with excess blasts-1 (RAEB-1)	Cytopenia(s) <5% Blasts No Auer rods $<1 \times 10^9$/L monocytes	Unilineage or multilineage dysplasia 5–9% Blasts No Auer rods
Refractory anemia with excess blasts-2 (RAEB-2)	Cytopenia(s) 5–19% blasts +/– Auer rods $<1 \times 10^9$/l monocytes	Unilineage or multilineage dysplasia 10–19% blasts +/– Auer rods
Myelodysplastic syndrome with isolated del(5q)	Anemia Usually normal or increased platelet count <1% blasts	Normal to increased megakaryocytes with hypolobated nuclei <5% blasts Isolated del(5q) genetic abnormality No Auer rods
Myelodysplastic syndrome, unclassified (MDS-U)	Cytopenias <1% blasts	Dysplasia in less than 10% of cells in one or more myeloid cell lines when accompanied by a typical cytogenetic abnormality in MDS <5% blasts

Adapted from S.H. Swerdlow et al. (eds.), WHO Classification of Tumours of Haematopoietic and Lymphoid Tissues, 4th ed. Lyon: IARC Press, 2008.

Figure 8.7. Karyotypic analysis revealing monosomy 7 in a patient with MDS. The arrow highlights the missing seventh chromosome.

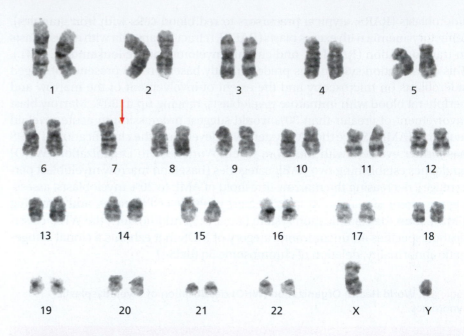

Karyotypic analysis is an important adjunct in the diagnostic evaluation of MDS. Chromosomal abnormalities are a characteristic feature of many patients with myelodysplastic syndromes, and certain clonal anomalies are associated with a particularly poor prognosis, such as "complex karyotypes" (more than 3 abnormalities), deletion of chromosome 5 (monosomy 5), and deletion of chromosome 7 (monosomy 7, shown in Figure 8.7).

Prognosis

The prognostication of MDS has also evolved in recent decades. The most commonly used tool is the International Prognostic Scoring System (IPSS) developed by Peter L. Greenberg and colleagues. Using three criteria with a weighted scoring system, four risk groups (low, intermediate-1, intermediate-2, and high) were established. These criteria included the number of cytopenias, the percentage of marrow involvement with myeloblasts, and cytogenetic risk, with the latter two criteria carrying greater weight in establishing prognosis, regarding leukemic evolution and overall survival (OS). As an example, median OS for those with low-risk MDS is calculated to be 5.7 years, versus that of high-risk MDS at 0.4 years. In 2012, the IPSS was revised as the IPSS-R, with further delineation of specific cytopenias and refinement of cytogenetic criteria. The IPSS-R has five risk-group categories (very low, low, intermediate, high, and very high; Table 8.2). A specific category of MDS, not included in the derivation of the IPSS scoring systems, is therapy-related MDS. Patients with therapy-related MDS develop myelodysplasia after prior exposure to alkylating agents or radiation and are considered to have very poor-risk disease, because their disease frequently displays very poor-risk cytogenetics.

The case of Hank Johnson, an older male with dyspnea on exertion and progressive cytopenias

Hank was a 78-year-old gentleman at the time he presented with shortness of breath. He had a history of coronary artery disease that had required bypass surgery more than 10 years earlier. He had only intermittently followed up with his primary care physician for clinical assessment in the years in between. He had been independent at home, where he worked on his cars and took care of

Table 8.2. Revised International Prognostic Scoring System (IPSS-R) in myelodysplastic syndromes.

Variable	Score							
	0	0.5	1.0	1.5	2.0	3.0	4.0	
Cytogenetics	Very good		Good		Intermediate	Poor	Very poor	
Marrow blasts	≤2%		>2% to <5%		5–10%	>10%		
Hemoglobin (g/dl)	≥10		8 to <10	<8				
Platelets (1000 s/μl)	≥100	50 to <100	<50					
Neutrophils (1000 s/μl)	≥0.8	<0.8						

Risk category	Risk score
Very low	≤1.5
Low	>1.5 to 3
Intermediate	>3 to 4.5
High	>4.5 to 6
Very high	>6

Note: Scores for each variable are added together to determine the risk score and risk category.

Adapted from P.L. Greenberg et al., *Blood* 120:2454–2465, 2012. With permission from the American Society of Hemotology.

his elderly wife. However, he noticed that he was increasingly short of breath on any exertion, with increased bruising on his arms and legs. Given his concern that this might be related to his heart disease, he presented to his primary care physician, who ordered a complete blood count evaluation and detected profound pancytopenia (cytopenias in all three cell types: white blood cells, red blood cells, and platelets).

Hank was referred to a hematologist, who found his WBC to be 1000/μl (normal range, 4500–11,000/μl), his hemoglobin level (HGB) to be 7.5 g/dl (normal range, 13.5–17.5 g/dl), and his platelet count to be 15,000/μl (normal range, 150,000–400,000/μl). His absolute neutrophil count was also suppressed, at 350/μl (normal range, 1800–7700/μl). Given his respiratory symptoms, which had become severe at home, he was admitted to the hospital. He underwent a bone marrow biopsy to evaluate his pancytopenia, which revealed dysplasia involving all cell lineages, as well as an increase in myeloblasts to 12% (normal is less than 5%). Chromosomal analysis revealed a deletion of chromosome 7, which is known to be a high-risk karyotypic feature. Based on IPSS-R criteria, Hank was diagnosed with very high-risk MDS.

After transfusion of red cells and platelets, Hank experienced clinical improvement, including improvement of his shortness of breath and decreased bruising, and was discharged. Unfortunately, given his age, he was not a candidate for stem cell transplantation, and after a discussion of therapeutic options, he was initiated on the **hypomethylating agent 5-azacitidine** on an outpatient basis. He tolerated this therapy quite well, and after approximately six weeks, his cytopenias improved and he gradually became transfusion-independent. Concurrently, his energy, appetite, and breathing also improved, and he returned to his normal routine. A bone marrow biopsy performed after three cycles (three months of therapy) revealed a lesser degree of dysplasia and a decrease in myeloblast involvement to 2%. Hank went on to receive ten additional cycles of 5-azacitidine, with stable peripheral blood counts, good clinical status, and continued therapeutic response.

Unfortunately, after more than a year of treatment with 5-azacitidine, Hank developed gradual and progressive cytopenias, and disease progression was suspected. A bone marrow biopsy was performed, revealing the return of marked dysplasia and an increase in myeloblasts to 14%. The decision was made to proceed with a second-line therapy, and **lenalidomide** was started at 10 mg daily. He tolerated this treatment well, but his cytopenias persisted, he became transfusion dependent, and his functional status once again deteriorated. After six weeks of treatment with lenalidomide, myeloblasts appeared in the peripheral blood and quickly increased in proportion and number. Given the progression to leukemia, he was taken off of lenalidomide. He was referred for hospice care, and ultimately died from progressive disease 18 months after his initial diagnosis.

Treatment

The therapy for MDS can be as heterogeneous as its classification—that is, depending on the nature, subclass, and prognostic risk of MDS, the therapeutic interventions can vary significantly. Indeed, the classification of MDS into low-, intermediate-, and high-risk prognostic categories aids in the determination of appropriate modalities of treatment. The goals of therapy can also vary and include controlling MDS-related symptoms, improving cytopenias, decreasing the risk of leukemic progression, and enhancing survival.

An important determination in patients with MDS, especially those with high- or very high-risk disease (see "Prognosis," above; and see Table 8.2), is whether they are potential candidates for hematopoietic stem cell transplantation (HSCT; for more description about HSCT, see Case 4-1). Allogeneic HSCT is the only known potentially curative modality for MDS, but it leads to long-term survival in only approximately one-third of patients. The timing of transplantation appears to be important for successful therapy. Studies have demonstrated that early HSCT, as opposed to transplant at the time of disease progression, may be beneficial in those with higher-risk disease. Nevertheless, given the logistics and timing of transplantation, many HSCT-eligible patients do receive some treatment prior to HSCT. This usually involves hypomethylating therapy (described below), but in more robust patients with a high marrow myeloblast burden (>15%), intensive induction chemotherapy (as would be used for AML) can be employed prior to HSCT for purposes of **cytoreduction** (reducing the number of cancer cells).

There are a range of non-transplant therapies available for patients with MDS, but none are curative. Many patients with lower-risk disease do not require therapy, but rather need close monitoring for disease progression. Supportive care can be an important aspect of therapy. For those with refractory cytopenias, hematopoietic growth factors that stimulate blood cell production (for example, erythropoietin stimulates red blood cell production) can be used as effective palliative adjuncts, in addition to regular transfusions of red cells and/or platelets. However, such erythropoiesis stimulating agents (ESAs) tend to be more effective in transfusion-independent patients and in those with suppressed serum levels of erythropoietin.

The most commonly employed class of epigenetic-targeted agents in MDS includes the hypomethylating agents decitabine and 5-azacitidine, which are thought to release the abnormal block in myeloid differentiation that is seen in cases of MDS. These agents are effectively employed and recommended for patients with refractory, clinically significant cytopenias (such as neutropenia and thrombocytopenia), and for those with intermediate- or high-risk disease (see Table 8.2). Hypomethylating agents exert their effect by impacting epigenetic regulation of genes essential for myeloid differentiation. Levels of gene expression can be impacted at the DNA level by addition of methyl groups to the promoter regions of key genes, leading to suppression of transcription. DNA methylation is catalyzed by the DNA methyltransferase (DNMT) family of enzymes. Aberrant hypermethylation and resultant inactivation of promoter regions of tumor suppressor genes are common in myeloid malignancies such as MDS. In addition, suppression of the expression of key genes necessary for the developmental maturation of hematopoietic cell lineages can result in cytopenias. Decitabine and 5-azacitidine are inhibitors of DNMT, and in this fashion are thought to target the abnormal hypermethylation that promotes MDS development and progression. DNMT inhibitors release the differentiation block, thereby allowing myeloid maturation and differentiation and subsequent clinical response.

DNMT inhibitors can indeed be very effective therapy in MDS, but response can be delayed, often occurring months after initiation of therapy. The response can range from improvement in cytopenias to transfusion independence to complete marrow responses. As an example, in a large study of patients with advanced MDS, 5-azacitidine was well tolerated and found to prolong overall survival when compared with conventional care regimens for MDS (median OS 24 versus 15 months). As a result of these extremely promising studies, 5-azacitidine and decitabine have become standard therapies for patients with higher-risk disease.

Interstitial deletions involving the long arm of chromosome 5 are a common cytogenetic anomaly in MDS, with a deleted segment extending from bands 5q31 to 5q32.3 as the most common cytogenetic abnormality in MDS. Patients with deletion 5q (del 5q) MDS often have a distinct clinicopathological presentation, with characteristic megakaryocytes seen in the marrow, a hypoproliferative anemia, and a propensity for dependence on red blood cell transfusions. 5q deletions can occur in isolation or can occur concurrently with other karyotypic abnormalities. The molecular mechanism of disease in del 5q syndrome appears to be deletion of the ribosomal subunit 14 gene (*RPS14*), which interrupts ribosomal biosynthesis and leads to apoptosis of affected cells.

An important therapeutic agent in MDS, specifically del 5q MDS, is the **immunomodulatory** agent lenalidomide. Immunomodulatory agents stimulate or suppress immune system functions (see Case 10-2). Lenalidomide, used effectively in other hematologic malignancies, such as multiple myeloma, is thought to exert its effect through a variety of mechanisms. These include augmenting the immune system by enhancing the functional capacity of T lymphocytes and amplifying co-stimulatory immune pathways, as well as exerting anti-angiogenic effects (preventing the development of new blood vessels; see Chapter 10). Lenalidomide has been studied for treatment of MDS, including in those with lower-risk disease and red cell transfusion dependence. Intriguingly, in the del 5q subset of patients, the responses to lenalidomide can be profound and appear to be related to up-regulation of tumor suppressor genes on the remaining chromosome 5q. Indeed, the majority of patients with del 5q MDS responded with red cell transfusion independence of prolonged duration in clinical studies. Additionally, achieving prolonged red cell transfusion independence was associated with significant reduction in leukemic progression and death. Therefore, lenalidomide is now recommended for and increasingly used in the treatment of patients with del 5q MDS. Lenalidomide is also effective in achieving red cell transfusion independence for a smaller subset of non-del 5q patients (26% in one study), where its mechanism of action is distinct from that in 5q-deleted MDS.

Future prospects and challenges in MDS

The treatment of MDS remains a challenge, although there has been marked improvement in effective therapies over the last decade. Epigenetic-targeted therapy with hypomethylating agents and immunomodulatory therapy with lenalidomide currently have important roles in the management of many patients with MDS. Clinical trials are studying novel targeted agents such as inhibitors of HDAC and other pathways. Novel combinations of therapies are also under study in clinical trials. In time, there is emerging hope for additional effective therapies that will improve the care and outcomes for patients with MDS.

Take-home points

✓ Myelodysplastic syndromes (MDS) are heterogeneous clonal malignancies of myeloid precursors that are characterized clinically by progressive cytopenias, bone marrow failure, and progression to acute leukemia.

✓ MDS prognostic risk is derived from the number of cytopenias, the cytogenetic risk, and degree of marrow involvement with myeloblasts.

✓ Allogeneic stem cell transplantation is the only potentially curative modality in the treatment of MDS.

✓ Other therapeutic and supportive therapies used in the management of MDS include transfusions and erythropoiesis stimulating agents (ESAs), epigenetic-targeted therapy with hypomethylating agents, and lenalidomide.

✓ Hypomethylating agents inhibit DNA methyltransferases (DNMTs), and in this fashion, are thought to target the aberrant hypermethylation that arrests the normal differentiation of myeloid cells in the marrow of patients with MDS.

✓ The hypomethylating agent 5-azacitidine has been shown to improve survival outcomes in patients with higher-risk MDS, and hypomethylating agents are now standard-of-care therapies.

✓ The immunomodulator lenalidomide has been shown to be particularly effective in achieving red cell transfusion independence in the subset of MDS patients with a deletion in chromosome 5q.

Discussion questions

1. How is MDS diagnosed and classified?

2. What specific feature about MDS might make hypomethylating therapy a reasonable approach?

3. What is the postulated mechanism of action of 5-azacitidine and decitabine in myelodysplastic syndromes?

4. What guides the treatment choice for patients with MDS? When would therapy with DNMT inhibitors or lenalidomide be appropriate?

Case 8-2	Cutaneous T-cell Lymphoma

Introduction

The majority of non-Hodgkin lymphomas (NHILs) derive from B lymphocytes, but approximately 10% arise from T lymphocytes and have a unique biology, requiring distinct approaches to therapy (see Case 5-2 for more background regarding NHL). The classification of NHLs is described in Table 5.5. Currently there are approximately two dozen distinct subtypes of T-cell NHL recognized, and they may occur predominantly in either nodal or extranodal (outside of lymph nodes) sites, or circulate in the blood (in which event they are also called leukemic) (Table 8.3). Among extranodal T-cell lymphomas, the skin is the most commonly involved site. Unlike lymphomas primarily involving lymphoid organs and other extranodal sites where B-cell lymphomas are far more common, **cutaneous T-cell lymphomas** (CTCLs) predominate among the cutaneous lymphomas, constituting about 75% of cutaneous lymphomas, with the remainder derived from B cells.

The most common CTCL is **mycosis fungoides** (MF), which primarily involves the skin but may secondarily involve lymph nodes, blood, and organs. This disease typically has an indolent natural history with a favorable prognosis, and approximately 80% of affected patients will be alive five years after diagnosis. A small subset of patients with MF, however, will develop a diffuse **erythroderma** (redness of the skin) with leukemic involvement of the peripheral blood, a condition known as **Sézary syndrome** (SS).

SS is highly aggressive and carries an inferior prognosis with five-year OS of only 25%. MF and SS are both considered incurable but highly treatable

Topics bearing on this case

Non-Hodgkin lymphomas

Cutaneous lymphoma

HDAC inhibition

Table 8.3. Subtypes of T-cell non-Hodgkin lymphomas.

Predominantly nodal	Predominantly extranodal	Predominantly leukemic
Peripheral T-cell lymphoma, not otherwise specified	Extranodal natural killer (NK)/T-cell lymphoma, nasal type	T-cell prolymphocytic leukemia
Angioimmunoblastic T-cell lymphoma	Enteropathy-associated T-cell lymphoma	T-cell large granular lymphocytic leukemia
Anaplastic large-cell lymphoma (ALCL), anaplastic lymphoma kinase (ALK) positive	Hepatosplenic T-cell lymphoma	Chronic NK-cell lymphoproliferative disorder
	Subcutaneous panniculitis-like T-cell lymphoma	Aggressive NK-cell leukemia
	Mycosis fungoides	Adult T-cell leukemia/lymphoma
	Sézary syndrome	
	Primary cutaneous anaplastic large-cell lymphoma	
	Primary cutaneous aggressive epidermotropic CD8-positive cytotoxic T-cell lymphoma	
	Primary cutaneous gamma-delta T-cell lymphoma	
	Primary cutaneous small/medium CD4-positive T-cell lymphoma	
	Hydroa vacciniforme-like lymphoma	

cancers, characterized by a chronic relapsing–remitting course that may require multiple lines of therapy during a patient's lifetime. The goal of treatment for both MF and SS is to induce remission of disease in the skin and other sites, prolong length of life, and improve and preserve quality of life.

Epidemiology

MF accounts for approximately 44% of all CTCL cases, or approximately 4% of all cases of NHL. The median age at diagnosis is 60 years and it occurs twice as commonly in men as in women. The median life expectancy of patients with MF is close to 20 years from diagnosis, but there is significant variation depending on the stage of the cancer. Blacks are more commonly affected than whites. SS accounts for only 4% of CTCLs and occurs exclusively at advanced stage with a more aggressive natural history. Unlike MF, patients with SS are more commonly white.

Diagnosis, workup, and staging

MF most commonly presents with slowly progressive skin lesions in non-sun-exposed areas. The lesions may occur in patches, plaques, tumors, or, less commonly, diffuse erythroderma. Patches are defined as flat skin lesions of any size without elevation or significant thickening (Figure 8.8A). Increased or decreased pigmentation may be seen, along with scaling or crusting. A plaque is a skin lesion of any size that does have elevation or thickening, and also may have alterations in pigmentation, scaling, crusting, or ulceration (see Figure 8.8B). Tumors are nodules in the skin of greater than 1 cm in depth (Figure 8.8C).

The lesions are usually quite itchy, and ulceration or scratching may lead to infections. Disease of the scalp may result in hair loss (**alopecia**). Unlike MF, SS does not initially present with isolated patches and plaques but instead with diffuse erythroderma of the skin, thickening of the skin (known as lichenification), and a characteristic wrinkled appearance. Nails may similarly become thickened. These clinical features may progress rapidly, over weeks to months, and may be associated with intense itching. Infections related to skin breaks from frequent scratching frequently occur.

Figure 8.8. Skin photographs of mycosis fungoides, presenting as (A) patches, (B) plaques, and (C) tumors.

(A)

(B)

(C)

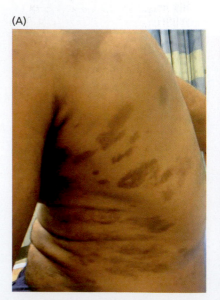
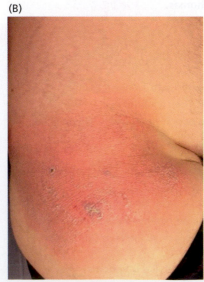
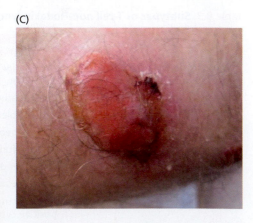

Diagnosis of MF is made when a skin biopsy reveals infiltration of the skin by small- to medium-sized atypical lymphocytes having cerebriform (convoluted, or folded) nuclei, with preferential involvement of the epidermis, a condition known as epidermotropism (Figure 8.9). By immunohistochemistry, the cells stain positive for proteins typically found on the surface of T cells, including CD2, CD3, CD4, CD5, and CD7. Notably, healthy T cells express these same proteins, but one clue favoring the diagnosis of a malignancy is the loss of one or more of the normal markers. For example, T cells that are positive for CD2, CD3, and CD5 but negative for CD7 indicate malignancy. CD30 is frequently seen in MF as well, where it may serve as a therapeutic target in relapsed/refractory disease (see Case 11-2, Hodgkin lymphoma). Because it may be difficult to distinguish malignant from healthy T cells, **polymerase chain reaction** (PCR) amplification can be used to assess for clonal T-cell gene rearrangements, which will be identified in nearly all cases of CTCL.

Patients with SS exhibit **lymphocytosis** (an increase in the number of lymphocytes) in the peripheral blood due to circulating leukemic disease. Cells in the blood will have convoluted cerebriform nuclei, called Sézary cells, with the same immunophenotype as the lymphoma cells found in the skin (Figure 8.10). Flow cytometry is most commonly used to evaluate the circulating lymphocytosis, and the finding of a CD4+ T-cell population that has lost a normal T-cell antigen, such as CD7, is highly suggestive of SS in the proper clinical context. As in MF, a clonal T-cell receptor gene rearrangement, verified by PCR, is also present.

Prognosis and treatment selection in MF is largely a function of stage as determined by the TNMB system, which considers extent of involvement of the skin (T), lymph nodes (N), visceral organ sites (M), and blood (B). Staging is performed with a detailed skin examination. CT scans are recommended when there is clinical suspicion for nodal or visceral involvement. A complete blood count (CBC) to assess for lymphocytosis is performed in all patients, and flow cytometry of the blood for further characterization is performed in the setting of an increased lymphocyte count on the CBC. The TNMB scores are listed in Table 8.4, and the stage groupings based on TNMB scores are in Table 8.5.

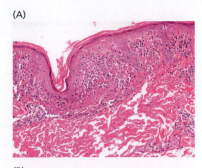

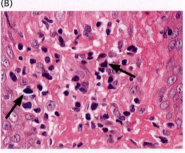

Figure 8.9. Morphologic findings in mycosis fungoides (hematoxylin and eosin staining). (A) There is an infiltrate of hyperchromatic lymphoid cells (cells with dark nuclear staining) centered in the epidermis (the uppermost layer of the skin). (B) On higher magnification, the lymphoid cells are highly atypical, with enlarged, hyperchromatic nuclei containing marked nuclear irregularities (*arrows*).

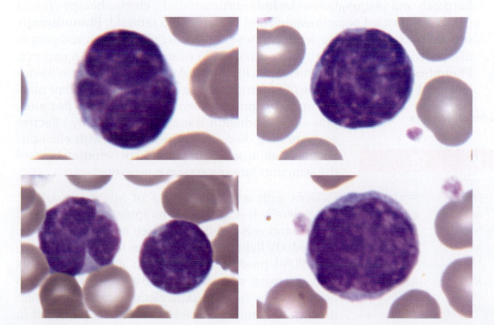

Figure 8.10. Circulating cells in the peripheral blood in Sézary syndrome. SS cells are characterized by highly convoluted, folded or clefted nuclei, also described as "cerebriform."

Table 8.4. TNMB classification of mycosis fungoides.

Skin	
T1	Limited patches, papules, and/or plaques covering <10% of the skin surface
T2	Patches, papules, and/or plaques covering ≥10% of the skin surface
T2a	Patch only
T2b	Patch +/− plaque
T3	One or more tumors (>1 cm in diameter)
T4	Erythroderma covering ≥80% of the body surface area
Node	
N0	No abnormal lymph nodes on physical examination
N1	Abnormal lymph nodes; grade 1 on biopsy
N2	Abnormal lymph nodes; grade 2 on biopsy
N3	Abnormal lymph nodes; grade 3–4 on biopsy
Nx	Abnormal lymph nodes; no biopsy confirmation
Organs	
M0	No visceral organ involvement
M1	Visceral involvement (must have pathology confirmation)
Mx	Abnormal visceral site; no histologic confirmation
Blood	
B0	≤5% of peripheral blood lymphocytes are atypical (Sézary) cells
B1	>5% of peripheral blood lymphocytes are atypical (Sézary) cells but do not meet the criteria of B2
B2	≥1000/μl Sézary cells or CD4/CD8 ratio ≥10 or ≥40% CD4$^+$/CD7$^-$ or ≥30% CD4$^+$/CD26$^-$ cells

Based on these stage groupings, patients with early-stage disease have a favorable prognosis, with stages IA, IB, and IIA having 10-year overall survival of approximately 93%, 86%, and 72%, respectively. Patients with more advanced disease classified as stages IIB–IIIB have a 10-year survival of 51%, and for IVA–IVB, it is only 24%.

Principles of MF and SS management

Treatment of MF is determined by stage (Table 8.6). Early-stage MF (stages IA–IIA) is treated with local skin-directed therapies. Effective topical agents for patch and plaque disease include corticosteroids, chemotherapy creams (nitrogen mustard or carmustine), and bexarotene, a retinoid. **Phototherapy** using ultraviolet light also demonstrates excellent activity in patch and plaque disease with complete remissions achieved in over 80% of stage IA patients. Phototherapy may be administered with ultraviolet light B (UVB) or ultraviolet light A combined with the photosensitizing agent psoralen (together making up the regimen known as PUVA: P, for psoralen, plus UVA). Radiation therapy is also highly effective at controlling discrete lesions, usually with electron-beam radiation, which has limited penetration beyond the skin, thus limiting systemic toxicity.

Patients with advanced stages of MF, stages IIB–IV, are generally less amenable to applications of topical creams and gels given the surface area needing to be covered. UV light therapy can cover the entire skin surface. PUVA produces responses in 30–70% of patients with extensive skin disease. Advanced MF may also be amenable to total-skin electron-beam radiation therapy. Systemic therapies may also be employed in

Table 8.5. Stages for mycosis fungoides and Sézary syndrome based on TNMB.

Stage	T	N	M	B
IA	T1	N0	M0	B0–1
IB	T2	N0	M0	B0–1
IIA	T1–2	N1–2	M0	B0–1
IIB	T3	N0–2	M0	B0–1
IIIA	T4	N0–2	M0	B0
IIIB	T4	N0–2	M0	B1
IVA$_1$	T1–4	N0–2	M0	B2
IVA$_2$	T1–4	N3	M0	B0–2
IVB	T1–4	N0–3	M1	B0–2

advanced-stage disease, either as initial therapy or after failure of ultraviolet light or radiation therapy. Systemic retinoids (such as oral bexarotene) act by binding the retinoid X receptor, which promotes cell differentiation and induces apoptosis, and produces responses in about half of patients (see Chapter 7 for discussion of retinoids and differentiation therapy). Potential toxicities of systemic retinoids include decreased thyroid function and hyperlipidemia (elevated levels of lipids in the blood). HDAC inhibitors are available in both oral and intravenous formulations, and produce responses in approximately one-third of subjects, with potential side effects including fatigue, nausea, diarrhea, low platelets, and abnormal heart rhythms (see the case and discussion below). Interferons result in objective responses in the majority of patients and are administered as a subcutaneous injection three times a week. Interferon acts by up-regulation of a T-cell immune response against the malignant cells, and side effects may be significant, including flu-like symptoms, fatigue, headaches, loss of appetite, and depression.

Patients with advanced MF, and particularly patients with SS, may receive a specialized systemic method of administering PUVA that uses a process called extracorporeal photopheresis (ECP). In ECP, the patient's blood is removed via a catheter, exposed to psoralen and UVA light, and then returned to the patient. This procedure induces remission in over half of patients, though best responses may take months of therapy to become manifest. These responses may occur more quickly if combined with a retinoid or interferon.

For patients with multiple relapses of MF, or for patients with SS, systemic chemotherapy will often be required, usually in the form of single agents. Options include methotrexate, pralatrexate, gemcitabine, pentostatin, and pegylated liposomal doxorubicin, as well as the anti-CD30 antibody–drug conjugate brentuximab vedotin, the anti-CD52 monoclonal antibody **alemtuzumab**, and the anti-CCR4 monoclonal antibody mogamulizumab. Selected patients with highly refractory or aggressive disease may be considered for allogeneic stem cell transplant.

Finally, supportive care is a critical component in the management of MF and SS. Moisturizers are employed to minimize drying and cracking, so as to help reduce discomfort, itching, and risk of infection. The itching may be severe and disabling in advanced MF; it may be addressed with antihistamines, with refractory itching addressed with gabapentin, mirtazapine, or aprepitant. Oral corticosteroids may be required for highly resistant itching but are used with caution as they further increase the risk of infection. Patients must be monitored for secondary skin infections and treated with antibiotics accordingly. For patients with multiply recurrent infections, bleach baths may be used to minimize skin colonization by bacteria.

The case of Jon Carter, a 62-year-old man with skin lesions

Jon is a 62-year-old previously healthy architect who presented to a dermatologist complaining of a diffuse erythematous rash affecting his entire body. The rash began in a mild form several months earlier and subsequently progressed to involve his entire skin surface. The rash was intensely itchy and uncomfortable. He initially felt it might be an environmental allergic reaction or an allergic reaction to a household product, so he changed his clothes detergent, soap, and shampoo, but there was no resolution of his symptoms. He used topical and systemic antihistamines to reduce the itching, but experienced only little relief.

Table 8.6. Treatment options for mycosis fungoides or Sézary syndrome.

Stages	Treatment
IA–IIA	Topical corticosteroids
	Topical bexarotene
	Topical nitrogen mustard
	Phototherapy (UVB or PUVA)
	Local electron-beam radiation
IIB–IVB	PUVA
	Total skin electron-beam radiation
	Oral bexarotene
	Oral methotrexate
	HDAC inhibitor (vorinostat, romidepsin)
	Interferon alfa
	Extracorporeal photopheresis (ECP)
	Pegylated doxorubicin
	Gemcitabine
	Pentostatin
	Combination chemotherapy regimens

The dermatologist found Jon to have diffuse erythroderma covering over 80% of the skin surface. The skin was brightly erythematous (red) and had some thickening and scaling, as well as excoriations from scratching the intensely itchy skin. On further physical examination, the dermatologist found enlarged lymph nodes affecting the neck, armpits, and groin. A skin biopsy was performed and showed diffuse infiltration of the upper dermis of the skin by medium-sized atypical lymphocytes with convoluted-appearing, "cerebriform" nuclei and further infiltration of the epidermis. These findings were highly suspicious for mycosis fungoides. By immuno-histochemistry, the malignant-appearing cells stained positive for mature T-cell markers CD2, CD3, CD4, and CD5, but lost normal expression of CD7, consistent with malignant T cells. Malignancy was confirmed by a PCR that was positive for a clonal T-cell receptor gene rearrangement. A CBC showed an elevated white blood cell count of 35,000 with 70% lymphocytes. Jon had no anemia or thrombocytopenia. Examination of the blood under the microscope revealed a large proportion of atypical lymphocytes with cerebriform nuclei, consistent with circulating CTCL cells (Sézary cells; see Figure 8.10). Flow cytometry showed the cells to be positive for CD23, CD4, and CD5, and negative for CD7 and CD26. A diagnosis of Sézary syndrome was made.

A PET/CT scan of the chest, abdomen, and pelvis confirmed lymphadenopathy involving the cervical, axillary, and inguinal nodes, as well as mediastinal, mesenteric, and retroperitoneal lymph nodes. The nodes measured up to 2.5 cm. No masses in other organs were demonstrated.

Jon was initiated on ECP treatment, which was performed on two consecutive days every two weeks. After one month of treatment, he noted improvement, with a decrease in the redness and thickening of the skin. His skin improvement correlated with a decrease in the amount of circulating Sézary cells. However, given the degree of persistent disease after four months, oral bexarotene was added, and his disease continued to improve. He received six total months of ECP with bexarotene, and achieved a partial response with near complete resolution of his intense itching. Unfortunately, the disease progressed six months after completion of initial therapy. At that time he was treated with the HDAC inhibitor **romidepsin**. Romidepsin was administered intravenously on days 1, 8, and 15 of a 28-day cycle. His itchiness began to subside within one month, and after two months, his erythroderma and peripheral adenopathy had dramatically improved. The romidepsin was continued, but Jon was counseled that the response was unlikely to be complete or to last for a lengthy period of time, so additonal therapies would likely be required in the future. For the moment, Jon is grateful that the itching has resolved.

HDAC inhibition in cutaneous MF and SS

Histone acetylation by HATs inhibits histone interactions with DNA, relaxing the compact nature of chromatin to expose genes that can then undergo transcription. HDACs oppose this HAT activity, and by deacetylating histones, HDACs restore the electrostatic attraction between positively charged lysine residues on histones and negatively charged phosphate groups on DNA, thus silencing the transcription of the affected genes. HDACs have been shown to affect the transcription of critical pro-apoptotic genes *Bim*, *Bfm*, and *TRAIL*, as well as the gene encoding the tumor suppressor p53. HDAC inhibitors are epigenetic agents that swing

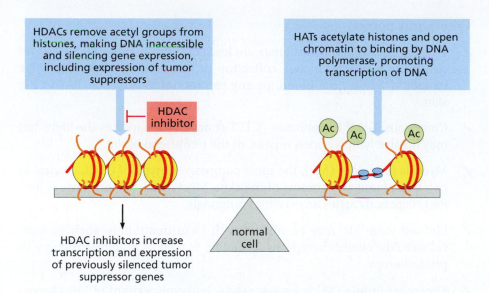

HDACs remove acetyl groups from histones, making DNA inaccessible and silencing gene expression, including expression of tumor suppressors

HATs acetylate histones and open chromatin to binding by DNA polymerase, promoting transcription of DNA

HDAC inhibitor

Ac Ac Ac

normal cell

HDAC inhibitors increase transcription and expression of previously silenced tumor suppressor genes

Figure 8.11. Mechanism of action of HDAC inhibitors. Histone deacetylases (HDACs) and histone acetyltransferases (HATs) regulate gene expression by modifying acetylation of histones. Acetylation of histones allows binding of DNA polymerase and promotes transcription of DNA. Histone deacetylase removes acetyl groups from histones, silencing gene expression. HDAC inhibitors swing the balance to HAT activity, promoting gene expression. This appears to work in CTCL by promoting transcription of previously silenced tumor suppressor genes.

the balance toward HAT activity, increasing histone acetylation and resulting in transcriptional activation of genes involved in apoptosis and growth inhibition (Figure 8.11).

Vorinostat is an oral HDAC inhibitor that was evaluated in a Phase II clinical trial for relapsed or refractory MF patients with stage IB or higher disease. The trial produced responses in 30% of subjects, all partial responses. The average duration of remission in responding patients was 11 months. Clinical toxicities included fatigue, diarrhea, nausea, and loss of appetite, and laboratory toxicities included mild anemia and thrombocytopenia. Romidepsin appears to be a slightly more potent intravenously administered HDAC inhibitor with an overall response rate of 34%, including 6% complete responses, and an average remission duration of 15 months. Toxicities were similar to vorinostat but with more prominent fatigue, thrombocytopenia, and a prolonged QT interval on electrocardiograms, which predisposes to dangerous arrhythmias (abnormal heart rhythms). Vorinostat and romidepsin were FDA approved for relapsed or refractory CTCL in 2006 and 2009, respectively. A third HDAC inhibitor, belinostat, was FDA approved in 2014 to treat systemic T-cell lymphomas that were relapsed or refractory after front-line chemotherapy.

Future prospects and challenges in cutaneous T-cell lymphomas

The majority of patients with CTCL will enjoy an excellent prognosis and often a normal life expectancy. The small subset of patients with advanced MF and SS, however, continue to have a shortened life expectancy despite the wide range of treatment options available and the emergence of HDAC inhibitors. A number of additional novel agents have demonstrated encouraging activity in cutaneous and systemic T-cell lymphomas that will hopefully yield ongoing improvement in prognosis for high-risk patients. These include the anti-CD30 antibody–drug conjugate brentuximab vedotin for cases of MF or SS that express CD30 (see Case 11-2), the anti-CCR4 monoclonal antibody mogamulizumab, proteasome inhibitors (see Case 9-2), the immunomodulator lenalidomide (see Case 10-2), PD-1 (immune checkpoint) inhibitors (see Case 12-3), and duvelisib, an inhibitor of the delta and gamma isoforms of PI3 kinase.

Take-home points

✓ T-cell non-Hodgkin lymphomas are less common than B-cell NHLs and constitute a heterogeneous collection of diseases, which may involve lymphoid tissues, the blood, or any extranodal location, including the skin.

✓ Cutaneous T-cell lymphomas (CTCLs) occur primarily in the skin, but may involve lymph nodes, organs, or the peripheral blood.

✓ Mycosis fungoides (MF), the most common subtype of CTCL, occurs in patches, plaques, or tumors of the skin. Most often, MF presents at limited stage and carries an excellent prognosis.

✓ Limited-stage MF may be treated with local treatments, such as topical steroids, chemotherapy, or retinoids, as well as radiation therapy or phototherapy.

✓ Sézary syndrome (SS) is an aggressive, leukemic variant of MF, characterized by diffuse erythroderma and thickened skin, but not plaques or patches.

✓ Advanced MF and SS carry a poor prognosis and require systemic treatment. Options include systemic histone deacetylase (HDAC) inhibitors, chemotherapy, interferon, and alemtuzumab, as well as extracorporeal photopheresis in the setting of SS.

✓ HDAC inhibitors are epigenetic agents that regulate gene transcription. In CTCL they appear to reactivate expression of tumor suppressor genes.

✓ Essential supportive care in MF and SS includes moisturizing the skin, treatment of severe itching, and prevention and treatment of cutaneous infections.

Discussion questions

1. What are signs and symptoms of MF and SS?

2. What features distinguish MF lymphocytes from normal T cells? What molecular tests are used to confirm the diagnosis?

3. How is MF staged?

4. What are the goals of treatment in MF?

5. What are treatment options for limited and advanced-stage MF?

6. What is Sézary syndrome?

7. How do HDAC inhibitors work in CTCL?

Chapter summary

As Cases 8-1 and 8-2 demonstrate, targeting epigenetic activities is feasible and effective for some cancers. The lack of specificity of gene regulation using epigenetic therapies likely contributes to the limited use of this approach. For example, it seems unlikely that an HDAC inhibitor might be developed that can specifically (and only) induce expression of genes involved in apoptosis and preferentially have activity in cancer cells. It remains to be seen whether the other activities of HDAC inhibitors can be exploited, including acetylation of nonhistone targets like proteins involved in DNA repair or induction of apoptosis-related protein expression, and whether these activities can potentiate the impact of conventional DNA-damaging therapies. Combinations of epigenetic and other therapies represent a potentially important therapeutic avenue.

Our growing understanding of the biology of miRNAs and lncRNAs could open up entire new fields of epigenetic therapies. **Antisense oligonucleotides** act like miRNAs by binding to complementary sequences of a target mRNA, creating double-stranded RNA that can be targeted for degradation. Antisense oligos are FDA-approved for the treatment of cytomegalovirus retinitis and a form of familial hypercholesterolemia but not as a cancer therapy, to date. Limitations of using antisense molecules include the need to modify nucleotides to avoid the ubiquitous nucleases within human cells and the difficulty of transporting these negatively charged oligonucleotides across cell membranes. It is possible that miRNAs or lncRNAs could have the advantage over antisense oligonucleotide strategies of being expressed from viral or plasmid (circles of DNA that can replicate independently of chromosomes) vectors (vehicles that carry DNA into cells) that might enter cells more readily than naked oligonucleotides. Targeting or exploiting these particular RNA molecules has not yet been validated for human cancer therapy.

Chapter discussion questions

1. What are the key elements that regulate gene expression?

2. What are the mechanisms by which epigenetic modifiers impact protein expression?

3. How might epigenetic therapies be incorporated with other cancer treatments?

4. Consider how you might design a miRNA cancer therapy.

Selected references

Introduction

Chi P, Allis CD, Wang GG et al. (2010) Covalent histone modifications: miswritten, misinterpreted and mis-erased in human cancers. *Nat. Rev. Cancer* 10:457–469.

Esteller M (2008) Epigenetics in cancer. *N. Engl. J. Med.* 358: 1148–1159.

Ponting CP, Oliver PL & Reik W (2009) Evolution and functions of long noncoding RNAs. *Cell* 136:629–641.

Ventura A & Jacks T (2009) MicroRNAs and cancer: short RNAs go a long way. *Cell* 136:586–591.

Case 8-1

Adès L, Itzykson R & Fenaux P (2014) Myelodysplastic syndromes. *Lancet* 383:2239–2252.

Bennett JM, Catovsky D, Daniel MT et al. (1982) Proposals for the classification of the myelodysplastic syndromes. *Br. J. Haematol.* 51:189–199.

Cutler CS, Lee SJ, Greenberg P et al. (2004) A decision analysis of allogeneic bone marrow transplantation for the myelodysplastic syndromes: delayed transplantation for low-risk myelodysplasia is associated with improved outcome. *Blood* 104:579–585.

Fenaux, P, et al. (2009) Efficacy of azacitidine compared with that of conventional care regimens in the treatment of higher-risk myelodysplastic syndromes: a randomised, open-label, phase III study. *Lancet Oncol.* 10:223–232.

Greenberg PL, Mufti GJ, Hellstrom-Lindberg E et al. (2009) Treatment of myelodysplastic syndrome patients with erythropoietin with or without granulocyte colony-stimulating factor: results of a prospective randomized phase 3 trial by the Eastern Cooperative Oncology Group (E1996). *Blood* 114:2393–2400.

Greenberg PL, Tuechler H, Schanz H et al. (2012) Revised international prognostic scoring system for myelodysplastic syndromes. *Blood* 120:2454–2465.

Kantarjian H, Issa J-P, Rosenfeld CS et al. (2006) Decitabine improves patient outcomes in myelodysplastic syndromes: results of a phase III randomized study. *Cancer* 106:1794–1803.

List AF, Dewald J, Bennett J et al. (2006) Lenalidomide in the myelodysplastic syndrome with chromosome 5q deletion. *N. Engl. J. Med.* 355:1456–1465.

List AF, Kurtin S, Roe DJ et al. (2005) Efficacy of lenalidomide in myelodysplastic syndromes. *N. Engl. J. Med.* 352:549–557.

Raza A, Reeves JA, Feldman EJ et al. (2008) Phase 2 study of lenalidomide in transfusion-dependent, low-risk, and intermediate-1 risk myelodysplastic syndromes with karyotypes other than deletion 5q. *Blood* 111:86–93.

Tefferi A & Vardiman JW (2009) Myelodysplastic syndromes. *N. Eng. J. Med.* 361:1872–1885.

Vardiman JW, Harris NL & Brunning RD (2002) The World Health Organization (WHO) classification of the myeloid neoplasms. *Blood* 100:2292–2302.

Case 8-2

Duvic M, Talpur R, Ni X et al. (2007) Phase 2 trial of oral vorinostat (suberoylanilide hydroxamic acid, SAHA) for refractory cutaneous T-cell lymphoma (CTCL). *Blood* 109:31–39.

Olsen E, Vonderheid E, Pimpinelli N et al. (2007) Revisions to the staging and classification of mycosis fungoides and Sézary syndrome: a proposal of the International Society for Cutaneous Lymphomas (ISCL) and the cutaneous lymphoma task force of the European Organization of Research and Treatment of Cancer (EORTC). *Blood* 110:1713–1722.

Prince HM, Whittaker S & Hoppe RT (2009) How I treat mycosis fungoides and Sézary syndrome. *Blood* 114:4337–4353.

Whittaker SJ, Demierre MF, Kim EJ et al. (2010) Final results from a multicenter, international, pivotal study of romidepsin in refractory cutaneous T-cell lymphoma. *J. Clin. Oncol.* 28:4485–4491.

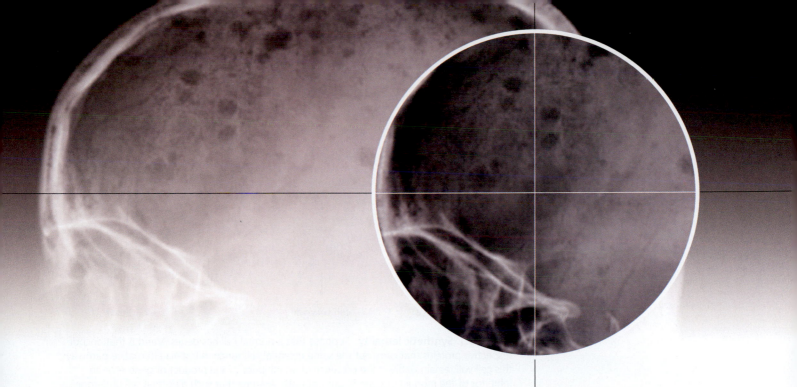

Chapter 9

Targeting Protein Degradation and DNA Repair

Introduction

DNA repair and protein **homeostasis** are metabolic processes that are essential features of cell life. Many pathways of DNA repair contribute to the maintenance of genome integrity. Cell proliferation, maintenance, and survival require appropriate levels of protein synthesis and degradation. The discovery that some cancer cells are addicted to particular pathways of DNA repair has revealed a vulnerability that can be exploited to the detriment of those malignant cells. The finding that some cancers are especially sensitive to inhibition of particular pathways of protein degradation provides the basis for a targeted therapy for those cancers. This chapter provides examples of developments in both approaches to killing cancer cells: compromising essential pathways of DNA repair, and disrupting pathways of protein degradation.

Two genes, A and B, distinct and encoding different products, are **synthetic lethals** if a disabling mutation in *either* gene A or gene B allows survival but disabling mutations in *both* genes, A and B, result in death (Figure 9.1). Cancer is a genetic disease that arises from an accumulation of mutations, some of which are essential drivers of the oncologic state. Some of these mutations will result in cancer cells with an altered requirement for specific biochemical activities, and it is likely that there will be genes that could be synthetic lethal partners in the cancerous cell. As illustrated in Figure 9.1, a cell without

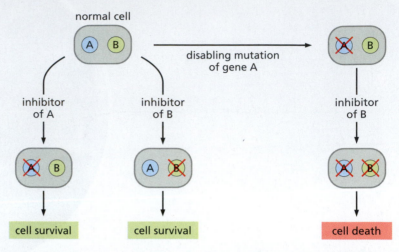

Figure 9.1. Synthetic lethality. Suppose that a normal cell has genes A and B that encode respective proteins that carry out the same essential cell function but via alternative pathways; this cell will remain viable in the presence of an inhibitor of the product of gene A or an inhibitor of the product of gene B, but not both. Assume that such a normal cell undergoes a disabling mutation of gene A. Having no functional product of gene A, the cell must rely on gene B to encode a product capable of conducting the essential cell function, and it will die if an inhibitor of the product of gene B is added. Note that this dependency on the product of gene B applies only if gene A carries disabling mutations at *both* loci.

disabling mutations in either member of a potentially synthetic lethal partnership survives. However, in a cell with a disabling mutation in either gene of the set, and therefore an absence of the functional product encoded by the altered gene, drugs that inhibit the product of the intact gene cause death of the cell. Identification of cancer-relevant genes and their synthetic lethal interactors provides a path that has great potential for the discovery of highly targeted anti-cancer drugs. Such drugs will be much less toxic to normal cells than to cancer cells that are dependent on the products of targeted genes with which the disabled cancer-relevant genes have a synthetic lethal relationship. This is true because normal cells have a functional version of the potential synthetic lethal that compensates if the partner gene is effectively inactivated by an inhibitor.

Conventional screens for anti-cancer drugs usually seek to identify compounds that are cytotoxic for cells whether they are normal or cancer cells. A compelling feature of synthetic lethality, the cell death that results when synthetic lethal partners are disabled, is that compounds developed to induce it afford an intrinsic selectivity for killing cancer cells while sparing normal cells. However, the challenge in developing synthetic lethality for cancer therapy lies in the great difficulty posed by the need to identify the synthetic lethal interactors of cancer-associated genes. Although it is suspected that many synthetic lethal interactors exist, at present only a small number have been identified.

Paramount in cellular homeostasis is the maintenance of an appropriate repertoire of proteins, wherein each type is present and at a level appropriate for existing conditions. The qualitative and quantitative repertoire of proteins is dependent on both protein synthesis and protein degradation. Which proteins are synthesized and in what quantities are determined by a complex of regulatory factors including epigenetic states (see Chapter 8), transcription factors, and regulators of translation, including microRNAs. The key regulatory pathway for protein degradation is the **ubiquitin-proteasome pathway** (UPP). In this highly conserved pathway, proteins are marked for degradation

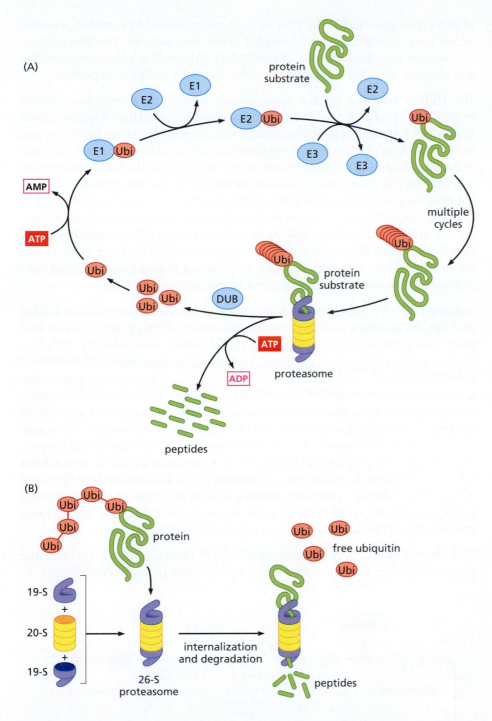

(A)

protein substrate

E1

E2

E2

E2 Ubi

Ubi

E1 Ubi

E3

E3

AMP

Ubi

multiple cycles

ATP

Ubi

Ubi

Ubi

Ubi

Ubi

DUB

protein substrate

Ubi Ubi

ATP

ADP

proteasome

peptides

(B)

Ubi Ubi

Ubi Ubi

Ubi

Ubi

protein

Ubi Ubi

Ubi free ubiquitin

Ubi

19-S

+

20-S

+

19-S

internalization and degradation

peptides

26-S proteasome

Figure 9.2. The Ubiquitin-proteasome pathway (UPP).
(A) Overview of the UPP. After cellular proteins undergo multiple cycles of conjugation with the small protein ubiquitin, they enter the proteasome and are degraded to small peptides. Initially in the UPP, ubiquitin ligases perform an ATP-dependent stepwise process, in which ubiquitin is conjugated to the targeted protein substrate. The resulting ubiquitinated protein becomes the recipient of many additional ubiquitin molecules by multiple cycles of ATP-dependent ubiquitination and then enters the proteasome, where it is unfolded and degraded to peptides. (Not shown: prior to degradation, de-ubiquitinating enzymes trim ubiquitins from the protein and perform other functions.) (B) Proteasome composition. The 26-S proteasome is comprised of a 19-S regulatory subunit and a 20-S catalytic subunit. The 19-S subunit binds polyubiquitinated proteins, strips their ubiquitins, and, in an ATP-requiring process, unfolds and translocates the protein into the 20-S catalytic subunit, where the protein is hydrolyzed into peptides. The proteasome includes three distinct types of protease activity: trypsin-like, chymotrypsin-like, and caspase-like. The amino acid threonine is present in the catalytic sites of all three types of protease activities and is targeted by many proteasome inhibitors. E1, ubiquitin-activating enzyme; E2, ubiquitin-conjugating enzyme; E3, ubiquitin ligase; DUB, de-ubiquitinating enzyme; Ubi, ubiquitin; 26-S proteasome, a large protein complex that unfolds and hydrolyzes polyubiquitinated proteins. (B, modified from J.A. Marteijn et al., *Leukemia* 20:1511–1518, 2006. Reprinted by permission from Macmillan Publishers Ltd.)

by polyubiquitination, a process in which many molecules of the small protein ubiquitin are conjugated to a targeted protein, via a lysine residue (Figure 9.2). The UPP is complex and is mediated by an enzymatic cascade that requires the sequential action of three different classes of enzyme: an activating enzyme, E1; a conjugation enzyme, E2; and a ligase, E3. The enzymes in these classes effect the addition of ubiquitin to a protein. Several cycles of this sequence of reactions yield a polyubiquitinated protein, which is flagged for disposal by the proteasome, where it is unfolded and hydrolyzed into peptides (see Figure 9.2A). Proteasomes are cylindrical, multi-enzyme complexes found in all eukaryotic cells. They are located in the nucleus and cytoplasm and are responsible for the disposal of misfolded proteins and obsolete proteins. The proteasome is comprised of a 20-S core cylindrical complex with **proteolytic** activity and capped by two 19-S regulatory units, for a total size of 26 S (see Figure 9.2B). The 19-S subunit recognizes ubiquitinated proteins, which are then unfolded and delivered to the 20-S subunit. The 20-S subunit has several enzymatic sites, categorized as trypsin-like, chymotrypsin-like, and caspase-like, and it degrades proteins into 3–25 amino acid peptides.

The UPP is essential for the maintenance of cellular homeostasis. While proteasome-mediated protein degradation was discovered first, it is now known that the UPP also plays an indispensable role in many signaling pathways and in the trafficking of some proteins to particular cellular compartments. Because of the large number and great variety of roles they play in the biology of normal and cancer cells, many components of the UPP are being investigated as potential drug targets. Interference with protein degradation, a hallmark activity of this pathway, compromises a broad diversity of essential cell functions, including cell division and cell survival. As illustrated in Case 9-2, proteasome inhibitors have already demonstrated their therapeutic utility for some cancers and are in clinical use. However, thus far, only certain hematopoietic (blood cell) cancers have been successfully targeted by inhibitors of the UPP.

Although the UPP is the main pathway for degradative clearance of misfolded proteins, an auxiliary pathway for removal of misfolded proteins is activated when the proteasome is overloaded or inhibited. The auxiliary pathway, known as the **aggresome pathway**, conveys large aggregates of misfolded polyubiquitinated proteins (called aggresomes) to lysosomes, where the aggresomes are proteolytically degraded (Figure 9.3). The assembly into aggresomes depends on histone deacetylase 6 (HDAC6), a member of a family of enzymes that catalyzes the removal of acetyl groups from histones. The transport of aggresomes to lysosomes on microtubules is driven by dynein motors. Aggresome-mediated protein degradation is disrupted by agents that inhibit tubulin or HDAC6. Combinations of proteasome inhibitors and inhibitors of HDAC6 have shown synergistic activity against multiple myeloma, making aggresome inhibitors of great interest as companion drugs for proteasome inhibitors, and clinical trials of these agents are in progress.

This chapter provides two examples of cancer therapies based on targeting vulnerabilities in the macromolecular metabolism of some cancer cells. In

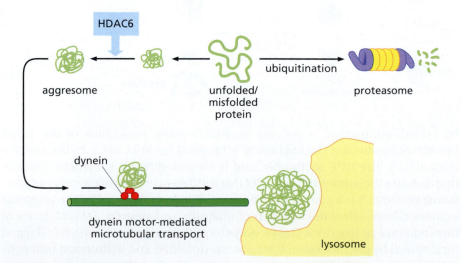

Figure 9.3. The aggresome pathway. Ubiquitinated misfolded proteins are degraded by the proteasome or form aggregates when the UPP is overloaded. Interaction of these aggregates with HDAC6 enables the formation of aggresomes that are transported on microtubules by the dynein/dynactin motor complex to lysosomes, where they are degraded. (From T. Simms-Waldrip et al., *Mol. Genet. Metab.* 94:283–286, 2008. With permission from Elsevier.)

Case 9-1, the principle of synthetic lethality is the basis of a therapy that targets repair-associated DNA synthesis, and in Case 9-2, disruption of an essential pathway of protein degradation leads to the death of myeloma cells.

Take-home points

✓ Disturbance or disruption of DNA repair or protein degradation can compromise cell survival, and cancer therapies targeting these processes are in use.

✓ Two genes are synthetic lethals if disabling either gene allows survival but disabling *both* genes results in cell death.

✓ Therapies based on synthetic lethality offer the promise of selectively targeting cancer cells.

✓ The ubiquitin-proteasome pathway (UPP) has many potential drug targets. One of these, the proteasome, has been successfully targeted with an inhibitor that is in clinical use.

✓ Targeting of both proteasome- and aggresome-mediated pathways of protein degradation may have future clinical applications.

| Case 9-1 | **Hereditary Breast Cancer** |

Introduction

Breast cancer is the most common cancer in American women. The epidemiology, diagnosis, workup, and management of breast cancer were described in Case 6-1. Molecular characterization has allowed for subgrouping this heterogeneous disease into categories that predict disease behavior and response to therapy. Although most breast cancers are sporadic and have no known familial predisposition to development of the disease, approximately 5–10% of breast cancer cases are hereditary. Another 20% are associated with familial predisposition, likely related to a combination of factors, including low-penetrance genes and environmental factors. Women harboring specific gene mutations have well-characterized risks of developing breast and other cancers. Most hereditary cases are due to mutations in *BRCA1*, *BRCA2*, and *PALB2*, and rare cases involve mutations in *PTEN*, *p53*, *CDH1*, and *STK11* (Table 9.1). The *BRCA* gene mutations are of particular interest to researchers developing targeted therapies that rely on synthetic lethality.

When normal, the genes *BRCA1* and *BRCA2* encode tumor suppressor proteins that repair damaged DNA. The BRCA1 protein is a component of a multi-subunit protein complex, the BRCA1-associated genome surveillance complex (BASC). Through BASC, BRCA1 repairs **double-strand DNA breaks** via **homologous recombination**. BRCA1 has also been associated with another form of DNA repair termed **mismatch repair**. Further, BRCA1 interacts with RNA polymerase II and histone deacetylase (HDAC) complexes. BRCA1 thus plays important roles in control of transcription and DNA double-strand break repair, as well as other functions. The BRCA2 protein interacts with single-strand DNA and the recombinase RAD51. Together with BRCA1 and PALB2 (partner and localizer of BRCA2), BRCA2 completes DNA double-strand break repair by recruiting RAD51 for the final steps of homologous

Topics bearing on this case

Synthetic lethality

poly(ADP-ribose) polymerases (PARP)

Double-strand breaks (DSB)

Base excision repair (BER)

Homologous recombination

BRCA1 and *BRCA2*

Table 9.1. Genetic mutations associated with hereditary breast cancer syndromes.

Gene(s)	Gene product function	Syndrome	Major clinical manifestations	Estimated risk of breast cancer development
BRCA1 BRCA2	DNA double-strand break repair, DNA mismatch repair	Hereditary breast and ovarian cancer (HBOC) syndrome	Autosomal dominant (AD); breast cancer; ovarian cancer; fallopian tube cancer; peritoneal cancer; male breast cancer; prostate cancer (BRCA2)	55–65% (BRCA1); 45–47% (BRCA2)
PALB2	Interacts with BRCA2 in DNA damage repair		Autosomal recessive (AR); biallelic loss causes Fanconi anemia; monoallelic loss associated with breast and pancreatic cancers	35%
PTEN	Phosphatase	Cowden syndrome	AD; skin manifestations (trichilemmomas, oral fibromas and papillomas, keratoses); dysplastic gangliocytomas of the cerebellum; breast cancer	25–50%
TP53	Tumor suppressor; binds DNA in the setting of DNA damage	Li–Fraumeni syndrome	AD; pre-menopausal breast cancer; childhood soft-tissue sarcoma; brain tumors; leukemia; adrenocortical carcinoma	49%
CDH1	Cell adhesion molecule	Hereditary diffuse gastric cancer syndrome	AD; diffuse gastric cancer in 40–83%; lobular breast cancer	30–50%
STK11 (LKB1)	Serine/threonine kinase	Peutz–Jeghers syndrome	AD; gastrointestinal hamartomas (benign focal tumors); hyperpigmented (darker, melanocytic) lesions on the lips, perioral region, buccal mucosa (inner cheek)	32–54%
ATM	Repair of DNA damage	Ataxia-telangiectasia	AR; cerebellar ataxia; ataxic gait; telangiectasias in sun-exposed areas; lymphomas; leukemias; breast cancer in heterozygotes	15%
CHEK2	DNA damage repair	Hereditary diffuse gastric cancer syndrome	AD; breast cancer—male and female; prostate cancer; colon cancer; thyroid cancer; kidney cancer	20–34%

recombination (Figure 9.4). When either of the *BRCA1* or *BRCA2* genes is mutated or altered, a loss of the remaining wild-type allele (loss of heterozygosity) will impair the DNA repair response, allowing the cell to accumulate additional genetic alterations that may lead to malignant transformation.

Poly(ADP-ribose) polymerases, abbreviated as PARPs, are a group of proteins that play a central role in repair of **single-strand DNA breaks** (SSBs). PARPs detect SSBs, bind DNA, and synthesize a poly(ADP-ribose) chain that signals other critical DNA repair enzymes to travel to the site of DNA damage. This mode of DNA repair is termed **base-excision repair** (BER). Normal cells have both BER and homologous recombination pathways (mediated by BRCA proteins) to repair damaged DNA (Figure 9.5). Cells with mutant BRCA proteins can compensate for nonfunctioning homologous recombination through intact BER or other DNA repair mechanisms. Similarly, cells with intact BRCA pathways can overcome loss of BER function via PARP inhibitors by repairing DNA damage through homologous recombination. However, cancer cells with absent BRCA function that are treated with PARP inhibitors have neither homologous recombination nor BER pathways to repair DNA damage, leading to cell death. Exploitation of the tumor-specific BRCA defect with PARP inhibition allows for selective tumor cytotoxicity, or a synthetic-lethality therapeutic approach. Case 9-1 highlights the link between genetic mutations and breast cancer development and management.

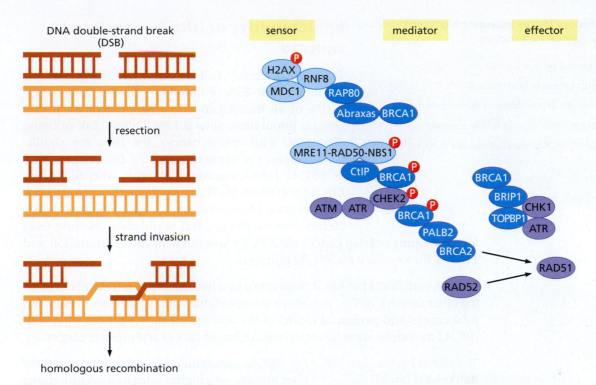

Figure 9.4. Roles of BRCA1 and BRCA2 in DNA double-strand break repair. In response to DNA double-strand breaks (DSBs), sensors (*light blue*) detect the damage, and signaling mediators recruit or activate effectors that repair the damage and activate cell cycle checkpoints. BRCA1-containing macro-complexes (*dark blue*) are crucial mediators of the DNA damage response. The BRCA1–PALB2–BRCA2 complex is important in mediating RAD51-dependent homologous recombination (HR). CHEK2-dependent phosphorylation of serine 988 (S988) in BRCA1 appears to be required for the BRCA1–PALB2–BRCA2 effector complex, which is important in RAD51-mediated HR. DNA damage is also recognized by ataxia telangiectasia-mutated (ATM) and ATR kinases, which phosphorylate BRCA1 and BRCA1-associated proteins, and mediate signaling to form macro-complexes and activate cell cycle checkpoints. Thus, multiple gene products mutated in hereditary breast cancer syndromes interact in this DNA repair process. (From R. Roy, J. Chun and S.N. Powell, *Nature Rev. Cancer* 12:68–78, 2012. Reprinted by permission from Macmillan Publishers Ltd.)

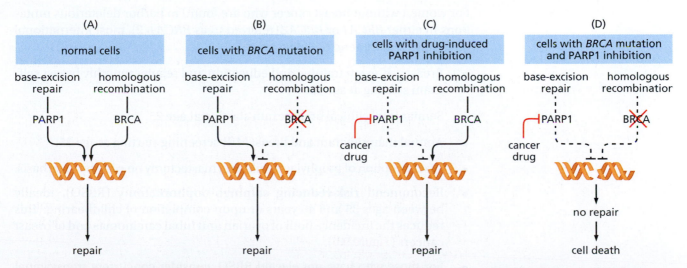

Figure 9.5. Mechanism of cell death from synthetic lethality in the BRCA context. (A) In normal cells, both base-excision repair (BER) and homologous recombination DNA repair pathways are functional and can repair damaged DNA. (B) In cells with *BRCA1* or *BRCA2* mutations and consequent nonfunctional homologous recombination, BER and other DNA-repair processes can compensate and still repair damaged DNA. (C) In cells with functional homologous recombination (at least one functioning copy of *BRCA1* and *BRCA2*) but loss of BER due to PARP1 inhibition, homologous recombination is intact and can repair damaged DNA. (D) In the cancer cells of mutation carriers, treatment with a PARP inhibitor combined with *BRCA* mutation renders the cell unable to repair DNA damage by homologous recombination or base-excision repair, with consequent cell death. (From J.D. Iglehart and D.P. Silver, *N. Engl. J. Med.* 361:198–191, 2009. Reprinted with permission from the Massachusetts Medical Society.)

Table 9.2. Family history factors indicating higher risk of having a harmful *BRCA1/2* mutation.

Breast cancer diagnosed before age 50
Family member with cancer in both breasts
Both breast and ovarian cancers among family members
Multiple breast cancer cases among family members
Two or more types of *BRCA1/2*-related cancers in a single family member
Male breast cancer
Ashkenazi Jewish ethnicity

Epidemiology of *BRCA*-associated cancers

Together, harmful *BRCA1* and *BRCA2* mutations account for 20–25% of hereditary breast cancers and 5–10% of all breast cancers. Whereas women in the general population have a 12% lifetime risk of being diagnosed with breast cancer, the risks are significantly higher for women harboring *BRCA1* (55–65%) or *BRCA2* (45%) mutations. Women with *BRCA1* or *BRCA2* mutations also have significantly increased risk of developing ovarian cancer. Compared with the 1.4% lifetime risk in the general population, lifetime risks for developing ovarian cancer are 39% for women with *BRCA1* mutation and 11–17% for women with *BRCA2* mutation.

Patients with *BRCA1* or *BRCA2* mutations also have an increased risk of developing other cancers. *BRCA1* mutations are associated with higher rates of fallopian tube cancer and peritoneal (lining of the abdominal cavity) cancer. Men with *BRCA2* mutations are at increased risk for breast cancer and prostate cancer.

The risk of harboring a *BRCA1* or *BRCA2* mutation is highest among people of **Ashkenazi Jewish** descent. Other groups with higher rates of such mutations compared with the general populations include people of Norwegian, Dutch, and Icelandic origin. Testing for *BRCA1* or *BRCA2* mutations is not routinely performed on the general population but is recommended for individuals considered to be at high risk based on family history (Table 9.2). Before genetic testing, patients should be counseled about the medical implications and psychological risks and benefits of a positive or negative test result. Additionally, patients should be made aware of the risk and consequences of passing a mutation onto their children.

Recommendations for *BRCA1/2* asymptomatic carriers

For women without breast cancer who are found to harbor deleterious mutations in either *BRCA1* or *BRCA2* (referred to as *BRCA1/2*), most international guidelines recommend the following:

- Breast self-exam training and education and regular monthly breast self-exam starting at age 18.

- Semiannual clinical breast exam starting at age 25.

- Annual mammogram and breast MRI screening starting at age 25.

- Discuss option of prophylactic bilateral mastectomy on case-by-case basis.

- Recommend **risk-reducing salpingo-oophorectomy** (RRSO), ideally between ages 35 and 40 years or upon completion of childbearing. This reduces the incidence both of ovarian and tubal carcinomas and of breast cancers (Figure 9.6).

- For those who have not elected RRSO, consider concurrent transvaginal ultrasound and CA-125 tumor marker testing every six months starting at age 30 or 5–10 years earlier than the earliest age of first diagnosis of ovarian cancer in the family. Note that CA-125 is a blood test with poor sensitivity to detect ovarian cancer as a screening measure in *asymptomatic* women at *average* risk of developing ovarian cancer, but the blood test may have value in high-risk, *BRCA* mutation-positive women.

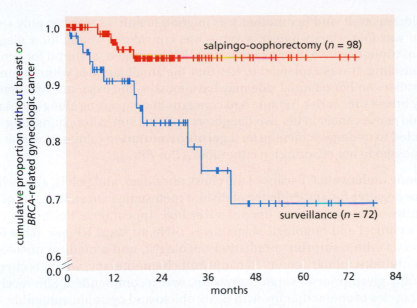

Figure 9.6. Effect of prophylactic bilateral salpingo-oophorectomy on breast and gynecologic cancer incidence in *BRCA1/2* mutation carriers. In a study of 170 women over age 35 with *BRCA* mutations, 98 had risk-reducing salpingo-oophorectomy (RRSO) and 72 did not. Patients were followed and screened for breast and gynecologic malignancies. The Kaplan–Meier curve demonstrates the proportion of women free of breast or gynecologic cancers during the observation period. Women who had RRSO had a 75% decrease in risk of developing breast or gynecologic cancers (hazard ratio, 0.25; 95% confidence interval, 0.08–0.74; p = 0.006). (From N.D. Kauff et al., *N. Engl. J. Med.* 346:1609–1615, 2002. Reprinted with permission from the Massachusetts Medical Society.)

- Consider **chemoprevention** options for breast and ovarian cancer (for example, tamoxifen taken daily for five years or, in post-menopausal women, exemestane taken daily for five years).

- Consider investigational imaging and screening studies.

The case of Yvonne Charles, a young mother with a breast lump

Yvonne was 37 years old and gave birth to her third child just five weeks before presenting for evaluation of a breast mass. She had noticed both breasts enlarging, as expected during and for some time after pregnancy, but in the couple of months previous to presenting, her left breast appeared particularly lumpy. Her obstetrician noted *"peau d'orange"* skin changes on the inferior portion of her left breast. *Peau d'orange* translated means "peel of the orange," which denotes the swollen, dimpled appearance of the skin. The manifestation of *peau d'orange* can be caused by an underlying lymphatic obstruction stemming from breast cancer (Figure 9.7). A mammogram of Yvonne's breasts revealed bilateral dense tissue (as would be expected in a young, lactating woman) but no suspicious findings to indicate breast cancer (Figure 9.8). However, because of the *peau d'orange*, her obstetrician's concern persisted and more tests were ordered. A breast ultrasound and a needle biopsy of the abnormal area (Figure 9.9) revealed a high-grade, invasive ductal carcinoma. Immunohistochemical tests were negative for the hormone receptors for estrogen and progesterone (ER and PR), as well as HER2. The **triple-negative** staining for these receptors categorized Yvonne's tumor in the "basal-like" group, associated with the most aggressive natural history among the five breast cancer subtypes (see Case 6-1). Ultrasound of her left axillary (armpit) lymph nodes to capture the lymphatic drainage from the left breast did not reveal any enlarged or abnormal lymph nodes.

Yvonne's family history was noteworthy, because her paternal grandmother had been diagnosed with breast cancer at age 50. There were no other cases of cancer in the family, as far as Yvonne knew. Her father, an only

Figure 9.7. *"Peau d'orange"* appearance of skin overlying an invasive breast cancer. The term means "orange skin" in French and describes skin that has the dimpled texture of an orange. (Photo courtesy of Michaela Higgins, M.D.)

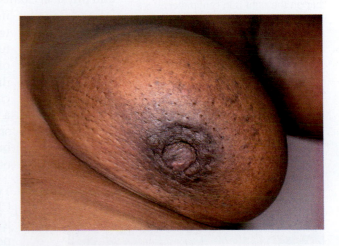

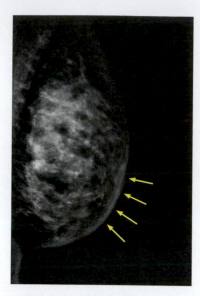

Figure 9.8. Left breast mammogram. The image shows heterogeneously dense glandular tissue with an area of mild skin thickening (*yellow arrows*).

Figure 9.9. Ultrasound evaluation of area of clinical concern in the left breast. In the six o'clock position, a 15-mm irregular hypoechoic mass is identified (*dashed yellow circle*). Ultrasound-guided biopsy of this area was performed (indicated as a line on the inset diagram of the breast position). Scale bar on the right is in centimeters.

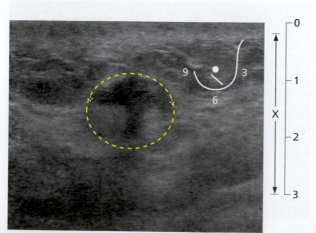

child, was well, and her mother was in good health, as was her only sister, who was four years younger. Yvonne was considered young for a diagnosis of triple-negative breast cancer, and this prompted a referral for genetic counseling. It was explained to her that the small families without female members on her paternal side may have masked the existence of a hereditary breast cancer risk. Yvonne had concerns about passing along genes that could cause cancer in her two daughters and one son. After counseling, she elected to undergo screening for a **germline mutation** (a mutation that can be passed to her offspring) in either *BRCA1* or *BRCA2*.

Yvonne underwent CT scans of the chest, abdomen, and pelvis, all of which were negative for any evidence of distant metastatic cancer. Because of the skin involvement causing the *peau d'orange* appearance, her breast cancer's clinical and radiologic staging was T4N0, or stage III (see Table 6.3). She met with a surgeon, a radiation oncologist, and a medical oncologist in a multidisciplinary clinic. Neoadjuvant chemotherapy, which is chemotherapy given prior to surgical excision, was recommended. She received four cycles of chemotherapy with doxorubicin and cyclophosphamide, followed by four cycles of an additional chemotherapy drug, paclitaxel, and all cycles were well tolerated. After the first cycle of therapy, Yvonne did notice softening of her left breast and improvement in her *peau d'orange*. Upon completion of her neoadjuvant chemotherapy, she no longer had an appreciable breast mass. A left mastectomy and sentinel lymph node biopsy were planned.

During genetic testing, a deleterious mutation in *BRCA1* was identified. Yvonne decided to undergo a prophylactic (preventative) right mastectomy at the same time as her modified left mastectomy and sentinel lymph node biopsy, with immediate implant reconstructions. She recovered well from surgery. Pathologic examination of her left breast and sentinel lymph node did not identify any viable cancer cells, denoting a **pathologic complete response** (pCR) to chemotherapy. Due to the large size of the original breast tumor that caused changes in the overlying skin, Yvonne received adjuvant post-mastectomy radiation. No adjuvant hormonal therapy was recommended because her cancer did not express receptors for estrogen or progesterone.

Yvonne's family was invited for genetic counseling, due to her *BRCA1* mutation. Her father tested positive for the same *BRCA1* mutation, but both her mother and sister tested negative. Yvonne and her husband elected to defer any testing of their young children until adulthood.

Yvonne continued with post-treatment follow-up, including periodic history and physical exams. Because she had undergone double mastectomy, she did not require mammograms. One year after her double mastectomy, Yvonne elected to have her ovaries and fallopian tubes removed (salpingo-oophorectomy) to reduce the risk of ovarian cancer development. Three years after her initial diagnosis, at age 40, Yvonne complained to her primary care physician of a persistent cough. A chest X-ray indicated a possible mass in her lung, and a chest CT scan confirmed small, bilateral masses that were suspicious for metastatic breast cancer (Figure 9.10). Other scans did not reveal any other sites of disease.

Yvonne discussed the findings with her medical oncologist. Chemotherapy options were considered, as well as a clinical trial of a novel class of treatments, poly(ADP–ribose) polymerase (PARP) inhibitors. Yvonne consented to enroll in a clinical trial of the PARP inhibitor olaparib, hoping that participation in a study of a novel treatment could lead to improved therapies for other patients, potentially including her own children if they inherited the *BRCA1* mutation. After three months on olaparib, her lung lesions diminished in size but did not disappear entirely. Besides mild fatigue, which she attributed to caring for her three young children, Yvonne tolerated olaparib very well. She has been on olaparib therapy for three years and counting.

Neoadjuvant therapy for breast cancer

Chemotherapy is given to some women with breast cancer early in treatment to reduce the risk of recurrence. Large studies have shown that chemotherapy given before surgery (neoadjuvant treatment) or afterward (adjuvant treatment) results in equivalent outcomes in terms of disease-free and overall survival (OS). The advantages of neoadjuvant treatment are that it may downsize a tumor and allow for breast conservation in cases where a mastectomy would otherwise have been necessary, and it provides the opportunity to learn about the tumor's chemosensitivity *in vivo*. In young women with breast cancer, or others in whom a hereditary breast cancer syndrome is suspected, the time taken to deliver neoadjuvant chemotherapy allows for concurrent genetic testing and the availability of results, which can influence the surgery. For example, in Yvonne's case, the detection of a mutation confirmed that the **contralateral** (opposite side) breast was also at high risk of developing cancer in her lifetime, so whether to undergo prophylactic mastectomy of that breast became a crucial decision that she needed to make before her cancer operation.

Approximately 25% of the time after neoadjuvant chemotherapy is administered, no remaining cancer cells are identified in the breast or axilla at the time of surgery. This is termed a pathologic complete response (pCR), and portends a very good prognosis. Such a response is most likely to occur in HER2-positive tumors when treated with combination chemotherapy and HER2-targeting drugs, such as trastuzumab (see Case 11-3) or pertuzumab; or in high-grade triple-negative tumors like Yvonne's.

Management of localized *BRCA1/2*-related breast cancer

At this time, the standard treatment of *BRCA1/2*-related breast cancers does not differ from the treatment of spontaneous breast cancers in noncarriers. Therefore, women with *BRCA1/2* mutations and high-risk early breast cancer should be offered the same neoadjuvant and adjuvant systemic treatment options as their peers. Where available, participation in clinical trials of novel therapies for this small population should be encouraged. Increased sensitivity to platinum-based chemotherapy is frequently reported among *BRCA1/2* carriers, and for the treatment of metastatic cancer, many oncologists would therefore select a single-agent platinum compound for this population, such as cisplatin or carboplatin. Although there are some data that these agents are also effective in early breast cancer, there are no large prospective studies comparing outcomes with current standard-of-care combination chemotherapy regimens, such as doxorubicin/cyclophosphamide/paclitaxel (see Case 6-1). Therefore, outside of clinical trial protocols, cisplatin and carboplatin are rarely used for treatment of early breast cancer.

Management of metastatic *BRCA1/2*-related breast cancer

Similar to localized disease, the standard management of metastatic *BRCA1/2*-related breast cancer is the same as metastatic disease in noncarriers. The use of

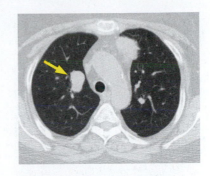

Figure 9.10. Chest CT image from Yvonne's case. An axial view (horizontal slicing through the body, viewed looking up from the patient's feet; the patient's right side is on the viewer's left). The yellow arrow points to a round lesion that is consistent with a metastatic nodule.

PARP inhibitors for the treatment of *BRCA1/2* mutation-related breast cancer is the subject of ongoing research and excitement.

There are now more than ten PARP inhibitors in development. The oral drug olaparib (trade name Lynparza) is a PARP inhibitor developed for the treatment of solid tumors. It is the first of this drug class to receive FDA approval for use in *BRCA* mutation-positive ovarian cancer. FDA approval was given on December 19, 2014. In a randomized Phase II study, subjects with recurrent ovarian cancer received olaparib or placebo. Olaparib was associated with an overall response rate of 34%, and an 8-month improvement in progression-free survival compared with placebo. Importantly, the FDA also approved the genetic test BRACAnalysis CDx as a companion diagnostic to detect *BRCA* mutations, highlighting the importance of codevelopment of accurate diagnostics and effective therapeutics. In 2018, olaparib was FDA approved for metastatic breast cancer patients who previously received chemotherapy and carry a germline *BRCA1/2* mutation.

Accumulating evidence suggests that PARP inhibitors may have a wider application in the treatment of cancers with defective DNA damage repair pathways, which include prostate, lung, endometrial, and pancreatic cancers. In a recent Phase II trial, the efficacy and safety of olaparib in a spectrum of *BRCA1/2*-associated metastatic cancers was evaluated. Responses to olaparib were observed across multiple tumor types, producing an overall response rate of 26%. The most common adverse events associated with PARP inhibitor treatment are fatigue, nausea, and vomiting.

Future prospects and challenges in *BRCA*-mutant cancers

Cancers caused by harmful genetic mutations pose unique challenges. The importance of cancer screening for carriers of such mutations cannot be overemphasized. For patients carrying a germline *BRCA* mutation but who are asymptomatic and cancer-free, the data for prophylactic treatments, such as mastectomy and salpingo-oophorectomy, are compelling. Nevertheless, whether a patient should undergo prophylactic treatment requires careful consideration. The resultant effects on body image and fertility, let alone the medical risks of surgery, are not to be taken lightly. For *BRCA* mutation-associated breast cancer, the standard treatment is the same as for sporadic breast cancer. The emergence of olaparib and other PARP inhibitors points the way toward targeted therapies based on synthetic lethality.

Take-home points

✓ Individuals who harbor a deleterious mutation in either the *BRCA1* or the *BRCA2* gene are at increased risk of breast and other cancers.

✓ *BRCA1*-related breast cancers are most likely to be "triple-negative."

✓ Neoadjuvant chemotherapy can be given before primary breast surgery to downstage the tumor and provide evidence of chemosensitivity.

✓ *BRCA*-related breast cancers are currently treated as per guidelines for spontaneous breast cancers; however, there is evolving literature to suggest that they may be particularly sensitive to platinum chemotherapy agents and poly(ADP-ribose) polymerase (PARP) inhibitors.

✓ PARP inhibitors may be particularly toxic to *BRCA*-deficient cells in a manner that is based on synthetic lethality.

✓ Olaparib is the first PARP inhibitor to be approved for use in *BRCA*-related ovarian cancer and breast cancer.

Discussion questions

1. What clinical recommendations would you give to a healthy individual who has recently learned that she harbors a deleterious *BRCA1* mutation?

2. Why is hormone therapy not typically used for breast cancer patients with harmful *BRCA* mutations?

3. Explain the concept of synthetic lethality.

| Case 9-2 | Multiple Myeloma |

Introduction

Multiple myeloma (MM) is a disease characterized by proliferation of malignant plasma cells in the bone marrow (Figure 9.11). Normal plasma cells in the body are the final differentiation step in lymphocyte development and are responsible for the production and secretion of antibodies, which our bodies use to fight infection. Myeloma cells accumulate in the bone marrow, an occurrence that may result in decreased red cell production (anemia). Another characteristic feature of myeloma is activation of **osteoclasts**, which are cells that resorb bone, leading to destructive bony lesions known as **osteolytic metastases** (Figure 9.12) and elevated calcium levels in the blood (**hypercalcemia**). The proliferating myeloma cells produce abnormal levels of a monoclonal immunoglobulin, called a paraprotein or monoclonal protein (**M protein**), which may cause injury to kidneys, resulting in renal failure. MM belongs to a spectrum of diseases known as plasma cell dyscrasias or **monoclonal gammopathies**, which include monoclonal gammopathy of undetermined significance (MGUS), solitary **plasmacytoma** (an isolated myeloma tumor), Waldenström macroglobulinemia, and amyloid light-chain (AL) amyloidosis.

Asymptomatic (or smoldering) multiple myeloma—that is, MM without the adverse clinical features of hypercalcemia, renal failure, anemia, or bone disease—may be monitored without active therapy. The features of active myeloma requiring therapy follow established criteria, abbreviated "**CRAB**" (hypercalcemia, renal insufficiency, anemia, and bone disease).

Epidemiology and etiology

MM is the second most common hematologic malignancy in the United States, with over 30,000 new cases diagnosed annually and over 12,000 deaths. It is a disease of older adults, with median age at diagnosis of 66 years and 30% of cases occurring over the age of 75. MM is generally preceded by monoclonal

Topics bearing on this case

Proteasome inhibitors

Ubiquitin-proteasome pathway (UPP)

Osteolytic metastases

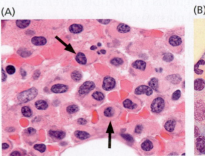 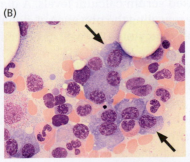

Figure 9.11. Morphologic findings in multiple myeloma. The bone marrow core biopsy (A; hematoxylin and eosin stain) and aspirate smear (B; Wright–Giemsa stain) both contain clusters of plasma cells (*arrows*) with eccentric nuclei, clumped chromatin, and moderate amount of cytoplasm. (Photographs courtesy of Dr. Aliyah Sohani.)

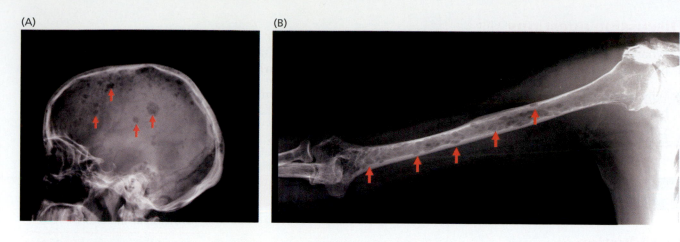

(A)

(B)

Figure 9.12. Lytic bone lesions in multiple myeloma. Multiple myeloma is frequently characterized by lytic bony lesions (arrows mark some of the innumerable lesions), as seen in the skull (A) and in the humerus (B). Bone lesions are part of the "CRAB" diagnostic criteria in multiple myeloma (hypercalcemia, renal dysfunction, anemia, bone lesions).

gammopathy of undetermined significance (MGUS), a common asymptomatic condition defined by the presence of monoclonal gammopathy (abnormal proliferation of lymphoid cells that produce immunoglobulins) alone. MGUS progresses to MM at a rate of roughly 1% of affected patients per year. It is estimated that MGUS is present in 3 to 4% of the general population. Men are affected by MM more frequently than women (1.6:1 ratio), and disease prevalence in patients of African descent is twice that of individuals of European descent.

The etiology and the mechanisms of myeloma progression are still largely unknown. Several studies have shown an association between high body mass index and risk of myeloma. Specifically, it has been observed that obese individuals tend to have higher levels of cytokines, such as interleukin-6 (IL-6), or of insulin-like growth factor (IGF-1), produced by adipocytes (fat cells), which are potent growth factors for myeloma cells. Occupational exposure to pesticides, organic solvents (benzene, petroleum derivatives, and styrene), or chronic radiation has been associated with development of MM. An association between myeloma risk and autoimmune diseases, such as rheumatoid arthritis or pernicious anemia, or certain chronic infections, such as HIV and hepatitis C, has been proposed, suggesting an immune-mediated mechanism for malignant transformation. No consistent associations have been observed with any particular diet, alcohol consumption, or smoking.

Diagnosis, workup, and staging

The most common clinical finding at diagnosis of MM is anemia, present in approximately three-quarters of patients, and it often results in fatigue. Bone pain related to skeletal involvement is present in over half of patients, while approximately one-quarter of patients will have unintentional weight loss or increased serum calcium levels. A diagnosis of multiple myeloma requires a biopsy revealing the presence of clonal plasma cells in bone marrow, or a solitary plasmacytoma, as well as detection of a monoclonal paraprotein in the blood or urine (Table 9.3; Figure 9.13). Myeloma is then classified as symptomatic if it meets any of the CRAB criteria for organ impairment. Smoldering myeloma is diagnosed if no CRAB criteria are present and the patient has either greater than 3 g/dl of monoclonal protein in the blood or more than 10% clonal plasma cells in the bone marrow. The presence of a monoclonal paraprotein at less than 3 g/dl in the serum with fewer than 10% clonal plasma cells is characterized as MGUS.

Table 9.3. Diagnostic criteria for multiple myeloma.

Multiple myeloma, symptomatic	Multiple myeloma, asymptomatic (smoldering)	Monoclonal gammopathy of uncertain significance (MGUS)
Monoclonal protein in blood or urine; bone marrow with clonal plasma cells, or plasmacytoma; CRAB criteria*	Monoclonal protein in blood measuring >3 g/dl and/or bone marrow with >10% clonal plasma cells; no CRAB criteria present	Monoclonal protein in blood measuring <3 g/dl; bone marrow with 10% clonal plasma cells in bone marrow; no CRAB criteria present; no evidence of other B-cell lymphoproliferative disease

*CRAB criteria: hypercalcemia, renal insufficiency, anemia, or bone lesions

Metaphase cytogenetic studies and interphase fluorescence *in situ* hybridization (FISH) analysis are performed on myeloma cells from bone marrow aspirates to evaluate for the presence of chromosomal abnormalities, such as deletion of chromosome 17 (containing the *TP53* gene) as well as the t(4;14) or t(14;16) translocations, which are associated with higher-risk disease and an inferior prognosis (Table 9.4).

Hematologic evaluation includes a complete blood count with differential, a comprehensive serum metabolic panel for the detection of hypercalcemia and renal insufficiency, and tests for β-2-microglobulin, C-reactive protein, and lactate dehydrogenase (LDH). The amount of β-2-microglobulin correlates with the burden of disease and is used for staging and prognosis. Myeloma proteins are assessed using **serum protein electrophoresis** (SPEP) with **immunofixation**, which identifies normal and abnormal immunoglobulins in the blood, as well as quantitation of immunoglobulin levels; serum free light-chain assay; and a 24-hour urine collection to quantitate total urinary protein and detect an immunoglobulin light chain known as the Bence Jones protein, using urine electrophoresis with immunofixation. About 15% of patients will have light-chain only disease, where disease is detectable primarily by the serum free light-chain assay or by urine studies.

The initial radiographic evaluation of newly diagnosed MM includes a complete skeletal survey, which includes X-rays of the chest, spine, arms, legs, skull, and pelvis. Approximately 80% of MM patients will have evidence of

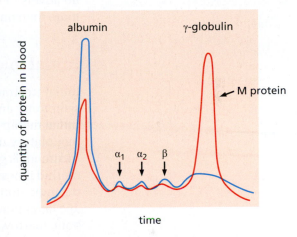

Figure 9.13. Serum protein electrophoresis in multiple myeloma. Heights of peaks on the graph represent quantities of specific proteins in the blood. Blue represents normal serum with normal serum proteins. The red curve represents a multiple myeloma patient with the presence of an abnormal paraprotein also called the M protein (*arrow*) and a decrease in the normal albumin level.

Table 9.4. Risk stratification of multiple myeloma based on genetics.

Risk	Karyotype or FISH
Standard-risk	Hyperdiploidy t(11;14) t(6;14)
Intermediate-risk	t(4;14) Deletion 13 or hypodiploidy by conventional karyotyping
High-risk	17p deletion t(14;16) t(14;20) High-risk gene expression profiling signature

Table 9.5. International Staging System (ISS) in multiple myeloma.

Stage	Criteria
I	Serum β-2-microglobulin <3.5 mg/l and serum albumin ≥3.5 g/dL
II	Serum β-2-microglobulin <3.5 mg/l but serum albumin <3.5 g/dl or serum β-2-microglobulin 3.5 to <5.5 mg/l, irrespective of serum albumin
III	Serum β-2-microglobulin ≥5.5 mg/l

bone involvement on skeletal survey, with 67% of all patients having lytic bony disease and 20% having osteoporosis or fractures. However, the sensitivity of X-rays is limited, and it may miss 10–20% of early lytic lesions. CT, MRI, and fluorodeoxyglucose positron-emission tomography (FDG-PET) are more sensitive than the traditional skeletal survey and better capture early bone disease, the extent of bone disease, and other disease outside of the bone marrow.

MM is staged according to the International Staging System (ISS), which stratifies patients according to serum β-2-microglobulin and albumin levels into three groups (Table 9.5). At the time this system was created, patients with stage I disease had a median overall survival of 62 months; stage II disease, median survival 44 months; and stage III disease, 29 months. Combining ISS staging and FISH findings can further provide prognostic information, with median survival ranging from 68 months in patients with ISS stage I to II and no adverse FISH results to 19 months in patients with ISS stage II to III and more than one adverse FISH finding.

Initial management of multiple myeloma

Initial treatment for MM generally consists of induction therapy to alleviate symptoms and reduce disease burden, followed by a maintenance phase. Initial therapy will often include a combination of drugs, including immuno-modulatory drugs, proteasome inhibitors, and steroids. Selection of precise treatment is guided by the patient's age and fitness, as well as presence of co-morbid illnesses such as renal insufficiency. In patients who are medically fit, the induction phase is typically followed by a remission consolidation phase. The latter phase consists of high-dose melphalan chemotherapy to ablate the bone marrow, including the MM, and infusion of previously collected autologous stem cells to repopulate the healthy bone marrow, a process known as an autologous stem cell transplant. Conventionally, clinical trials of high-dose chemotherapy and autologous stem cell transplantation in MM have enrolled patients 65 years of age or younger. In the United States, greater emphasis is placed on the patient's "physiologic" age and co-morbidities rather than chronologic age. Allogeneic stem cell transplant, where stem cells are taken from a donor, remains investigational in the initial treatment of patients, unlike autologous stem cell transplantation, which is considered routine. Supportive treatments with bisphosphonates are also a key component of MM care. Bisphosphonates such as zoledronic acid help strengthen bone, decrease bone resorption, reduce pain from lytic bony lesions, prevent fractures, and also improve survival (see Case 10-2). The RANK ligand inhibitor denosumab inhibits osteoclast maturation and is also available to reduce bone resorption and treat hypercalcemia in patients with multiple myeloma.

While MM is not curable, it is highly treatable, with the majority of patients enjoying symptomatic improvement and significant disease response. However, since MM is not curable, the disease course invariably involves relapse, retreatment, response, subsequent relapse, and so forth.

The case of Malik Jackson, a 39-year-old man who presented with anemia and multiple lytic bone lesions

Mr. Jackson was 39 years old when he was found to have an IgA kappa monoclonal protein of 2.37 g/dl during workup for a mild anemia. At the time, he had no chronic medical problems, was a nonsmoker, and exercised regularly, but he felt a decrease in endurance while running and an increase in overall fatigue. His primary care physician ordered blood tests, including a complete blood count, chemistry panel, and thyroid evaluation. Blood tests indicated that Malik was anemic, with a hematocrit of 32% (normal range, 41–53%), and that he had normal kidney function. Further testing by his physician indicated that he was not deficient in any of the factors causing anemia that can be readily treated with supplements, such as iron, folic acid, or vitamin B_{12}. Malik was referred to a hematologist for evaluation of his anemia.

The hematologist noted that the original chemistry panel indicated elevated immunoglobulins, a potential indicator of a monoclonal gammopathy, and therefore, the physician pursued a diagnostic workup for multiple myeloma. SPEP with immunofixation revealed an M protein at 2.37 g/dl (see Figure 9.13). Serum β-2-microglobulin was measured at 3.8 mg/l. Multiple lytic bone lesions were seen on skeletal survey. Given the strong suspicion for a diagnosis of MM, a bone marrow biopsy was performed, which showed 34% clonal plasma cells, confirming the diagnosis.

Malik was diagnosed with multiple myeloma and treated with induction chemotherapy of the three drug regimen RVD, which stands for Revlimid (lenalidomide), Velcade (**bortezomib**), and dexamethasone. He completed eight cycles of RVD and achieved a very good partial response (>90% reduction in monoclonal protein). He then proceeded to high-dose chemotherapy with autologous stem cell support. His M protein, serum β-2-microglobulin, and bone marrow plasma cell fraction all normalized.

Unfortunately, three years after high-dose chemotherapy, he presented with back pain and was found to have a 7.2-cm lytic soft tissue mass involving the left fourth anterior rib. The mass was biopsied under CT guidance, and pathology revealed homogeneous clonal plasma cells (a plasmacytoma), consistent with relapsed multiple myeloma. Given his disease progression, he was then treated with carfilzomib/lenalidomide/dexamethasone treatment (CRd), and achieved a second remission.

Proteasome inhibition and multiple myeloma

The introduction of newer, more effective treatments, such as proteasome inhibitors, has transformed the treatment of MM by significantly improving overall survival following disease relapse. For example, one study found that patients who relapsed after the year 2000 had an overall survival of two years compared with one year in patients who relapsed prior to this modern era. Indeed, overall survival in MM from time of diagnosis has improved over the years, with median OS of 6.1 years in patients diagnosed from 2006 to 2010 compared with 2.5 years in patients diagnosed before 1996. The increase in OS is largely due to novel agents, such as proteasome inhibitors and immunomodulatory drugs (see Case 10-2 for further discussion), as well as use of high-dose melphalan and autologous stem cell transplant.

The regulated degradation of proteins is key to normal cell homeostasis. This is accomplished by the proteasome, a cylindrical, multi-enzyme complex found in all eukaryotic cells (see Figure 9.2). Proteins are flagged for disposal by the proteasome by the attachment of ubiquitin polypeptides through the ubiquitin-proteasome pathway.

While normal cells are universally dependent on proteasome function, malignant cells are even more dependent on proteasome function and thus more vulnerable to its inhibition. Bortezomib reversibly binds to the 20-S subunit of the proteasome. Inhibition of the proteasome by bortezomib in cell culture leads to apoptosis across multiple types of cancer cell lines. The effectiveness of bortezomib is particularly profound in MM, where cells produce large amounts of monoclonal immunoglobulin, a defining characteristic of this malignancy. Early clinical trials across different tumor types showed that bortezomib was unusually active in MM.

In MM cells, treatment with bortezomib results in the accumulation of unfolded proteins, activating the unfolded protein response. Bortezomib interferes with the clearance of proteins involved in cell cycle regulation and apoptosis, such as cyclins and **caspases**, leading to cell cycle arrest. Bortezomib also increases the levels of the NFκB inhibitor, IκB, thereby inhibiting the activity of NFκB, a transcription factor that contributes to the generation of anti-apoptosis factors and cell survival. The unfolded protein response is a cellular stress response that ultimately triggers the induction of cell death via apoptosis.

Importantly, bortezomib can synergize with other anti-myeloma drugs, allowing the use of lower doses to achieve increased cytotoxic effects. This cooperativity is leveraged in several triplet regimens commonly used in MM, such as the RVD regimen that was initially prescribed for Mr. Jackson.

Proteasome inhibitors, including bortezomib, play a core role in the treatment of MM at all stages of illness. Bortezomib was first FDA approved in May 2003 for patients with relapsed disease based on the results of two Phase II clinical trials, including the SUMMIT trial, which showed an overall response rate of 27% to bortezomib as a single agent. These patients had multiple lines of prior treatment, with a median of six prior therapies. A Phase III trial, the APEX trial, demonstrated higher response rates (overall response rate 38% versus 18%) and superior overall survival (1 year OS 80% versus 66%) with single-agent bortezomib compared with high-dose dexamethasone in patients who relapsed after one to three prior types of systemic myeloma treatment.

Following demonstration of efficacy in the relapsed setting, bortezomib was incorporated into initial induction treatment regimens to be used prior to autologous stem cell transplant. A **meta-analysis** (a statistical method of combining results from different studies) of several Phase III trials showed that bortezomib-based induction regimens compared with non-bortezomib regimens had superior response rates, median progression-free survival (36 months versus 29 months), and three-year overall survival (80% versus 75%). Bortezomib was combined with lenalidomide and dexamethasone (creating the RVD regimen) in newly diagnosed patients and resulted in an encouraging overall response rate of 100%. In patients who are not considered eligible for intensive therapy with autologous stem cell transplant, bortezomib has also played a central role. For example, the VISTA trial randomized older patients to the combination of VMP (bortezomib, melphalan, and prednisone) versus MP and showed that the patients randomized to the bortezomib-containing arm lived longer (three-year OS of 69% versus 54%). Finally, bortezomib has also shown efficacy as maintenance therapy after autologous stem cell transplant.

Common, notable side effects of bortezomib treatment include thrombocytopenia (low blood platelet count) and **peripheral neuropathy** (damage of nerves in the peripheral nervous system). Grade 3 or 4 thrombocytopenia occurred in 30% of patients in the SUMMIT and CREST trials of bortezomib (Table 9.6). Unlike thrombocytopenia due to traditional chemotherapy, the thrombocytopenia with bortezomib is unique and may be due to reversible effects on megakaryocyte function rather than direct cytotoxicity. Bortezomib-induced thrombocytopenia is not cumulative. Peripheral neuropathy also occurred in 35% of patients in these trials and was grade 3 or 4 in 13% of patients

Table 9.6. Grading of common adverse events related to bortezomib.

Adverse event	Grade 1	Grade 2	Grade 3	Grade 4	Grade 5
Thrombocytopenia (low platelets)	75,000/µl to the lower limit of normal (typically 150,000/µl)	50,000/µl to 75,000/µl	25,000/µl to 50,000/µl	<25,000/µl	N/A
Peripheral neuropathy	Asymptomatic; loss of deep tendon reflexes	Moderate symptoms limiting instrumental activities of daily living (ADL), such as preparing meals	Severe symptoms limiting self-care ADL, such as bathing, dressing, feeding	Life-threatening consequences; urgent intervention indicated	Death

Standardized grading allows for comparison of adverse event severity between patients and statistical analysis across patient cohorts in clinical trials.

(see Table 9.6). This neuropathy is generally reversible if the drug is discontinued. Subcutaneous administration, rather than intravenous administration, and weekly instead of twice weekly dosing, have been associated with significantly less peripheral neuropathy with preserved efficacy. Finally, herpes zoster (also known as shingles) has been seen in increased frequency in bortezomib-treated patients. In the Phase III APEX study, herpes zoster occurred in 13% of patients compared with 5% in the dexamethasone arm. This risk can be mitigated with the routine use of the antiviral medication acyclovir (or equivalent) prophylaxis.

Next-generation proteasome inhibitors include **carfilzomib**, ixazomib, and oprozomib. The latter two are under investigation in clinical trials. The FDA approved carfilzomib in July 2012 for patients with relapsed disease who received at least two prior therapies, including bortezomib and an immuno-modulatory drug (for example, lenalidomide). Carfilzomib irreversibly binds to the proteasome and does not bind to other proteases. This is in contrast to bortezomib, which binds the proteasome reversibly and also inhibits other serine proteases (possibly accounting for its neurotoxicity). A Phase II trial of single-agent carfilzomib in heavily pre-treated MM patients who received a median of five prior lines of treatment showed an overall response rate of 24%. Notably, peripheral neuropathy was uncommon, and grade 3 or 4 neuropathy occurred in only 1% of patients.

The ASPIRE trial examined the combination of carfilzomib with lenalidomide and dexamethasone (CRd) compared with lenalidomide and dexamethasone in relapsed MM. Patients were eligible to participate if they received one to three prior lines of therapy. Prior lenalidomide and bortezomib or lenalido-mide treatment was permitted if there was no disease progression on these treatments; the majority of patients (80%) had not received prior lenalidomide therapy. The overall response rate was significantly higher in the carfilzomib arm compared with the control arm, 87% versus 67%. The median progression-free survival was 26 versus 18 months, favoring carfilzomib. This duration of response in the treatment arm was unprecedented, and serious adverse events were uncommon, though grade 3 or 4 dyspnea (shortness of breath, 3%), hypertension (4%), and heart failure (4%) were higher in the carfilzomib group.

Future prospects and challenges in multiple myeloma

Proteasome inhibitors are an important class of agents in oncology, best illustrated by their transformational role in multiple myeloma. Treatment with proteasome inhibitors, such as bortezomib and more recently carfil-zomib, especially in combination with other agents, has led to significant improvement in disease responses and overall survival. Ongoing trials with these drugs and newer proteasome inhibitors, such as ixazomib and oprozomib, are expected to yield additional, more effective, better tolerated, and more convenient treatment options for MM. Whether the combinatorial targeting of HDAC6 activity for the aggresome with proteasome inhibitors will provide additional benefit is the subject of ongoing studies as well.

Take-home points

✓ Multiple myeloma (MM) is the second most common hematologic malignancy in the United States.

✓ MM is not curable.

✓ MM is a monoclonal gammopathy, an abnormal proliferation of B lymphocyte-lineage cells producing high levels of antibodies that are known as the M protein.

✓ Adverse clinical features of MM include the CRAB criteria: hypercalcemia, renal insufficiency, anemia, and bone disease.

✓ Treatment of MM includes chemotherapy, autologous stem cell transplant, bisphosphonates, and proteasome inhibitors.

✓ The proteasome plays a key role in cellular homeostasis, and malignant cells, especially those of MM, are vulnerable to proteasome inhibition.

✓ Proteasome inhibitors, such as bortezomib and carfilzomib, are highly active in MM both as single agents and when combined with other anti-myeloma therapies, such as lenalidomide and dexamethasone. Treatment of MM with proteasome inhibitors has led to remarkable improvements in progression-free and overall survival.

✓ Side effects of bortezomib include thrombocytopenia, peripheral neuropathy, and risk of herpes zoster. Subcutaneous administration of bortezomib significantly improves tolerability with decreased risk of peripheral neuropathy.

Discussion questions

1. What are the clinical features of MM that warrant initiation of therapy?

2. What is the mechanism of action in a proteasome inhibitor?

3. What features might make MM more sensitive to proteasome inhibition compared with other cancers?

4. What are the main side effects of proteasome inhibitors?

Chapter summary

An appreciation of the principle of synthetic lethality and an understanding of the ubiquitin-proteasome pathway have enabled the design of therapeutic strategies that disrupt cancer cell metabolism. The success of these approaches encourages consideration of the development of additional drugs based on synthetic lethality and the creation of drugs that respond to the diverse spectrum of targets presented by the UPP. Some of the prospects and challenges accompanying further drug development in these areas are identified and briefly discussed below.

In addition to the proteasome complex, the remarkably complex UPP (see Figure 9.2) includes many distinct components, including three different classes of enzymes—E1s, E2s, and E3s—for ubiquitination and deubiquitinases (DUBs) for polyubiquitin trimming. Each of these components is potentially "druggable" by a small molecule. Case 9-2 illustrates the successful targeting of the proteasome by the inhibitor bortezomib. Other components of the UPP, and components of the auxiliary aggresome pathway, present an unusually rich variety of potentially druggable enzymes for drug discovery. Furthermore, there is optimism that selective inhibition of components of these pathways may allow a focused attack on cancer cells while perhaps minimizing the impact on normal cells.

The breast cancer case presented in this chapter illustrates the effectiveness of a synthetic lethality approach based on the loss of BRCA-mediated pathways. Loss of BRCA makes cancer cells critically dependent on PARP-mediated base excision repair. The realization that PARP, a druggable target, is a synthetic lethal interactor with BRCA-mediated DNA repair enabled the discovery of small-molecule inhibitors that could be used as drugs. The large number of genetic mutations typically found in cancer cells suggests that there are likely to be a great many opportunities for the development of therapies based on synthetic lethality. However, the major roadblock to the more rapid development of drugs designed to exploit synthetic lethality is the difficulty of identifying synthetic lethal interactors. Consequently, a great deal of effort is being invested in the development of systematic approaches to the discovery of synthetic lethal interactors.

High-throughput screening provides a general approach to the systematic discovery of synthetic lethals. This entails screening large collections of compounds against large libraries of many different cell lines, each bearing knockdowns or knockouts of a different and distinct gene. In many cases, due to metabolic redundancy, there will be different parallel pathways, each of these generating the same or an equivalent essential product or supporting an equivalent essential pathway. High-throughput screening allows testing of many compounds to identify those that block a parallel process whose absence, combined with the absence of the phenotype dictated by a particular gene, will result in the death of the cell. This approach requires large libraries of many thousands of cell lines, each line bearing a loss of function in both alleles of a particular gene. **RNA interference**-mediated knockdowns, or, better, **CRSPR/Cas9**-mediated knockouts, are the key technologies for engineering the creation of libraries of human cell lines with known and distinct gene knockdowns or knockouts.

Take-home points

✓ Drug discovery based on interference with protein degradation has led to proteasome inhibitors for multiple myeloma.

✓ The diversity of enzyme isoforms and groups making up the ubiquitin-proteasome pathway and the components of the auxiliary aggresome pathway provide a wide diversity of targets for drug discovery.

✓ Drug discovery based on the principle of synthetic lethality has yielded PARP inhibitors for some types of breast and ovarian cancers.

✓ Although synthetic lethality holds promise for the discovery of additional cancer-targeted therapies, synthetic lethal interactors have been very difficult to identify.

✓ High-throughput screening of large numbers of compounds on large libraries of appropriately engineered cell lines provides a systematic approach to the identification of druggable synthetic lethal interactions.

Chapter discussion questions

1. Consider Figure 9.1 and draw a similar figure that involves three independent genes, A, B, and C, each of which encodes a product that enables the same essential process. Assume that a mutation disabling gene A enabled a normal cell to become a cancer cell. Also assume that distinct inhibitors that block the products of each of these genes are available. Be sure your diagram shows which inhibitor or combination of inhibitors would be required to kill the cancer cell.

2. Why are there so few drugs based on synthetic lethality, and what might be done to identify more?

3. Consider E1, E2, E3, and the proteasome (see Figure 9.2). Would inhibition of any, some, or all of these classes of the UPP shut down all ubiquitin-dependent protein degradation? Justify your choice.

Selected references

Introduction

Bedford L, Lowe J, Dick LR et al. (2011) Ubiquitin-like protein conjugation and the ubiquitin-proteasome system as drug targets. *Nat. Rev. Drug Discov.* 10:29–46.

de la Cruz FF, Gapp BV & Nijman SM (2015) Synthetic lethal vulnerabilities of cancer. *Annu. Rev. Pharmacol. Toxicol.* 55:513–531.

Kaelin WG Jr (2009) Synthetic lethality: a framework for the development of wiser cancer therapeutics. *Genome Med.* 1:99.

Pal A, Young MA & Donato NJ (2014) Emerging potential of therapeutic targeting of ubiquitin-specific proteases in the treatment of cancer. *Cancer Res.* 74:4955–4966.

Simms-Waldrip T, Rodriguez-Gonzalez A, Lin T et al. (2008) The aggresome pathway as a target for therapy in hematologic malignancies. *Mol. Genet. Metab.* 94:283–286.

Case 9-1

Ashworth A (2008) A synthetic lethal therapeutic approach: poly(ADP) ribose polymerase inhibitors for the treatment of cancers deficient in DNA double-strand break repair. *J. Clin. Oncol.* 26:3785–3790.

Byrski T, Gronwald J, Huzarski T et al. (2010) Pathologic complete response rates in young women with *BRCA1*-positive breast cancers after neoadjuvant chemotherapy. *J. Clin. Oncol.* 28:375–379.

Fong PC, Boss DS, Yap TA et al. (2009) Inhibition of poly(ADP-ribose) polymerase in tumors from *BRCA* mutation carriers. *N. Engl. J. Med.* 361:123–134.

Iglehart JD & Silver DP (2009) Synthetic lethality—a new direction in cancer-drug development. *N. Engl. J. Med.* 361:189–191.

Kauff ND, Satagopan JM, Robson ME et al. (2002) Risk-reducing salpingo-oophorectomy in women with a *BRCA1* or *BRCA2* mutation. *N. Engl. J. Med.* 346:1609–1615.

Kaufman B, Shapira-Frommer R, Schmutzler RK et al. (2015) Olaparib monotherapy in patients with advanced cancer and a germline *BRCA1/2* mutation. *J. Clin. Oncol.* 33:244–250.

Ledermann J, Harter P, Gourley C et al. (2014) Olaparib maintenance therapy in patients with platinum-sensitive relapsed serous ovarian cancer: a preplanned retrospective analysis of outcome by *BRCA* status in a randomised phase 2 trial. *Lancet Oncol.* 15:852–861.

Roy R, Chun J, Powell SN et al. (2012) BRCA1 and BRCA2: different roles in a common pathway of genome protection. *Nat. Rev. Cancer* 12:68–78.

Case 9-2

Attal M, Harousseau J-L, Stoppa A-M et al. (1996) A prospective, randomized trial of autologous bone marrow transplantation and chemotherapy in multiple myeloma. *N. Engl. J. Med.* 335:91–97.

Greipp PR, San Miguel J, Durie BG et al. (2005) International staging system for multiple myeloma. *J. Clin. Oncol.* 23:3412–3420.

Herndon TM, Deisseroth A, Kaminskas E et al. (2013) U.S. Food and drug administration approval: carfilzomib for the treatment of multiple myeloma. *Clin. Cancer Res.* 19:4559–4563.

Marteijn, JA, Jansen JH & van der Reijden BA (2006) Ubiquitylation in normal and malignant hematopoiesis: novel therapeutic targets. *Leukemia* 20:1511–1518.

McCarthy PL, Owzar K, Hofmeister CC et al. (2012) Lenalidomide after stem-cell transplantation for multiple myeloma. *N. Engl. J. Med.* 366:1770–1781.

Moreau P, Richardson PG, Cavo M et al. (2012) Proteasome inhibitors in multiple myeloma: 10 years later. *Blood* 120:947–959.

Palumbo A & Anderson K (2011) Multiple myeloma. *N. Engl. J. Med.* 364:1046–1060.

Rajkumar SV, Dimopoulos MA, Palumbo A et al. (2014) International Myeloma Working Group updated criteria for the diagnosis of multiple myeloma. *Lancet Oncol.* 15:e538–e548.

Richardson PG, Weller E, Lonial S et al. (2010) Lenalidomide, bortezomib, and dexamethasone combination therapy in patients with newly diagnosed multiple myeloma. *Blood* 116:679–686.

Siegel RL, Miller KD & Jemal A. (2018) Cancer statistics, 2018. *CA Cancer J. Clin.* 68:7–30.

Summary

Ashworth A & Lord CJ (2018) Synthetic lethal therapies for cancer: what's next after PARP inhibitors? *Nat. Rev. Clin. Oncol.* 15:564–576.

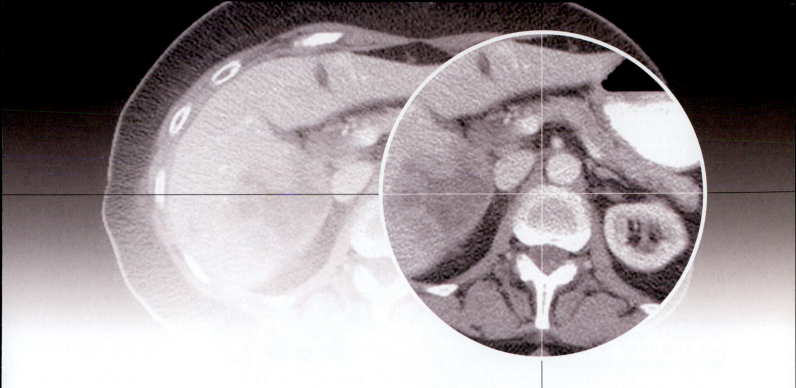

Chapter 10

Interference with the Tumor Microenvironment

Introduction

Cancers are communities in which malignant cells interact with many different cell types as well as with noncellular tissue components (Figure 10.1). Collectively, these elements make up what is called the tumor microenvironment (TME). While the cancer cell is the instigating element in the TME, nonmalignant cell types that make important contributions include:

- *Cancer-associated fibroblasts* *(CAFs)*. These are populations of fibroblasts that become functionally distinct from their normal counterparts when imported into the TME. Their growth within the TME is stimulated by cytokines such as IL-6 or CXCL-12, as well as fibroblast growth factor (FGF) and platelet-derived growth factor (PDGF). In turn, CAFs secrete a variety of cytokines and growth factors that support tumor maintenance and growth. They are also an important source of vascular endothelial growth factor (VEGF), which is required for the generation of tumor blood vessels. Furthermore, CAFs secrete pro-inflammatory factors such as chemokines and cytokines that recruit immune and inflammatory cells to the TME. CAFs produce the collagen that is a major component of the **extracellular matrix** (ECM).

- *Cells of immunity and inflammation.* When in the TME, some cell types that normally protect against invading microbes will actually enhance cancer survival, growth, and invasion. For example, many classes of immunocytes are present in the TME and include lymphocytes, macrophages, neutrophils, and populations of immature myeloid cells, called

Figure 10.1. The Tumor Microenvironment (TME). *Several of the major cell types comprising the TME are shown. The many types of T lymphocytes that may be present in the TME, such as regulatory T cells (T$_{reg}$ cells), cytotoxic T cells (T$_C$ cells), and helper T cells (T$_H$ cells), are represented by the lymphocyte icon. However, some immune cells that are present in the TME, such as neutrophils, dendritic cells, and natural killer (NK) cells, are not shown in the figure. CAF, cancer-associated fibroblast; TAM, tumor-associated macrophage; MDSC, myeloid-derived suppressor cell. (Adapted from F. Klemm and J.A. Joyce,* Trends Cell Biol. *25:198–213, 2015. With permission from Elsevier.)*

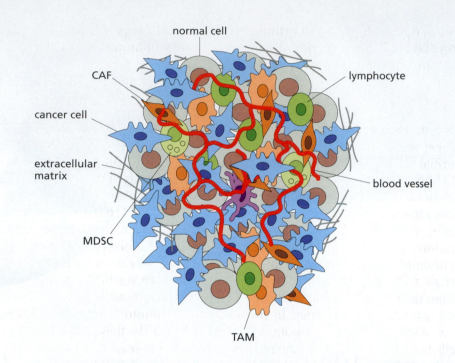

myeloid-derived suppressor cells (MDSCs). The TME recruits MDSCs, a heterogenous immunosuppressive population of immunocytes, from bone marrow. In the TME, they support tumor growth by suppressing anti-tumor immune responses against the tumor.

- *Vascular cells.* A growing tumor will rapidly utilize oxygen and outgrow its blood supply, leading to lower relative oxygen concentrations (hypoxia) in the TME. Endothelial cells (cells that form the inner lining of blood vessels), pericytes (long, contractile cells that wrap around capillaries), and mural cells (nonendothelial cells of the basement membrane of capillaries) are recruited to the TME or generated within it. These vascular cells support the assembly and maintenance of the network of tumor blood vessels that are indispensable for the growth and survival of the tumor.

Cancer cells, key elements of the TME, are the primary organizers responsible for setting in motion the recruitment of the various noncancerous cell populations that join them to compose the TME. The other cells types can in turn support cancer cell survival and tumor growth. An important way in which cancer cells recruit noncancerous cells into the TME is by the secretion of **chemotactic factors** that attract populations of fibroblasts, bone marrow-derived cells, endothelial cells, and immune cells. **Chemokines**, a subclass of cytokines whose principal function is to initiate and direct the chemotaxis (the directional movement) of their target cell population, induce the movement of cells bearing chemokine-specific receptors up concentration gradients of these factors and into the TME. Specificity of chemokine receptors for their chemokine ligands enables the summoning of only particular cell types from the vast and complex repertoire present in the body. For example, it is the generation of the chemokine CCL2 (C-C motif chemokine ligand 2, also referred to as MCP-1, or monocyte chemotactic protein-1) in the TME that attracts monocytes to the TME by stimulating the CCL2 receptors CCR2 and CCR4. Another example is provided by CXCL12 (also known as SDF-1, or stromal cell-derived factor-1), a chemokine that recruits mesenchymal stem cell populations in the bone marrow to the TME, where they can differentiate into endothelial cells and CAFs. Endothelial cells are essential for the formation of a system of blood vessels within the tumor. CAFs contribute to the extracellular matrix of the tumor and play a role in organizing and maintaining the immunosuppressive state present in the TMEs of many cancers.

The immunosuppressive TME

An immunosuppressive microenvironment is important for the survival of cancer cells, whose altered repertoire of expressed proteins compared with normal cells could be **immunogenic** and capable of provoking destruction by the immune system. The risk of destruction by the agents of the immune system poses great risk to cancers, and the TME is a key mediator of immune evasion, a hallmark of cancer. This is not due to an exclusion of cells of the immune system from the TME. Paradoxically, the TME recruits a variety of cells of the immune system including macrophages, T cells, and a heterogeneous group of immune-related cells collectively termed myeloid-derived suppressor cells (MDSCs). However, many of these immune cells are versions that *suppress* rather than conduct the cytotoxic attack that is usually associated with T cells and activated macrophages. In many cases, the immunosuppressive activities of these immune cell populations result in a TME that protects cancer cells from immune attack. The regulation of immune responses depends on the interplay of activating and inhibitory populations of cells and a variety of cytokines, some of which are inhibitory. For example, within the TME, there are regulatory T cells (T_{reg} cells) that use a variety of mechanisms to inhibit the induction of immune responses. Among **tumor-associated macrophages** (TAMs) in the TME, there are subpopulations that exist in states of activation that enable their ability to mount cytotoxic immune responses and other subpopulations in alternative suppressive states that inhibit such responses. Suppressive TAMs can secrete cytokines, such as IL-10, that inhibit T-cell activation and cytotoxic activity. Typically, the T-cell and macrophage populations of the TME are enriched for T_{reg} cells and alternative macrophages. Unfortunately, cytotoxic TAMs do not dominate the TME, and there is great interest in gaining an understanding of the factors and conditions within the TME that would favor dominance by this class of macrophages. Efforts to discover drugs that would tip the balance in favor of cytotoxic TAMs are being pursued.

Angiogenesis in the TME

In addition to containing a repertoire of cells that collaborate to produce an immunosuppressive environment, which enables immunoevasion and provides conditions that may facilitate metastasis, the TME makes a determinative contribution to tumor growth and survival by providing a **vasculature** (a network of blood vessels) that delivers nutrients and removes waste. Tumors are critically dependent on the generation of a vasculature, and angiogenesis is well established as a hallmark of cancer (Figure 10.2). Angiogenesis, the ability to induce the production of new blood vessels from existing vessels, is an essential feature of cancer because in the absence of a vascular system, diffusion can supply oxygen and nutrients and remove waste from a tumor only over a modest distance, just a millimeter or so. Without a vascular system, solid tumors are quite limited in their growth and can achieve a volume of little more than a cubic millimeter before cells in the tumor's interior become hypoxic, starved for nutrients, and poisoned by their own waste. However, the hypoxia experienced by cancer cells beyond the reach of the vasculature can induce an angiogenic switch that leads to neovascularization (new blood vessels) within the tumor.

Normally, a balance between pro-angiogenic molecules and angiostatic molecules keeps the vasculature from extending new vessels. A shift in this

Figure 10.2. Steps in angiogenesis. (A) A growing tumor becomes hypoxic in the absence of blood vessels but with rapid consumption of oxygen. (B) Hypoxia induces the expression of VEGF and other angiogenic factors, stimulating the sprouting of a capillary. (C) With vascular branching and elongation of the blood vessel, a lumen forms and the vessel matures, with endothelial cells lining the inside of the vessel and pericytes along its periphery.

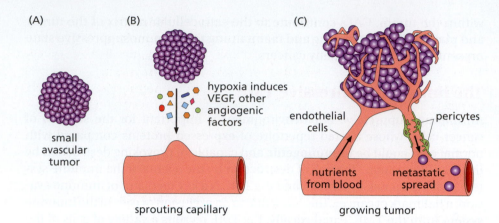

balance in favor of angiogenesis activates the movement of endothelial and other cells to the site of excess angiogenic factor (such as vascular endothelial growth factor, or VEGF). Once there, the endothelial and other cells assemble into new vessels. In the normal physiological process of wound healing, angiogenesis eventually stops because the balance between pro-angiogenic and angiostatic factors has been restored (Table 10.1). In contrast, this balance is not restored in the tumor microenvironment, due to a continuing abundance of pro-angiogenic signals. As a result, the tumor vasculature is much less orderly than normal vasculature. Endothelial cell connections are haphazard, resulting in malformed and leaky vessels, making blood flow irregular and slow, and perhaps facilitating **intravasation** (invasion of cancer cells through the blood vessel basement membrane and into the circulation) and dissemination of cancer cells. The suboptimal blood flow provided by the tumor vasculature leads to areas of hypoxia in the tumor, stimulating increased production of angiogenic factors and inducing even more angiogenesis.

Myeloid cells recruited from the circulation and bone marrow secrete angiogenic factors that drive formation of the tumor vasculature. In addition to being immunosuppressive, subpopulations of TAMs secrete VEGF, the most important of the many angiogenic signals listed in Table 10.1. Additionally, TAMs and tumor-associated neutrophils (TANs) secrete proteases, notably MMP-9 (matrix metalloproteinase-9), that release VEGF sequestered by the ECM.

Programs of drug development and discovery have sought inhibitors that interfere with the angiogenesis-dependent neovascularization that is essential for tumor growth and maintenance. The promise of this approach has been explored in more than 1000 clinical trials testing a variety of anti-angiogenic compounds and approaches.

Table 10.1. A partial list of activators and inhibitors of angiogenesis.

Activators	Inhibitors
VEGFs (vascular endothelial growth factors)	Tsp-1 and Tsp-2 (thrombospondin)
aFGF (acidic fibroblast growth factor)	Angiostatin
bFGF (basic fibroblast growth factor)	Endostatin
Angiopoietin	TIMPs (tissue inhibitors of metalloproteinase)
MMP-9 and other MMPs (matrix metalloproteinases)	(many others)

Adapted from R.A. Weinberg, The Biology of Cancer, 2nd ed. New York: Garland Science, 2014.

Targeting and disturbing the TME

A useful adjunct to attacking cancer cells would be to target and disturb the TME. Two strategies in current use are illustrated in the cases in this chapter. Case 10-1 demonstrates that antibodies can be used to block the binding of angiogenic ligands to their receptors, employing an anti-VEGF antibody to inhibit ligand–receptor interactions between VEGF and the

VEGF receptor (VEGF-R), to treat colon cancer. VEGF-Rs are receptor tyrosine kinases, which can also be targeted using small molecule inhibitors (specific VEGF-R inhibitors are not discussed here, but kinase inhibitors are discussed in Chapter 4). Case 10-2 concerns the anti-myeloma drug lenalidomide, a small molecule and member of the thalidomide family, which impacts the tumor microenvironment in many ways including immunomodulation and angiogenesis inhibition.

Take-home points

✓ Cancerous tumors are composed of both malignant and nonmalignant cells embedded in an extracellular matrix (ECM).

✓ Nonmalignant cell populations make essential contributions to the growth and survival of cancers.

✓ The nonmalignant cells of the tumor microenvironment (TME) are diverse and include cancer-associated fibroblasts (CAFs); cells of the vasculature including endothelial cells and pericytes; and immune cells, including macrophages, T cells, and a variety of others.

✓ The installation and maintenance of a vascular system is essential for tumor growth.

✓ Due to the activity of alternatively activated macrophages, regulatory T cells (T_{reg}s), myeloid-derived suppressor cells (MDSCs), and others, the TME is often immunosuppressive and prevents the elimination of cancer cells by the immune system.

Case 10-1   Colorectal Cancer

Topics bearing on this case

Adenomatous polyps and cancer progression

Mismatch repair (MMR) and microsatellite instability (MSI)

Angiogenesis and anti-angiogenesis therapies

Introduction

Colon and rectal cancers (CRCs) are typically grouped together, given their contiguous anatomy and similar histology. The vast majority of CRCs arise from proliferation of glandular cells lining the bowel wall, which then form precancerous polyps, or **adenomas**. These adenomas can acquire mutations and ultimately transform into adenocarcinomas (Figure 10.3). Precancerous

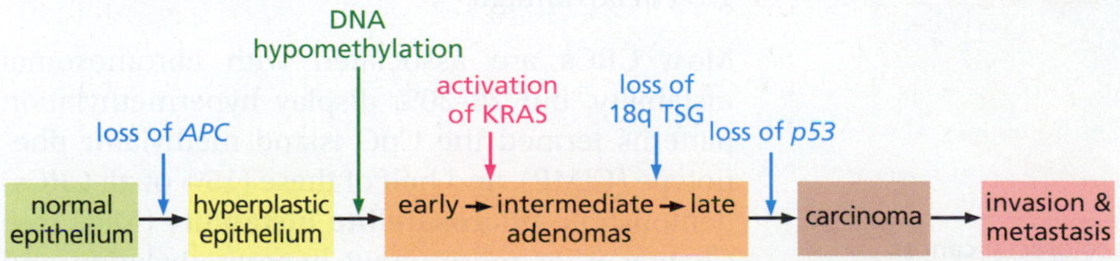

Figure 10.3. Multistep tumorigenesis in colorectal cancer (CRC). Common, successive mutations lead to the development of CRC. The mutation of *APC* (either inherited or somatic) leads to hyperproliferative epithelium, with DNA hypomethylation then leading to an early adenoma. Mutation in *KRAS* and subsequent loss of chromosome 18q (containing the tumor suppressor gene *SMAD4*) leads to progression to intermediate and late adenomas, whereupon inactivation of p53 leads to the development of frank adenocarcinoma. TSG, tumor suppressor gene. (From R.A. Weinberg, The Biology of Cancer, 2nd ed. New York: Garland Science, 2014.)

Figure 10.4. Types of intestinal polyps. Polyps can be (A) sessile (flat) or (B) pedunculated (connected by a stalk). (C) Adenocarcinomas invade into the intestinal wall. (From A.T. Skarin, Dana-Farber Cancer Institute Atlas of Diagnostic Oncology, 4th ed. Mosby Elsevier, 2010. With permission from Elsevier.)

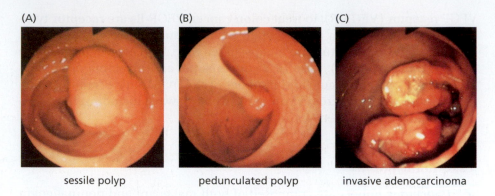

(A)   (B)   (C)

sessile polyp   pedunculated polyp   invasive adenocarcinoma

polyps occur in 30–50% of adults. Specifically, they are identified in 25% of people aged 50 and 45% of people aged 70. Fewer than 1% of polyps become cancer. They are more likely to transform if they are greater than 1.5 cm, sessile (flat) rather than pedunculated (connected by a stalk), and villous (hair-like) rather than tubular (tube-shaped) (Figure 10.4). It generally takes at least five years for a polyp to become a cancer.

Molecular alterations and CRC

The development of colon polyps results from dysregulation of cell proliferation within the glandular crypts lining the colon mucosa. A series of genetic alterations (either inherited or acquired, or a combination) leads to transformation of a benign polyp to a malignant tumor (see Figure 10.3). An acquired mutation in the *APC* gene is found in 80% of tumors and results in an important loss of cell growth regulation, leading to tumorigenesis. The importance of the RAS signaling pathway in CRC tumorigenesis is underscored by the fact that activating mutations in *KRAS* and *BRAF* occur in 50% and 10% of sporadic CRCs, respectively (Figure 10.5). These activating mutations appear to be mutually exclusive in an individual cancer, because if the KRAS/BRAF pathway is already activated by one mutation, a second mutation would not confer an additional selective growth advantage.

Most CRCs are associated with chromosomal instability, but 20–30% display hypermethylation patterns termed the CpG island methylator phenotype (CIMP), and half of these (15% of all CRCs) demonstrate **microsatellite instability** (MSI) (see Chapter 8 for more about hypermethylation and epigenetic modifications). Cancers with MSI display alterations of specific repeating units of DNA in the genome called microsatellites and have defective DNA mismatch repair (MMR). Defects in MMR lead to increased rates of mutation throughout the

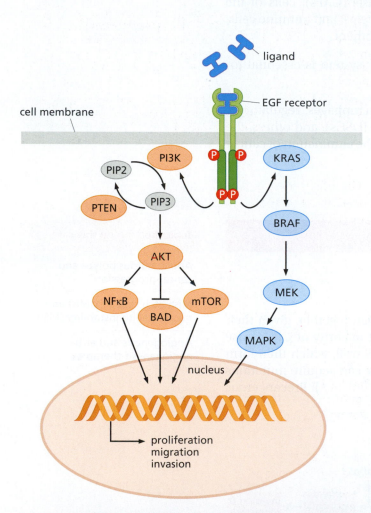

Figure 10.5. EGFR/RAS/RAF signaling in colorectal cancer. *APC* mutation may be one of the inciting events in colon tumorigenesis, but the signaling pathway downstream of the epidermal growth factor receptor (EGFR) drives the growth of established CRCs. Activating mutations in *KRAS* and *BRAF* occur in 50% and 10% of sporadic CRCs, respectively, driving MAPK signaling and cell proliferation. Targets for therapy along this pathway include the EGFR and BRAF.

Table 10.2. Inherited syndromes related to CRC.

Syndrome	Gene	Major clinical manifestations
Familial adenomatous polyposis (FAP)	*APC*	Accounts for 1% of CRCs. Autosomal dominant (AD) with 100% cancer penetrance. Thousands of adenomatous polyps develop by age 20 in 75% of carriers and lead to cancer. Total proctocolectomy (removal of entire colon and rectum) with ileoanal anastomosis (joining of the ileum component of small intestine with the anus) is treatment of choice. Associated with congenital hypertrophy of the retinal pigment epithelium (a layer of the retina of the eye), desmoid tumors (noncancerous growth of the connective tissue of the abdomen, called Gardner syndrome), and brain tumors (Turcot syndrome).
Hereditary non-polyposis colon cancer (also known as Lynch syndrome)	*MLH1, MSH2, MSH6, PMS2*	Accounts for 3–5% of CRCs with 25–75% cancer penetrance (proportion of carriers that exhibit the characteristic phenotype). Autosomal dominant resulting from germline mutations in mismatch repair genes, which lead to DNA instability and errors in DNA replication. Median age of presentation <50 years. Ascending (right side) colon tumors are more common than descending (left) colon, unlike sporadic CRCs. Additionally associated with stomach, pancreas, ureter and kidney, biliary tract (gallbladder and bile ducts), brain (glioblastoma), endometrial, and ovarian cancer. Screening colonoscopy for HNPCC patients is recommended at age 21 and every five years thereafter. These cases generally have better prognosis than sporadic CRCs.
MutYH-associated polyposis (MAP)	*MUTYH* (also termed *MYH*)	Accounts for <1% of CRCs. Germline mutation in base-excision repair gene *MUTYH* results in predisposition to polyposis. Only known autosomal recessive colon cancer syndrome.

genome, and in cancer-associated genes in particular. Germline mutations of MMR genes, such as *MLH1*, *MSH2*, and *MSH6,* are associated with the genetic syndrome **hereditary nonpolyposis colorectal cancer** (HNPCC), or **Lynch syndrome** (Table 10.2).

Epidemiology and etiology

CRC is the third most common malignancy in the United States, with an estimated 140,250 new cases in 2018 (approximately 70% of these arising in the colon and 30% in the rectum). At presentation, 18% of patients will have metastatic disease and 35% of all patients will ultimately develop metastases. CRC will account for 50,630 deaths in 2018, the second leading cause of cancer-related death in the United States. It is the third most common cancer cause of death in men (after prostate and lung cancers) and women (after breast and lung cancers).

Risk factors associated with CRC include family history, sedentary lifestyle, and age (Table 10.3). Although there are many established genetic syndromes that lead to increased risk of developing CRC, the vast majority of CRCs (over 70%) are sporadic (see Table 10.2). In addition, patients with **inflammatory bowel disease** (ulcerative colitis or Crohn's disease) have a markedly increased risk of developing CRC. The risk in ulcerative colitis correlates with extent, duration, and severity of disease; the risk of developing CRC is 10% at 10 years with ulcerative colitis and 20% at 20 years. A total colectomy eliminates the risk, and is often performed prophylactically in patients with ulcerative colitis.

Table 10.3. Risk factors associated with CRC development.

Family history of CRC or adenomatous polyps (first-degree family member increases risk 1.7-fold)
Western or urbanized society
Diet with high quantity of red meat or processed meat
Diabetes mellitus
Increased bowel anaerobic flora
Inflammatory bowel disease (Crohn's disease, ulcerative colitis)
Smoking
Alcohol consumption
Ureterosigmoidostomy (a procedure for bladder cancer in which the bladder is removed and urine is redirected by attaching the ureters into the colon; uncommon in current practice)
Streptococcus bovis bacteremia
Prior pelvic radiation
Sedentary lifestyle and/or obesity
Race
Age
Acromegaly (hypersecretion of growth hormone)

Diagnosis, workup, and staging

Approximately 30% of CRC cases are identified with routine screening of asymptomatic patients. For patients at average risk, screening with colonoscopy begins at age 50. Screening colonoscopy (that is, testing of asymptomatic patients) has been demonstrated to decrease death from CRC, similar to screening tests for breast cancer (Case 6-1) and lung cancer (Case 4-3). Colonoscopy is both diagnostic and therapeutic in that it allows for removal of precancerous lesions. If there are no polyps or cancers identified, a colonoscopy can then be performed every 10 years until life expectancy is less than 10 years. Most guidelines recommend stopping by age 85.

Virtual colonography (also called CT colonography because it uses computed tomography scanning instead of direct visualization with a camera) appears to be just as good (if not better) for lesions that are at least 1 cm in size, and is indicated for patients who have incomplete colonoscopies or cannot safely tolerate the procedure. The procedure is diagnostic but not therapeutic—that is, if a lesion is identified, a conventional colonoscopy is then required for biopsy or removal.

Some patients with CRC will present with symptoms of anemia, such as fatigue or pallor, resulting from occult (hidden, or microscopic) blood loss. The blood loss is due to friable (fragile) edges of the tumor that can bleed into the gastrointestinal tract and lead to iron deficiency. Patients may also present with symptoms related to the location of their tumor. For example, tumors located close to or involving the rectum may result in change in bowel habits, rectal pain, tenesmus (feeling the need to pass stools, which can cause cramping and pain), bright red blood with bowel movements, or change in the caliber (width) of stool. More advanced left-sided (descending colon) tumors may cause obstructive symptoms such as abdominal distention, pain, nausea, and vomiting. This is rare for tumors arising from the right side (ascending) of the colon, where stool is liquid and is less likely to be blocked by an intraluminal tumor.

A comprehensive medical and family history, physical exam, and laboratory evaluation should be performed for a patient suspected to have CRC. The laboratory evaluation should include a complete blood count, iron studies, and blood chemistries with liver function tests and a test for **carcinoembryonic antigen** (CEA), which is a tumor marker associated with some intestinal cancers. Because the CEA test has poor **specificity** and **sensitivity**, it is not helpful as a screening or diagnostic tool. However, in patients with CRC, a preoperative CEA level is prognostic (poor outcomes for patients with a level >5 ng/ml) and can be followed during post-treatment surveillance as an indicator of cancer recurrence.

Following diagnosis, staging studies are performed to evaluate the extent of CRC. The typical sites of metastasis include lymph nodes, liver, peritoneum, and lung. Colorectal cancer can be locally advanced and invade into nearby structures, or can be disseminated via hematogenous (blood stream) or lymphatic routes. The colon is drained by the portal system, which is a network of blood vessels leading into the portal vein, which delivers blood to the liver, and

(A)

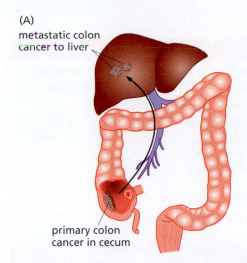

metastatic colon cancer to liver

primary colon cancer in cecum

(B)

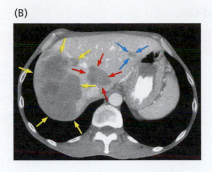

Figure 10.6. Colorectal cancers often metastasize to the liver.
(A) Schematic diagram of colon cancer metastasis. The normal venous drainage of the colon is via the mesenteric veins, which coalesce into the portal system and drain into the liver, where liver enzymes detoxify chemicals and metabolize drugs. Thus, a typical first metastatic site for CRCs is the liver. (B) CT scan of a colon cancer patient with three distinct liver lesions (denoted by different-colored arrows).

therefore the liver is usually the first site of distant metastasis (Figure 10.6). Staging studies include CT scans of the chest, abdomen, and pelvis. Bone scans and brain imaging are not routinely performed for CRC as metastases to these areas are rare. PET scan is reserved for patients who have lesions on CT scans that are considered "indeterminate." A PET scan may help distinguish between metastatic disease and noncancerous lesions such as scar tissue or postsurgical changes.

Staging of CRC employs the TNM system of the American Joint Committee on Cancer (AJCC) (Table 10.4). As with most cancers, survival correlates with stage at diagnosis, as indicated in the table.

Management of localized CRC

Fortunately, 80% of CRC patients present with nonmetastatic disease and are candidates for surgical resection of the primary tumor and regional lymph nodes. The colon can be divided into right (or ascending) colon, transverse colon, left (or descending) colon, and sigmoid colon, which connects to the rectum. For colon cancers, surgery typically removes the affected segment of colon; that is, a right hemicolectomy, transverse colectomy, left hemicolectomy, or sigmoidectomy is performed. Preservation of the circular muscle band in the colon that controls bowel movement, the anal sphincter, is desired if possible. Depending on the distance from the anus, surgical resection of a rectal cancer is either sphincter-sparing (called a low anterior resection) or sphincter-taking (called an abdominoperineal resection with resulting colostomy, a created opening in the abdominal wall through which stool exits and is collected in an applied bag). Lymph node resection is considered adequate if 12 or more nodes are removed.

Patients with locally advanced rectal cancers (T3–T4 and/or N1–N2) are at higher risk for recurrence within the pelvis and are often treated with combined chemotherapy and radiation prior to surgical resection to attenuate this risk.

Adjuvant chemotherapy is designed to eradicate any microscopic metastases that are left behind after surgery. It is indicated for node-positive (stage III) disease and considered, but not absolutely indicated, for stage II CRC patients with high-risk features (T4, poorly differentiated, invasion into lymphatic channels or blood vessels, presentation with obstruction or perforation, or inadequate lymph node sampling). The standard of care consists of six months of FOLFOX, which is comprised of 5-fluorouracil administered by

Table 10.4. AJCC staging classification of CRC.

Primary tumor (T)		Regional lymph nodes (N)		Distant metastasis (M)	
Tx	Primary tumor cannot be assessed	Nx	Nodes cannot be assessed	M0	No distant metastasis
T0	No evidence of primary tumor	N0	No regional node metastasis	M1	Distant metastasis
Tis	Carcinoma in situ	N1a	Metastasis in 1 regional node		
T1	Tumor invades submucosa	N1b	Metastasis in 2–3 regional nodes		
T2	Tumor invades muscularis propria	N1c	No regional lymph nodes positive but tumor deposits are present in subserosa, mesentery, or pericolic tissues		
T3	Tumor invades through muscularis propria into pericolorectal tissues				
T4a	Tumor penetrates to the surface of the visceral peritoneum	N2a	Metastasis in 4–6 regional nodes		
T4b	Tumor directly invades or is adherent to other organs or structures	N2b	Metastasis in 7 or more regional nodes		

Stage grouping	T	N	M	5 year overall survival (%)	
				Colon	Rectum
Stage I	T1, T2	N0	M0	74	74.1
Stage IIa	T3	N0	M0	66.5	64.5
Stage IIb	T4a	N0	M0	58.6	51.6
Stage IIc	T4b	N0	M0	37.3	32.3
Stage IIIa	T1 or T2	N1 or N1c	M0	73.1	74
	T1	N2a	M0		
Stage IIIb	T3 or T4a	N1 or N1c	M0	46.3	45
	T2 or T3	N2a	M0		
	T1 or T2	N2b	M0		
Stage IIIc	T4a	N2a	M0	28.0	33.4
	T3 or T4a	N2b	M0		
	T4b	N1 or N2	M0		
Stage IV	any T	any N	M1	5.7	6.0

a continuous intravenous infusion over two days plus leucovorin and oxaliplatin administered on the first of the two days; the chemotherapy cycles are repeated every two weeks. For patients with stage III disease, the addition of adjuvant chemotherapy can improve overall survival by 15–25%. This is generally a well-tolerated regimen; however, 15% of patients can be left with a lasting peripheral neuropathy (numbness, tingling, and/or difficulty with dexterity especially in fingers and toes).

Following treatment for CRC, patients are followed closely for recurrence to allow for timely administration of palliative chemotherapy if necessary or, in the case of limited disease, allow for consideration of an aggressive surgical approach to possibly cure the patient. Patients are followed with blood tests including a CEA test and annual chest, abdomen, and pelvis CT scans for three to five years. Patients require a colonoscopy one year following surgical resection and, in the presence of recurrent polyps, yearly colonoscopies. After a normal colonoscopy, these exams can be performed every two to three years.

After treatment for CRC, patients should be counseled regarding **secondary cancer prevention**. Many physicians recommend a daily low-dose aspirin to reduce the risk of polyp growth and a second occurrence of CRC. A healthy diet of nonprocessed foods and regular exercise may also reduce risk of recurrence or development of second primaries.

Management of metastatic CRC

If metastatic disease manifests as a few tumors and is limited to the liver or to the lung, then an aggressive surgical approach may lead to cure in 20–30% of carefully selected patients. However, for the majority of cases, metastatic CRC is considered incurable.

For incurable CRC, palliative chemotherapy is the primary treatment and is designed to both help patients feel better and live longer. The chemotherapy agent 5-fluorouracil (5-FU) acts as the backbone to many palliative regimens and can be used alone or in combination with other drugs. As a single agent, 5-FU can improve median overall survival to 12 months (over 6 months for supportive care alone). With the introduction of newer agents, survival rates have improved. In 2000, when the chemotherapy agents irinotecan and leucovorin were added to 5-FU in the regimen known as IFL, median overall survival improved to 15 months compared with 13 months with 5-FU plus leucovorin. In 2004, new chemotherapy combinations established standards of care for metastatic CRC. The use of FOLFOX further improved survival to 20 months compared with 15 months with IFL. Thus, iterative improvements in chemotherapy regimens have extended survival for metastatic CRC from 6 months with supportive care only to approximately 20 months with chemotherapy. With the introduction of targeted agents (described below), median survival for metastatic CRC continues to improve. For many patients, the added months to years of survival can afford precious quality time.

The case of Mary Maspin, a woman with blood in her stool and an unintentional 20-pound weight loss

At the age of 64, Mary had a history of hypertension (high blood pressure) and hypothyroidism (underactive thyroid); she had previously declined screening colonoscopies. She was seen by her primary care physician in her annual follow-up and was noted to have an unintentional 20-pound weight loss over the previous eight months. She reported blood in her stool, which she attributed to hemorrhoids, as well as irregular bowel patterns and diarrhea. She also reported increased fatigue and shortness of breath with exertion. On examination, she had some mild tenderness to palpation in the left lower quadrant of her abdomen, and the edge of her liver, something that cannot normally be felt, was palpable 3 cm below the rib cage. Her blood tests revealed iron deficiency anemia. She was referred to a gastroenterologist and underwent colonoscopy, which revealed a nearly circumferential tumor in the sigmoid colon involving 60% of the luminal wall (Figure 10.7). The lesion was friable (easily crumbling on contact). The colonoscope was able to pass by the tumor, and on examination of the remainder of the colon, three 1-cm polyps were found and resected, with pathology examination demonstrating tubular adenomas. Biopsy of the sigmoid lesion was consistent with a moderately differentiated adenocarcinoma. Mismatch repair protein expression was normal.

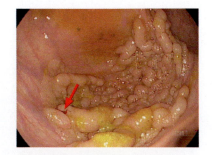

Figure 10.7. Mary's tumor found at colonoscopy. The relatively smoother contour of the colon can be seen at the periphery of the image. The tumor is nearly circumferential. The surface of the tumor is friable (fragile), and contact with the colonoscope can lead to bleeding (*red arrow*), which explains the frequent presentation of CRC patients with anemia. (Photo courtesy of David Forcione, M.D.)

Staging CT scans of her chest, abdomen, and pelvis showed multiple masses in both lobes of the liver, ranging in size from 1 to 5 cm. Additionally, she had bilateral subcentimeter suspicious nodules in her lungs. A CEA test measured at 36 ng/ml (normal range, <3.4 ng/ml). A CT-guided biopsy of one of the peripheral liver lesions confirmed metastatic colon cancer.

Mary reviewed treatment options with her oncologist. She was not considered a candidate for surgical removal of the liver lesions. Given the extent of her tumor burden as well as her symptoms, including fatigue and weight

Figure 10.8. Mary's response to FOLFOX and bevacizumab. (A) CT scan images of Mary's liver prior to treatment. The liver contains multiple metastases that are indicated with blue arrows. (B) CT scan images after four months of therapy. The liver metastases are reduced in size and number (*red arrows*).

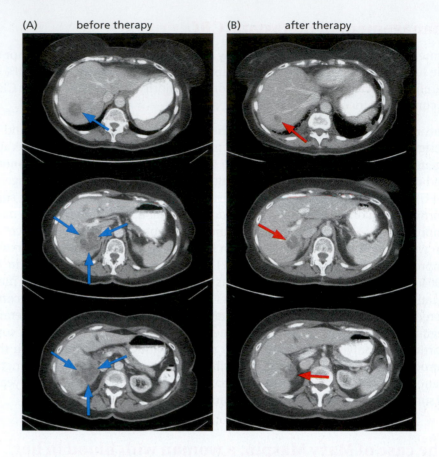

loss, her doctor recommended initiating chemotherapy with palliative intent. They reviewed chemotherapy options and agreed upon FOLFOX based on the efficacy and tolerability data. They further discussed the addition of targeted therapies, with either VEGF inhibitors such as bevacizumab or EGFR inhibitors such as **cetuximab** and **panitumumab**. Mary and her medical oncologist decided on treatment with FOLFOX plus bevacizumab.

After two months of therapy, restaging scans indicated a decrease in size of the hepatic metastases and lung nodules, and her CEA decreased to 12 ng/ml (Figure 10.8). After four months of therapy, she had further decline in her CEA to 8 ng/ml and ongoing improvement in her CT scans. However, she began to develop a persistent sensory neuropathy, which was starting to interfere with her ability to perform fine motor tasks such as buttoning buttons and which eventually affected her balance. The oxaliplatin was dose-reduced and then stopped altogether. She was maintained on 5-FU, leucovorin, and bevacizumab alone. A year after her diagnosis of metastatic colon cancer, Mary felt generally well and was able to attend her son's wedding.

Targeted therapies and metastatic CRC

Targeted therapies for CRC generally inhibit angiogenesis or the EGF-R/RAS/RAF pathway. Angiogenesis, the development of new blood vessels, is critical for the survival of growing tumors. Once a tumor surpasses a critical size (thought to be 1 mm), the diffusion of nutrients is insufficient to supply the tumor and this triggers the creation or recruitment of new blood vessels. The relative hypoxia of a growing tumor induces hypoxia-inducible factor (HIF), a transcription factor that can promote expression of VEGF. VEGF binds to VEGF receptors (VEGF-Rs) and acts as a specific **mitogen** (a stimulant for mitosis) for endothelial cells.

Strategies for inhibiting angiogenesis primarily target the VEGF/VEGF-R signaling axis. Bevacizumab is a humanized recombinant monoclonal antibody that binds and inhibits the activity of VEGF-A, the primary VEGF family member involved in angiogenesis. Other family members include VEGF-B, -C, and -D and placenta growth factor (PGF), which play roles in angiogenesis and formation of lymphatic channels. Bevacizumab is delivered by intravenous infusion. Early clinical trials indicated an improvement in response rate and survival when bevacizumab was combined with chemotherapy. In a randomized, Phase III clinical trial of patients given chemotherapy with irinotecan, 5-FU, and leucovorin (the IFL regimen) with or without bevacizumab, the group receiving bevacizumab had a significant improvement in median overall survival (OS) from 16 months to 20 months. The most common adverse event on bevacizumab was hypertension, which was manageable. These results led to FDA approval of bevacizumab for colorectal cancer on February 26, 2004, marking the first anti-angiogenesis drug available to treat cancer. Later studies demonstrated similar OS improvement when bevacizumab was combined with other chemotherapy regimens for colon cancer, such as FOLFOX (21 months), making bevacizumab-based combinations with chemotherapy the standard approach for initial therapy in metastatic CRC.

A different anti-angiogenesis approach is with small-molecule inhibitors of VEGF-R, which is a receptor tyrosine kinase. FDA-approved kinase inhibitors with activity against VEGF-Rs include sunitinib, pazopanib, and axitinib (for more about kinase inhibitors in general, refer to Chapter 4). **Regorafenib** is an oral small-molecule tyrosine kinase inhibitor of VEGF-R, as well as other kinases including BRAF, that confers a modest (less than two-month) improvement in OS and is FDA approved for previously treated metastatic CRC.

The EGF-R/RAS/RAF pathway provides other targets in treatment of CRC. FDA-approved treatments to target EGF-R include the monoclonal antibodies cetuximab and panitumumab, in combination with chemotherapy. In a large study comparing cetuximab with bevacizumab along with chemotherapy in untreated metastatic CRC, there was no statistically significant difference in OS (approximately 29 months). It should be noted that survival rates cannot be compared between trials (for example, FOLFOX–bevacizumab OS was 21 months in one study but 29 months in another) because of differences in subject populations and era of treatment (that is, subjects in later studies may initiate treatment with lower disease burden due to earlier detection of metastatic disease, so their survival may appear to be longer than subjects in earlier studies). EGF-R antibodies are effective only in patients with wild-type KRAS, because inhibitory antibodies to the upstream EGF-R signal would not be able to turn off the downstream signal propagated by constitutively active mutant KRAS.

Future prospects and challenges in metastatic CRC treatment

The management of CRC has seen steady improvements in the past 20 years, with improvements in surgical standards, tactical employment of radiation in rectal cancer, and improvements in systemic therapy for metastatic disease. Chemotherapy treatments improved median overall survival from 6 months on best supportive care to over 20 months, during a flurry of development of combination chemotherapy regimens in the late 1990s and early 2000s. The further improvements in survival using targeted therapies against angiogenesis or EGF-R signaling have created new standards of care for metastatic CRC. Despite this remarkable progress, some key questions remain. The development of prognostic biomarkers to identify those patients at highest risk, as well as improved adjuvant therapy approaches, would help further prevent the development of recurrent or metastatic disease. For metastatic CRC, the mechanisms of resistance to current systemic strategies have not been well elucidated. Ongoing work seeks to identify other feasible targets for therapy; for example, although *BRAF* is mutated in 5–10% of CRCs, BRAF inhibitors have not yet demonstrated significant activity in CRC patients. Hope for future developments rests in ongoing molecular characterization of resistant cancers and the immunomodulatory environment, and the deployment of novel targeted therapies.

Take-home points

✓ Colorectal cancer (CRC) is a common cancer that is largely preventable with appropriate screening.

✓ Patients with a family history suggestive of a genetic CRC syndrome should be screened earlier than the normal population.

✓ For localized disease, resection of the affected section of colon is the mainstay of treatment; adjuvant chemotherapy is recommended for stage III and highly selected stage II patients.

✓ Chemotherapy has increased overall survival (OS) of metastatic CRC from 6 months to over 20 months.

✓ The VEGF antibody bevacizumab, when combined with chemotherapy, improved OS in metastatic CRC. Contemporary trials have demonstrated OS nearing 30 months, with bevacizumab or the EGF-R targeting antibody cetuximab.

Discussion questions

1. Patients without polyps on screening colonoscopy often do not need another exam for 10 years, but those with polyps are examined every three to five years. Why do you think this might be?

2. Explain the rationale for targeting angiogenesis as a cancer therapy. Is the therapy cytotoxic or cytostatic?

3. Why would CRC patients with activated *RAF* mutations not respond to the EGF-R antibody cetuximab?

Topics related to this case

Bone marrow microenvironment

Immunomodulatory drugs (IMiDs)

Osteoclast-targeted therapy

Case 10-2   Multiple Myeloma

Introduction

Multiple myeloma (MM) is the second most common hematological malignancy in the United States. It is primarily a disease of older adults with median age at diagnosis of 66 years with 30% of patients over the age of 75. Further information about epidemiology, diagnosis, and management of MM can be found in Case 9-2. Management of the disease is based on clinical presentation as well as age and performance status at the time of diagnosis. As discussed in Case 9-2, in patients who are medically fit, routine initial therapy includes induction therapy with immunomodulatory drugs, proteasome inhibitors, and steroids, followed by a remission consolidation treatment consisting of high-dose chemotherapy to ablate the bone marrow, including the MM, and infusion of previously collected autologous stem cells to repopulate the healthy bone marrow, a process known as an autologous stem cell transplantation. Many MM patients, however, are not eligible for transplant due to advanced age or co-morbid conditions. In this case, we will focus on the management of transplant-ineligible patients treated with immunomodulatory drugs.

Principles of management for transplant-ineligible patients with newly diagnosed MM

Over the past 10 years, the availability of effective new drugs with acceptable toxicity, such as the **immunomodulatory drugs** (IMiDs) and the proteasome inhibitors, has modified the traditional treatment paradigms for multiple myeloma, resulting in improvement in the quality and the duration of life. Despite these advances, however, data from the Surveillance, Epidemiology, and End Results (SEER) database demonstrate that little improvement has been achieved among patients older than 60 years as suggested by an apparent lag in improvement in survival compared with younger patients. The survival difference may reflect increased co-morbidities in the older patient population, or fundamental differences in the tumor biology of elderly patients, as well as the lack of participation in clinical trials.

Historically, the combination of melphalan and prednisone (MP) was the standard initial treatment for transplant-ineligible patients; this regimen produces an overall response rate (ORR) of 50–60%. With the advent of newer, more potent anti-myeloma therapies, trials have been designed that incorporate both IMiDs (thalidomide and lenalidomide) and proteasome inhibitors (bortezomib) in combination with MP; each of them improved overall survival over MP alone. The "Rd" regimen of continuous lenalidomide plus low-dose dexamethasone was evaluated as a "chemotherapy-sparing" approach and produced an encouraging ORR of 75% and four-year OS of 59%. The safety profile with continuous Rd was manageable, with the most commonly reported severe side effects being neutropenia, infections, anemia, and **thromboembolism** (a clot that is carried through the circulation and plugs a blood vessel, a category that includes deep vein thrombosis, or DVT, and pulmonary embolus, or PE). The effectiveness combined with tolerability compared with chemotherapy-based regimens established Rd as the standard of care for initial therapy of multiple myeloma patients who are not candidates for high-dose chemotherapy with autologous stem cell transplantation.

To address the discrepancies in survival between older and younger patients with MM, ongoing trials are looking at three drug combinations including bortezomib, lenalidomide, and dexamethasone at reduced doses and attenuated schedules in this population as well. Interim analysis of results from a Phase II study of modified lenalidomide, bortezomib, and dexamethasone (RVD "lite") demonstrated an ORR of 90% after four cycles with excellent tolerability. Ongoing work continues to better incorporate the IMiDs and proteasome inhibitors into treatment regimens that balance efficacy and tolerability.

The role of IMiDs in multiple myeloma therapy

Historically, the combination of vincristine, doxorubicin (Adriamycin), and dexamethasone (VAD) was used as the standard induction chemotherapy for MM, but VAD has since been replaced by newer, more targeted drugs—IMiDs and proteasome inhibitors. **Thalidomide** was the first IMiD introduced. The drug was originally marketed in 1957 as an effective treatment for nausea and morning sickness during pregnancy. It was subsequently discovered that the drug had teratogenic properties (the capacity to cause malformations during development, resulting in birth defects), causing limb malformation in the children of women who had taken thalidomide during their pregnancies as a treatment for morning sickness, and this drug was removed from the market in countries where it was available (thalidomide had not been FDA approved in the United States).

In 1994, thalidomide was recognized to be a potent inhibitor of angiogenesis. The drug showed efficacy as an inhibitor of angiogenesis in animal models and was subsequently explored in clinical trials for a variety of different types of cancer. Two studies combined thalidomide with dexamethasone as initial therapy for MM and achieved rapid responses in two-thirds of patients. A subsequent randomized Phase III trial showed higher response rates for thalidomide/dexamethasone versus dexamethasone alone, resulting in FDA approval of thalidomide for MM on May 26, 2006. The necessary dose of dexamethasone was then explored in a Phase III trial that demonstrated that low-dose dexamethasone combined with thalidomide was associated with a better safety profile and improved OS compared with higher-dose dexamethasone combined with thalidomide. In the Phase III study comparing thalidomide and dexamethasone versus dexamethasone alone, 17% of patients receiving thalidomide developed DVTs, and so prophylaxis with blood-thinning agents including aspirin, warfarin, or subcutaneous heparin is needed when patients are treated with IMiD therapy.

Lenalidomide is a second-generation IMiD that demonstrates greater potency than thalidomide against MM cells, and is not associated with the thalidomide side effects of sedation, peripheral neuropathy, and severe constipation. In two large, randomized Phase III trials, lenalidomide with high-dose dexamethasone produced a superior response and delayed time to progression compared with dexamethasone and placebo. An analysis of over 1000 patients showed that the combination of lenalidomide and dexamethasone had an acceptable safety profile with fewer than 10% of patients experiencing pneumonia and deep vein thrombosis when aspirin prophylaxis was included. Lenalidomide (Revlimid) was then evaluated in combination with both bortezomib (Velcade) and dexamethasone (RVD) and demonstrated a very encouraging ORR of 97% in a Phase II trial with a one-year PFS of 83%. At present, either Rd or RVD remains a highly effective lenalidomide-based combination therapy for the initial treatment of multiple myeloma.

The case of Tom Brown, a 70-year-old male with a history of monoclonal gammopathy and severe low back pain

Tom Brown was an otherwise healthy 70-year-old man who was found to have monoclonal gammopathy of uncertain significance (see Case 9-2 for more details about MGUS) on routine screening two years prior to presentation. At the time of his MGUS diagnosis, a bone marrow biopsy was performed and

(A)   (B)

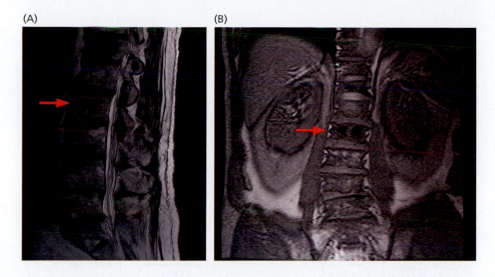

Figure 10.9. Magnetic resonance imaging (MRI) of lumbar spine. Two views of the same MRI are provided: (A) sagittal—a view of Tom standing and turned towards the left, (B) coronal—a view of Tom standing and facing the reader. The MRI demonstrates abnormal lesions involving the second lumbar (L2) vertebral body (*arrow*).

showed approximately 1% monoclonal plasma cells. He was followed by a local hematologist and had been doing well until this latest time of presentation.

Prior to presentation, Tom was mowing his lawn and strained his back when operating the lawn mower. He was subsequently found to have compression fractures at T12 (twelfth thoracic) and L2 (second lumbar) vertebral bodies (Figure 10.9). A **kyphoplasty** and biopsy for these compression fractures was performed. Kyphoplasty is a procedure whereby a vertebral body that has collapsed due to a compression fracture is re-expanded using a catheter and a specialized cement is then injected to maintain normal height of the bone. A vertebral biopsy taken at that time showed the presence of IgG κ-restricted plasma cells. Tom then underwent a bone marrow biopsy, which showed 80% plasma cells. Karyotype and FISH were performed and were normal. A serum protein electrophoresis (SPEP) showed an IgG of 1907 mg/dl with a monoclonal component (M-spike) of 1.07 g/dl of IgG kappa, and a free kappa light-chain level of 2320.0 mg/l. Mr. Brown's blood tests revealed that he was not anemic and had normal kidney function and calcium levels. His β-2-microglobulin was elevated at 2.99 μg/ml, and his albumin measured 4.7 g/dl. Mr. Brown was diagnosed with International Staging System stage I multiple myeloma (see Table 9.5).

Mr. Brown's only other co-morbid conditions were hyperlipidemia (high level of lipids such as cholesterol in the blood) and gout, for which he was taking simvastatin and allopurinol, respectively. He was married and had two adult children. He had no family history of hematologic malignancy.

Based on his age and treatment preferences, Tom was initiated on combination therapy with modified lenalidomide, bortezomib, and dexamethasone (RVD "lite") with a plan to receive a total of eight cycles followed by lenalidomide maintenance.

In addition, **zoledronic acid** was administered once monthly, concurrent with chemotherapy. Tom achieved remission and continues on maintenance lenalidomide along with monthly zoledronic acid.

The bone marrow microenvironment in multiple myeloma

Plasma cells are terminally differentiated antibody-producing activated B cells. Maturation of B cells involves migration from the bone marrow (BM) to the secondary lymphoid organs and ultimately back to the BM. The BM

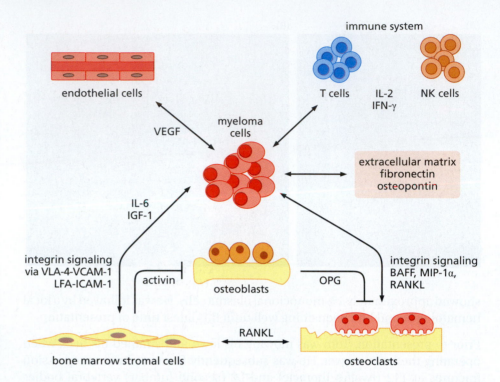

Figure 10.10. Myeloma cell interactions within the bone marrow microenvironment. The bone marrow microenvironment is composed of extracellular matrix proteins and cells, including endothelial cells, hematopoietic stem cells, bone marrow stromal cells, osteoclasts, and osteoblasts. Myeloma cells physically interact with extracellular matrix proteins and accessory cells to gain growth, survival, and drug resistance advantages.

microenvironment provides a sanctuary for myeloma cells, supporting their proliferation and survival. Myeloma cells benefit not only from the normal lymphocytes within the BM microenvironment but they also take advantage of the environment in a manner that promotes tumorigenesis and neovascularization, alters bone metabolism, and promotes drug resistance.

The BM microenvironment is composed of ECM proteins, hematopoietic stem cells, bone marrow stromal cells (BMSCs), and endothelial cells, as well as osteoclasts and **osteoblasts** (Figure 10.10). MM cell adhesion induces BMSC secretion of chemokines, cytokines, and growth factors. For example, accessory cells (BMSCs, endothelial cells, osteoclasts, and osteoblasts) secrete factors including interleukin-6 (IL-6), insulin-like growth factor-1 (IGF-1), vascular endothelial growth factor (VEGF), tumor necrosis factor-α (TNF-α), fibroblast growth factor (FGF), stromal cell-derived factor-1α (SDF-1α), and B-cell-activating factor (BAFF); all are capable of promoting activation of survival factors such as NF-κB and induction of the phosphatidylinositol 3-kinase (PI3K)/AKT, STAT3, and mitogen-activated protein kinase (MAPK) signaling pathways.

Lymphoid and myeloid cells are also part of the BM microenvironment and can modulate MM survival. Myeloid cells, such as macrophages, mast cells, and neutrophils, control both pro- and anti-inflammatory responses and regulate antigen presentation.

Bone metabolism

Osteolytic bone lesions, bone pain, increased risk of pathological fractures, and generalized bone loss (or osteoporosis) are defining features of myeloma. Typically, the severity of bone destruction correlates with tumor burden and prognosis. The pathophysiology of osteoclast-mediated diseases suggests that myeloma cell growth is mediated by cell-to-cell interaction and through release of factors from the BM microenvironment. As myeloma burden increases, an imbalance between osteoblast and osteoclast activities develops, with suppression of bone formation by osteoblasts and activation of osteoclasts. Osteoclast activity may contribute to myeloma cell survival, growth, and resistance to apoptosis. Osteoclast activity that drives osteolytic lesions is regulated by the RANK ligand (RANKL), which plays an essential role in osteoclast formation, function, and survival. Myeloma cells induce RANKL expression in BMSCs, contributing to enhanced osteoclastogenesis in myeloma bone disease. The endogenous soluble decoy receptor known as osteoprotegerin (OPG) inhibits RANKL–RANK interaction to prevent osteoclast formation. Myeloma cells inhibit production and induce degradation of OPG. These effects result in an increased RANKL-to-OPG ratio, favoring osteoclast formation and activation.

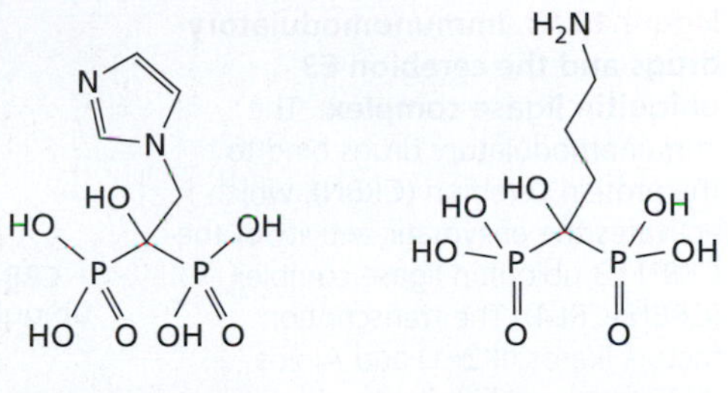

Figure 10.11. Structure of bisphosphonates. The anionic bisphosphonate forms of these compounds are employed to inhibit osteoclast-mediated bone destruction.

Controlling progression of myeloma bone disease may have direct consequences affecting survival and health-related quality of life for myeloma patients. Bisphosphonates, such as pamidronate and zoledronic acid, inhibit osteoclast activity and can palliate pain and prevent bone-related complications. In addition to playing an important supportive role, bisphosphonates may have a direct anti-tumor effect (Figure 10.11). A clinical trial comparing zoledronic acid and oral clodronic acid in MM patients found that zoledronic acid reduced mortality by 16% and increased median OS. **Denosumab**, a monoclonal antibody that inhibits RANKL and downstream osteoclast activity, showed promising activity in MM in a Phase II trial. A Phase III trial demonstrated that denosumab was noninferior to zoledronic acid in prevention of skeletal-related events in newly diagnosed MM patients, leading to FDA approval of denosumab for this indication on January 5, 2018.

In addition to anti-myeloma therapy, the use of either pamidronate or zoledronic acid is recommended monthly for patients with multiple myeloma and lytic bone disease.

Immunomodulatory drugs and the bone marrow microenvironment

Until very recently, the mechanism by which thalidomide and its IMiD derivatives, lenalidomide and pomalidomide, work was not well understood, though the agent was felt to have pleiotropic (multiple) effects on the myeloma microenvironment, and myeloma cells themselves (Figure 10.12). Such effects include direct cytotoxic action on tumor cells, inhibition of tumor cell–stromal cell interactions, inhibition of angiogenic activity (new blood vessel formation), up-regulation of interleukin-2 (IL-2) production by activating T cells, induction of oxidative stress, inhibition of pro-inflammatory cytokines such as TNF-α, and stimulation of natural killer (NK) cells.

The very property that was responsible for its devastating teratogenic effects inspired the idea that thalidomide might be useful in controlling angiogenesis in MM. It was believed that disruption of the interaction between plasma cells and bone marrow stromal cells would lead to reduced induction of VEGF and IL-6 and, consequently, decreased tumor growth and survival. Despite the efficacy of the drug, resistance patterns quickly suggested that inhibition of angiogenesis was not the primary mechanism by which the IMiDs worked.

Figure 10.12. Chemical structures of the immunomodulatory drugs thalidomide, lenalidomide, and pomalidomide.

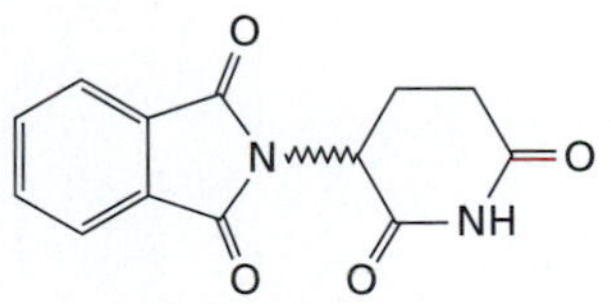

thalidomide

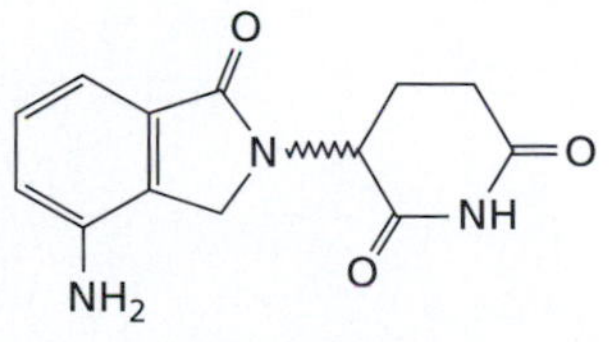

lenalidomide

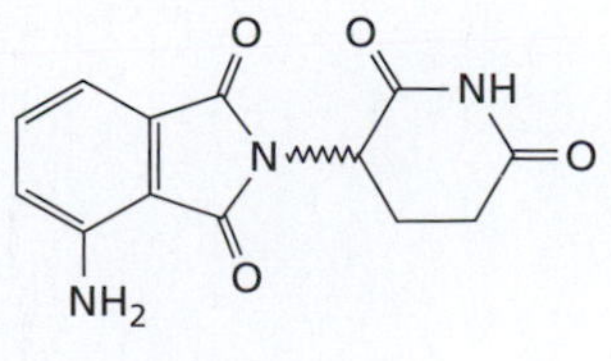

pomalidomide

Figure 10.13. Immunomodulatory drugs and the cereblon E3 ubiquitin ligase complex. The immunomodulatory drugs bind to the protein cereblon (CRBN), which activates the enzymatic activity of the CRBN E3 ubiquitin ligase complex (CRBN–CRL4). The transcription factors Ikaros (IKZF1) and Aiolos (IKZF3) are modified with ubiquitin (Ubi) molecules, targeting them for proteolysis.

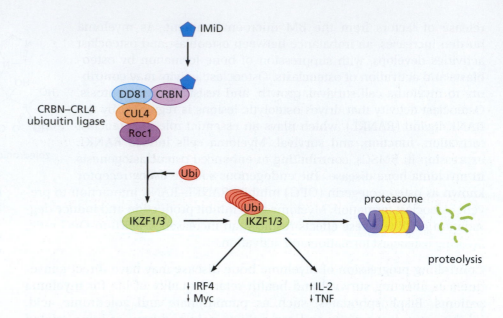

This hypothesis that angiogenesis inhibition was not the primary mechanism of IMiDs in MM was validated in 2010 by the discovery that thalidomide binds to the protein cereblon (CRBN). CRBN forms an E3 ubiquitin ligase complex with the proteins cullin-4A (CUL4A), damaged DNA binding protein 1 (DDB1), and the regulator of cullin-1 (ROC1). This complex, known as the CRBN–CRL4 ubiquitin ligase, attaches ubiquitin to substrate proteins. Once ubiquitinated, these proteins undergo proteolysis (hydrolysis of a protein into amino acids) (Figure 10.13).

The CRBN–CRL4 ubiquitin ligase pathway has been further elucidated by the discovery that the transcription factors and zinc finger proteins IKZF1 (Ikaros) and IKZF3 (Aiolos) are selectively bound by CRBN. Binding of lenalidomide to CRBN activates CRBN's E3 ligase activity, resulting in selective ubiquitination and proteasomal degradation of IKZF1 and IKZF3, which appears to increase IL-2 expression, resulting in an anti-myeloma T-cell response in lenalidomide-sensitive myeloma cells. Further, IMiD-induced CRBN silencing leads to down-regulation of IRF4, a transcription factor, which leads to myeloma cell death via decreased expression of IRF4 target genes that promote proliferation and inhibit apoptosis (including *MYC* and *CDK6*). Though we now have a greater understanding of the mechanism of action of IMiDs in multiple myeloma, the CRBN–CRL4 ligase complex is only one target, and other mechanisms continue to be explored.

Pomalidomide is a third-generation IMiD with potent anti-myeloma effects. Several studies have looked at pomalidomide in combination with low-dose dexamethasone in the relapsed population, culminating in the approval of pomalidomide for relapsed or refractory MM. A randomized Phase II study compared pomalidomide plus low-dose dexamethasone with single-agent pomalidomide in patients with relapsed or refractory myeloma, and showed an improved ORR and PFS favoring the combination. Refractoriness to lenalidomide and bortezomib did not affect outcomes with pomalidomide. A Phase III clinical trial has compared pomalidomide plus low-dose dexamethasone with high-dose dexamethasone alone, and also shows improved outcomes in patients treated with combination therapy. Pomalidomide was FDA approved as a treatment for relapsed and refractory MM in 2013.

Future prospects and challenges in treating multiple myeloma

Recent studies of the cereblon–Ikaros/Aiolos–IRF4/Myc signaling pathway have begun to elucidate the mechanisms by which thalidomide and the other IMiD drugs work; however, there are still gaps in our understanding. Questions, such as why the combination of IMiDs and proteasome inhibitors is particularly potent despite the fact that IMiDs—which work through proteasomal degradation of Ikaros and Aiolos—are paradoxically effective in combination with proteasome inhibitors, require further investigation.

The elucidation of this pathway opens up the opportunity to develop more specific and targeted therapies. Currently, there are few available drugs that directly target the ubiquitin system. Recent advances in our understanding of these pathways provide optimism that there may be several opportunities to target this system. Specific targets include the E3 ubiquitin ligases as well as CRBN itself. Even more targeted therapies may enhance efficacy while avoiding the potential teratogenicity of thalidomide and its analogs.

Take-home points

✓ Microenvironment targets in multiple myeloma (MM) include osteoclast inhibition, angiogenesis inhibition, and immunomodulation.

✓ The immunomodulatory drugs (IMiDs), including thalidomide, lenalidomide, and pomalidomide, are highly effective in the treatment of MM.

✓ The mechanism of action of IMiDs has been elucidated and is now understood to involve the binding of the IMiDs to cereblon (CRBN), which causes modulation of the activity of the CRBN–CRL4 ligase complex to increase ubiquitination and proteasomal degradation of Ikaros and Aiolos.

✓ The efficacy of the IMiDs is augmented by using them in combination with proteasome inhibitors.

✓ Continuous IMiD therapy or maintenance therapy is associated with overall survival benefits in the transplant-ineligible MM population and is considered the standard of care for all patients treated for MM.

✓ Ongoing clinical trials are being designed to incorporate IMiDs into multi-drug regimens that balance efficacy and safety in the transplant-ineligible MM population.

✓ Inhibiting osteoclast activity in bone using bisphosphonates or denosumab decreases skeletal events, such as fractures, a major cause of morbidity in MM patients, as well as improving overall survival.

Discussion questions

1. Identify the clinical factors that define Mr. Brown's condition as symptomatic multiple myeloma versus MGUS.

2. What are the standard treatment options for transplant-ineligible patients with newly diagnosed MM?

3. What are the roles of the different IMiDs in MM treatment?

4. Discuss the mechanism by which the immunomodulatory drugs work.

5. What are the risks of therapy with IMiDs?

Chapter summary

The tumor microenvironment (TME) makes an indispensable contribution to tumor survival and growth, and provides multiple opportunities for therapeutic intervention. The Cases in this chapter demonstrate the efficacy of therapies targeting the cells and the intercellular collaborations of the TME. There is a need to develop a variety of agents and approaches that transform the cellular and molecular properties of the TME to favor tumor stasis and elimination rather than proliferation and growth. It will be especially important to develop effective strategies for converting immunosuppressive microenvironments to ones that support and mount immune responses that are tumoricidal. Immunological strategies discussed in Chapter 12 and other approaches under development may be useful in addressing this problem. TME-directed therapies now in existence and those yet to be developed confront the challenge of making rational determinations of which drugs will be effective in each clinical situation. This has provoked great interest in developing predictive biomarkers, and the identity, physiology (for example, cytokine and chemokine profiles), and gene signatures of cellular components of the TME are being investigated with the goal of identifying clinically useful markers.

Tumor responses to therapeutic agents can be modified by the TME. A desmoplastic (that is, dense and stiff) stroma, largely the result of the nature and amount of ECM produced by CAFs, can hamper the distribution of therapeutic agents in such regions of the tumor, effectively lowering the dose received by those areas and facilitating their escape from the cytotoxic or cytostatic effects of the treatment. In this regard, the chaotic and leaky vasculature of the TME provides an inefficient conduit for delivery of conventional or targeted therapies to all areas of the tumor, resulting in a lowered treatment efficacy. Additionally, expression of the multi-drug resistance (MDR) transporter is induced by the binding of tumor cell integrins to components of the ECM such as fibronectin. MDR activity lowers the toxicity of cytotoxic drugs by drug efflux from the tumor. This results in less effective killing of cancer cells and lowers the efficacy of chemotherapeutic treatments. Integrin–ECM interactions can also initiate signaling via FAK (focal adhesion kinase), PI3K/AKT, and NF-κB pathways, any one of which can directly or indirectly trigger anti-apoptotic pathways, making it harder for cytotoxic drugs to kill cancer cells. Furthermore, signaling by these pathways provides yet another avenue for resistance to tyrosine kinase inhibition with small-molecule inhibitors or specific antibodies.

The growing understanding of the roles of the TME's nonmalignant cell fraction provides a variety of avenues for therapeutic intervention. Some of these are briefly described below (Figure 10.14).

- *Inhibit formation of the tumor vasculature.* Case 10-1 explored inhibition of tumor vasculature formation in CRC patients.

- *Immunomodulatory strategies that facilitate immune attack and decrease immunosuppression.* Case 10-2 discussed the application of an immunomodulatory strategy to treat transplant-ineligible multiple myeloma patients that employs members of the thalidomide family of inhibitors. Approaches to the release of immunosuppression of anti-cancer immune responses include blockade of receptors that inhibit T-cell activation by

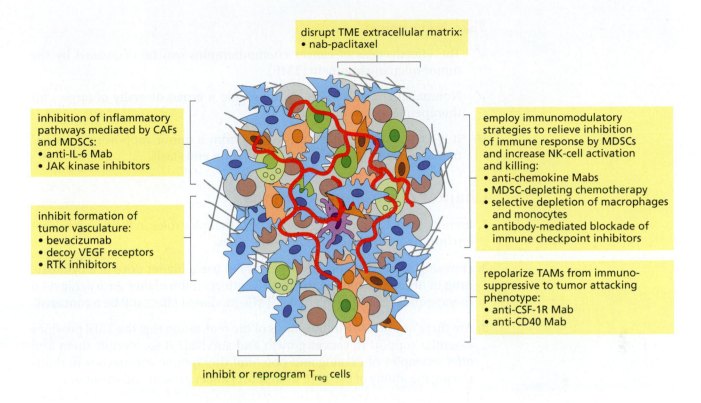

disrupt TME extracellular matrix:
• nab-paclitaxel

inhibition of inflammatory
pathways mediated by CAFs
and MDSCs:
• anti-IL-6 Mab
• JAK kinase inhibitors

inhibit formation of
tumor vasculature:
• bevacizumab
• decoy VEGF receptors
• RTK inhibitors

employ immunomodulatory
strategies to relieve inhibition
of immune response by MDSCs
and increase NK-cell activation
and killing:
• anti-chemokine Mabs
• MDSC-depleting chemotherapy
• selective depletion of macrophages
  and monocytes
• antibody-mediated blockade of
  immune checkpoint inhibitors

repolarize TAMs from immunosuppressive to tumor attacking
phenotype:
• anti-CSF-1R Mab
• anti-CD40 Mab

inhibit or reprogram T$_{reg}$ cells

the checkpoint inhibitors to be discussed in Chapter 12. The blockade of chemokine receptors instrumental in directing the importation and retention of suppressive myeloid populations into the TME is another approach that might have clinical utility.

- *Inhibition of supporting inflammatory pathways.* CAFs and MDSCs are important sources of inflammatory cytokines and chemokines. Inhibition of the IL-6-mediated pathways by anti-IL-6 monoclonal antibodies and by use of the small-molecule kinase inhibitor roxolitinib to interfere with JAK1/JAK2-mediated signaling has shown promise.

- *Repolarization of suppressive macrophages and myeloid cells.* Conversion of immunosuppressive macrophage/myeloid phenotypes to ones that are not antagonistic to immune responses would be highly beneficial. Exploration of approaches to effecting such reprogramming include binding of agonistic (activating) anti-CD40 monoclonal antibodies (MAbs) to the activating ligand, CD40, present on macrophages and monocytes, favoring their conversion to an activated state. Also, inhibition of the influence of colony-stimulating factor-1 (CSF-1) by blockade of the CSF-1 receptor (CSF-1R) with an anti-CSF-1R monoclonal antibody can promote reprogramming of macrophages and monocytes away from a suppressive phenotype.

- *Disruption of the TME extracellular matrix.* Nanoparticle albumin-bound paclitaxel, or nab-paclitaxel, is a protein-bound form of the chemotherapeutic agent paclitaxel. Nab-paclitaxel has been shown to collapse the tumor stroma in pre-clinical models of pancreatic cancer, although the precise mechanism is unclear.

Future efforts incorporating these approaches and others that may emerge will be aggressively pursued to reduce the ability of the TME to support the proliferation and survival of cancer cells.

Figure 10.14. Therapeutic approaches to modifying the tumor microenvironment. A variety of strategies are outlined for the transformation of the TME from a cancer-supportive environment to a nonsupportive or hostile one. CSF-1R: colony stimulating factor-1 receptor; JAK: Janus activated kinase; Mab: monoclonal antibody; nab-paclitaxel: nanoparticle albumin-bound paclitaxel. (Adapted from F. Klemm and J.A. Joyce, *Trends Cell Biol.* 25:198–213, 2015.)

Take-home points

✓ The effectiveness of cancer chemotherapies can be decreased by the tumor microenvironment (TME).

✓ Nonmalignant cells of the TME provide a broad diversity of targets for therapeutic attack on cancer.

✓ It is essential to reprogram the TME from a tumor-supporting environment to one that favors tumor elimination and stasis.

Chapter discussion questions

1. Identify three types or categories of cells that play roles in the tumor micro-environment, and briefly outline their roles.

2. How might the TME make it necessary to use a higher dose of a cytotoxic drug in order to achieve a therapeutic effect? If the higher dose achieves a therapeutic effect, why might this TME-mediated effect still be a concern?

3. Are there any positive implications of the realization that the TME provides essential support to cancer growth and survival? If so, identify them and offer examples of existing and potential therapeutic approaches to mini-mizing the ability of the TME to support cancer growth and maintenance.

Selected references

Introduction

Balkwill FR, Capasso M & Hagemann T (2012) The tumor microenvironment at a glance. *J. Cell. Sci.* 125:5591–5596.

Klemm F & Joyce JA (2015) Microenvironmental regulation of therapeutic response in cancer. *Trends Cell Biol.* 25:198–213.

Case 10-1

Goldberg RM, Sargent DJ, Morton RF et al. (2004) A randomized controlled trial of fluorouracil plus leucovorin, irinotecan, and oxaliplatin combinations in patients with previously untreated metastatic colorectal cancer. *J. Clin. Oncol.* 22:23–30.

Hurwitz H, Fehrenbacher L, Novotny W et al. (2004) Bevacizumab plus irinotecan, fluorouracil, and leucovorin for metastatic colorectal cancer. *N. Engl. J. Med.* 350:2335–2342.

Lengauer C, Kinzler KW & Vogelstein B (1997) Genetic instability in colorectal cancers. *Nature* 386:623–627.

Saltz LB, Clarke S, Díaz-Rubio E et al. (2008) Bevacizumab in combination with oxaliplatin-based chemotherapy as first-line therapy in metastatic colorectal cancer: a randomized phase III study. *J. Clin. Oncol.* 26:2013–2019.

Saltz LB, Cox JV, Blanke C et al. (2000) Irinotecan plus fluorouracil and leucovorin for metastatic colorectal cancer. Irinotecan Study Group. *N. Engl. J. Med.* 343:905–914.

Siegel RL, Miller KD & Jemal A (2018) Cancer Statistics, 2018. *CA Cancer J. Clin.* 68:7–30.

Tournigand C, André T, Achille E et al. (2004) FOLFIRI followed by FOLFOX6 or the reverse sequence in advanced colorectal cancer: a randomized GERCOR study. *J. Clin. Oncol.* 22:229–237.

Zauber AG, Winawer SJ, O'Brien MJ et al. (2012) Colonscopic polypectomy and long-term prevention of colorectal-cancer deaths. *N. Engl. J. Med.* 366:687–696.

Case 10-2

Krönke J, Udeshi ND, Narla A et al. (2014) Lenalidomide causes selective degradation of IKZF1 and IKZF3 in multiple myeloma cells. *Science* 343:301–305.

Kumar S, Flinn I, Richardson PG et al. (2012) Randomized, multicenter, phase 2 study (EVOLUTION) of combinations of bortezomib, dexamethasone, cyclophosphamide, and lenalidomide in previously untreated multiple myeloma. *Blood* 119:4375–4382.

Lu G, Middleton RE, Sun H et al. (2014) The myeloma drug lenalidomide promotes the cereblon-dependent destruction of Ikaros proteins. *Science* 343:305–309.

Morgan GJ, Davies FE, Gregory WE et al. (2010) First-line treatment with zoledronic acid as compared with clodronic acid in multiple myeloma (MRC Myeloma IX): a randomised controlled trial. *Lancet* 376:1989–1999.

Palumbo A, Hajek R, Delforge M et al. (2012) Continuous lenalidomide treatment for newly diagnosed multiple myeloma. *N. Engl. J. Med.* 366:1759–1769.

San Miguel J, Weisel K, Moreau P et al. (2013) Pomalidomide plus low-dose dexamethasone versus high-dose dexamethasone alone for patients with relapsed and refractory multiple myeloma (MM-003): a randomised, open-label, phase 3 trial. *Lancet Oncol.* 14:1055–1066.

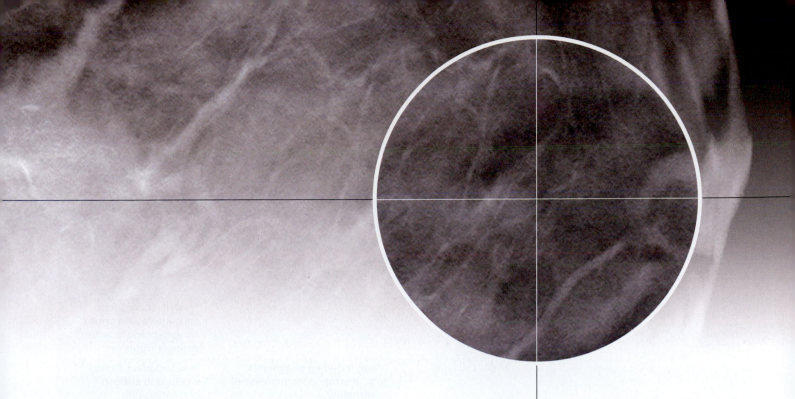

Chapter 11

Immunological Strategies: Monoclonal Antibodies

Introduction

The immune response to cancer is the outcome of a network of collaborative interactions among the cells and soluble factors of the adaptive and innate immune systems. In successful anti-tumor responses this collaboration results in the death of the cancer cell. The major immune effectors of cancer cell death are **cytolytic cells** and antibody-dependent processes. The cytolytic cell populations involved include **cytotoxic T cells**, **natural killer (NK) cells**, and **macrophages**. Cancer cells can therefore be recognized, targeted, and killed by the immune system. Immunotherapies mediated by **antibodies** and **T cells** are the dominant ones in clinical use. Both of these immunotherapies are antigen-triggered manifestations of **adaptive immunity**. **Antigens** are molecules that specifically bind to antibodies, or that contribute structural elements that specifically bind to T-cell receptors. Although adaptive immune responses are triggered by antigen recognition, they are dependent upon interactions between the adaptive and innate arms of the **immune system** (Table 11.1). As summarized in the table, **B cells**, which are the only cells capable of making antibodies, and T cells are the cellular agents of the precise antigen specificity and **immunological memory** that are hallmarks of adaptive immunity. **Innate immunity** is mediated by a large and diverse repertoire of cell types and molecules, whose actions are triggered by recognition of either **pathogen-associated molecular patterns** (PAMPs) or **damage-associated molecular patterns** (DAMPs). PAMPS are microbial components or products, and are generated during infections. In general, DAMPS are of host origin and may arise in one of two ways—as by-products of infection-related damage to cells and tissues, or as a

Table 11.1. Innate and adaptive immunity: a comparison.

Characteristic	Innate immunity	Adaptive immunity
Specificity	Recognition of general molecular patterns or motifs present in pathogens but absent from hosts; also damage- or death-associated molecules	Highly specific, capable of distinguishing between molecules or regions of molecules that differ only slightly
Receptors	Many families with few members of highly conserved receptors	Extremely diverse repertoires of T-cell and B-cell antigen receptors
Response time after initial encounter	Minutes to hours	Days
Memory	Nonexistent or very weak	Present. Subsequent encounters with same antigen produce faster, stronger responses
Major cell types	Many, including phagocytes (e.g., macrophages, monocytes, neutrophils), natural killer cells, other leukocytes, and some non-hematopoietic cells	Few, including T cells, B cells, and antigen-presenting cells

consequence of cell or tissue injury caused by trauma, such as a blow, laceration, or burn. Cells implicated in innate immunity are indispensable first responders to infection and injury. In addition, innate immune cells, particularly **dendritic cells** (DCs), bear ligands that are essential for the adaptive immune system's initial recognition of and response to antigens.

Many immune-based therapeutic approaches have been devised and are increasingly finding their way into clinical use. Consequently, **immunotherapy** has now taken its place alongside surgery, radiation, and chemotherapy as a major category of cancer therapy. Immunotherapeutic agents currently in clinical use include: (a) monoclonal antibodies (MAbs), (b) monoclonal **antibody–drug conjugates** (ADCs), (c) **immune checkpoint inhibitors**, (d) **adoptive T-cell transfer** strategies, including **chimeric antigen receptor (CAR) T cells**, and (e) both **prophylactic** and **therapeutic cancer vaccines**. This chapter discusses the use of monoclonal antibody-based immunotherapies, and presents cases that illustrate the clinical use of these therapies. Chapter 12 explores approaches to cancer treatment that manipulate host adaptive immune responses; these include **vaccination**, adoptive T-cell transfer, and release of checkpoint inhibition of immune responses.

Antibodies are the most familiar elements of the immune system. As shown in Figure 11.1, they consist of two identical immunoglobulin light chains and two identical immunoglobulin heavy chains. In addition to the roles of antibodies as agents of immune defense, the specificity and high affinity of **antigen–antibody reactions** have found many applications, including the identification and characterization of tissues, cells, and molecules. They have a long history of use as agents for the immunotherapy of infectious disease, and are now finding increasing application in the treatment of cancer.

The introduction of an appropriately **immunogenic antigen** into a vertebrate stimulates B cells that have receptors specific for the antigen to divide and mature into **plasma cells**, which secrete antibodies specific for the immunizing antigen (Figure 11.2A). The B-cell population of an individual is made up of many distinct populations. These are referred to as **clones**, because

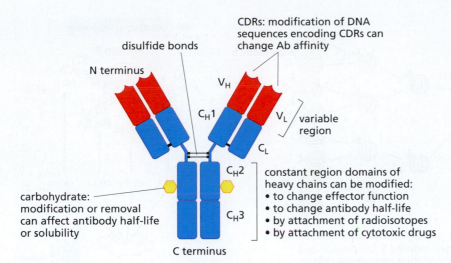

N terminus

disulfide bonds

CDRs: modification of DNA sequences encoding CDRs can change Ab affinity

V_H

C_H1

V_L / variable region

C_L

C_H2

constant region domains of heavy chains can be modified:
• to change effector function
• to change antibody half-life
• by attachment of radioisotopes
• by attachment of cytotoxic drugs

carbohydrate: modification or removal can affect antibody half-life or solubility

C_H3

C terminus

Figure 11.1. Antibody structure and its manipulation. The basic structure of an antibody consists of a pair of disulfide-linked immunoglobulin heterodimers. Each of the heterodimers is identical to the other and contains a heavy-chain immunoglobulin molecule with disulfide links to a light-chain immunoglobulin molecule. Note that both heavy and light chains bear regions of amino acid sequence known as constant regions (labeled C_H for heavy and C_L for light) that do not vary greatly from one member of the immunoglobulin class to another. The same is true of the constant regions of light-chain types. In contrast, other regions (called V regions) show great variation from one antibody to another. Most of this variation is concentrated in three noncontiguous regions known as complementarity-determining regions (CDRs), located in the V regions of the heavy and light chains. The CDRs of the heavy and light chains of each heterodimer constitute an antigen-binding site, endowing the antibody illustrated with two identical antigen-binding sites, each having the same antigen specificity and binding affinity for antigen. As indicated by the covalently attached carbohydrate, antibody molecules are glycosylated. The properties of antibodies can be manipulated by the indicated modifications of antibody structure.

all of the members of an individual population arose from a single cell and are genetically identical to each other. However, each B-cell clone differs to a greater or lesser extent from all other B-cell clones with regard to the DNA sequences that encode their immunoglobulin molecules, particularly in relation to those portions of the sequence that encode the binding site of the antibody and therefore determine its antigen specificity. The total repertoire of **B-cell clones** in an individual is vast, and contains a sufficient diversity of binding sites to ensure that one or more of that individual's resident clones will recognize any antigen encountered. Provided that conditions are appropriate, it will then be triggered to proliferate, greatly increasing the number of cells that recognize the antigen. The subsequent differentiation of a large fraction of this population into plasma cells—the B-cell type that secretes antibody—results in the body accumulating antibodies that react with the immunizing antigen.

As shown in Figure 11.2B, the antibody response usually involves induction of antibody secretion from many different B-cell clones, resulting in the production of **polyclonal antibodies**. These are a complex mixture of many different antibodies that vary in composition and whose constituent antibodies may react with different regions of the inducing antigen. Even when the same antigen and immunization conditions are used, the exact composition of a polyclonal antibody varies from animal to animal and from day to day. Therefore, when generated for research or diagnostic use, each batch of a polyclonal antibody has to be characterized and standardized.

In contrast to a polyclonal antibody, a monoclonal antibody comes from a single clone of B cells, and is uniform with regard to structure and reactivity. This is a very useful attribute for a substance intended for use as a drug. Furthermore,

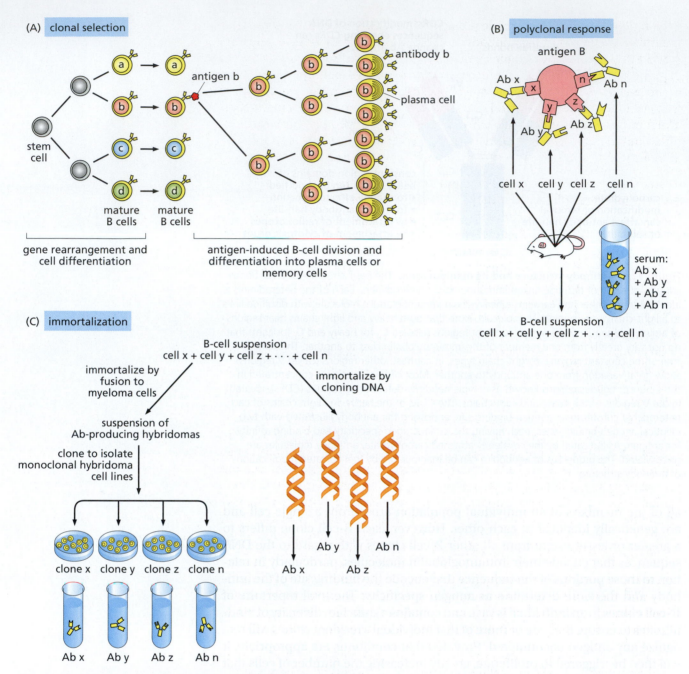

Figure 11.2. Primer on the origin, generation, and immortalization of monoclonal antibodies. (A) Clonal emergence and selection. B cells differentiate from hematopoietic stem cells during B-cell development. The random rearrangements of portions of the genome encoding the variable heavy-chain (V_H) and variable light-chain (V_L) regions of the antibody molecule endow each B cell with a unique V_H and V_L region, making each B cell emerging from development unique. An emergent B cell's antigen receptors, its complement of cell surface antibodies, have antigen-binding sites that are identical to each other but different from those of other emerging B cells, and therefore have the potential to bind antigens that are different from those bound by other emerging B cells. Each emerging B cell is the founder of a unique clone of B cells. Antigen recognition (binding of an antigen to the antigen-binding site of the B-cell receptor) triggers activation of the immune system for proliferative expansion of the B-cell clone and the differentiation of some members of the clone into terminally differentiated B cells called plasma cells that produce and secrete large amounts of antibody. Furthermore, because the antibody is the product of a single clone of B cells, it is a monoclonal antibody. (B) Polyclonal response. Injection of a complex foreign antigen into a mouse usually induces the response of many B-cell clones (a polyclonal response) to its various antigenic determinants (epitopes). Each of the different antibodies induced by the antigen is the product of a single clone, and hence is a monoclonal antibody. Under normal conditions, all of these different monoclonal antibodies pool in the bloodstream and tissue fluids, constituting a polyclonal antibody against an antigen (antigen B in the diagram). (C) Immortalization. The ability to produce a particular monoclonal antibody in unlimited quantities at any time is achieved in either of two ways: (1) B-cell populations secreting the desired antibody can be fused with myeloma cells to produce hybrid cells, called hybridomas. Hybridomas have the immortal lifespan and proliferative capacity of the cancerous myeloma parent, as well as the capacity to produce the desired antibody of the normal B-cell parent. (2) Clone antibody genes of B-lineage cells that encode the desired monoclonal antibody and then amplify the DNA. The DNA can be incorporated into a vector and used to express the desired antibody.

cloning of the genes that encode a useful MAb makes it possible to produce unlimited quantities of the antibody and to produce that antibody at any time and in the quantity that might be needed in the future. The development of highly efficient technologies for the generation of partially or fully human MAbs represented a vital technical advance, because nonhuman immunoglobulins can elicit severe allergic reactions in humans. Such technologies include genetically engineering rodents and other animals to produce fully human MAbs. Alternatively, recombinant DNA technologies can be used to produce fully human MAbs completely *in vitro*, without the need to use animals at all.

Many MAbs have received U.S. Food and Drug Administration (FDA) approval for use in cancer therapy, and MAbs are the immunological agents most widely used for cancer treatment (Table 11.2). Most of these antibodies are directed against **tumor-associated antigens** (TAAs), rather than **tumor-specific antigens** (TSAs), which are found exclusively on the cancer cells. TAAs are expressed by cells of the cancer, but can also be found on healthy cells of the same or related tissues. For example, normal lymphocytes of the B-cell lineage as well as many **B-cell malignancies** (e.g., non-Hodgkin lymphoma) express CD20, a cell-surface protein that is characteristic of the B-cell lineage. Thus when it is present on cancer cells, CD20 is a TAA. On the other hand, many cervical carcinomas express proteins encoded by **human papillomavirus** (HPV), the oncogenic virus that is the etiologic agent of most cases of cervical cancer. These HPV-encoded proteins are TSAs, present in cells of HPV-induced cancers but not in normal cells uninfected by HPV. Immunotherapies directed at TSAs are particularly appealing because they offer an ideal means of focusing an attack on cancer cells while sparing normal cells. At first glance, it might appear that TAAs are not attractive targets for immunotherapy, because an immunological attack on TAAs would kill both normal cells and cancer cells that express the antigen. However, clinical experience has demonstrated that some TAAs are quite useful targets for immune attack. Indeed, the majority of the monoclonal antibodies that are used in cancer immunotherapy (see Table 11.2) target the much more common and readily identifiable TAAs, rather than the less common and much less easily identifiable TSAs.

There are several mechanisms by which MAbs kill or inhibit the growth of cancer cells, and in some cases a particular antibody may exert its anti-cancer effects in more than one way (Figure 11.3). The main mechanisms used by MAbs to kill cancer cells or inhibit their growth are briefly reviewed below.

Although antibodies themselves are not cytotoxic, their binding to the surface of a cell can target it for killing by cellular and noncellular agents of the immune system. For example, when the Fc receptor of a natural killer (NK) cell is engaged by the Fc region of a monoclonal antibody bound to a cancer cell, the NK cell will kill that cancer cell. This interaction whereby antibody binding to a cancer cell directs the cellular immune response is known as antibody-dependent cell-mediated cytotoxicity (ADCC), and may be mediated by NK cells, macrophages, or monocytes. The antigen–antibody complex that is formed by the reaction of the anti-cancer monoclonal antibody may also trigger activation of the **complement pathway**. This component of the immune system assists in the clearance of pathogens from the body, and can kill cells by a process known as **complement-dependent cytotoxicity** (CDC). A major mechanism of CDC arises from the generation, as a consequence of complement activation, of a protein complex dubbed the **membrane attack complex** (MAC), which proceeds to punch holes in the membranes of targeted cells, thereby causing their death.

Antibody blockade of essential ligand–receptor interactions provides an effective means of killing or inhibiting the growth of cancer cells that are

Table 11.2. Monoclonal antibodies approved by the U.S. Food and Drug Administration (FDA) for cancer treatment.

International name	Trade name	Specificity and characteristics	First approved indication	First FDA approval
Rituximab	Rituxan	Anti-CD20 Chimeric IgG1	Non-Hodgkin lymphoma	1997
Trastuzumab	Herceptin	Anti-HER2 Humanized IgG1	Breast cancer	1998
Alemtuzumab	Campath	Anti-CD52 Humanized IgG1	Chronic lymphocytic leukemia	2001
Cetuximab	Erbitux	Anti-EGFR Chimeric IgG1	Colorectal cancer	2004
Ibritumomab	Zevalin	Anti-CD20 Murine IgG1	Non-Hodgkin lymphoma	2002
Bevacizumab	Avastin	Anti-VEGF Humanized IgG1	Colorectal cancer	2004
Panitumumab	Vectibix	Anti-EGFR Human IgG2	Colorectal cancer	2006
Ofatumumab	Arzerra	Anti-CD20 Human IgG1	Chronic lymphocytic leukemia	2009
Brentuximab vedotin	Adcetris	Anti-CD30 Chimeric IgG1/ immunoconjugate	Hodgkin lymphoma	2011
Ado-trastuzumab emtansine	Kadcyla	Anti-HER2 Humanized IgG1/ immunoconjugate	Breast cancer	2013
Ramucirumab	Cyramza	Anti-VEGF Human IgG1	Gastric cancer	2014
Obinutuzumab	Gazyva	Anti-CD20 Humanized EgG1 Glycoengineered	Chronic lymphocytic leukemia	2013
Ipilimumab	Yervoy	Anti-CTLA-4 Human IgG1	Metastatic melanoma	2011
Nivolumab	Opdivo	Anti-PD-1 Human IgG4	Advanced melanoma	2014
Pembrolizumab	Keytruda	Anti-PD-1 Humanized IgG4	Advanced melanoma	2014
Atezolizumab	Tecentriq	Anti-PD-L1 Humanized IgG1	Urothelial carcinoma	2016
Durvalumab	Imfinzi	Anti-PD-L1 Human IgG1	Urothelial carcinoma	2017

Due to rapid advances, this should not be considered a comprehensive list. Adapted from Janice M. Reichert/Antibody Society Web site.

dependent upon these interactions. The anti-cancer effect can be achieved by using antibodies directed against either the ligand, such as a growth or survival factor, or the receptors for these ligands on the surface of the cancer cell. In some instances, the engagement of receptors may trigger intracellular pathways that lead to cell death.

Cancers are dependent on interactions with normal cells in the human body that together with the cancer cells form a tumor microenvironment. These include **fibroblasts**, **endothelial cells**, and a variety of other cell types. MAbs that compromise the function of normal host cells in the tumor

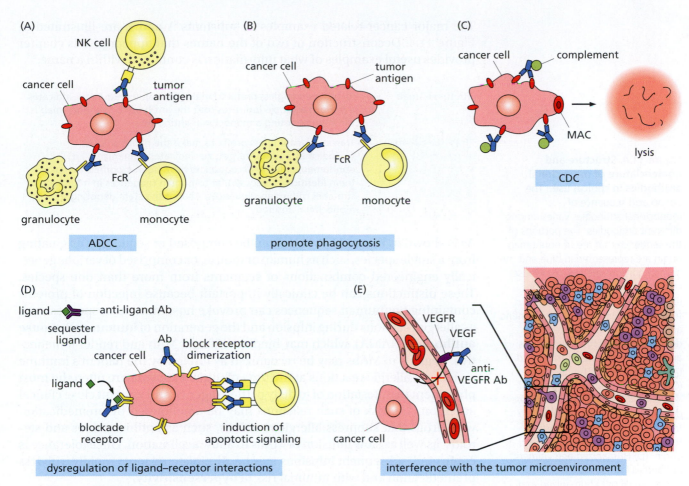

Figure 11.3. Antibodies kill or inhibit cancer cells in many different ways. (A) Antibody-dependent cell-mediated cytotoxicity (ADCC). Antibodies specific for tumor antigens bind to Fc receptors on NK cells, monocytes, or granulocytes, and trigger the killing of the cancer cell. (B) Phagocytosis. The Fc regions of antibodies bound via their antigen-binding sites to tumor antigens interact with Fc receptors on monocytes or granulocytes to trigger phagocytosis of the cells. (C) Complement-dependent cytotoxicity (CDC). Antigen–antibody complexes formed by the reaction of antibodies with TSAs or TAAs on cancer cells trigger their complement-mediated lysis by the membrane attack complex (MAC). (D) Dysregulation of ligand–receptor interactions. Direct tumor cell death can be caused by such ligand–receptor disruptions as sequestration of ligand by anti-ligand antibody or blockade of the receptor by anti-receptor antibodies targeting the binding site. In addition, inappropriate engagement of receptors by antibodies can induce apoptosis. Antibodies that have been modified to bear toxic payloads (i.e., cytotoxic drugs or radioisotopes) can also kill cancer cells (*see* Figure 11.5). (E) Interference with the tumor microenvironment. Disruption of processes that are essential for support of the tumor inhibits the growth or survival of cancer cells. In the example shown, the activity of VEGF (vascular endothelial growth factor), a factor that is essential for the growth of blood vessels, is blocked by an anti-VEGF antibody. (A–D, adapted from G.J. Weiner, *Nat. Rev. Cancer* 15:361–370, 2015. Reprinted by permission from Macmillan Publishers Ltd. E, adapted from F.R. Balkwill et al., *J. Cell Sci.* 125:5591–5596, 2012.)

microenvironment, or their ability to interact with cancer cells, can inhibit the growth of the cancer.

Table 11.2 lists a bewildering collection of names that at first glance appear quite strange (e.g., trastuzumab, rituximab, pembrolizumab). In fact, these generic names of MAbs are quite rational, and are structured so as to provide the following information: (a) the source species (mouse, human, or other); (b) whether the amino acid sequence is that of a human antibody or contains immunoglobulin sequences derived from another species (e.g., mouse); and (c) information about the disease, process, molecule, or physiological system that is targeted by the antibody. Generic names are composed as schematized below:

Prefix a distinctive but meaningless syllable	Substem A designating target	Substem B designating nature of antibody sequence	Suffix "mab" as stem

The major cancer-related examples of substems A and B are illustrated in Figure 11.4. Deconstruction of two of the names that appear in this chapter provides useful examples of what information is contained within a name.

Ri- tu- xi- -mab	Ri- is a meaningless prefix to make the name unique; tu- indicates a tumor target; xi- indicates that the antibody is chimeric; -mab is the stem denoting a monoclonal antibody.
Tras- tu- zu- -mab	Tras- is a meaningless prefix to make the name unique; tu- indicates a tumor target; zu- indicates humanized, i.e., nonhuman amino acid sequences have been modified to make them identical or very similar to human sequences in homologous stretches of human sequences; -mab is the stem denoting a monoclonal antibody.

Figure 11.4. Structure and nomenclature of monoclonal antibodies in clinical use. The amino acid sequence of monoclonal antibodies varies among different antibodies. The portions of the sequence that are of nonhuman origin are represented in blue and the human portions are red. (A) Mouse or nonhuman heavy (H) and light (L) chains are completely derived from nonhuman species. (B) Chimeric antibodies have variable regions derived from a nonhuman species, and constant regions of the H and L chains that are of human origin. (C) Humanized antibodies have only their CDRs (complementarity-determining regions) derived from foreign species; the rest of the H and L chains are of human origin. (D) Humanized and chimeric antibodies have variable regions of one chain (H or L) humanized and those of the other chain chimeric. (E) Fully human antibodies, as expected, have no components from other species. The systemized nomenclature denoting nonhuman, in this case mouse, and the targeted tissue or process are specified in the table.

As is shown in Figure 11.4, MAbs can be composed of sequences originating from a single species, such as human or mouse, or comprised of various genetically engineered combinations of segments from more than one species. These distinctions can be clinically important because injection of proteins containing nonhuman sequences can provoke host immune responses such as allergic reactions during infusion and the generation of **human anti-mouse antibodies** (HAMA), which may bind the infused MAb and render it ineffective. Regions of MAbs may be recognized as foreign by the patient's immune system, so allergic reactions at the time of initial exposure are often the most prominent manifestation of toxicity of these agents, and warrant close clinical attention. The risk of such reactions may be ameliorated by premedication with drugs that suppress allergic reactions, such as antihistamines and steroids, as well as slow infusion of MAbs to allow acclimation. Once tolerance is confirmed, subsequent infusions may be administered more rapidly with less premedication and with minimal risk of **hypersensitivity**.

Antibodies provide a means of precisely and selectively targeting highly cytotoxic drugs at cancer cells. By attaching drugs to antibodies, it is possible to generate antibody–drug conjugates (ADCs) that are specific for TSAs or TAAs found on the surface of cancer cells. Tumor cells bind and internalize the ADC, at which time the drug is enzymatically cleaved from the antibody and the drug is released into the cancer cell to exert its cytotoxic effect. This approach

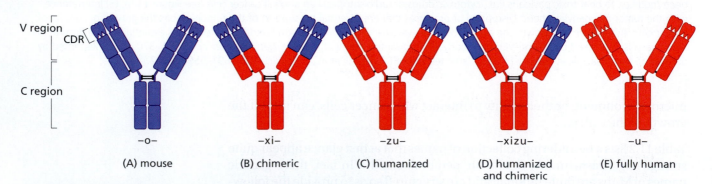

V region				
CDR				
C region				
–o–	–xi–	–zu–	–xizu–	–u–
(A) mouse	(B) chimeric	(C) humanized	(D) humanized and chimeric	(E) fully human

substem designating target of monoclonal antibody		substem designating nature of monoclonal antibody sequence	
ci	circulatory system	o	mouse
li	immune system	u	human
tu	tumor	zu	humanized
kin	cytokine	xi	chimeric
so	bone	xizu	chimeric and humanized

specifically delivers the cytotoxic drug into cells that display the tumor antigen, sparing normal cells that do not bear the targeted antigen. For students of mythology, this mechanism may be reminiscent of the Trojan horse being welcomed inside the city walls of Troy only to release enemy soldiers who would be the city's downfall. Although straightforward in principle, many technical requirements needed to be met before the promise of ADCs was transformed into clinical reality. First, the synthesis of the ADC must generate a product that leaves the antigen-binding sites of the antibody intact and undiminished in affinity. Second, the immunogenicity of the chemically modified antibody must be sufficiently low for it not to elicit unacceptable levels of allergic reactions. Third, its **pharmacokinetics** (i.e., *in vivo* stability of the linkage, tissue distribution, and *in vivo* half-life of the ADC) must be acceptable.

The pharmacokinetic requirements for the *in vivo* stability of the ADC are particularly challenging, because the requirements of the covalent link between antibody and drug are demanding and contradictory. During its journey to the surface of the tumor cell, this linkage must be very stable and resist cleavage in the bloodstream and tissue fluids. However, once the ADC has bound to the surface of the tumor cell and gained access to its interior, the antibody–drug linkage must undergo intracellular cleavage, thereby releasing its cytotoxic cargo and allowing the latter to move by diffusion to its intracellular target. As shown in Figure 11.5, the ADC has to complete a number of steps in order to deliver its cytotoxic cargo to the target molecule. Strikingly, even if the ADC completes each of these steps with an efficiency of 50%, only 1.6% of the drug will reach the target. Since the amount of drug delivered is so small, only the most powerful cytotoxic agents are employed. ADCs also have the attractive property of significantly increasing the potency of antibodies without loss of the cancer-killing attributes of their unconjugated counterparts, as ADCs are still able to mediate cell death by CDC and by ADCC. Furthermore, if the targeted tumor-cell antigen is a receptor of growth or survival ligands, the ADC will block the receptor and inhibit this essential receptor–ligand interaction.

The cases that follow illustrate the contribution of MAb-based immunotherapies to the treatment of non-Hodgkin lymphoma (NHL), Hodgkin lymphoma, and breast cancer.

Take-home points

✓ The immune system is capable of recognizing cancer cells and using cytolytic cell populations and antibody-dependent processes to kill them.

✓ Monoclonal antibodies offer the advantages of reproducibility, uniformity, and unlimited quantity, and with the use of recombinant DNA methodologies, they can be generated in fully human versions.

✓ Monoclonal antibodies are the most widely used immunotherapeutic agents, and exert their anti-cancer effects through a variety of mechanisms, including (a) targeting cancer cells for cytotoxic attack, (b) blockade of ligand–receptor interactions, and (c) rendering the cancer microenvironment less supportive for cancer cell growth and survival.

✓ Antibody–drug conjugates (ADCs) can target a drug to a particular cancer cell population and deliver a cytotoxic payload, thus deploying both immunotherapy and chemotherapy with the same molecule.

✓ A broad diversity of approaches to cancer immunotherapy have been developed, and several of these have entered clinical use.

Figure 11.5. Nature of an antibody–drug conjugate (ADC) and a typical pathway of delivery. (A) Antibody–drug conjugate. A cytotoxic drug is covalently attached to the constant region of an antibody molecule that is specific for a tumor-specific or tumor-associated antigen. Successful ADCs have antibody–drug linkages that do not disturb the antigen-binding site of the antibody and are stable in the extracellular environment of the cancer cell, but undergo cleavage inside the cell. The drugs employed are highly potent and act in subnanomolar concentrations. (B) The path of an ADC from injection to the induction of cell death involves six steps: Step 1, injected ADC reaches the tumor site; Step 2, ADC binds to target tumor cell; Step 3, ADC is internalized into an endosome; Step 4, the endosome fuses with a lysosome, and enzymatic digestion releases the drug; Step 5, the drug reaches the cytoplasm; Step 6, the drug reaches the target and exerts its biological effect, resulting in death of the cell. (C) Efficiency. The diagram represents a hypothetical calculation of how much of an ADC-borne drug reaches the target and has a biological effect. The calculation assumes a 50% yield at each step, and concludes that only 1.56% of the total amount of the drug that was given to the patient has a biological effect.

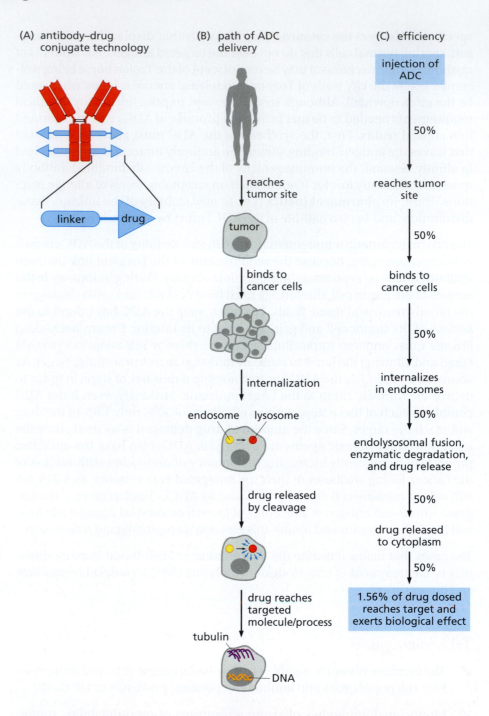

(A) antibody–drug conjugate technology

(B) path of ADC delivery

(C) efficiency

Topics bearing on this case

Diffuse large B-cell lymphoma

Monoclonal antibody treatment of lymphoma

Antibody-dependent cell-mediated cytotoxicity (ADCC)

Case 11-1 Non-Hodgkin Lymphoma

Introduction

Non-Hodgkin lymphoma (NHL) comprises a heterogeneous collection of cancers derived from either B or T lymphocytes. There are currently over 60 recognized subtypes of NHL, each characterized by a unique biology and often distinct natural history and therapy (Table 11.3). Understanding the discrete biology of a given histology is therefore critical to understanding prognosis and treatment selection for a lymphoma patient. Approximately 90% of NHLs arise from B cells, with around 10% arising from T cells. Among **B-cell lymphomas**, the clinical course can often be categorized as disease with an

Table 11.3. Subtypes of non-Hodgkin lymphomas.

Indolent non-Hodgkin lymphomas	
B-cell lymphomas	**T-cell lymphomas**
Follicular lymphoma (grades I–II)	T-cell large granular lymphocyte leukemia
Small lymphocytic lymphoma/chronic lymphocytic leukemia	Mycosis fungoides
Marginal zone lymphomas	
Lymphoplasmacytic lymphoma (Waldenström macroglobulinemia)	
Hairy cell leukemia	
Aggressive non-Hodgkin lymphomas	
B-cell lymphomas	**T-cell lymphomas**
Diffuse large B-cell lymphomas	Peripheral T-cell lymphoma, not otherwise specified
Follicular lymphoma (grade III)	Anaplastic large-cell lymphomas, ALK+ and ALK−
Mantle cell lymphoma	Angioimmunoblastic T-cell lymphoma
B-cell prolymphocytic leukemia	Extranodal NK/T-cell lymphoma
	Enteropathy-associated T-cell lymphoma
	Subcutaneous panniculitis-like T-cell lymphoma
	Hepatosplenic T-cell lymphoma
	Adult T-cell leukemia/lymphoma
	T-cell prolymphocytic leukemia
Highly aggressive non-Hodgkin lymphomas	
B-cell lymphomas	**T-cell lymphomas**
Burkitt lymphoma	Precursor T-cell lymphoblastic lymphoma
Precursor B-cell lymphoblastic lymphoma	

indolent, aggressive, or highly aggressive natural history. Indolent lymphomas have an untreated natural history on the order of years or decades, whereas aggressive and highly aggressive B-cell lymphomas would be associated with a life expectancy of weeks or even days without effective treatment. In general, indolent lymphomas are highly treatable but not considered curable with conventional therapy. Indolent lymphomas are treated as needed over the course of the patient's lifetime, with the goal of preventing the disease from ever becoming life-threatening or impairing quality of life. However, in the case of aggressive lymphomas, the goal of therapy is to cure, because unless they are successfully eradicated these lymphomas will be uniformly fatal.

Diffuse large B-cell lymphoma (DLBCL) is the most common lymphoma and the prototype for aggressive NHL. DLBCL is a biologically heterogeneous disease in which numerous different genetic and molecular aberrations may contribute to **lymphomagenesis**, which is the growth and development of lymphoma. The most common recurring chromosomal defect is rearrangement of the *BCL6* gene; this gene is critical for formation of the germinal-center reaction in healthy lymph nodes. The germinal-center reaction is a process characterized by B-cell proliferation, antibody class switching, and strikingly high rates of mutation that drive increases in antibody affinity. Other common abnormalities

Figure 11.6. Molecular subtypes of diffuse large B-cell lymphoma (DLBCL) as determined by mRNA expression patterns using gene expression profiling. Each column represents a patient tumor sample, and each row represents a gene that is differentially expressed across different subtypes. The color scale indicates relative expression, with red denoting high-level expression and green denoting low-level expression. Distinct molecular subtypes of DLBCL include activated B-cell (ABC)-like, germinal-center (GC)-like, and primary mediastinal B-cell lymphoma (PMBCL). (From M. Roschewski et al., *Nat. Rev. Clin. Oncol.* 11:12–23, 2014. Reprinted by permission from Macmillan Publishers Ltd.)

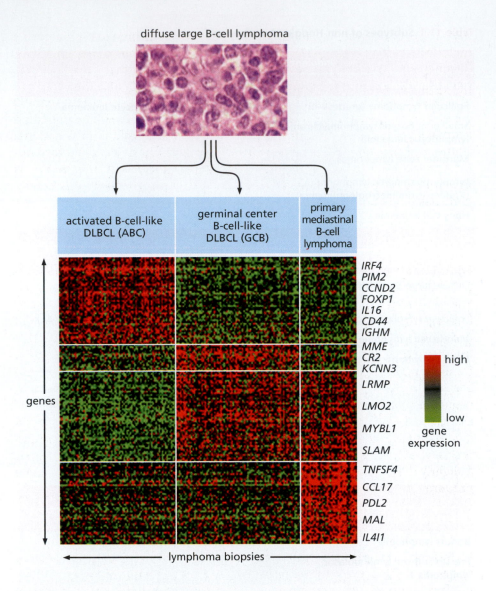

include rearrangement of *BCL2* (a critical anti-apoptotic gene), *MYC* (a driver of proliferation), mutation or loss of the tumor suppressor gene *p53*, and up-regulation of pathways, including B-cell receptor signaling, NF-κB, PI3 kinase, and JAK/STAT signaling. Transcriptional profiling has been used to divide DLBCLs into molecular subsets with shared features based on their resemblance to normal stages of B-cell development, or cell of origin (Figure 11.6). These subtypes are **activated B-cell (ABC)-like** and **germinal-center B-cell (GCB)-like**, and a distinct molecular entity from these subsets can also be identified, namely **primary mediastinal B-cell lymphoma** (PMBCL).

GCB-like DLBCL relies on constitutive expression of *BCL6* and may have activating mutations of *EZH2*, whereas ABC-like DLBCL is characterized by chronic active B-cell receptor (BCR) signaling, activating MYD88 mutations and resultant activation of the NF-κB pathway. PMBCL, on the other hand, is characterized by activation of the JAK/STAT signaling pathway. Since gene expression profiling is not widely available for clinical management, lymphoma biopsies can be stained with antibodies against specific markers on their cell surface or in their cytoplasm, which can help to differentiate between GCB-like DLBCL, ABC-like DLBCL, and PMBCL.

The molecular subtypes have prognostic relevance, with GCB-like DLBCL and PMBCL carrying a distinctly more favorable prognosis compared with

ABC-like DLBCL. Additional genetic features that contribute to the prognosis include activation of the *MYC* gene (by either translocation or amplification), which predicts particularly aggressive behavior and an inferior cure rate, particularly when combined with abnormalities of *BCL2*. Clinical features can also predict inferior outcome in DLBCL; these include age greater than 60 years, stage 3 or 4 disease, involvement of multiple sites in the body other than normal lymphoid organs (lymph nodes, spleen, and tonsil), poor functional status, and elevation of lactate dehydrogenase (LDH) as determined by laboratory testing. These five risk factors constitute the widely used "**International Prognostic Index**," or IPI, and are summed to create low-, intermediate-, and high-risk groups.

Epidemiology

DLBCL is the most common subtype of NHL, accounting for approximately 30% of the 74,000 cases of NHL that are diagnosed every year in the United States. The vast majority of cases are sporadic, although the level of risk is increased in certain populations. These include patients with congenital immunodeficiency syndromes, suppression of the immune system by HIV infection, or medical immunosuppression, such as following a solid organ transplantation. The risk is also increased in patients with autoimmune diseases, such as systemic lupus erythematosus and rheumatoid arthritis. In aggregate, however, all of these risk factors constitute a small minority of DLBCL patients. DLBCL occurs at a median age in the mid-60s, but can affect any age group, from children to the very elderly. The disease affects both men and women, with a slight male predilection. Of note, PMBCL occurs in a slightly different demographic than the rest of DLBCLs, with a slight female predominance and a median age in the mid-30s.

Diagnosis, workup, and staging

DLBCL typically presents with symptoms related to rapidly growing lymph nodes or masses. The growing nodes may be felt or seen by patients, or may produce symptoms by encroaching on nerves or by interfering with the normal functions of tissues or organs in the chest or abdomen. Nerve compression or invasion by tumor may result in pain, and symptoms of disease in the chest may include shortness of breath because of compression or cancer invasion of the lungs by cancer, whereas symptoms in the abdomen may be caused by compression and/or obstruction of the bowel or other intra-abdominal organs. DLBCL most commonly occurs in lymphoid organs, such as the lymph nodes, spleen, and tonsil tissues, but can also involve any sites outside the lymph-node tissues. These so-called extranodal sites include the gastrointestinal tract, lungs, liver, kidneys, skin, bone, bone marrow, soft tissues, and central nervous system, among others. Approximately 40% of DLBCLs will also present with systemic symptoms known as "B" symptoms, which include unexplained fevers, drenching night sweats, and unintentional weight loss. Occasionally, patients will have decreased blood cell counts (red cells, platelets, and healthy white blood cells) due to infiltration of the bone marrow.

Physical examination may be normal, but will often be notable for enlarged lymph nodes or masses, or obvious enlargement of the spleen or tonsillar tissue. Affected lymph nodes are usually firm to the touch but are generally not tender.

Diagnosis is made by biopsy of an enlarged lymph node or mass. The malignant lymphocytes efface the normal lymph node architecture in a *diffuse*

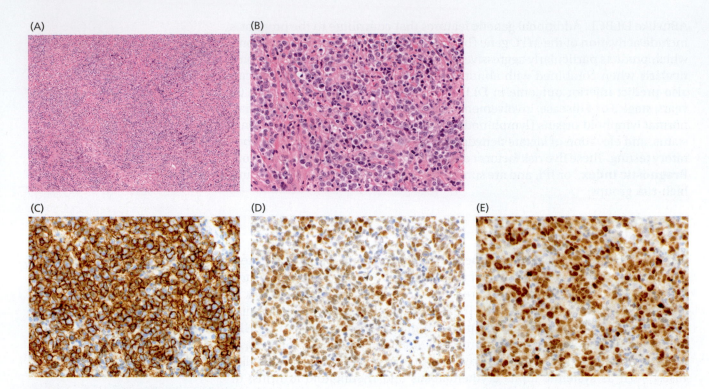

Figure 11.7. Histologic and immunophenotypic features of DLBCL. Low-power magnification shows a diffuse infiltrate of atypical lymphocytes without formation of nodules or follicles (A, hematoxylin and eosin). At higher magnification, the tumor cells are found to be large, containing nuclei that exceed twice the size of background normal small lymphocytes, suggestive of large-cell lymphoma (B, hematoxylin and eosin). Membranous CD20 expression is present as indicated by immunohistochemistry (C), confirming the large cells to be of B-cell lineage and supporting a diagnosis of DLBCL. Further immunophenotyping of this tumor showed it to be negative for CD10 (not shown), and positive for both BCL6 (D) and MUM1 (E), consistent with an ABC-like DLBCL.

pattern and are much *larger* than normal lymphocytes. Stains with antibodies to characterize these cells prove them to be of *B-cell* origin due to uniform expression of B-cell markers, such as CD20, CD19, and CD79B. This microscopic appearance leads directly to the name *diffuse large B-cell lymphoma*. DLBCL can be further classified based on the cell of origin distinction (GCB-like versus ABC-like) using selected antibodies against CD10 and BCL6, which typically characterize GCB-like tumors, as opposed to MUM-1, which is commonly expressed in ABC-like DLBCL (Figure 11.7).

In addition to the use of antibody stains, also known as immunohistochemistry, genetic analyses are often undertaken, either by performing a karyotype on the fresh biopsy tissue or by evaluating for selected chromosomal rearrangements of genes, such as *BCL2*, *BCL6*, and *MYC*, using fluorescence *in situ* hybridization (FISH). DLBCL is staged with the Ann Arbor classification (see Figure 5.13).

Unlike most cancers, stage in DLBCL and other lymphomas has only modest prognostic influence. The reason for this is that DLBCL is potentially curable regardless of stage. Whereas many other cancers have historically relied upon surgical removal of a localized tumor as the greatest predictor of cure, DLBCL is usually highly sensitive to chemotherapy, which means that it can potentially be cured independent of extent of involvement throughout the body. Staging is performed using a full body positron-emission tomography computerized tomography (PET-CT) scan, which incorporates anatomical CT staging with fluorodeoxyglucose (FDG) imaging (see Chapter 1 for further information on imaging techniques for staging). PET imaging in DLBCL is particularly helpful for demonstrating areas that may not appear abnormal on a CT scan, such as bone, or for identifying disease in otherwise normal sized lymph nodes (Figure 11.8). In addition, evaluation of the bone marrow with a needle biopsy performed on a posterior hip bone may be required at diagnosis. Laboratory tests, including a complete blood count and measurement of chemistries (electrolytes, and liver and renal function), are evaluated at diagnosis. The LDH is also checked, as it carries prognostic relevance in the IPI. Human immunodeficiency virus (HIV) is often tested, as this is a risk factor for developing DLBCL, and identifying DLBCL in an HIV-infected patient would necessitate treating the infection and the lymphoma concurrently, with combination antiviral therapy for the HIV and chemotherapy and/or radiation for the lymphoma. Following staging and laboratory evaluation, the patient's IPI score can be calculated in order to determine which risk group they fall into based on their number of adverse risk factors at diagnosis. For patients treated with standard chemotherapy, the IPI historically predicted cure rates ranging from approximately 25% in the highest-risk group to 75% in the lowest-risk group. However, modern advances in therapy have increased the likelihood of cure in all risk groups.

Principles of DLBCL management

Prior to the introduction of combination chemotherapy, DLBCL was an incurable disease with a poor prognosis. In 1976, a four-drug chemotherapy regimen called "CHOP" was developed, and was found to cure approximately 50% of patients with aggressive lymphomas. CHOP is an acronym for the following four chemotherapy drugs with activity against lymphoma cells: cyclophosphamide, hydroxydaunomycin (doxorubicin), Oncovin (vincristine), and prednisone. Following this major advance, numerous additional chemotherapy regimens were tested with the goal of further improving the cure rate, largely by giving more chemotherapy drugs and administering them at higher doses. However, when these strategies were compared head to head, it was found that CHOP cured exactly the same number of patients as the more intensive chemotherapy regimens, but with much less toxicity. The CHOP regimen remained the standard initial therapy for DLBCL for over 20 years.

The case of Linda White, a 61-year-old school teacher with pain in her shoulder

Linda was in her usual state of excellent health when she noted shoulder pain. The pain developed over a two-month period, and was worse on exercise but also noticeable at rest. She visited her primary care physican, who performed an X-ray which was unremarkable. Exercises for possible rotator-cuff injury were recommended. When the pain continued, Linda went back to her doctor, and magnetic resonance imaging (MRI) of the shoulder was performed. The MRI demonstrated a partial tear of the supraspinatus muscle in the rotator cuff, but also showed an abnormality in the head of the humerus, suggestive of an invasive malignancy. A needle biopsy was performed under radiographic guidance, with the finding of a mixed population of atypical lymphocytes and histiocytes, but the amount of tissue on the needle biopsy was insufficient to allow a definitive diagnosis. Ultimately, Linda underwent a surgical biopsy of the suspicious bone lesion. The biopsy revealed diffuse large B-cell lymphoma (DLBCL), which was GCB-like. Linda's staging PET/CT scan demonstrated the disease in the humeral head, as well as disease not easily visualized by the CT scan, involving multiple vertebral bodies, the right scapula, and the left ninth rib. A bone marrow biopsy was performed, which showed normal bone marrow without DLBCL involvement. Based on the extranodal involvement, Linda's stage was IV. Her IPI risk score was 2, with her risk factors determined by her age and the stage of disease.

Linda was treated with six cycles of CHOP with the addition of the monoclonal antibody rituximab (R-CHOP). Within days of initiation of chemotherapy, her shoulder pain had dramatically improved, and by the end of the second

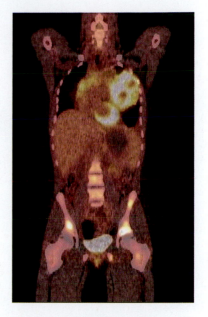

Figure 11.8. Staging PET-CT scan. The gray-scale images denote the CT scan, and the yellow-scale images denote the superimposed PET scan, which highlights areas of increased metabolic activity. This scan shows bright metabolically active disease involving the left lung, mediastinum, vertebral bodies, and pelvic bones. Physiologic uptake is seen in the tonsils, heart, and bladder. This is stage IV disease based on involvement of extranodal locations.

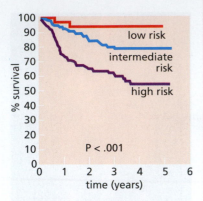

Figure 11.9. Modern Kaplan–Meier survival curves in DLCBL treated with R-CHOP. (Modified from L.H. Sehn et al., *Blood* 109:1857–1861, 2007.)

week of a three-week chemotherapy cycle it had completely resolved. A PET-CT scan, performed after completion of six cycles of R-CHOP, showed complete remission in all sites of her disease. She was followed with surveillance scans every six months for two years, after which the intervals were lengthened. It is now six years since Linda completed her lymphoma therapy, and she is almost certainly cured of her disease.

CHOP chemotherapy for a patient like Linda White, with two adverse risk factors on the IPI score, would predict a cure rate of approximately 50%. After CHOP had remained standard therapy for DLBCL for approximately 30 years, the next major advance occurred in 1997 with the FDA approval of the anti-CD20 monoclonal antibody rituximab. Rituximab binds to the CD20 antigen on the surface of malignant B cells and normal B cells, and kills cells predominantly via ADCC. In ADCC, immune effector cells target the antibody-coated lymphoma cells for cellular destruction, as shown in Figure 11.3.

When rituximab is given alone in relapsed DLBCL, responses are typically short-lived, but when it is combined with CHOP (R-CHOP) for the initial treatment of DLBCL, several large randomized trials have demonstrated a significant improvement in the cure rate for both younger and older patients, and for patients with low- or high-risk IPI scores. The introduction of rituximab heralded the era of targeted therapy for lymphoma, and had a far-reaching impact in improving outcomes of countless patients with DLBCL and virtually all other B-cell non-Hodgkin lymphomas and leukemias. In DLBCL, the cure rate increased to 66% on average, with cure rates based on IPI scores ranging from 55% in the highest-risk patients to greater than 90% in the lowest-risk individuals (Figure 11.9). It is a remarkable fact that the majority of even high-risk patients with this previously incurable disease may now be cured with modern chemoimmunotherapy.

Management of relapsed DLBCL

Patients who relapse after R-CHOP therapy have only a modest chance of being cured by additional chemotherapy. Standard therapy consists of an alternative combination chemotherapy regimen followed by high-dose chemotherapy with autologous stem cell transplantation, assuming that the patient is young and fit enough to undergo such intensive treatment. The primary predictor of outcome at relapse is the duration of initial remission after R-CHOP. Patients who relapse more than one year after completing R-CHOP have a cure rate of approximately 50%, compared with less than 20% in patients who relapse within less than one year. Patients who are not eligible for high-dose chemotherapy, due to inability to achieve a response to second-line salvage chemotherapy, advanced age, or comorbid disease, are given palliative treatment with lower-intensity chemotherapy regimens.

Future prospects and challenges in DLBCL

Given the disappointing outcomes for DLBCL patients at relapse, several novel treatments are being explored in DLBCL to improve the efficacy of therapy, particularly by targeting the underlying disease biology (Figure 11.10). These treatments include drugs that may have selective activity in ABC-like DLBCL by acting through distinct mechanisms to decrease NF-κB activation, which is a dominant biologic driver in this DLBCL subtype. Such drugs include ibrutinib, which targets Bruton's tyrosine kinase (BTK, see Case 4-2), a critical pathway step in chronic active B-cell receptor (BCR) signaling, and lenalidomide (see Case 10-2), which inhibits IRF4, leading to down-regulation of BCR-dependent NF-κB.

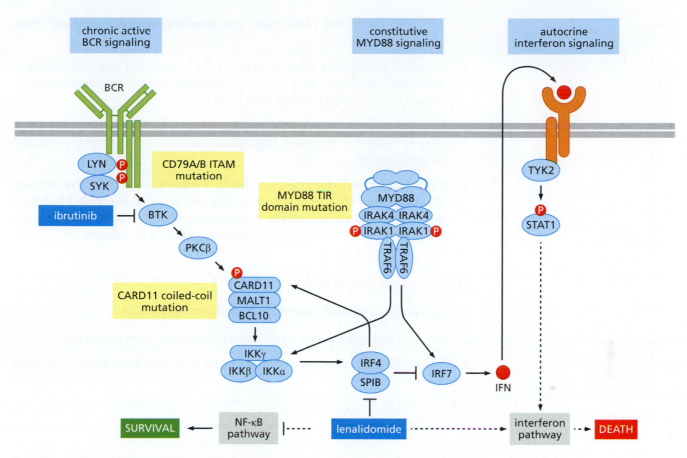

Figure 11.10. Potential molecular targets for DLBCL therapy. Oncogenic mutations in ABC-like DLBCL activate B-cell receptor (BCR) and myeloid differentiation primary response 88 (MYD88) signaling, which results in constitutive NF-κB activity and lymphoma survival. Interferon regulatory factor (IRF4) and the transcription factor Spi-B (SPIB) lie at the nexus of both pathways, promoting ABC-like DLBCL survival. Lenalidomide down-regulates IRF4 and SPIB. Ibrutinib inhibits Bruton's tyrosine kinase (BTK), which interrupts chronic active BCR signaling. Both of these actions lead to decreased NF-κB activity and represent novel therapeutic targets in ABC-like DLBCL. (Modified from Y. Yang et al., *Cancer Cell* 21:723–737, 2012. With permission from Elsevier.)

Modest activity has been demonstrated for these agents in relapsed ABC-like DLBCL, but clinical trials are now evaluating the addition of each of these agents to standard R-CHOP with the goal of curing more patients with ABC-like DLBCL with their initial therapy. The goal is ultimately to cure all patients with their initial treatment, just like Linda White. An ongoing challenge will be to identify mechanisms of resistance to biologically targeted therapies so as to ultimately select the optimal targeted agent to combine with chemotherapy at the time of every patient's diagnosis. Ongoing clinical trials are seeking to address this important challenge.

Take-home points

✓ Diffuse large B-cell lymphoma (DLBCL) is an aggressive but highly treatable and often curable disease with modern therapy.

✓ Patients with DLBCL may be risk stratified based on clinical and biologic factors.

✓ DLBCL is a biologically heterogeneous disease with multiple different underlying genetic lesions that contribute to pathogenesis. Distinct biologic subtypes of DLBCL include germinal-center B-cell-like (GCB-like), activated B-cell-like (ABC-like), and primary mediastinal B-cell lymphoma (PMBCL).

✓ Standard therapy for DLBCL includes the anti-B-cell targeted antibody rituximab combined with CHOP chemotherapy, which cures approximately two-thirds of patients with DLBCL.

✓ The majority of patients who relapse after R-CHOP will not be cured despite intensive therapies, highlighting the need for novel biologically targeted agents.

✓ Current targeted therapies are being explored with selective activity within distinct biological subsets of disease, and will likely lead to different targeted therapies used in combination with chemoimmunotherapy and based on the biologic profile of each patient's lymphoma.

Discussion questions

1. Can you name some distinct biologic subsets of DLBCL?

2. How is DLBCL staged, and what impact does stage have on prognosis?

3. What is the goal of initial treatment of DLBCL, and how is this currently achieved?

4. How is DLBCL treated at relapse?

Topics bearing on this case

Hodgkin Reed-Sternberg (HRS) cell

Initial chemotherapy for classical Hodgkin lymphoma

Antibody-drug conjugate (ADC) therapy for Hodgkin lymphoma

Case 11-2 | Hodgkin Lymphoma

Introduction

Hodgkin lymphoma (HL) is derived from B lymphocytes and, unlike most cancers, preferentially affects young people. Thomas Hodgkin first described this entity in 1832, although the lymphoid derivation remained elusive until 1994, when molecular techniques defined the cell of origin to be germinal-center B cells, which express virtually no B-cell surface proteins due to crippling mutations. This discovery changed "Hodgkin disease" to "Hodgkin lymphoma." There are two distinct categories within Hodgkin lymphoma, namely **classical Hodgkin lymphoma** (CHL), which constitutes 95% of all cases of Hodgkin lymphoma, and **nodular lymphocyte-predominant Hodgkin lymphoma** (NLPHL), which accounts for only 5% of cases. Within classical Hodgkin lymphoma there are four recognized histologic subtypes: nodular sclerosis, which accounts for the majority of cases, followed by mixed cellularity, lymphocyte rich, and lymphocyte depleted. The classical subtypes of Hodgkin lymphoma all share the same underlying malignant cell, known as the **Hodgkin Reed–Sternberg** (HRS) cell, and are treated according to common principles. NLPHL, on the other hand, has a different malignant cell, known as an LP cell, and has a distinct natural history and approach to therapy.

Hodgkin lymphoma is unique under the microscope in that the malignant HRS cells represent only about 1% of the overall tumor cellularity (Figure 11.11). The majority of cells in Hodgkin lymphoma actually consist of inflammatory cells, including healthy lymphocytes (mostly T cells), eosinophils, neutrophils, histiocytes, plasma cells, and others. The presence of an apparent robust host inflammatory response surrounding the tumor cells suggests a histologic paradox in Hodgkin lymphoma. Why are the tumor cells not being destroyed by the potent immune onslaught surrounding them? The answer to this question speaks to the unique underlying biology of this malignancy.

The HRS cell carefully choreographs the immune micro-environment, such that the surrounding host inflammatory cells are rendered incapable of attacking the tumor cells. Furthermore, through cell–cell interactions and cytokine production, the immune microenvironment actually supports HRS cell growth and survival (Figure 11.12). Direct cell–cell interactions include tumor cell expression of FAS ligand, PD-1 ligand, and galectin-1, which directly inhibit the anti-tumor activity of infiltrating T lymphocytes. Secretion of the cytokines IL-10 and TGF-β by HRS cells further impairs the efficacy of cytotoxic T lymphocytes. Finally, down-regulation of human leukocyte antigen (HLA) class I and class II molecules helps to cloak HRS cells from recognition by cytotoxic T lymphocytes.

T-cell engagement of CD40, CD30, and occasionally LMP1 on the surface of HRS cells directly activates the NF-κB pathway, which is one of the most potent drivers of HRS cell growth and survival. Additional signaling from the infiltrating immune cells comes from secretion of IL-13, which binds to the IL-13 receptor on tumor cells and activates the JAK-STAT pathway, which is also a dominant contributor to HRS cell survival. HRS cells also secrete TARC, a chemokine that attracts a subtype of lymphocytes called **regulatory T cells** (T_{reg} cells), which work to decrease the efficacy of cytotoxic T lymphocytes, further promoting immune escape.

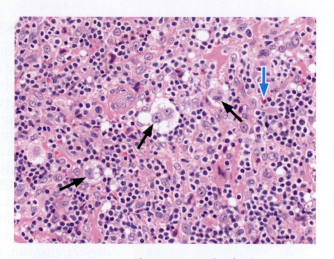

Figure 11.11. Histologic appearance of classical Hodgkin lymphoma. This contains infrequent HRS cells with multiple nuclei and prominent eosinophilic nucleoli (*black arrows*). The majority of the cells in the tumor are polyclonal host inflammatory cells, including small lymphocytes and occasional eosinophils (*blue arrow*).

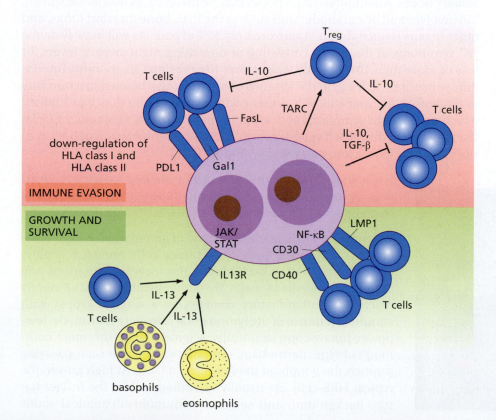

Figure 11.12. Hodgkin lymphoma (HL) interactions with the immune microenvironment. Hodgkin Reed–Sternberg (HRS) cells evade immune escape by inducing anergy (unresponsiveness to antigens) in the surrounding T lymphocytes via interactions with programmed death ligand 1 (PD-L1), FAS, and Gal-1 on the cell surface with their respective ligands on T cells. Secretion of TARC, IL-10, and TGF-β further suppresses an activated immune response. IL-10 secretion by regulatory T cells in the microenvironment also suppresses infiltrating cytotoxic T lymphocytes, and down-regulation of major histocompatibility complex (MHC) classes I and II helps to cloak the HRS cell from immune recognition. Infiltrating immune cells also promote HRS cell growth and survival via activation of the NF-κB and JAK/STAT pathways by both direct cell–cell and cytokine stimulation.

Epidemiology

There are approximately 8500 new cases of HL diagnosed every year in the United States, and about 1050 deaths. The discordance between incidence and fatality rate reflects the fact that CHL is one of the most curable of all malignancies, representing a major success story in the history of modern medicine, as this disease had previously been uniformly fatal. The median age of diagnosis is 35 years, but this belies two incidence peaks for this disease, with the primary peak occurring in young people in the late teenage years and early 20s, and the second peak occurring in older adults over the age of 60 years. The vast majority of CHL cases are sporadic, although there is an approximately tenfold increased risk in the setting of HIV infection, as well as an increased risk in patients with suppressed immune systems due to medications, such as drugs used to treat autoimmune diseases or to prevent rejection after a solid organ transplant. Increased risk in the setting of compromised immunity in part reflects the pathogenic role that the **Epstein–Barr virus** (EBV) plays in a subset of patients with CHL. EBV may be detected in a proportion of HRS cells, where it plays a role in promoting tumor growth and survival by NF-κB activation. Patients with underlying HIV, those with immune suppression, and the elderly are at increased risk of having EBV-associated HL. People with normal immune systems may also develop EBV-associated CHL, and the risk of developing CHL is approximately fourfold after prior **infectious mononucleosis**, a viral infection that is also caused by EBV. Given the low overall incidence of HL, the risk of developing HL after infectious mononucleosis remains quite low at approximately 1 in 1000.

Diagnosis, workup, and staging

The majority of patients present with painless lymph node enlargement, with the neck being the most common location, followed by the mediastinum and axillary nodes. Abdominal lymph nodes may be involved, as well as the spleen, but involvement of extranodal sites such as the liver, bone marrow, lungs, and other organs is uncommon. Approximately 30% of patients will have systemic "B" symptoms at diagnosis, consisting of drenching night sweats, fevers, or loss of more than 10% of body weight over the previous six months. Intense diffuse pruritus (itching) in the absence of rash may also be seen in a minority of patients. Some patients will present with very large mediastinal tumors, which may lead to symptoms of chest pressure, shortness of breath, or facial swelling due to decreased venous return caused by compression of the superior vena cava (SVC), known as SVC syndrome (Figure 11.13).

HL is staged using the Ann Arbor staging system. A slight majority of patients present at limited stage (stage I–II), with the remainder having advanced-stage disease (stage III–IV).

Diagnosis of HL is made by biopsy of an enlarged lymph node or mass. Given the paucity of tumor cells amidst the polyclonal inflammatory background, fine-needle aspirations and other small-needle biopsies are often misinterpreted as merely reactive or inflammatory changes, so a surgical biopsy with complete removal of the lymph node is usually critical. On low-power microscopy in nodular sclerosis CHL, the most common subtype, dense bands of sclerosis may be seen coursing through the lymphoid tissue (Figure 11.14). At high power, the typical HRS cells are usually identified amidst the inflammatory background, and selected immunohistochemical stains

Figure 11.13. Plain chest X-ray of a young patient with a large mediastinal mass. The yellow arrows point to the lateral borders of the mass, and the red arrow indicates the normal border of the heart.

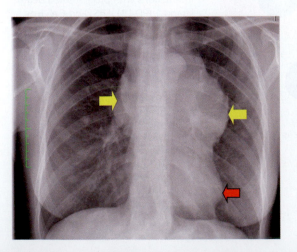

show the malignant cells to be positive for CD30 as well as CD15, but negative for the typical B-cell marker CD20.

As with the previously discussed DLBCL, stage is less relevant for prognosis in HL than in traditional organ-based cancers because HL is highly curable at any stage using chemotherapy. However, stage does assist in treatment selection. A full body PET-CT scan is employed for radiographic staging. A bone marrow biopsy is not usually required in HL, unlike non-Hodgkin lymphoma, as the PET-CT scan is excellent at identifying bone marrow disease in HL and may even be superior to biopsies, as the involvement can be patchy and thus can be missed on needle biopsies. Laboratory tests that encompass a complete blood count and measurement of chemistries, including electrolytes as well as liver and renal function, are evaluated at diagnosis. HIV is often tested, as this is a risk factor for developing HL and would have treatment implications. In general, the cure rate for advanced-stage CHL is approximately 80%, whereas the cure rate for limited-stage disease is greater than 90%.

Principles of CHL management

Historically, the initial treatment for HL was radiation therapy, which was found to cure over 80% of patients with limited-stage HL, but advanced-stage HL remained incurable, as did limited-stage disease that relapsed after radiation therapy alone. Combination chemotherapy was then developed for HL and was tested in patients with advanced-stage disease (whereas now more than 50% of these patients could be cured). Development of combination chemotherapy for HL dramatically decreased the incidence of death from HL both in the United States and worldwide (Figure 11.15).

The initial chemotherapy for HL was called MOPP [mechlorethamine, Oncovin (vincristine), procarbazine, and prednisone], which was shown to cure just over 50% of patients with advanced-stage disease. However, MOPP had several adverse effects, including making patients feel quite sick, carrying a high risk of inducing infertility, and causing a low but significant rate of secondary bone marrow injury (**myelodysplasia**) and acute leukemia that occurred years after patients were cured of their HL. The chemotherapy ABVD (adriamycin, bleomycin, vinblastine, and dacarbazine) was therefore developed for HL. ABVD is better tolerated and does not cause significant rates of either infertility or secondary myelodysplasia or leukemia. ABVD was compared head to head with MOPP in a randomized clinical trial, and ABVD came

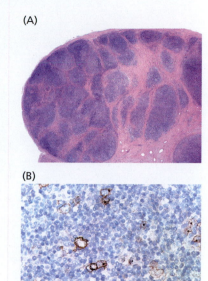

Figure 11.14.
Immunohistochemistry of Hodgkin lymphoma. (A) Low-power view of nodular sclerosis classical Hodgkin lymphoma shows dense bands of sclerosis dividing the lymphoid tissue into nodules (hematoxylin and eosin). (B) At higher power, a stain for CD30 highlights membranous CD30 expression on scattered malignant HRS cells.

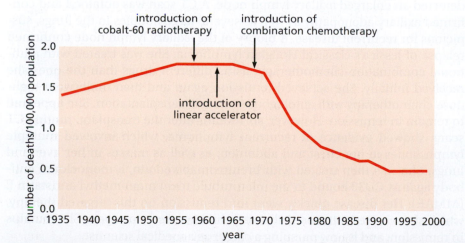

Figure 11.15. Incidence of death from Hodgkin lymphoma in the U.S. population. Minimal overall impact is seen from the development of radiation therapy (initially radioactive cobalt, and subsequently the linear accelerator), but a major impact is seen after the development of combination chemotherapy in 1968. (From V.T. DeVita Jr., *N. Engl. J. Med.* 348:2375–2376, 2003. Reprinted with permission from the Massachusetts Medical Society.)

out ahead, as it cured more patients while causing less toxicity. Thus ABVD became, and remains, the standard chemotherapy regimen in CHL. Given the efficacy of chemotherapy in advanced-stage disease, the treatment was then evaluated in combination with radiation therapy for limited-stage disease, where it was found to cure more patients than radiation alone. Given concerns about late toxicities of radiation, including secondary cancers, heart disease, and lung disease, more recent clinical trials have evaluated chemotherapy alone for limited-stage CHL. These clinical trials revealed that although there are slightly more relapses (about 5% more) when radiation is omitted, the majority of patients are cured with either approach. Many relapsed patients can be cured with second-line therapy after relapse, making the overall survival of these two strategies largely overlapping. The choice of chemotherapy plus radiation therapy versus chemotherapy alone is made on a patient-by-patient basis that weighs the risks and benefits of radiation. A young woman with a non-bulky mediastinal mass who would be at increased risk for radiation-induced breast cancer and heart disease would probably be considered a candidate for chemotherapy alone. However, if she had a very large tumor or disease limited to the cervical or inguinal lymph node regions, she might be a good candidate for chemotherapy followed by radiation therapy.

The case of Jessica Romero, a 22-year-old graduate student with shortness of breath

Jessica is a vigorously healthy graduate student in cell biology who first noted shortness of breath while running. An avid athlete, she began to become winded earlier than she was accustomed to while running. She initially attributed this to greater stress from her studies, but her exercise tolerance continued to decrease and she developed a dry cough. Jessica visited her primary care physician who performed a chest X-ray. This showed a 10 cm mass in Jessica's anterior mediastinum, which was then confirmed on a subsequent CT scan. She was referred to a thoracic surgeon who performed a mediastinoscopy and biopsy of the anterior mediastinal mass. The pathology showed classical Hodgkin lymphoma of the nodular sclerosis subtype. Jessica's staging PET-CT scan showed stage II disease, involving the mediastinal and hilar lymph nodes. She was treated with four cycles of ABVD chemotherapy, followed by 30 Gray of radiation therapy, and she achieved a complete response. It was felt that Jessica was highly likely to be cured.

Approximately one year after completing therapy, Jessica felt an abnormal lump under her right arm while showering. She visited her oncologist, who detected an enlarged axillary lymph node. A CT scan was obtained and confirmed axillary adenopathy as well as several large nodules in the lungs, suspicious for recurrent disease. A biopsy of the axillary lymph node confirmed relapse of Jessica's classical Hodgkin lymphoma. She was treated with additional combination chemotherapy, using different drugs than the ones she received initially. She achieved remission again and then underwent high-dose chemotherapy with autologous stem cell transplantation. She appeared to remain in remission. However, six months after the transplant, routine CT scans showed evidence of recurrent lymphoma, which involved multiple lymph nodes in her chest and abdomen, as well as masses in her liver and lungs. She was then treated with brentuximab vedotin, a monoclonal antibody against CD30 bound to the microtubule toxin monomethyl auristatin E (MMAE). Her disease quickly went into remission on this targeted therapy, which she received intravenously every three weeks for one year. She remains in remission, and is now pursuing a career as a medical scientist.

Treatment of relapsed Hodgkin lymphoma

Although the majority of patients with CHL are cured with initial chemotherapy, with or without radiation, a small but significant minority will relapse or be refractory to their initial treatment. Standard therapy for young fit patients is second-line chemotherapy with a non-cross-resistant regimen when compared with the initial ABVD, followed by high-dose chemotherapy with autologous stem cell support. This strategy cures 50–60% of patients who relapse after upfront ABVD therapy. Patients who have relapsed after high-dose chemotherapy, or those who were too old or too ill for such intensive treatment, have been treated with lower-intensity chemotherapy regimens for palliative control of their disease. Given the high levels of CD30, a cell surface antigen, expressed on HRS cells, intense interest developed in the use of a monoclonal anti-CD30 antibody to target the HRS cells, in much the same way as rituximab had revolutionized the management of B-cell lymphomas by binding to CD20 on non-Hodgkin lymphoma cells. However, when an anti-CD30 antibody was developed and given to patients with relapsed CHL, no treatment responses occurred. This probably reflects the biology of CHL, wherein the immune effector cells in the tumor microenvironment are shut down by the HRS cells, and are thus unable to kill the HRS cell by ADCC after binding the anti-CD30 antibody. Another factor was that the CD30 antigen was noted to be internalized into the cell upon receptor engagement by the antibody. This led to a redesign of the molecule, whereby the highly potent microtubule toxin, **monomethyl auristatin E** (MMAE), was covalently bound to the anti-CD30 monoclonal antibody. The anti-CD30 MAb conjugated to MMAE forms the ADC known as **brentuximab vedotin**. Now, upon antigen–receptor engagement, the ADC brentuximab vedotin would be internalized into the cell where the toxin, MMAE, would be released via endosomal cleavage into the cell and lead directly to cell death.

Brentuximab vedotin was studied in a Phase II clinical trial of CHL patients who had not been cured with initial chemotherapy, and who had subsequently relapsed after high-dose chemotherapy with autologous stem cell transplantation. In this clinical trial, nearly all patients showed a reduction in the size of their tumor, and 75% of patients achieved a response to therapy defined as a reduction in tumor size of more than 50% (Figure 11.16). Among the one-third of patients who achieved a complete response, half were sustained five years later. Brentuximab vedotin received FDA approval on August 19, 2011, for patients with relapsed Hodgkin lymphoma.

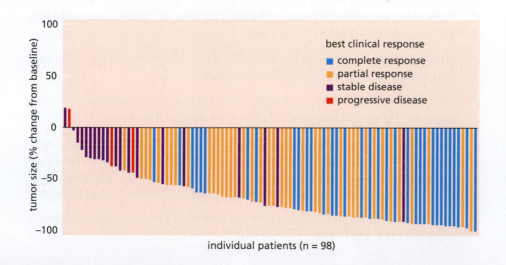

Figure 11.16. **Waterfall plot of change in tumor size in patients treated with brentuximab vedotin on the Phase II clinical trial.** Each bar represents an individual patient's total tumor measurement and change from the x-axis denotes percentage increase or decrease from baseline. All but two patients showed a reduction below baseline in their tumor measurement, with 34% of patients achieving a complete response. (From A. Younes et al., *J. Clin. Oncol.* 30:2183–2189, 2012. Reprinted with permission. © 2012 American Society of Clinical Oncology. All rights reserved.)

Future prospects and challenges in CHL

Ongoing lines of investigation include the addition of brentuximab vedotin to standard chemotherapy as the initial therapy for CHL, with the goal of increasing the already high cure rate with initial therapy. A Phase I study that combined brentuximab vedotin with ABVD found an increase in pulmonary toxicity with the combination (lung injury is also a problem in ABVD from bleomycin), so the combination of brentuximab vedotin with AVD, without bleomycin, was evaluated and found to be safe and preliminarily effective. A Phase III study recently compared brentuximab vedotin plus AVD with standard ABVD as initial treatment for advanced-stage disease. Brentuximab-AVD did modestly improve PFS over standard ABVD, but with no difference in overall survival, and at the cost of increased toxicity in the experimental arm due primarily to increased neutropenic fever and peripheral neuropathy. Studies of limited-stage disease are evaluating whether the addition of brentuximab can further decrease reliance on radiation therapy for young patients, and whether the use of an early response as indicated by PET-CT scan may be another mechanism whereby treatment intensity, including radiation, can be reduced in patients with early evidence of highly chemosensitive disease. Additional strategies that are becoming important are drugs that interrupt the immune suppression induced by HRS cells on the surrounding cells, known as immune checkpoint inhibitors (see Chapter 12), thus restoring antitumor immunity within the richly inflammatory CHL microenvironment. Two inhibitors of the immune checkpoint PD-1, nivolumab and pembrolizumab, are now FDA approved for the treatment of relapsed classical Hodgkin lymphoma.

Take-home points

✓ Hodgkin lymphoma (HL) is a highly treatable and usually curable cancer that most commonly affects young men and young women.

✓ The tumor in HL is notable for a dominant population of polyclonal host inflammatory cells surrounding the malignant HRS cells, which constitute only about 1% of the overall tumor cellularity.

✓ The immune microenvironment is carefully choreographed by the HRS cell, which inhibits activity of the surrounding immune effector cells. The surrounding immune cells also provide pro-growth and pro-survival signals to the HRS cell.

✓ Treatment of advanced-stage disease consists of combination chemotherapy, usually the ABVD regimen.

✓ Limited-stage HL may be treated with a combination of chemotherapy and radiation, or with chemotherapy alone.

✓ Relapsed HL may be cured with additional chemotherapy, including high-dose chemotherapy and autologous stem cell support.

✓ The antibody–drug conjugate (ADC) of the anti-CD30 monoclonal antibody bound to the microtubule toxin MMAE, brentuximab vedotin, is highly effective in chemotherapy-resistant and newly-diagnosed Hodgkin lymphoma, where it induces remissions in the majority of patients.

✓ Immune checkpoint inhibitors of the protein PD-1 are highly active in relapsed classical Hodgkin lymphoma.

Discussion questions

1. What makes HL unique among other cancers when it is studied under the microscope?

2. What is the standard therapy for advanced-stage HL?

3. What may influence the decision to include radiation as a component of therapy for limited-stage HL?

4. What is the standard therapy for a young fit patient with relapsed HL?

5. Describe the mechanism of action of brentuximab vedotin.

Case 11-3   **Breast Cancer**

Topics bearing on this case

HER2-positive breast cancer

Monoclonal antibody therapy for HER2+ breast cancer

Antibody-drug conjugate (ADC) therapy for HER2+ breast cancer

Breast cancer is the most common malignancy among women in the United States. The presence of estrogen receptor (ER), progesterone receptor (PR) and/or **human epidermal growth factor receptor 2** (HER2) on breast cancer cells provides important targets for therapy. The epidemiology, diagnosis, management, and genetic alterations of breast cancer have been described in Case 6-1. One gene of particular importance in breast cancers is *HER2* (also called *ErbB2* or *Neu*), which encodes a transmembrane tyrosine kinase, HER2, in the epidermal growth factor receptor (EGFR) family. HER2 is expressed on the surface of 25% of breast cancers. Amplification of *HER2* is associated with an aggressive breast cancer subtype. Although HER2 can be expressed on normal breast epithelial cells, its overexpression (by 10- to 100-fold) on the cell surface of breast cancer cells due to amplification of the *HER2* gene makes anti-HER2 strategies an attractive approach. Targeted therapy approaches to HER2-overexpressing breast cancers could thus focus on small-molecule tyrosine kinase inhibitors, such as lapatinib, or on the MAb, trastuzumab.

Trastuzumab (Herceptin) is a humanized IgG1 MAb that binds to the extracellular domain of HER2/Neu. The mechanism by which trastuzumab causes the death of HER2-overexpressing breast cancer cells has not been clearly defined. IgG antibody-coated breast cancer cells may bind the Fc γ receptors on cytotoxic NK cells and macrophages, causing ADCC. Trastuzumab may cause the internalization and degradation of HER2 on the cell surface of the breast cancer cells, reducing HER2-based proliferative and survival signaling. Trastuzumab may interfere with the ability of HER2 to form heterodimers with its primary signaling partner, HER3, causing a decrease in PI3-kinase and downstream Akt signals that convey protection from apoptosis and promote cell survival (Figure 11.17).

For patients with localized or locally advanced disease, trastuzumab combined with adjuvant chemotherapy has been found to reduce the recurrence rate by 50% and mortality by 30% in randomized controlled trials involving 3752

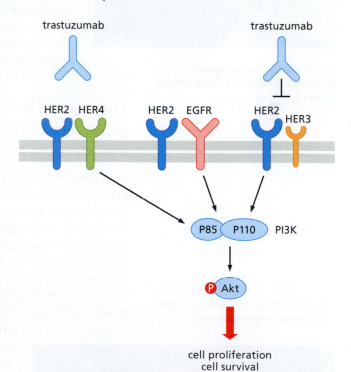

Figure 11.17. HER2 can bind to other members of the epidermal growth factor receptor (EGFR) family, including EGFR, HER3, and HER4. The activation of PI3-kinase then phosphorylates Akt. The downstream effect of activating this pathway is an increase in cell proliferation and protection from apoptosis.

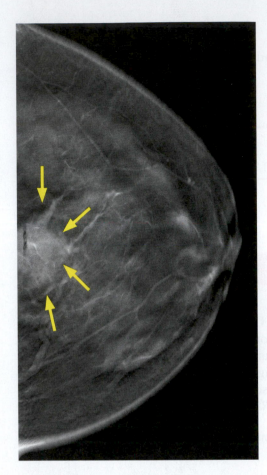

Figure 11.18. Mammogram demonstrating a left-sided breast tumor (*arrows*).

women whose cancers overexpressed HER2 and who received standard adjuvant chemotherapy with doxorubicin, cyclophosphamide, and paclitaxel. On this basis, on November 16, 2006, trastuzumab was given FDA approval for use as adjuvant therapy. In metastatic HER2-positive breast cancer, trastuzumab produces a 14% response rate when used as a single agent. However, when combined with chemotherapy for metastatic disease, trastuzumab increases response rates by 18–27%, prolongs disease-free survival by 3–5 months, and increases overall survival by 5–9 months. Trastuzumab was given FDA approval for use in metastatic HER2-positive breast cancer on September 25, 1998.

Trastuzumab is generally well tolerated. As with other antibodies, an initial infusion reaction may occur in 40% of patients. Diarrhea can occur in 25% of patients. Heart failure is a rare but significant adverse event, affecting 4% of patients, although the rate can rise to 20% in patients who have also received anthracyclines, a category of cytotoxic chemotherapy agents that includes doxorubicin, and which are also associated with increased risk of heart failure. Prior to starting trastuzumab therapy, patients should undergo baseline cardiac evaluation, including a history, physical examination, electrocardiogram, and an echocardiogram or other evaluation of the ejection fraction (a measure of the "squeeze" of the left ventricle of the heart). Acute symptoms of trastuzumab-related heart failure include shortness of breath and peripheral edema. Trastuzumab should be discontinued if heart failure is suspected. Although many patients show symptom improvement with time off therapy, some can have progressive heart failure that requires more intense management.

The case of Julie Rosen, a 53-year-old woman with a growing breast mass and a family history of breast cancer

Julie Rosen is a small business owner who first noted a firm mass in her left breast on self-examination, as well as a lump approximately 2 cm in diameter in her armpit. In retrospect, she had noted an increase in the breast mass for one year. A mammogram demonstrated asymmetric increased density in the upper outer quadrant of the left breast (Figure 11.18). Julie is otherwise in good health and is taking no prescription medications. She is married and has no children. Her menstrual cycles started at the age of 13 years, and she underwent menopause at the age of 50 years. Notably, her mother was diagnosed with breast cancer at the age of 48 years, and three female first cousins were also diagnosed with breast cancer before the age of 50 years. Her family ethnic background includes Ashkenazi Jewish heritage on both maternal and paternal sides.

Figure 11.19. CT scan of the chest demonstrating left axillary lymph node enlargement (*arrow*).

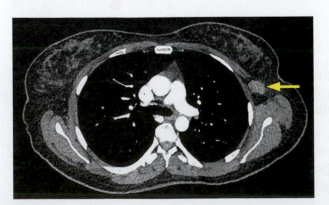

A core biopsy of the left breast mass was performed, which revealed invasive ductal carcinoma, moderately differentiated, grade 2. A CT scan to evaluate the axillary mass revealed a 4 cm left breast mass consistent with the biopsy-proven cancer, as well as enlarged left axillary and subpectoral (under the chest muscles) lymph nodes (Figure 11.19). There was no evidence of distant metastasis. A bilateral breast MRI was performed that confirmed the large breast mass with extension down the duct toward the nipple, as well as significant skin thickening, but no chest wall involvement. Again, the axillary and subpectoral lymphadenopathy was noted.

Julie underwent a modified radical mastectomy. Pathology revealed invasive ductal carcinoma that was poorly differentiated, grade 3, and 7 cm in its greatest dimension. On immunohistochemistry, the tumor was negative for estrogen and progesterone receptors (ER and PR), but positive for HER2. Lymphovascular invasion was present, but the nipple and overlying skin were negative for carcinoma. All resection margins were negative for cancer; however, 12 out of 20 resected lymph nodes did have cancer involvement. Julie ultimately had genetic testing for the *BRCA* mutation due to her family history, but tested negative for *BRCA1* and *BRCA2* mutations (see Case 9-1).

Given her node-positive, HER2-positive, ER/PR-negative breast cancer, Julie was considered to be at very high risk for recurrence of breast cancer (*see* Case 6-1 for a discussion of the relative benefits of hormone therapy and chemotherapy in terms of decreasing the likelihood of recurrence postoperatively). She began standard adjuvant chemotherapy with doxorubicin, cyclophosphamide, and paclitaxel, along with trastuzumab, a MAb against HER2. She also received adjuvant radiation therapy to the left breast to reduce the likelihood of a local recurrence.

Trastuzumab is typically administered starting with the initiation of paclitaxel chemotherapy, which lasts for 12 weeks. In February 2013, while continuing on trastuzumab therapy, Julie developed new headaches and vertigo, prompting a CT scan of her head. Unfortunately, the imaging revealed brain metastases in the cerebellum (Figure 11.20). She had no other evidence of distant spread on CT scans of her chest, abdomen, and pelvis, or on a bone scan.

Julie underwent surgery to remove three cerebellar metastases via a left suboccipital microsurgical craniotomy. Pathology confirmed breast carcinoma metastasis (i.e., stage IV) that was poorly differentiated, estrogen receptor/progesterone receptor-negative (ER/PR-negative), and HER2-positive. Julie received postoperative whole brain radiation therapy to reduce the likelihood of local recurrence in the brain. She continued on trastuzumab therapy.

Four months later, routine restaging CT scans indicated a new liver lesion and enlarged retroperitoneal lymph nodes (Figure 11.21A). A CT-guided biopsy of the liver lesion confirmed metastatic breast cancer. Julie and her medical oncologist reviewed her treatment options, including chemotherapy, other HER2-targeted options, and clinical trials. Because all of her primary and metastatic biopsies proved to be HER2-positive, it was agreed to start her on the ADC, **ado-trastuzumab emtansine**. After four treatments with ado-trastuzumab emtansine given every three weeks by intravenous infusion, the liver lesion and retroperitoneal lymphadenopathy were found to have gone

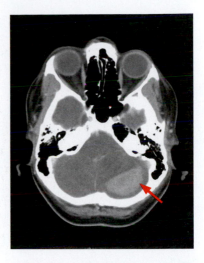

Figure 11.20. CT scan of the brain demonstrating left cerebellar metastasis (*arrow*).

(A)   (B)

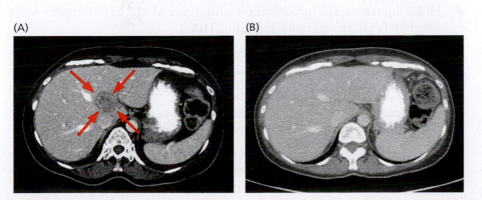

Figure 11.21. Response to treatment. (A) CT scan demonstrating a liver metastasis (*arrows*). (B) CT scan after 4 cycles of ado-trastuzumab, demonstrating complete resolution of the liver metastasis.

into remission (Figure 11.21B). Julie has continued on this therapy for over a year without other evidence of progression, with good energy levels, and with few side effects attributable to ado-trastuzumab emtansine.

The advantage of trastuzumab is its relative selectivity for HER2-overexpressing breast cancer cells. Ado-trastuzumab emtansine (hereafter referred to as *ado-trastuzumab*, and also known as T-DM1) is an antibody–drug conjugate (ADC) that combines trastuzumab and the microtubule inhibitor, emtansine. In addition to the afore-mentioned mechanisms by which trastuzumab kills cancer cells, ado-trastuzumab binds to HER2, and the conjugate is internalized into the breast cancer cell, where the link between antibody and emtansine is enzymatically broken down. Emtansine can then disrupt microtubule assembly and thereby cause selective cell death.

To test the efficacy of ado-trastuzumab, 991 women with HER2-positive breast cancer who had previously received trastuzumab and a taxane (a class of antimicrotubule agents that inhibit mitosis) were randomized to receive ado-trastuzumab or lapatinib (a small-molecule inhibitor of HER2) with capecitabine (an antimetabolite chemotherapy agent). The response rate, progression-free survival, and overall survival all favored ado-trastuzumab, which was generally better tolerated than lapatinib/capecitabine therapy. The median overall survival for patients who received ado-trastuzumab was 31 months, compared with 25 months for those who were given lapatinib/capecitabine. On the basis of this study, ado-trastuzumab was given FDA approval on February 22, 2013, for use in HER2-positive metastatic breast cancer.

Future prospects and challenges in breast cancer

The management of breast cancer is constantly evolving, with the defining of molecular subtypes and a growing understanding of the potential targets for intervention. The vast number of affected patients globally continues to warrant ongoing investigation. Although hormone therapy and HER2-targeted therapies have roles for breast cancers that express ER, PR, or HER2, there remain clear roles for surgery, radiation therapy, and cytotoxic chemotherapy. Future strategies may move away from cytotoxic chemotherapies and toward the targeting of other active proliferative and survival signaling pathways, and the optimal combination or sequencing of existing and future therapies will pose important questions for clinical trials.

Take-home points

✓ Some breast cancers overexpress HER2.

✓ HER2 increases cell proliferation and survival via heterodimers with other cell surface receptors, including HER3.

✓ Trastuzumab, a monoclonal antibody against HER2, improves response rates and survival in adjuvant and metastatic settings for HER-positive breast cancer.

✓ Trastuzumab is generally well tolerated, but heart failure is an important concern, especially in patients who have received anthracyclines.

✓ Ado-trastuzumab emtansine is an antibody–drug conjugate that demonstrates the power of using a cytotoxic therapy that is delivered to the target cell via the monoclonal antibody component.

✓ Use of immune strategies, such as monoclonal antibodies and antibody–drug conjugates, may be more broadly applied to other diseases that overexpress HER2 or other cell surface receptors.

Discussion questions

1. What factors in Julie's history increase or decrease her risk of developing breast cancer?

2. How is it suspected that trastuzumab kills breast cancer cells, and how might you envisage that this differs from small-molecule kinase inhibitors? (For a detailed discussion of kinase inhibition as a strategy, see Chapter 4.)

3. How do antibody–drug conjugates work?

Chapter summary

The cases presented in this chapter allow an exploration of the rational and targeted application of monoclonal antibodies to the treatment of a variety of cancers. Cytotoxic MAb therapies work by one or a combination of the following mechanisms:

1. Binding of the antibody to a cell surface antigen enables ADCC and triggers complement-mediated lysis of antibody-targeted cells.

2. Highly cytotoxic drugs appropriately conjugated to antibodies against cell surface antigens specifically target cells for cytotoxic effects.

3. Antibody blockade of essential receptor–ligand interactions compromises cell proliferation and survival.

In Case 11-1, the effective treatment of non-Hodgkin leukemia with rituximab, a MAb that binds to CD20, a cell surface antigen, illustrates the use of antibodies to provoke killing of antibody-targeted cells by ADCC and by complement lysis. In Case 11-2, Hodgkin lymphoma is treated with an antibody–drug conjugate, brentuximab vedotin. This employs the specificity of an antigen–antibody reaction to selectively deliver a cytotoxic cargo, monomethyl auristan (MMAE), an anti-mitotic drug that inhibits tubulin polymerization, to cells that bear CD30. In addition to the toxic effects of the drug, CD30-bearing cells are marked for ADCC and also become targets for complement-dependent lysis. In Case 11-3, a different ADC, namely ado-trastuzumab emtansine, is used in the treatment of breast cancer, an epithelial cancer. In this case, breast cancer cells with elevated levels of HER2/neu are attacked by the same antibody-mediated set of cytotoxic mechanisms as those targeted by rituximab or brentuximab vedotin. Like brentuximab vedotin, it also delivers a highly cytotoxic drug, in this case emtansine, which is also an inhibitor of microtubule formation. Additionally, ado-trastuzumab emtansine blockades the EGF receptor family member, HER2/neu, making targeted cells vulnerable to death due to the disruption of essential receptor–ligand interactions.

Thus far, the ability to derive or engineer MAbs specific for particular cancer-related antigens or processes, and to produce them in unlimited amounts, has been the most successful approach to cancer immunotherapy. Although powerful, MAbs are not normally deployed as single agents. As illustrated in these case studies, they contribute to therapeutic programs that include other agents and procedures, such as chemotherapy and radiation. The search for antigens that act as TAAs or TSAs continues, as do efforts to identify antibody-targetable processes essential for the survival and maintenance of cancer cells or the tumor microenvironment. As they are discovered, these will add to an already crowded pipeline of new antibodies and ADCs in development and in clinical trials. The successful candidates that emerge from clinical testing will increase and diversify the repertoire of antibody-based cancer immuno-therapies. Another important focus of current and future clinical studies will be the determination of how existing anti-cancer antibodies and ADCs can be optimally combined or sequenced with other drugs and tumor therapies.

Chapter discussion questions

1. Contrast a monoclonal antibody with a polyclonal antibody. Why might one prefer to use a monoclonal rather than a polyclonal antibody for cancer therapy?

2. Successful anti-cancer antibodies can cause the death of cancer cells. What are the major mechanisms of antibody-mediated cancer cell death? What agents play important roles in each of the mechanisms that you have identified?

3. In what ways are the mechanisms used by ADCs to attack a patient's cancer similar or identical to those of unconjugated monoclonal antibodies? What advantages do they offer over unconjugated antibodies? What are the challenges associated with the development of clinically useful ADCs?

Selected references

Introduction

Mellman I, Coukos G & Dranoff G (2011) Cancer immunotherapy comes of age. *Nature* 121:2436–2446.

Page DB, Postow MA, Callahan MK et al. (2014) Immune modulation in cancer with antibodies. *Annu. Rev. Med.* 65:185–202.

Scott AM, Wolchok JD & Old LJ (2012) Antibody therapy of cancer. *Nat. Rev. Cancer* 12:278–287.

Sievers EL & Senter PD (2013) Antibody-drug conjugates in cancer therapy. *Annu. Rev. Med.* 64:15–29.

Weiner GJ (2015) Building better monoclonal antibody-based therapeutics. *Nat. Rev. Cancer* 15:361–370.

Case 11-1

Abramson JS & Shipp MA (2005) Advances in the biology and therapy of diffuse large B-cell lymphoma: moving toward a molecularly targeted approach. *Blood* 106:1164–1174.

Coiffier B, Lepage E, Brière J et al. (2002) CHOP chemotherapy plus rituximab compared with CHOP alone in elderly patients with diffuse large-B-cell lymphoma. *N. Engl. J. Med.* 346:235–242.

Roschewski M, Staudt LM & Wilson WH (2014) Diffuse large B-cell lymphoma-treatment approaches in the molecular era. *Nat. Rev. Clin. Oncol.* 11:12–23.

Sehn LH, Berry B, Chhanabhai M et al. (2007) The revised International Prognostic Index (R-IPI) is a better predictor of outcome than the standard IPI for patients with diffuse large B-cell lymphoma treated with R-CHOP. *Blood* 109:1857–1861.

Siegel RL, Miller KD & Jemal A (2018) Cancer statistics, 2018. *CA Cancer J. Clin.* 68:7–30.

Case 11-2

Canellos GP, Anderson JR, Propert KJ et al. (1992) Chemotherapy of advanced Hodgkin's disease with MOPP, ABVD, or MOPP alternating with ABVD. *N. Engl. J. Med.* 327:1478–1484.

Connors JM, Jurczak W, Straus DJ et al. (2017) Brentuximab vedotin with chemotherapy for stage III or IV Hodgkin's Lymphoma. *N. Engl. J. Med.* [epub ahead of print].

Diefenbach C & Steidl C (2013) New strategies in Hodgkin lymphoma: better risk profiling and novel treatments. *Clin. Cancer Res.* 19:2797–2803.

Meyer RM, Gospodarowicz MK, Connors JM et al. (2012) ABVD alone versus radiation-based therapy in limited-stage Hodgkin's lymphoma. *N. Engl. J. Med.* 366:399–408.

Rancea M, Monsef I, von Tresckow B et al. (2013) High-dose chemotherapy followed by autologous stem cell transplantation for patients with relapsed/refractory Hodgkin lymphoma. *Cochrane Database Syst. Rev.* 6:CD00941.

Younes A, Gopal AK, Smith SE et al. (2012) Results of a pivotal phase II study of brentuximab vedotin for patients with relapsed or refractory Hodgkin's lymphoma. *J. Clin. Oncol.* 30:2183–2189.

Case 11-3

Plostker GL & Keam SJ (2006) Trastuzumab: a review of its use in the management of HER2-positive metastatic and early-stage breast cancer. *Drugs* 66:449–475.

Verma S, Miles D, Gianni L et al. (2012) Trastuzumab emtansine for HER2-positive advanced breast cancer. *N. Engl. J. Med.* 367: 1783–1791.

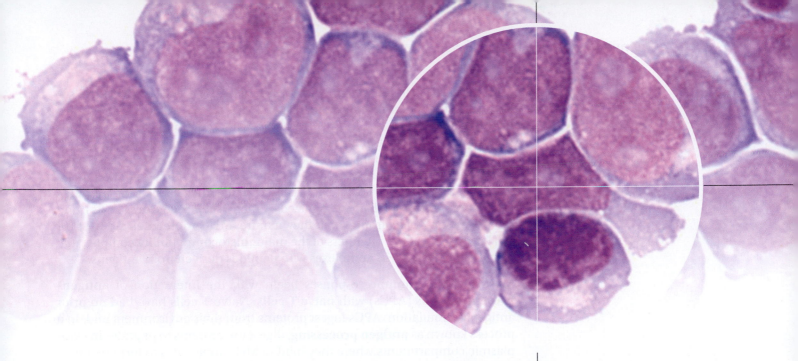

Chapter 12

Immunological Strategies: Vaccination and Adoptive T-Cell Transfer

Introduction

T cells within the immune system can be manipulated to prevent and treat cancer. At the time of writing, three specific types of T-cell manipulations are the most promising immunological strategies for the treatment of cancer, namely vaccination, **adoptive T-cell therapies**, and **checkpoint inhibitors** to blockade immune-response checkpoints. Vaccination involves specificity, amplification, and memory—three of the adaptive immune system's most striking and potent attributes. Adoptive T-cell transfer involves infusion of patients with populations of T cells directed against cancer cells. These T cells are either obtained from the patient or selected *in vitro* for anti-cancer activity, or they may be engineered to attack the patient's cancer. Checkpoint inhibitors target an important set of receptor–ligand interactions that negatively regulate immune responses. These negative regulators play an important role in controlling excessive or chronic immune responses; however, this normal and essential homeostatic mechanism can restrain or prevent the immune system from generating an effective response against cancer. Monoclonal antibody blockade of receptor–ligand interactions that mediate immune checkpoints has proved to be an effective approach for the immunotherapy of cancer. Because T cells and their interactions underlie all of the case studies presented in this chapter, an outline of a few key features of T-cell immunobiology will help the reader to appreciate the design of the immunotherapies described in the individual cases.

Most human T cells fall into one of two groups—CD4 T cells or CD8 T cells—based on whether the glycoprotein CD4 or CD8 is expressed on the T cell's surface. In general, CD4 T cells provide cytokines and membrane-bound ligands that are essential for the initiation, maintenance, and regulation of a wide variety of immune responses. This category includes **helper T cells** (T_H) that support immune responses, and regulatory T cells (T_{reg}) that suppress them. The **CD8** group includes cytotoxic T cells (T_C) that kill cells bearing **peptide–MHC complexes**, which are recognized by their **T-cell receptors** (TCRs), and the precursor T cells from which they mature. **MHC molecules** are host (or self) proteins that are members of the **m**ajor **h**istocompatibility **c**omplex that form the peptide–MHC complexes with peptides derived from antigens.

Most adaptive immune responses begin with the interaction of **antigen-presenting cells** (APCs) with **naive T cells**. Naive T cells have had no prior antigenic stimulation. APCs ingest proteins from their environment and, in a process known as **antigen processing**, digest the proteins to peptides in cytoplasmic compartments where they bind to MHC molecules to form peptide–MHC complexes that are exported to the surface of the APC and displayed on the plasma membrane. There are two molecular families of MHC. One of these, MHC-I, presents antigen-derived peptides to CD8 T cells; the other, MHC-II, presents antigen-derived peptides to CD4 T cells.

Unlike antibodies, TCRs cannot bind intact protein antigens. A TCR can only recognize a peptide–MHC complex whose three-dimensional conformation is complementary to the TCR's binding site. Therefore the **proteolysis** (breakdown of proteins into peptides) and the steps of antigen processing resulting in the display of antigen-derived peptides are essential for antigen-triggered T-cell responses. However, APCs do more than merely present antigens. They have other families of molecules on their surface known as **co-stimulatory molecules** that interact with receptors on the T cell. In addition to the recognition of peptide–MHC complexes by its TCRs, activation of a naive T cell requires interactions of co-stimulatory ligands and receptors. APCs bear the dual and essential responsibilities for both antigen presentation and co-stimulation. Although some cell types, such as appropriately activated macrophages and monocytes, can present antigens to T cells, a category of APCs known as dendritic cells are by far the most efficient and versatile inducers of immune responses in naive CD4 and CD8 T cells. Helper T cells collaborate with APCs to induce the activation and development of naive CD8 T lymphocytes into cytotoxic T lymphocytes (CTLs) that can kill target cells. Also, in most instances, the production of antibodies by B cells is dependent on signals from T_H cells. Interactions among the network of APCs, T cells, and B cells are reviewed in Figure 12.1.

It is apparent from Figure 12.1 that the initiation of an adaptive immune response is complex and requires the interaction of many cell types and receptor–ligand interactions. Induction of anti-cancer immune responses must successfully convene and activate this elaborate network of interactions. In addition to the complexities of the process, other factors can diminish or even prevent the induction of anti-cancer immune responses by tumor antigens. One of these is **tolerance**, namely the reluctance or inability of the patient's immune system to mount immune responses against self-antigens. Tolerance is a predictable obstacle to anti-tumor responses because most of the antigens present in cancer cells closely resemble or are identical to their normal counterparts. Consequently, the body's response to cancer must penetrate the formidable barrier of tolerance to self-antigens. In addition to tolerance, anti-cancer responses are confronted by a variety of inhibitory mechanisms that have evolved to restrain excessive or chronic immune responses. These intrinsic safeguards are necessary because

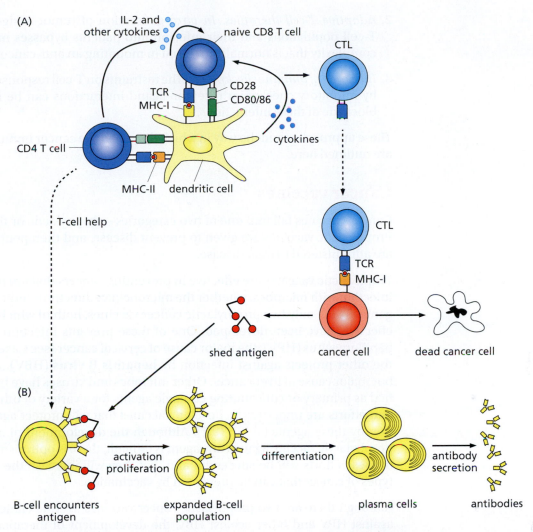

Figure 12.1. An overview of cell interactions required for an adaptive immune response to cancer. (A) Activation of CD4 T_H (helper T cell) and CD8 T cells. Proteins shed from tumor cells are internalized by an antigen-presenting cell (APC), such as the dendritic cell shown here, and processed to peptides, some of which are presented on major histocompatibility complex (MHC) class II molecules (MHC-II) to CD4 T cells and others on class I MHC molecules (MHC-I) to CD8 cells. Interaction of the T-cell receptors (TCRs) of a CD4 T_H with peptide–MHC-II complexes and co-stimulatory ligands on the APC can activate the CD4 T_H to secrete cytokines, such as interleukin-2 (IL-2). Depending on conditions, activated T_H cells may proliferate and differentiate into various subpopulations of T_H cells. The activation and differentiation of CD8 cytotoxic T lymphocytes (CTLs) with full cytotoxic capability requires binding of TCRs with peptide–MHC-I, and co-stimulation by interaction of co-stimulatory ligands on the APC with co-stimulatory receptors on the T cell and IL-2. In this case, CD80 co-stimulatory ligands on the APCs engage the CD28 receptor on the T cell. Killing a cancer cell requires the CTL's encounter with the cancer cell and engagement of the TCRs of the cytotoxic T cell (T_C) with peptide–MHC-I on the surface of the cancer cell. (B) Induction of antibody secretion by B cells. Two signals are required—one from the B-cell receptor (BCR) upon cross-linking by binding the tumor antigen, and the other being "T-cell help" (a combination of CD4 T_H membrane-bound ligands and T_H-produced cytokines). Activated B cells proliferate, with some of the population differentiating into antibody-secreting cells.

strong or chronic immune responses can be quite damaging to the patient if they are allowed to proceed unchecked. Inhibitory mechanisms include the following: immune suppression by T regulatory cells (T_reg cells); immunosuppressive cytokines, such as TGF-β and IL-10; and inhibitory immune checkpoints mediated by receptor–ligand combinations, whose engagement decreases the activation and effector functions of T cells. Three of the major strategies evolved to solve these problems are illustrated by case studies in this chapter. They are:

1. *Cancer vaccines.* Some cancer vaccines employ viral and microbial antigens and do not encounter barriers of tolerance.

2. *Adoptive T-cell therapies. In vitro* generation of tumor-antigen-specific T-cell populations and their infusion into patients bypasses much of the complexity that is normally involved in mounting an anti-cancer response.

3. *Blockade of immune checkpoints.* The restraints on T-cell responses imposed by inhibitory checkpoint receptor–ligand interactions can be relieved by blockade of these interactions.

These approaches and the way in which they facilitate cancer immunotherapy are outlined here.

Cancer vaccines

Cancer vaccines fall into one of two categories—prophylactic or therapeutic. Prophylactic vaccines are given to prevent disease, and therapeutic vaccines are administered to treat disease.

Prophylactic vaccines are effective in preventing cancers that are the result of infection with microbes, whether the microbes are directly or indirectly oncogenic. So far, only two prophylactic cancer vaccines, both of which are highly effective, have been developed. One of these prevents infection by human papillomavirus (HPV), the major cause of cervical cancer (see Case 12-1), and the other protects against infection by **hepatitis B virus** (HBV), an indirect but major cause of liver cancer. Other microbes and viruses have been identified as primary or contributing etiologic agents for a variety of other cancers, and efforts are underway to develop vaccines that will protect against infection by these agents (Table 12.1). Although the development of an effective prophylactic vaccine against any agent is difficult, there is optimism that some of these efforts will be successful and that they will increase the number of types of cancer that can be prevented by vaccination.

Although there are two prophylactic cancer vaccines in wide clinical use, one against HPV and other against HBV, the development of therapeutic cancer vaccines has been especially challenging. The first therapeutic cancer vaccine approved by the U.S. Food and Drug Administration (FDA) was sipuleucel-T for the treatment of metastatic **castration-resistant prostate cancer**. Sipuleucel-T extends median overall survival by 4 months. Although the therapeutic benefit of this vaccine has been modest, the design of sipuleucel-T illustrates an approach to the creation of a therapeutic cancer vaccine. This vaccination strategy involves obtaining peripheral blood mononuclear cells (PBMCs) from the patient and the subsequent *in vitro* culture of those PBMCs with a chimeric protein that contains the amino acid sequence of prostatic acid phosphatase (PAP) and **granulocyte-macrophage colony-stimulating factor** (GM-CSF). PAP is a tumor-associated

Table 12.1. Viral and microbial candidates for development of prophylactic cancer vaccines.

Infection-associated malignancy	Microbe
Stomach cancer; gastric lymphoma	*Helicobacter pylori*
Cervical, vaginal, and anal cancer	Human papillomavirus
Liver cancer	Hepatitis B virus, hepatitis C virus
Nasopharyngeal cancer; Burkitt lymphoma	Epstein–Barr virus
Kaposi sarcoma	Human herpesvirus 8
Adult T-cell leukemia/lymphoma	Human T-cell lymphotrophic virus 1

Adapted from D.M. Parkin, *Int. J. Cancer* 118:3030–3044, 2002; and I.H Frazer, D.R. Lowy and J.T. Schiller, *Eur. J. Immunol.* 37:S148–S155, 2007.

antigen and the GM-CSF promotes the uptake of the PAP by APCs in the PBMC population. When reinfused to the patient, the APCs present peptides derived from PAP to T cells, initiating an immune response that attacks prostate cancer cells as well as normal prostate epithelial cells that express the tumor-associated antigen. Other therapeutic cancer vaccines are in various stages of development, and efforts to create a therapeutic vaccine for the treatment of cervical cancer are briefly summarized in the conclusion to Case 12-1. The major challenge to a therapeutic vaccine is the induction of a strong and durable immune attack on the cancer that effects its sustained remission or elimination.

Adoptive T-cell therapies

Adoptive T-cell therapies involve four steps: the collection of T cells from the patient; *in vitro* selection or modification for reactivity with cancer cells; the expansion of the cancer-reactive T-cell population to large numbers; and their reinfusion into the patient. The use of autologous T cells avoids the serious and life-threatening immune responses (host versus graft or graft versus host) that occur when the graft and the host are not genetically identical. The broad outlines of adoptive T-cell therapies employing T cells with anti-tumor activity generated endogenously or created by genetic engineering are described here.

Adoptive T-cell therapy using endogenous T cells. Endogenous T cells with anti-tumor activity arise from the patient's immune response to their cancer and are found among the **tumor-infiltrating lymphocytes** (TILs) present in the patient's tumor. Consequently, tumor-reactive T-cell populations can be obtained from samples of the patient's cancer and expanded *ex vivo* with the aid of the cytokine, interleukin-2 (IL-2), an essential T-cell growth factor. The preferential *in vitro* expansion of T-cell populations with specificities for tumor antigens can be encouraged by inclusion of tumor cells or cells engineered to express antigenic determinants specific to the tumor. Once expanded, these populations of tumor-reactive T cells, which are a mixture of CD4 and CD8 T cells, are reinfused into the patient, where they proliferate, maintain their antigen specificity, and, most importantly, retain their anti-cancer activity. *In vitro* expansion bypasses whatever factors prevent or inhibit a robust expansion of T cells from recognizing tumor antigens in the patient. An important part of the therapy involves the deliberate depletion of the patient's lymphocyte population by treating with chemotherapy prior to infusion of the *in vitro*-expanded T-cell population. IL-2, a cytokine essential for T-cell proliferation, is often given with the cell infusion. Using TILs harvested from metastatic melanomas, expanded *in vitro,* and reinfused into the donor patients, this approach has achieved complete remission rates as high as 20% (Table 12.2).

Table 12.2. Selected early clinical trials of adoptive cell transfer (ACT) for cancer treatment.

Cells used for ACT	Year	Cancer	Patients	% of ORs	Target
TILs	1998	Melanoma	20	55%	Unknown/possibly tumor neoantigens
TILs	2011	Melanoma	93	56%	Unknown/possibly tumor neoantigens
TILs	2014	Cervical Cancer	9	33%	HPV antigens were probable targets
TILs	2014	Cholangiocarcinoma (bile duct carcinoma)	1	100%	Mutated HER2
CAR T cells	2011	CLL	3	100%	CD19
CAR T cells	2013	ALL	5	100%	CD19
CAR T cells	2014	ALL	30	90%	CD19
CAR T cells	2014	Lymphoma	15	80%	CD19

Abbreviations: ACT, adoptive cell transfer; ALL, acute lymphocytic leukemia; CAR T cells, chimeric antigen receptor T cells; CD19, B cell surface protein; CLL, chronic lymphocytic leukemia; HER2, human epidermal growth factor receptor 2; HPV, human papillomavirus; OR, objective response; TILs, tumor infiltrating lymphocytes.

Adapted from S.A. Rosenberg and N.P. Restifo, *Science* 348: 62–68, 2015.

Although most of the experience with adoptive T-cell therapies has been gained in melanoma patients, it has also shown promise for some other cancers. Strikingly, the likelihood of achieving a complete response to this immunotherapy appears to be less dependent on the bulk of the disease and the site of metastasis than would have been expected. However, the therapy (summarized in Figure 12.2) does depend upon the patient's generation of anti-tumor T cells, and requires that the size and location of the cancer make it possible to harvest tumor-infiltrating T-cell populations and expand anti-tumor T-cell populations *ex vivo*.

Adoptive T-cell therapy using host T cells engineered to recognize tumor antigens. A different approach employs genetic engineering to create antigen receptors that are specific for tumor-associated antigens (TAAs) or tumor-specific antigens (TSAs) known to be present on the patient's cancer. This is best accomplished by selecting an antigen that is present on the surface of all cancer cells and then deriving a monoclonal antibody specific for that antigen. Recombinant DNA techniques are used to clone the genes specifying the antigen-binding domains of the antibody's heavy and light chains. These genes are then employed as essential building blocks to construct the extracellular portion of a **chimeric antigen receptor** (CAR). This approach provides antigen receptors of known specificity that directly recognize their target antigen. In addition to bearing an extracellular antigen receptor of predefined

Figure 12.2. Generation and adoptive transfer of T cells recognizing antigens generated by cancer-specific mutations. (1) Samples of cancerous tissue and T cells are obtained from the patient. (2) and (3) DNA from normal and cancerous tissue is sequenced to identify cancer-specific mutations. (4) Tandem arrays of minigenes encoding cancer-specific peptides are constructed and transfected into the patient's APCs. (5) APCs present peptide–MHC-I and peptide–MHC-II (not shown). (6) Activated T cells are sorted for expression of activation markers. (7) The patient is lymphodepleted and activated T cells are introduced. (Adapted from S.A. Rosenberg and N.P. Restifo, *Science* 348:62–68, 2015.)

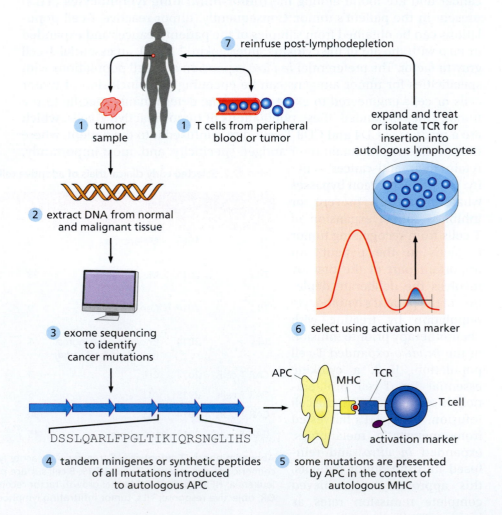

7 reinfuse post-lymphodepletion

1 tumor sample

1 T cells from peripheral blood or tumor

expand and treat or isolate TCR for insertion into autologous lymphocytes

2 extract DNA from normal and malignant tissue

3 exome sequencing to identify cancer mutations

6 select using activation marker

DSSLQARLFPGLTIKIQRSNGLIHS

APC   MHC   TCR   T cell   activation marker

4 tandem minigenes or synthetic peptides of all mutations introduced to autologous APC

5 some mutations are presented by APC in the context of autologous MHC

specificity, these constructs are designed to include domains that provide the intracellular signals associated with engagement of conventional T-cell receptors and co-stimulatory receptors. T cells engineered to have a CAR, known as CAR T cells, offer several advantages over dependence on conventional T-cell responses to cancer. These are readily appreciated by considering the immunobiology of T-cell responses.

The initiation of a conventional cytotoxic T-cell response in a population of naive CD8 T cells against cancer cells is complex and requires at least two different categories of receptor–ligand interactions, and the interactions of at least three different cell types (see Figure 12.1). Essential ligand–receptor interactions are (a) antigen-specific engagement of TCRs with peptide–MHC combinations derived from cancer cell antigens and (b) engagement of co-stimulatory receptors, such as **CD28** on T cells, with their co-stimulatory ligand partners on APCs. Cellular participants include (1) cancer cells as antigen donors and targets of the response, (2) T cells, and (3) dendritic cells as APCs. The complexity and unpredictability of the conventional T-cell response contrast sharply with that of CAR T cells that directly recognize the target antigen, bypassing APCs and antigen presentation as peptide–MHC complexes (Figure 12.3). Furthermore, because co-stimulatory signaling domains are a part of the CAR, there is no need for co-stimulatory receptors on T cells and their complementary ligands on APCs. Finally, the antigen specificity of the CAR is predetermined to be specific for a tumor-associated tumor antigen known to be present on the cancer cell. CAR T-cell responses involve only two cell types, namely the attacking CAR

Figure 12.3. Contrast between induction of conventional CTL-mediated cancer cell death and CAR T-cell responses. (A) Induction of conventional CTL-mediated cancer cell death. An antigen from a cancer cell is internalized and digested to peptides by a dendritic cell. Peptides forming peptide–MHC-I complexes of appropriate conformation are recognized by TCRs of naive CD8 T cells and induced to differentiate into mature CTLs. Interaction of co-stimulatory ligands on the dendritic cell with receptors on the T cells provides the necessary co-stimulatory signals, and cytokines such as IL-2 required for proliferation and development into mature CTLs come from a variety of cells. Recognition of peptide–MHC-I on cancer cells by TCRs of activated CTLs induces cytolysis of the cancer cell. Note that the interaction involves a minimum of three distinct cell types, namely cancer cells, dendritic cells, and CD8 T cells. In many cases CD4+ helper T cells are also involved. Also there are two distinct types of receptor–ligand interactions, namely MHC-I with TCRs of CD8 T cells, and co-stimulatory ligands with receptors. (B) CAR T-cell-mediated killing. Interaction of the CAR of a CD8 T cell with the targeted antigen on the cancer cell initiates all of the signaling cascades needed to induce killing by the CAR T cell. In CAR T-cell responses, only two cell types and one type of receptor–ligand interaction are required for cancer cell killing.

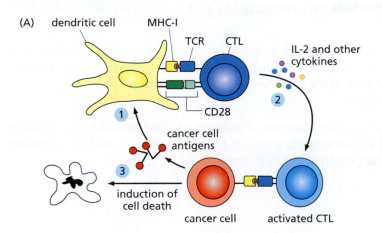

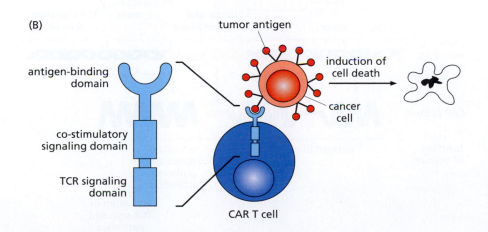

T cell and the targeted antigen-displaying cancer cell. Instead of several different receptor–ligand interactions, there is only one, namely the binding of the CAR on the T cell to the target antigen on cells displaying it.

The design and construction of CAR T cells is based on T-cell activation requiring three distinct activities: antigen recognition by the extracellular portion of the TCR complex; signals from the intracellular domains of the TCR complex; and signals from the T cell's co-stimulatory receptors. The CAR brings these three essential elements together into one engineered molecule. A key feature and advantage of CAR T cells is that their antigen recognition is provided by the antigen binding sites of antibody molecules, allowing antigen specificity to be predetermined and genetically engineered into their makeup. Although conventional T cells are limited to reactions against complexes of MHC molecules with peptides derived from proteins, the antibody-derived portions of CARs are not limited to antigenic determinants of proteins. It is also possible to design CAR T cells that can recognize carbohydrate or even lipid antigens. The process begins with the selection of an antigen that is either tumor associated or tumor specific. The next step is to derive a suitable monoclonal antibody against the chosen tumor antigen. One then proceeds to generate the required DNA sequences for the assembly of a CAR. These cloned sequences are assembled into an expression construct and transfected into T cells obtained from the patient. Figure 12.4 summarizes the derivation and assembly of DNA sequences encoding the following:

a. Antigen-binding domain. A fusion polypeptide known as a **single-chain variable fragment** (scFv), in which the variable regions of the **heavy chain** (V_H) and **light chain** (V_L), connected by a 10- to 25-amino-acid linker, bind antigen.

b. Hinge and transmembrane sequences. A transmembrane portion of a T-cell membrane protein, usually CD8 or CD28, which acts as a hinge, allowing the scFv of the CAR flexibility and connection to the cytoplasmic signaling domains.

c. Cytoplasmic signaling domains. A polypeptide that includes the signaling domains of a co-stimulatory molecule, such as CD28, and the signaling domain of the ζ chain of the TCR complex.

Figure 12.4. Engineering CAR T cells. Sequences encoding the indicated protein domains are ligated to generate a construct that expresses a chimeric antigenic receptor with the specificity to engage the targeted antigen and trigger T-cell activation.

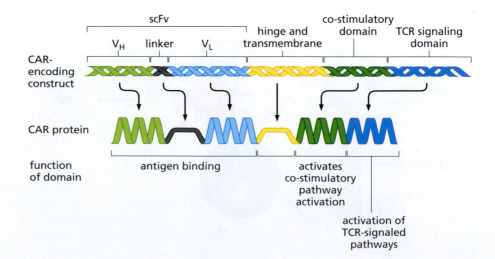

Binding of the antibody-based receptor portion of the CAR by the targeted antigen initiates many of the same T-cell-activating signaling cascades that are triggered by engagement of conventional T-cell receptors and co-stimulatory receptors, thereby triggering CAR T cells to kill the targeted cancer cells. They also kill normal host cells that bear the tumor-associated antigen, but spare those that do not.

Allogeneic bone marrow transplantation. Allogeneic (i.e., of the same species but genetically dissimilar) stem cell transplantation (SCT), also known as bone marrow transplant (BMT), is a type of adoptive cell transfer in which the recipient's hematopoietic system is replaced by transfusing stem cells from a donor whose MHC differs from the recipient. The procedure involves attenuation or ablation of the patient's adaptive immune system prior to the allogeneic transfer in order to minimize the risk of rejection of the infused donor cells by the patient. However, there is a major risk of graft-versus-host disease (GVHD), in which T cells in the donor transplant (the graft) recognize recipient tissues (the host) as foreign and mount an immune response, which can be serious and life-threatening. The anti-cancer effects of allogeneic BMT were recognized in 1979, when antileukemic activity was observed to accompany reconstitution of the hematopoietic system with allogeneic bone marrow. These therapeutic effects, mediated primarily by T cells but with some contribution by natural killer (NK) cells, can be quite robust and have curative potential for some hematologic cancers. The graft-versus-leukemia (GVL) phenomenon demonstrates the anti-cancer effects of adoptively transferred immune cells for the treatment of cancer in humans. However, the GVHD that accompanies the GVL effect makes allogeneic bone marrow transplantation a challenging clinical procedure, requiring immunosuppressive therapy that minimizes GVHD while still allowing anti-cancer GVL activity. It is these considerations that drive efforts to develop alternative adoptive transfer therapies that are not accompanied by GVHD. In addition to allogeneic SCT, cytokines, members of the thalidomide family, **imiquimod**, and **Bacillus Calmette–Guérin** (a strain of mycobacteria employed for vaccination against tuberculosis) are used as immunological approaches to cancer therapy for specific malignancies. These agents and other novel approaches recently added to the immunotherapy toolbox are briefly profiled later in the section "Other immunotherapies."

Blockade of immune checkpoints

As mentioned earlier, immune responses are the outcome of a complex and complicated network of interactions, of which some are stimulatory and others are inhibitory. As shown in Figure 12.1, the induction of an effective antitumor T-cell response requires TCR recognition of tumor-antigen-derived peptide–MHC complexes and a co-stimulatory signal delivered via the interaction of a co-stimulatory molecule, such as **CD80** or **CD86** on the APC with CD28 on the T cell. However, negative regulatory mechanisms have evolved to prevent excessive T-cell activation of naive T cells or chronic or inappropriately prolonged action of effector T cells. Excessive T-cell activation is discouraged by the appearance on the T-cell membrane of **CTLA-4**, a molecule related to CD28 that inhibits T-cell activation when engaged by CD80/86. The balance between stimulatory signals from CD28 and inhibitory ones from CTLA-4 determines the level of T-cell activation. The antagonism between CD28 and CTLA-4 protects against the damage (autoimmunity) that can be inflicted on healthy tissue by excessive T-cell activation. However, the CTLA4-mediated brake on T-cell activation makes it more difficult for tumor antigens to trigger an anti-cancer immune response. In peripheral tissues, long-term exposure to antigen induces CTLs and other effector T cells to express **programmed cell death protein 1** (PD-1), a cell surface receptor that inhibits T-cell responses when engaged by its ligands, **programmed death ligand 1** (PD-L1) or **programmed death ligand 2** (PD-L2). During the conditions of chronic inflammation that accompany prolonged immune responses, a number of cell types in peripheral tissues express PD-L1. Under normal circumstances, the inhibition or termination of prolonged T-cell responses is anti-inflammatory and beneficial. However, although regulation of the checkpoints of activation and persistence by CTLA-4 and PD-1, respectively, is an important mechanism of protection against *autoimmunity*, these checkpoints make it difficult to mount and sustain *anti-cancer* T-cell responses.

Release from the inhibitory effects of these checkpoints can be achieved by using monoclonal antibodies that block the interactions between inhibitory receptors and their ligands. Blockade by checkpoint inhibitors can facilitate anti-cancer responses at either of two phases. Anti-CTLA-4 antibodies prevent the engagement of CTLA-4 by CD80/86 and act to release an important brake on T-cell activation. The persistence of the response of effector T cells to antigens in peripheral tissues is prolonged and strengthened by blockade of the interaction of PD-1, with its ligands PD-L1 or PD-L2, by antibodies against PD-1 or its ligands (Figure 12.5). In addition to the CTLA-4- and PD-1-anchored checkpoints described here, there are several others, some of which are being used as targets for antibody-mediated checkpoint blockade. Checkpoint-blockading antibodies now constitute a powerful and growing class of immunotherapeutics.

Figure 12.5. Schematic diagram of blockade of CTLA-4 or PD-1 signaling by checkpoint-inhibiting monoclonal antibodies. (A) Activation phase. In lymph nodes, naive T cells are activated by engagement of their TCRs and co-stimulatory receptors by peptide–MHC complexes and co-stimulatory ligands, respectively, on dendritic cells. However, the degree of activation is reduced by engagement of CTLA-4, an inhibitory receptor that is also engaged by the otherwise co-stimulatory receptors, CD80/86. Antibodies against CTLA-4 that block its engagement of CD80/86 promote more robust activation of naive T cells. (B) Effector phase. In the periphery, antigen-experienced T cells exported from lymph nodes attack cancer cells bearing appropriate peptide–MHC complexes. However, the engagement of inhibitory PD-1 receptors on the T cell make it susceptible to inhibition by cancer cells or other cells of the tumor microenvironment bearing the ligands PD-L1 or PD-L2. As shown, anti-PD-1 or anti-PD-L1 antibodies that block the negative regulatory effects of these receptor–ligand interactions enable more potent T-cell responses against cancer cells. (Adapted from A. Ribas, *N. Engl. J. Med.* 366:2517–2519, 2012.)

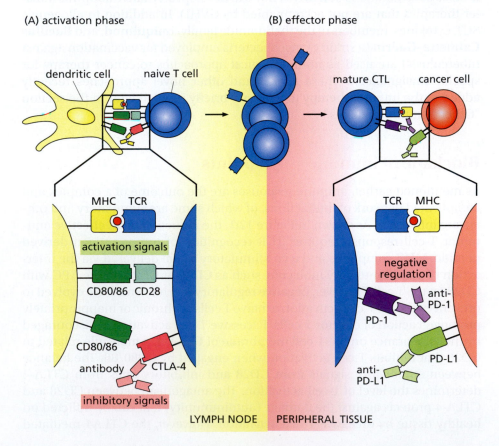

Other immunotherapies

Oncolytic viruses.   **Oncolytic virus** immunotherapy uses cytolytic viruses for selective killing of tumor cells and the induction of anti-tumor immune responses. Many types of viruses, including adenovirus, herpesvirus, coxsackievirus, measles virus, and others, are under investigation. Infection of tumor cells by an oncolytic virus directly and indirectly induces the production of cytokines, some of which, like interferon, have immunomodulatory actions. Virus-mediated lysis of tumor cells releases intracellular components that include a variety of inflammatory agents. Some of these, such as double-stranded RNA and other nucleic acids, are of viral origin, while others, including ATP, uric acid, and certain heat shock proteins, are of host cell origin. In addition, the lysis of the cancer cell releases potentially immunogenic neoantigens, making them available for recognition and targeting by the immune system. Oncolytic viruses can be targeted to the tumor by direct injection of the virus into a tumor site, exploiting the natural tropism of the virus for a particular cell surface molecule or a combination of both. With regard to viral tropisms, it is unusual to find a receptor for a given virus exclusively on the surface of the targeted cancer cell. However, there are many examples where receptors for a particular virus are overexpressed on cancer cells, thus making them more susceptible to infection by the virus. Once it has been determined that a surface protein is overexpressed on the surface of a particular cancer type, there is the possibility of genetically modifying the virus to recognize it.

One specific example of an oncolytic virus in clinical use is **Talimogene laherparepvec** (T-Vec), a genetically modified **herpes simplex virus 1** (HSV-1). T-Vec has been approved by the FDA for the treatment of melanoma. Genetic modifications were necessary because wild-type HSV-1 is an important human pathogen that causes skin lesions, infects neurons, and can enter prolonged latent phases from which it can emerge years later and cause a recurrence of disease. In the engineered version of HSV used for cancer therapy, the deletion of two genes disables the ability of the virus to interfere with antigen presentation, prevents the inhibition of viral replication by some key intracellular defenses, and renders it incapable of infecting neurons. In addition to these gene deletions, a gene encoding the cytokine GM-CSF is added. The result of these modifications is a virus that attracts the attention of the immune system by display of viral antigens and replicates free from the constraint of some important intracellular defenses. When it lyses the cell, it releases additional viral antigens and tumor neoantigens. The GM-CSF generated from T-Vec-infected cells increases the likelihood of induction of an adaptive immune response. The immune response against the oncolytic virus limits the spread of the virus, and the immune response to the tumor antigens increases the likelihood of systemic immune responses to the cancer at widely disseminated sites in the patient. There is considerable interest in employing oncolytic viruses in combination with other therapies, rather than limiting them to use as monotherapies. Early clinical trials indicate that the combination of T-Vec with a PD-1 checkpoint inhibitor increases the effectiveness of this oncolytic virus.

Bi-specific antibodies.   A **bi-specific antibody** (bsAb) is a single antibody engineered to bear two different binding sites, endowing one antibody with two different antigen specificities. Bi-specific antibodies, in which one binding site is specific for an antigen on the surface of a tumor cell and the other for CD3, a part of the T-cell receptor complex, can be used to bring a T cell together with a tumor cell. For example, bi-specific antibodies against CD3 and a tumor antigen can be used to bring a cytotoxic T cell and a tumor cell bearing that antigen together. Irrespective of the specificity of the T cell, engagement of its T-cell receptor complex by the bsAbs will trigger its activation and provoke killing of the tumor cell that it is forced to embrace by the bsAbs. The centerpiece of this approach is the convening of T cells with tumor targets, and this class of immunotherapeutic drugs has been labeled **T-cell-dependent bi-specific antibodies** (TBDs). This is a promising and rapidly diversifying technology that assumes many different forms. The bsAb can be a conventional one composed of two heavy and two light chains, or a variation of this structure. A highly stripped down version of the approach is the **bi-specific T-cell engager** (BiTE), which contains only the heavy- and light-chain variable regions of the TBD tethered to each other by a peptide linker. The FDA-approved **blinatumomab**, a BiTE that is specific for CD3 and CD19, a B-cell antigen, was first used for the treatment of **acute lymphoblastic leukemia**.

Cytokines. Cytokines are small proteins made by one cell that exert regulatory effects on other cells and, in some cases, also on the cytokine-producing cell. A subcategory of cytokines, known as chemokines, can direct the movement of cells that bear chemokine receptors. Although cytokines were first discovered in the immune system, they are now known to be made by and to affect the behavior of a wide variety of cell types. Interleukin-2 (IL-2) is a cytokine that promotes the growth of T cells and affects the activity of some other cells of the immune system. It has been employed in the treatment of metastatic melanoma and advanced kidney cancer. Interferon alfa (IFN-α) is a member of a large family of proteins, many of which affect the immune system, and was first recognized because of its antiviral activities. IFN-α was historically used to treat hairy cell leukemia, follicular lymphoma, and Kaposi sarcoma, among other cancers. However, due to the development of more effective and better tolerated therapies, it is rarely used today. Granulocyte-macrophage colony-stimulating factor (GM-CSF) stimulates the generation of granulocytes, monocytes, and macrophages in the bone marrow, and is often used to boost white cell counts in patients after the depletion of these populations by chemotherapy. In the periphery, GM-CSF promotes the differentiation of monocytes into dendritic cells—the immune system's most efficient antigen-presenting cell type, which is essential for the activation of naïve T cells. It also shows promise as an agent for boosting the response to other immunotherapies.

Thalidomide. Thalidomide and other structurally related compounds, such as lenalidomide and pomalidomide, can enhance immune responses. Thalidomide was originally developed and marketed in the 1950s as a sedative or anti-nausea therapy, particularly for pregnant women, but exposure of fetuses caused a severe birth defect involving limb malformation. It was not until the 1990s that the anti-cancer potential of thalidomide and its related compounds was appreciated and brought to clinical use. One mechanism of action of these drugs is the promotion of expansion of T cells and NK cells. Thalidomide and its relatives are used to treat multiple myeloma and several subtypes of non-Hodgkin lymphoma (see Chapter 10 for a detailed discussion of the use of thalidomide in immunomodulatory strategies to treat myeloma).

Imiquimod. Imiquimod, which is a small organic molecule, is a member of the imidazoquinoline family that stimulates the cells of the innate immune system in ways which promote the activation of adaptive immune responses. It has been approved by the FDA for the treatment of some forms of squamous cell carcinoma. The compound is formulated into a cream that is topically applied to early-stage skin cancers to stimulate a local immune response that can help to control the cancer.

Bacillus Calmette–Guérin.   Bacillus Calmette–Guérin (BCG) is a live attenuated strain of *Mycobacterium bovis* that has received FDA approval for the treatment of early-stage bladder cancer in immunocompetent patients with small tumor burdens. Although its mechanism of action is not understood, it is known to attract cells of the immune system to the bladder, and to activate those cells.

Indoleamine 2,3 dioxygenase.   **Indoleamine 2,3 dioxygenase (IDO) inhibitors** are small molecules that block IDO, an intracellular enzyme that degrades tryptophan, an amino acid which is essential for T-cell activity. IDO activity is a feature of many tumor microenvironments, and can be a contributing factor to the suppression of T-cell-mediated anti-cancer responses. Studies of tissue samples from patients with several types of cancer, including breast, cervical, and lung cancer, have shown that higher IDO levels are associated with poorer survival. Small-molecule inhibitors of IDO have been developed, and have shown promise in early-stage trials. Of particular interest is the finding that a combination of an IDO inhibitor with the checkpoint inhibitor, pembrolizumab, yielded encouraging results in a Phase II study of treatment for advanced melanoma.

The transforming growth factor beta pathway.   The **transforming growth factor beta (TGF-β) pathway** has many roles in development and homeostasis. In some contexts, TGF-β causes powerful immunosuppression. In the tumor microenvironment, TGF-β can contribute to the generation of an immunosuppressive environment by supporting Treg cells. This has encouraged the development of anti-TGF-β-receptor monoclonal antibodies to blockade the triggering of this pathway by its ligand, TGF-β. In addition, small-molecule inhibitors of the TGF-β receptor tyrosine kinase have been developed and have entered clinical trials.

The cases in this chapter illustrate immunotherapeutic approaches in which patient-derived T cells play essential roles in the treatment and even prevention of cancer. The first case illustrates the ability of prophylactic vaccination to markedly lower the incidence of cancers caused by an infectious agent that has been successfully targeted by an effective vaccine. In the next case, acute lymphoblastic leukemia (ALL) is treated with T cells bearing chimeric antigen receptors that have been genetically engineered to recognize and respond to a tumor-associated antigen that is found on the surface of ALL cells. The third case illustrates the ability of the immune system, freed of checkpoint inhibition of T-cell activity, to mount effective anti-cancer responses.

Take-home points

✓   The immune system is capable of recognizing cancer cells, and it can use T-cell-dependent processes to mount immune responses that kill cancer cells or provide a basis for prophylactic vaccination against infectious agents that cause cancer.

✓   Factors that restrain excessive immune responses, such as tolerance and homeostatic mechanisms, can diminish or prevent effective anti-cancer immune responses.

✓ Checkpoint inhibitors increase the potency of anti-cancer responses by blocking some key homeostatic mechanisms that restrain and diminish immune responses to cancer.

✓ Adoptive T-cell therapies involve the infusion of T cells with anti-tumor activities into patients with cancer. Infusion of *in vitro*-activated and expanded cancer antigen-specific host T-cell populations, chimeric antigen receptor-modified T cells (CAR T cells), and allogeneic bone marrow transfer are all examples of adoptive T-cell therapies.

✓ Vaccines for the prevention of cancer (prophylactic vaccines) and for the treatment of cancer (therapeutic vaccines) have been developed.

✓ A broad diversity of approaches to cancer immunotherapy, including oncolytic viruses, bi-specific antibodies and their derivatives, and a variety of small molecules and cytokines that increase immune responses, are also used in cancer therapy.

Topics bearing on this case
Viral oncogenesis
Cancer vaccine strategies
Tumor suppressors and inactivation of pRB and p53

Case 12-1   Cervical Cancer

Introduction

Worldwide, 20% of fatal cancers are caused by infectious agents, most of which are viruses. Cancers with a pathogenic link to viral etiologies cover a broad range of neoplastic diseases, including lymphomas and cancers of the cervix, liver, brain, and nasopharynx, among others (Table 12.3). In this case, we shall focus on human papillomavirus (HPV), a small DNA virus (Figure 12.6) that is responsible for almost all cervical cancer (over 99%). In addition to cervical cancer, HPV causes anal cancers (particularly in immunocompromised patients), genital warts, and a significant proportion of head and neck cancers.

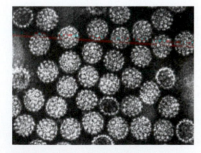

Figure 12.6. An electron micrograph of human papillomavirus (HPV), a small DNA virus. (Courtesy of D. DiMaio, Yale School of Medicine.)

Table 12.3. Viruses implicated in human cancer causation.

Virus	Virus family	Human malignancy
EBV (Epstein–Barr virus)	Herpesviridae	Burkitt lymphoma, Hodgkin lymphoma, pediatric leiomyosarcomas, nasopharyngeal carcinoma, plasmablastic lymphoma, post-transplant lymphoproliferative disease
HHV-8 (human herpesvirus-8)	Herpesviridae	Kaposi sarcoma, primary effusion lymphoma
HPV (human papillomavirus)	Papillomaviridae	Cervical carcinoma, anal cancer, squamous cell cancer of head and neck
MCCV (Merkel cell carcinoma virus)	Polyomaviridae	Merkel cell carcinoma
HTLV-1 (Human T-cell leukemia virus-1)	Retroviridae	Acute T-cell leukemia
HBV (hepatitis B virus)	Hepadnaviridae	Hepatocellular carcinoma
HCV (hepatitis C virus)	Flavivirus	Hepatocellular carcinoma

Adapted from Table 4.6, R.A. Weinberg, The Biology of Cancer, 2nd ed. New York: Garland Science, 2014.

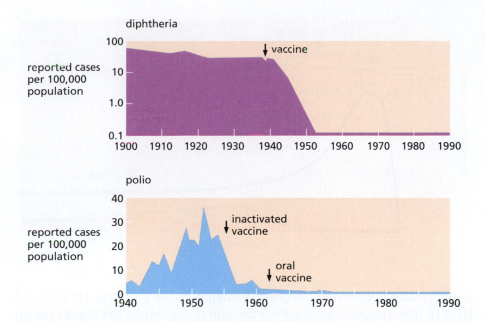

Figure 12.7. Examples of the effect of prophylactic vaccination. (From P. Parham, The Immune System, 3rd ed. New York: Garland Science, 2009.)

The optimal management of cervical cancer consists of prevention, even prior to the precancerous stage that is often detected with **Papanicolaou screening (Pap smears)** or by visual inspection with acetic acid (VIA). Vaccination is the most effective and economical medical intervention for the prevention of infectious diseases, and thus offers the opportunity to prevent cancers caused by infectious agents. It was vaccination that eliminated the once dreaded scourge of smallpox and led to dramatic reductions in the incidence of other infections (Figure 12.7).

In a disease caused by a sexually transmitted virus, one appealing strategy is to vaccinate target populations prior to the start of sexual activity. The development and clinical implementation of HPV vaccines are further described in the case below. Although this case differs from most of the other targeted agents described in this text that aim to disable cancer cells, the molecular approach to development of HPV vaccines may inspire other efforts aimed at *prevention*, rather than merely *control*, of a patient's malignancy.

Epidemiology and etiology

In the United States, 13,240 women will have been diagnosed with cervical cancer in 2018, with 4170 deaths attributed to the disease. Globally, cervical cancer causes 5% of cancer deaths in women and is the fourth most common cause of death from cancer in this group—with breast cancer being the most common cause.

Like breast cancer, cervical cancer has been known for thousands of years, and both diseases are mentioned on Egyptian papyri. The association between sexual intercourse and cervical cancer was recognized by the mid-nineteenth century, when the Italian physician Rigoni-Stern examined death certificates of widows, married women, female sex workers, virgins, and nuns. He found that this cancer was extremely rare in virgins and nuns, and in 1842, he concluded that there was a relationship between cervical cancer and sexual contact. Ultimately, the realization that an infectious agent was involved led to the identification in the early 1980s of HPV as the causal agent of almost all cervical cancer. Subsequent research led to the important discovery that only 15 "high-risk" types of the more than 150 types of HPV are implicated

Figure 12.8. Time course of cervical cancer development after HPV infection. (Adapted from M. Schiffman and P.E. Castle, *N. Engl. J. Med.* 353:2101–2104, 2005. With permission from the Massachusetts Medical Society.)

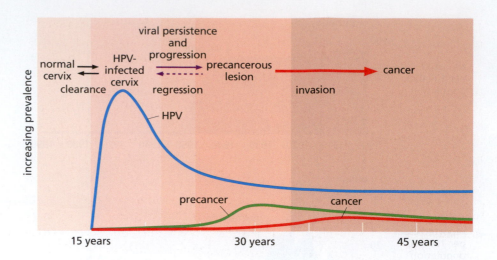

in the causation of cervical cancer. The major causal types are HPV-16 and HPV-18. The transmission of HPV is thought to depend on the tissue erosion and microtrauma of the overlying genital epithelia during sexual intercourse, allowing HPV access to the basal epithelial layer, and thereby enabling infection of the latter.

Cervical cancer arises in the epithelium of the cervix, which is the "neck" of the uterus that joins it to the uppermost portion of the vagina. Of the two most commonly found subtypes—squamous cell carcinoma and adenocarcinoma—squamous cell carcinoma is by far the most frequently occurring. The development of cervical cancer is very slow, requiring at least 10 years, and typically much longer (Figure 12.8).

Diagnosis, workup, and staging

The early stages of cervical cancer are asymptomatic, and when symptoms do eventually appear the most common ones are vaginal bleeding unrelated to menstruation, and, in some cases, pain associated with sexual intercourse (**dyspareunia**) (Table 12.4). Fortunately, unlike most cancers, reliable methods for early detection of cervical cancer at the precancerous stage are in routine clinical use. For many years, this has been accomplished by Pap smear screening, a minimally invasive procedure that involves collecting cells from the cervix and performing cytologic examination, looking for evidence of morphologic abnormalities typical of cells that have escaped normal growth controls. Detection of abnormal cells during this examination is followed by cervical biopsy for the definitive diagnosis and staging (Table 12.5) of a precancerous or cancerous state. Pap smear screening has reduced the incidence of invasive cervical cancer by as much as 80% in countries of the developed world. Introduction of the Pap smear led to a dramatic reduction in the incidence of invasive cervical cancer in the Western world, making the Pap smear one of the most successful public health innovations in the history of medicine. Unfortunately, invasive cervical cancer continues to be a significant cause of cancer morbidity and mortality in the developing world, where there is often limited access to Pap smears due to the scarcity of trained gynecologists and the lack of required cytological preparations and pathologic interpretations.

Visual inspection with acetic acid (VIA) is a simple, inexpensive test that can be performed by a health care worker and does not require a gynecologist or

Table 12.4. Symptoms of cervical cancer.

Presenting symptom
Abnormal vaginal bleeding
Dyspareunia (pain during sexual intercourse)
Pelvic pain
Vaginal discharge

special equipment for pathologic interpretation of cytology specimens. A dilute acetic acid solution (vinegar) is applied to the cervix by means of a swab. Abnormal cervical tissue temporarily turns white on routine visual inspection. Women with abnormal screening examinations can then be referred for further workup. A cluster randomized trial in Mumbai, India, where screening is not widely available, found a remarkable 31% reduction in cervical cancer mortality among women living in villages that had been randomized to VIA screening compared with those living in villages that received no VIA screening. This outcome has broad public health implications for women in developing countries.

Despite the success of screening by Pap smear or VIA, neither of these approaches is able to identify a significant number of women who will develop invasive cervical cancer. This is due in part to the subjectivity and relatively low sensitivity of these cytology-based approaches for detection of high-grade cervical cancer precursors. Since HPV genotypes associated with a high risk of cervical cancer are present and detectable by a PCR-based test, molecular testing for high-risk HPV genotypes has received FDA approval and in some settings may replace the Pap smear as first-line screening for cervical cancer. Women who test positively for high-risk HPV are referred for cervical cytology and possible biopsy.

Patients with abnormal cytology on Pap smear (or an abnormal VIA) must be evaluated by colposcopy, which provides an illuminated and magnified view of the cervix, together with directed biopsies. Application of a 3% acetic acid solution can enhance the vascular patterns that are seen in dysplasia (abnormal cell growth) or carcinoma. Patients with invasive cervical carcinoma should undergo directed physical examinations for palpable abnormalities, blood tests (including kidney tests, due to the possibility of a bulky tumor that can block and hence alter kidney function), and imaging. Pelvic imaging with computerized tomography (CT), magnetic resonance imaging (MRI), or combined positron-emission tomography and CT (PET-CT) are often performed, as well as lung imaging with a chest X-ray or CT scan to evaluate for distant metastasis. Surgical staging, in which the extent of disease is directly evaluated via surgical exploration and removal of suspicious lymph nodes, remains an option.

Cell biology of HPV

HPV is a small DNA virus composed of an 8-kb circular strand of DNA encased in a shell made up of two proteins, namely L1 (~55–60 kDa), the major **capsid protein**, and L2 (55–60 kDa). More than 150 types of HPV have been identified and sequenced. Among HPV types, viral genome sequences differ from one another by up to 1%, and sequences encoding L1 can differ by up to 30%. This degree of antigenic variation among HPV types necessitates the inclusion of the L1 protein from each HPV type in order to design a vaccine with broad protective immunity. All HPV types that have been studied to date replicate in the basal layer of the squamous epithelium, and about one-third of these infect the genital tract. Of those that infect the genital tract, 15 types can cause cervical cancer and are classified as high risk. The HPV genome encodes the structural proteins L1 and L2 as well as several small proteins that function inside infected and transformed cells to regulate the expression of the viral genome and manipulate the host cell biology to facilitate viral replication. Because their small genomes do not encode polymerases or other proteins essential for the synthesis of DNA, these viruses are completely dependent on the host cell for viral replication. Extensive studies of purified L1 found that, under appropriate conditions, it can assemble

Table 12.5. Staging of cervical cancer.

Stage	Description
0	Cervical intraepithelial neoplasia (CIN)
I	Tumor is confined to the cervix
II	Tumor extends beyond the cervix but not to the pelvic wall or the lower third of the vagina
IIIa	Tumor involves the lower third of the vagina without extension to the pelvic wall
IIIb	Tumor extends to the pelvic sidewall and/or impairs function of the kidney on the same side
IVa	Tumor invades the bladder or rectum
IVb	Tumor has metastasized to distant organs

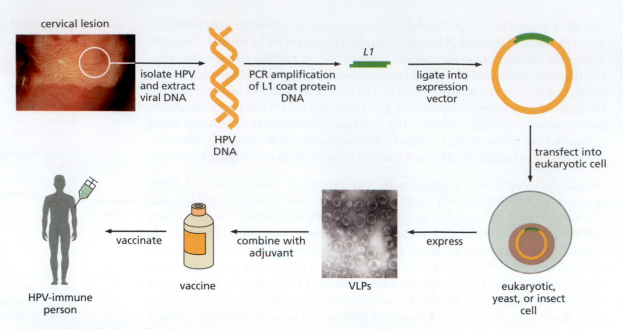

Figure 12.9. Production of anti-HPV vaccine. DNA from HPV virions was extracted and the sequence encoding the L1 protein was amplified by PCR using L1-specific primers, ligated into an expression vector, and transfected into cells for protein production. VLPs generated from the purified protein are mixed with a suitable adjuvant (a substance that enhances the immune response) to formulate a vaccine that is subsequently injected into humans. (Adapted from I.H. Frazer, G.R. Leggatt and S.R. Mattarollo, *Annu. Rev. Immunol.* 29:111–138, 2011. With permission from Annual Reviews.)

into **virus-like particles** (VLPs) that induce the production of neutralizing anti-HPV antibodies when injected into experimental animals or humans (Figure 12.9). These observations and the ability to produce large quantities of type-specific L1 proteins by recombinant DNA technology provide the rationale and technology for the manufacture of anti-HPV vaccines containing VLPs as the immunizing agent.

The carcinogenicity of HPV is a consequence of the life cycle of HPV, which is different from that of most viruses. Usually a virus infects a target cell and replicates in that same cell. Although HPV infects cells of the basal layer, it replicates only after these cells have divided to produce more differentiated daughter cells. Normally, even though the basal cells are capable of proliferation, the differentiated daughter cells are nondividing and would not be able to support HPV replication, because they do not enter S phase and therefore do not support either cellular or viral DNA synthesis. However, small proteins encoded by the HPV genome, most notably **E6** and **E7**, collaborate to prevent the infected differentiated cells that develop from initially infected basal cells from leaving the active growth-and-division cell cycle, and thus prevent them from undergoing cell death. E6 and E7 accomplish this by targeting two key tumor suppressor molecules, namely **pRB** and **p53**, interrupting the pathways regulated by these molecules (Figure 12.10). The sequestration of **E2F family** members necessary for passage of cells from G1 phase into S phase of the cell cycle is a major mechanism by which pRB regulates this critical step to proliferation. By causing the proteolysis of pRB via the ubiquitin-dependent proteasome pathway, E7 causes the release of E2Fs, entry into S phase, and cell proliferation. These actions of E7 result in increased levels of p53, which, if left undisturbed, would

Figure 12.10. HPV oncoproteins E6 and E7 collaborate to effect transformation. (Adapted from C.A. Moody and L.A. Laimins, *Nat. Rev. Cancer* 10:550–560, 2010. Reprinted by permission from Macmillan Publishers.)

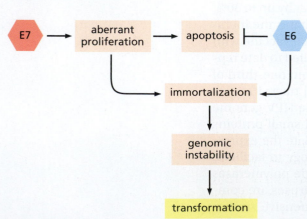

decrease proliferation and predispose the cells to apoptosis. However, E6 recruits other cellular proteins to bring about the degradation of p53 by the ubiquitin-dependent proteasome pathway. Through these and other mechanisms, E6 and E7 conspire to cause cells to enter the cell cycle, to remain in the cycle, to avoid apoptosis, and to suffer genetic instability that results in DNA damage and chromosomal aberrations. Typically, the HPV genome is initially carried in the cell as an extrachromosomal **episome**. In a small proportion of infected cells, much of the viral genome integrates into the host-cell genome, where it becomes an integral part of the latter, continuing to express the E6 and E7 oncoproteins. This integration ensures that the descendants of a cell that acquired this critical portion of the HPV genome continue to express the two key viral oncogenes—E6 and E7.

Note that the concerted and continuing action of these oncogenic proteins directly installs such hallmarks of cancer as sustaining proliferative signaling, resistance to cell death, replicative immortality (by induction of telomerase), and genome instability in members of the HPV-infected cell population. This sets the stage for the eventual acquisition of additional hallmarks of cancer, leading over the course of decades to the eventual development of cancer in approximately 1% of individuals who are infected with high-risk types of HPV.

Management of localized cervical cancer

In the earliest stages of cervical cancer, including carcinoma *in situ*, neoplastic cells are contained within the surface layer of the cervix and have not penetrated the basement membrane and entered the underlying tissues. Such early-stage disease can be treated successfully with curative outcome by surgical removal or other methods of ablation.

Larger tumors or those cancers with evidence of blood vessel or lymphatic channel invasion observed upon biopsy are managed with more extensive surgeries, from **cervicectomy** (also known as trachelectomy, or removal of the uterine cervix) to **hysterectomy** (removal of the entire uterus). Some patients also undergo removal of the lymph nodes in the pelvis that are the initial "landing spots" for cervical cancer that has spread beyond the uterus. An alternative, curative approach consists of concurrent administration of radiation therapy to the cervix and pelvis along with systemic chemotherapy. In this case, chemotherapy can render the cancer cells more susceptible to death by radiation ("radiation-sensitizing doses"), and typically require lower doses than are needed to kill cancer cells with chemotherapy alone ("cytotoxic doses").

Management of advanced cervical cancer

When cancer has spread beyond the cervix, either by invading nearby structures or by dissemination to distant organs and tissues (metastasis), successful therapy becomes increasingly difficult and the prognosis becomes increasingly grave. The role of extensive surgery in such patients is typically limited. Although combined chemotherapy and radiation may be appropriate for a fraction of patients with locally advanced (i.e., nonmetastatic) disease, systemic chemotherapy is the standard treatment. Chemotherapy combinations such as cisplatin with paclitaxel, or carboplatin with paclitaxel, are typically employed. To date, targeted agents have not changed contemporary practice with regard to advanced cervical cancer patients.

The case of Pamela Johnson, a 38-year-old woman with abnormal Pap smear findings

Pamela Johnson is a mathematics teacher at a suburban middle school near her home, where she lives with her husband, 11-year-old daughter, and 9-year-old son. During her annual physical examination, she had a routine Pap smear that showed evidence of cell dysplasia, which was classified as cervical intraepithelial neoplasia, grade II. The doctor advised her to undergo a loop electrosurgical excision procedure (LEEP), and explained that this procedure involved the use of an electrically heated wire to remove the affected area of the cervix. Pamela was told that she would be given a local anesthetic and would remain fully awake during the procedure, which would take less than half an hour, after which she could go home. She was told that this routine outpatient procedure would be expected to remove the precancerous lesion, thus preventing its possible progression to cancer.

When asked if she had any questions, Pamela was especially keen to be given advice about having her daughter vaccinated against HPV. In particular, she hoped that such a vaccination might protect her daughter from cervical cancer and spare her the need to have Pap smears. She wondered why parents were being encouraged to have their daughters vaccinated at such a young age, because she had never heard of a preteen with cervical cancer. Pamela also commented that someone in her book club had heard that some doctors were advising that boys should receive the vaccine, too, and she wondered if this was actually true or just a misunderstanding. In addition, she had heard that the vaccine contained virus-like particles and she was concerned that some children might develop infection from the vaccine. Pamela's questions and concerns are addressed in the following discussion of the first successful clinical trial of an HPV vaccine.

HPV vaccination and prevention of cervical cancer

When laboratory and clinical studies demonstrated that immunization of animals or humans with purified viral structural proteins (i.e., lacking viral DNA) generated an immune response that can prevent HPV infection, the groundwork was laid for the development of HPV vaccines. The discovery that the L1 capsid protein of HPV undergoes self-assembly to produce VLPs provided the approach to generating candidate vaccines (see Figure 12.9). Because VLPs lack viral nucleic acid, they are incapable of causing infection, but they offer the advantage of inducing protective immune responses because they display antigens in a fashion that is apparently identical to the way that these antigens are displayed by intact virus particles.

Large clinical trials, conducted by Merck and by GlaxoSmithKline, tested whether vaccination against oncogenic strains of HPV prior to infection would prevent infection and cancer of the cervix. Each company developed a vaccine that received approval from the FDA. Merck's vaccine, **Gardasil**, employs VLPs composed of L1 capsid proteins from four different types (6, 11, 16, and 18) of HPV, and is therefore termed a quadrivalent vaccine. The GlaxoSmithKline vaccine, **Cervarix**, deploys VLPs composed of capsid proteins from the two HPV types (16 and 18) that are known to be responsible for 70% of HPV-induced cancers, and is termed a bivalent vaccine. Here we discuss the clinical trial of the quadrivalent vaccine Gardasil. It should be noted that the HPV types 6 and 11 do not cause cervical cancer, but cause anal and genital warts, and are thus included in the vaccine to prevent these sexually transmitted diseases.

The clinical trial was set up to test the following hypothesis: "Compared with placebo, the vaccine would reduce the incidence of high-grade cervical intraepithelial neoplasia related to HPV-16 or HPV-18." Note that the parameter for determining vaccine efficacy—the clinical endpoint—was high-grade **cervical intraepithelial neoplasia** (CIN) containing DNA from HPV-16, HPV-18, or both. High-grade CIN, a precancerous lesion, rather than cervical cancer, was used as the endpoint because it would have taken many additional years to answer definitively the important question about carcinoma prevention, given the long period between the appearance of CIN and the development of invasive cancer (see Figure 12.8). Based on annual frequencies of 0.19% and 0.038% for HPV-16- and HPV-18-associated endpoints, respectively, in the general population, 12,167 women between the ages of 15 and 26 years were recruited as candidates for the study. To ensure genetic and cultural diversity, this population was recruited at 90 different sites in 13 countries, and included participants from Europe (65%), Latin America (26%), North America (7.5%), and the Asia-Pacific region (1.5%).

In this trial, the success of the vaccination was determined by its ability to prevent the development of HPV-16- or HPV-18-related precancerous lesions. This could only be established in a test population that (a) was not infected by either HPV-16 or HPV-18 at the time of enrollment, (b) did not already have cervical cancer, and (c) did not already show indications of developing cervical cancer.

To identify such individuals, following **randomization** (their random assignment to either a population that would receive vaccine or a control population that would not), a complete medical history was taken, each candidate was given a gynecologic examination, and a Pap smear was performed. Swabs of the anogenital areas, including the labia, vulva, and perianal regions, were also taken, and all of these samples were tested for HPV types 6, 11, 16, and 18, using sensitive and specific polymerase chain reaction (PCR) analyses. In addition, since infection with HPV usually provokes an immune response that generates circulating antibodies against infecting viruses, serologic tests of blood samples from the study population were tested for antibodies specific for HPV-16 or HPV-18.

Subjects were vaccinated with either the quadrivalent vaccine made up of HPV VLPs (see Figure 12.9) mixed with aluminum hydroxyl-phosphate, an **adjuvant** (an agent that increases the level of immune responses by recruiting innate immune responses), or sham-vaccinated with a visually indistinguishable formulation containing only the aluminum hydroxyl-phosphate. Vaccinations were administered on day 1, at 2 months, and at 6 months. Relatively few side effects were reported, the most common one being pain at the injection site. Serologic studies showed that the vaccine produced an immune response in most members of the treated groups. The vaccine showed an efficacy of 44% in preventing development of CIN2/3 or cancer that was positive for HPV-16 or HPV-18 (Figure 12.11). However, when the analysis was performed including all high-grade cervical lesions, regardless of HPV infection status or the HPV type found in the precancerous or cancerous lesion—in effect, a sample equivalent to an unselected general population—the efficacy was 17%. In

Figure 12.11. Efficacy of the quadrivalent anti-HPV vaccine. In a study of subjects without HPV-16 or HPV-18 infection at the time of vaccination, the vaccine decreased development of high-grade cervical lesions known to be precursors to cervical cancer by 44%. (From FUTURE II Study Group, *N. Engl. J. Med.* 356:1915–1927, 2007. With permission from the Massachusetts Medical Society.)

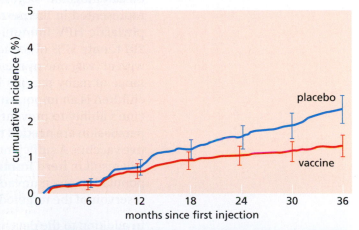

number at risk							
vaccine	6087	5918	5824	5733	5592	5427	2925
placebo	6080	5942	5832	5736	5586	5420	2994

cumulative number of subjects with an endpoint							
vaccine	0	13	36	53	61	67	76
placebo	0	17	43	72	91	109	132

terms of its effect on public health, if this anti-HPV vaccination was universally practiced, a 17% efficacy would prevent 45,900 of the 270,000 cervical cancer deaths that occur each year, worldwide.

The data from this trial led to FDA approval of Gardasil for vaccination of women between the ages of 13 and 26 years to prevent cervical cancer. Subsequently, data from a trial of Cervarix demonstrated similar efficacy, and this vaccine was granted FDA approval for vaccination of girls and women between the ages of 10 and 25 years, based on the ages of subjects enrolled in the clinical trial. Vaccination with either vaccine was associated with few adverse reactions, at low rates similar to those of other alum-adjuvant-containing vaccines. Since the approval of these vaccines for prevention of infection of women by HPV-16 and HPV-18, studies have shown that vaccination against HPV-16 and HPV-18 also protects against anal intraepithelial neoplasia (AIN), a precursor of anal cancer in both men and women. The quadrivalent vaccine, which also protects against infection by HPV-6 and HPV-11, has been approved for prevention of **anal condyloma** and for prevention of **genital warts** in males aged 9 to 26 years.

Future prospects and challenges in HPV vaccines

The HPV vaccines that were initially introduced contained types 16 and 18, which are responsible for approximately 70% of cervical cancer. Since its introduction, the repertoire of Gardasil has been expanded from four to nine HPV types. The FDA-approved nonavalent Gardasil-9 includes types 31, 33, 45, 52, and 58 in addition to the types 6, 11, 16, and 18 present in the original formulation. The inclusion of the additional five HPV types increases the coverage by 15–20%, and provides protection against up to 90% of cervical cancer. However, because existing vaccines do not include all of the HPV types responsible for the remaining 10–15% of cervical cancers (i.e., those not caused by HPV 16, 18, 31, 33, 45, 52, or 58), even those who have been immunized with an HPV vaccine should continue to undergo periodic screening for cervical cancer. Ultimately, perhaps future prophylactic vaccines will include all of the types of HPV that are known to be carcinogenic. Beyond broadening the spectrum of high-risk HPV types represented in the next generation of vaccines, the effectiveness of a prophylactic HPV immunization program depends on vaccination rates. By 2014, only 57% of teenage girls in the United States had been vaccinated with at least one of the recommended full course of three doses. The resistance of many parents in the United States to vaccinating their early teen children is an important challenge to realizing the full potential of the vaccine's ability to prevent cervical cancer. Better programs of education and persuasion are needed to overcome this behavioral barrier to a highly effective vaccine. A sufficient rise in the percentage of the population vaccinated will result in **herd immunity**, which is the protection of all members of the population, even unvaccinated ones, by reduction of the transmission and reservoir of the targeted pathogen.

In addition to the development of prophylactic vaccines with broader coverage, research is under way to develop therapeutic HPV vaccines that could be used to treat already established HPV-induced intraepithelial neoplasias and cancers. There have been encouraging clinical trials using vaccines containing peptides based on the sequences of E6 and E7 proteins from HPV-16 and HPV-18, the highest-risk HPV strains. These HPV-encoded proteins are particularly attractive targets because they are tumor-specific antigens that are not present in uninfected host cells.

Take-home points

✓ Virtually all cervical cancer is caused by sexual transmission of one of 15 carcinogenic types of human papillomavirus (HPV), with 70% of cases caused by either HPV-16 or HPV-18.

✓ HPV forces cell proliferation by encoding and expressing two proteins, E6 and E7, that inactivate p53 and pRB, two key tumor suppressors, thereby greatly increasing the probability that the cell will progress to a cancerous state.

✓ Development of anti-HPV vaccines to prevent cervical cancer confirmed the paradigm that cancers caused by infectious agents may be prevented by prophylactic vaccination.

✓ Testing prophylactic vaccines requires study populations that do not manifest the condition.

✓ The size of the population required to attain a statistically reliable estimate of vaccine effectiveness depends upon the frequency with which infection-dependent endpoints occur in the population.

✓ The anti-HPV vaccines Gardasil and Cervarix are approved by the FDA and are effective if given prior to onset of sexual activity, when patients may become exposed to HPV-16 or HPV-18.

✓ Since 10–15% of cervical cancers are due to infection by HPV types that are not included in the vaccine, even those who receive the vaccine should undergo periodic screening for cervical cancer.

Discussion questions

1. Identify and respond to the questions and concerns raised by Pamela Johnson during her conversation with the doctor.

2. What is HPV and what diseases does it cause? Which of these diseases have been shown to be prevented by vaccination?

3. How does the existence of different HPV types complicate the development of a vaccine that prevents cervical cancer?

4. Was the vaccine tested in trials actually shown to prevent cervical cancer? If not, why not?

5. Discuss the cellular and molecular mechanisms by which HPV causes the development of cervical cancer.

Case 12-2 | **Acute Lymphoblastic Leukemia**

Topics bearing on this case

Acute lymphoblastic leukemia (ALL)

Allogeneic hematopoietic stem cell transplantation (AHSCT)

Graft-versus-host disease (GVHD)

CAR T-cell therapy

Introduction and epidemiology

Leukemia is a general term for cancers derived from white blood cells (WBCs) that circulate in the bloodstream and bone marrow, although WBCs may also involve other parts of the body. White blood cells include both myeloid and lymphoid cells, either of which may yield leukemia (myelogenous or lymphocytic leukemias). Leukemias are classified as chronic or acute, based on their clinical behavior and natural history. Discussions of chronic myelogenous leukemia (CML) and chronic lymphocytic leukemia (CLL) can be found in Cases 4-1 and 4-2, respectively. A discussion of acute promyelocytic leukemia, a form of acute myelogenous leukemia (AML), can be found in Case 7-1. Acute lymphoblastic leukemia (ALL) is a highly aggressive form of acute leukemia that represents the most common cancer among children, and occurs less commonly among adults. Approximately 70–80% of ALL cases arise from B cells. Adolescent males may develop a distinct subtype of T-cell ALL known as T-cell lymphoblastic lymphoma, which typically presents as large mediastinal masses.

In the United States, 2500–3500 new cases of childhood ALL are diagnosed each year, representing an incidence of 2.8 cases per year per 100,000 members of the population, which accounts for approximately 30% of all pediatric cancers. The highest incidence is seen between the ages of 2 and 5 years, and this cancer is more common in boys than in girls. The frequency among Caucasians is twice that seen in black children, with the lowest incidence being found in children of Asian descent and the highest in Hispanic whites. Although frequencies of incidence differ among populations, in most cases ALL is not a familial cancer caused by inherited mutations in a susceptibility gene, and occurs sporadically.

Diagnosis, workup, and staging

In most cases, ALL presents with nonspecific symptoms, such as fever or infection, bleeding, bone pain, or lymphadenopathy (enlarged lymph nodes). Laboratory testing will virtually always reveal significant abnormalities, with either decreased WBCs, increased WBCs, decreased platelets (thrombocytopenia), or anemia (low red blood cells). Examination of the blood may show immature lymphocytes. Physical examination may reveal **pallor** (pale appearance) in the setting of anemia, or **petechiae** (tiny red spots on the skin from capillary hemorrhage) or **purpura** (larger purple spots) related to bleeding in the setting of thrombocytopenia. Patients may have lymphadenopathy or splenomegaly palpable on examination due to buildup of malignant B or T cells in those lymphoid organs. Headache, lethargy, double vision, or seizures may indicate central nervous system (CNS) involvement, which is a dangerous complication of ALL seen in approximately 5% of patients at diagnosis.

If ALL is suspected based on clinical and laboratory features, patients undergo detailed examination of the blood and bone marrow, which includes immunohistochemistry, flow cytometry, and cytogenetics. Figure 12.12 indicates the morphologic features of ALL. Given the increased risk of CNS involvement in this disease, all patients with ALL undergo a spinal tap (also known as **lumbar puncture**) to study the **cerebrospinal fluid** (which bathes the brain and spinal cord) for evidence of leukemic cells.

Leukemic cells in ALL are much larger than normal lymphocytes, and have abundant cytoplasm, dispersed chromatin, and prominent nucleoli, giving them a typical blast (immature) appearance. Examination of surface markers by means of flow cytometry or immunohistochemistry shows B-cell ALL to express the common ALL antigen CD10 as well as the immature lymphoid marker **TdT** and the B-cell marker CD19. T-cell ALL or lymphoblastic lymphoma will usually express TdT as well as typical T-cell antigens CD2, CD3, CD5, CD7, and either CD4 or CD8.

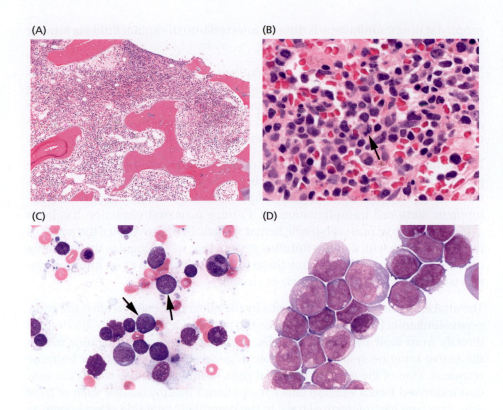

(A)

(B)

(C)

(D)

Figure 12.12. Morphologic features of ALL. (A) Low-power view of a bone marrow core biopsy involving ALL demonstrates increased marrow cellularity in between the bone trabeculae (hematoxylin and eosin). (B) At higher magnification, the cells that make up much of the cellularity are medium to large in size, with dispersed chromatin, consistent with primitive blasts. A mitotic figure is readily visible (*arrow*), indicating that the blasts are actively proliferating (hematoxylin and eosin). (C) Review of the bone marrow aspirate specimen shows numerous blasts (*arrows*) with dispersed chromatin and prominent nucleoli in a background of few normal marrow elements (Wright–Giemsa). (D) Examination of a cerebrospinal fluid specimen obtained by lumbar puncture demonstrates the presence of numerous blasts, indicative of CNS involvement in this case (Wright–Giemsa). Similar-appearing cells to those seen in the bone marrow aspirate and cerebrospinal fluid specimens are also typically seen in the peripheral blood at the time of diagnosis.

Cytogenetic analysis may reveal multiple chromosomal abnormalities in ALL. These abnormalities include variations in chromosome number, deletions, insertions, fusions, and translocations. Genetic findings associated with inferior prognosis in ALL include presence of the t(9;22) BCR-ABL translocation, also known as the Philadelphia chromosome (see Case 4-1), as well as abnormalities of *TP53*, *MLL* translocations, intrachromosomal amplification of chromosome 2p21, and hypodiploidy (fewer than 46 chromosomes). Favorable prognosis has been associated with hyperdiploidy (more than 50 chromosomes) and with the t(12;21) translocation, which results in the fusion of genes *ETV6* and *RUNX1*.

Management of ALL

Leukemia is a cancer of blood cells, and is therefore a systemic malignancy that must be treated with chemotherapy. The goal of treatment is cure, and the majority of pediatric ALL patients are cured of their disease. Treatment of ALL involves a complex multi-drug regimen with a duration of approximately 2 years, which includes three phases, namely induction, consolidation, and maintenance. By the completion of induction therapy, more than 90% of pediatric ALL patients achieve **complete remission** (CR). Consolidation therapy is instituted following induction therapy to eradicate potential **minimal residual disease** (MRD) left behind after induction therapy, and to discourage the emergence of drug-resistant leukemic cells. Typically, this phase employs a cocktail of drugs that differ from those used during induction. Maintenance therapy is the last phase and is less intensive than the previous two phases. Although leukemic involvement of the CNS is uncommon at the time of diagnosis, a high proportion of ALL patients will subsequently develop involvement within the CNS unless preventive therapy is employed. Traditional chemotherapy drugs administered intravenously have poor penetration into the CNS due to the protective effect of the blood–brain barrier, so prophylactic chemotherapy is directly injected into the spinal fluid via lumbar puncture during the course of therapy.

Management during treatment also requires attention to a variety of potential side effects. These include bleeding, thrombosis, **tumor lysis syndrome** (TLS), and infection. Because these events can be life-threatening, pediatric ALL patients should ideally be treated at high-volume cancer centers where there is expertise with identification and treatment of these potential toxicities.

Even after a CR is achieved, 15–20% of children will subsequently relapse. The goal for these patients continues to be cure of the disease, and involves reinduction therapy with intensive chemotherapy, often followed by allogeneic hematopoietic stem cell transplantation (SCT) once a second remission has been achieved. The goal of an allogeneic hematopoietic SCT is to replace the patient's immune system with a donor immune system that will recognize the leukemia cells as "foreign" and thus destroy those foreign cells, much as an immune system would attack invading bacteria or viruses.

The attack on the patient's cancer cells by the allogeneic cells is called the graft-versus-leukemia (GVL) effect. Because the patient's leukemia cells are derived directly from their own immune cells, the phenomenon of tolerance makes the native immune system less capable of generating an anti-cancer immune response. One of the greatest potential risks of this procedure is that the new donor-derived T cells will also attack the patient's healthy tissues, such as their skin, liver, and gastrointestinal tract, in the potentially fatal side effect known as graft-versus-host disease (GVHD). The risk of developing severe GVHD can be minimized by careful matching of the **human leukocyte antigen** (HLA) type of the donor to the recipient. All patients also require immunosuppressive medication in order to prevent GVHD, as well as to prevent rejection of the graft. Such immunosuppressive therapy results in an increased risk of potentially serious infections in the transplant recipient. Consequently, allogeneic hematopoietic SCT is considered to be a high-risk procedure with a significant rate of fatal outcomes, and therefore it is only performed when no other curative option is available.

The case of Roger Gonzales, a 5-year-old boy with fatigue, pallor, and bruising

Roger's mother became concerned when she observed that her usually rambunctious 5-year-old son had recently appeared to be fatiguing easily. Roger complained of feeling tired, and his mother also noticed that he appeared pale and had developed multiple bruises after playing with his friends. His mother, a nurse, brought him to his pediatrician, who noted the pallor, as well as bruising and petechiae of the skin. The pediatrician sent an urgent complete blood count (CBC), which showed an elevated WBC of 51,000/μl (normal range, 4500–11,000 cells/μl), with the majority of WBCs interpreted as immature cells, or blasts. The platelet count was markedly decreased at 12,000/μl (normal range, 250,000–450,000/μl), and Roger was anemic with a hemoglobin concentration of 6.5 g/dl (normal range, 13.5–17.5 g/dl). His bruising and petechiae were caused by the decreased platelet count, and his increasing fatigue was probably related to anemia. On closer questioning, Roger also reported pain in his bones, reflecting the increasing volume of leukemia cells packing the bone marrow cavities inside his bones. Roger's pediatrician urgently admitted him to the hospital with a presumed diagnosis of acute leukemia.

A pediatric oncologist was quickly consulted, and analyses of the circulating blood cells as well as a bone marrow biopsy confirmed a diagnosis of B-cell ALL. The t(9;22) translocation was not detected on cytogenetic analysis. Roger began induction therapy, which included the chemotherapy drugs vincristine, dexamethasone, L-asparaginase, and daunorubicin. A bone-marrow biopsy performed on day 15 of induction chemotherapy showed a decrease in lymphoblasts, but there was still persistent disease. A repeat bone-marrow biopsy at the conclusion of induction therapy showed complete remission.

Unfortunately, 9 months later, Roger developed fatigue, easy bruising, and fevers, and was found to have recurrent B-cell ALL in the blood and bone marrow. He received an initial reinduction cycle of vincristine, prednisone, pegylated (modified with polyethylene glycol to reduce clearance) asparaginase, and doxorubicin, as well as intrathecal chemotherapy injected into the CSF. A bone-marrow biopsy at the end of the reinduction cycle showed persistent ALL, denoting a reinduction failure. At this point, Roger's care team recommended participation in a clinical trial of genetically engineered T-lymphocytes targeted against his ALL cells, which had shown great promise in the initially treated patients, including those with chemotherapy-refractory disease. Roger's mother gave **informed consent** for him to participate in the clinical trial, and Roger also provided **assent** to participate. His own T cells were then collected from his bloodstream and were genetically modified using a viral vector to express chimeric antigen receptors (CARs) for CD19, a B-cell antigen on the surface of all B-ALL cells, and included the co-stimulatory domain 4-1BB. The cells were then expanded *ex vivo* to produce a sufficient quantity of anti-CD19-targeted T cells for reinfusion. Roger then received additional chemotherapy in preparation for CAR T-cell reinfusion, after which his modified CAR T cells were reinfused through a peripheral vein. There were no immediate infusion-related toxic effects. However, starting 5 days after the infusion, Roger developed high fever (Figure 12.13A) and confusion, which necessitated his transfer to the pediatric intensive care unit, where respiratory and cardiovascular compromise required use of a mechanical ventilator and blood pressure support. Laboratory findings included very high levels of inflammatory cytokines (Figure 12.13B) and cytokine receptors, as well as elevated ferritin (a protein that stores and releases iron), hepatic aminotransferases (liver enzymes used to monitor liver function), and triglycerides (fats). Treatment with **tocilizumab**, a monoclonal antibody that binds IL-6, a cytokine that plays a role in inflammatory and immune responses, was initiated.

Roger's fevers began to reduce and there was a decline in the levels of inflammatory cytokines, such as IL-6 and interferon gamma, consistent with a rapid response to the medication. There was a sharp rise in the WBC count, which

Figure 12.13. Fever and cytokine levels during CAR T-cell therapy. (A) Time course of Roger's temperature during the first 16 days post–CAR T-cell therapy. (B) Levels of Roger's inflammatory cytokines during the first 23 days post CAR T-cell therapy, expressed as fold change from baseline. (Adapted from S.A. Grupp et al., *N. Engl. J. Med.* 368:1509–1518, 2013.)

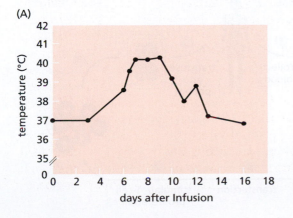

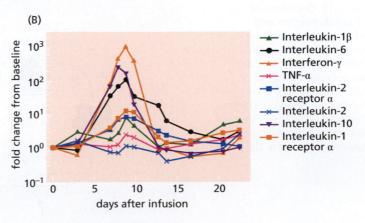

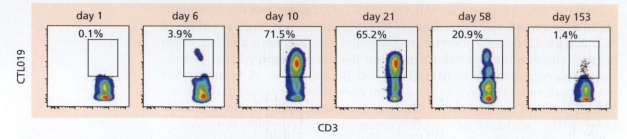

Figure 12.14. Level of CAR T cells during the first 153 days post–CAR T-cell infusion. Antibodies specific for the CAR (CTL019) and for CD3, a marker of the T-cell receptor complex, were used in flow cytometric analysis to determine the percentage of T cells bearing the CAR present in blood. Peak levels occurred between days 10 and 21, and then decreased. Note that at 153 days post infusion, CAR T cells are still present in the circulation. (Adapted from S.A. Grupp et al., *N. Engl. J. Med.* 368:1509–1518, 2013.)

consisted of 70% CAR T cells, reflecting the rapid *in vivo* expansion of the CAR T-cell population (Figure 12.14). A repeat bone marrow biopsy performed 2 weeks later showed a marked decline in malignant lymphoblasts. At 30 days post-infusion, Roger was in CR in the blood and bone marrow, including no evidence of MRD by flow cytometry or PCR for clonal immunoglobulin heavy-chain (IgH) rearrangement. At 6 months post-CAR T-cell infusion, samples of blood and bone marrow aspirates continued to show no evidence of disease. Roger has remained leukemia-free during the 1.5 years since his CAR T-cell-induced remission. He continues to have no detectable healthy B lymphocytes either, and has significantly decreased levels of circulating immunoglobulins, which require monthly infusions of **intravenous immune globulin** (IVIG) in order to prevent infections. Roger's energy levels are now fully restored, and his soccer coach predicts that he could be a future Olympian.

Treatment with CAR T cells

CAR T-cell therapy involves four phases: (1) genetic modification and *in vitro* expansion of the patient's T cells; (2) preparation of the patient and infusion of the modified T-cell population; (3) evaluation of the effects of CAR T-cell therapy; and (4) monitoring and management of the systemic effects of CAR T-cell therapy. These phases will now be discussed in turn, and an overview of CAR T-cell manufacture is presented in Figure 12.15.

Genetic modification and *in vitro* expansion of the patient's T cells. The patient's leukocytes are collected by **apheresis** and can be frozen for further processing and use in the future. Prior to modification, the T-cell fraction is

Figure 12.15. Overview of CAR T-cell derivation and transfusion. White blood cells are harvested from the patient, and T cells are purified and then transfected with a lentivirus or retrovirus vector, which inserts the desired chimeric antigen receptor into the T-cell DNA. The engineered T cells are expanded *in vitro* and the cells are then returned to the patient.

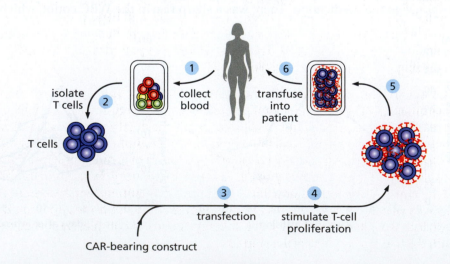

purified and then transduced with retroviral or **lentiviral vectors** containing genes for the expression of the CAR, and also includes co-stimulatory domains essential for full T-cell activation once antigen is bound. The engineered cells are then stimulated to proliferate by *in vitro* cultivation, with cytokines under conditions mimicking antigen stimulation.

Preparation of the patient and infusion of the modified T-cell population. Engraftment and proliferation of adoptively transferred T cells is greatly facilitated by lymphodepleting chemotherapy, which provides the necessary immune suppression to prevent the modified T cells from being rejected in the patient. The CAR T cells are then infused.

Evaluation of the effects of CAR T-cell therapy.   Effective CAR T-cell therapy kills both normal and neoplastic cells bearing the surface antigen. In the case presented above, the target antigen is CD19, an antigen found on virtually all B cells. The most important biomarker of the effectiveness of anti-CD19 CAR T cells is the elimination of CD19-positive cells from blood and bone marrow. The elimination of CD19-positive cells creates systemic **B-cell aplasia** (lack of B cells) and the concomitant inability of the patient to produce immunoglobulins. This condition persists as long as an active CAR T-cell population remains. The patient is monitored for persistence of the CAR T cells, for the presence of ongoing cancer, and for levels of immunoglobulins, which can be raised by intravenous administration of immunoglobulin (IVIG) if needed. The antigen-triggered activation and proliferation of the CAR T cells is highly inflammatory due to the direct and indirect effects of their production and secretion of large amounts of **inflammatory cytokines**, such as interferon gamma, IL-6, and others.

These high levels of inflammatory cytokines have systemic effects that include fevers, hypotension, hypoxia, abnormal blood chemistries, and liver malfunction. In some cases, neurological symptoms, such as disorientation, delirium, and seizures, are caused by the therapy. Most of these effects fall into one or more systemic syndromes, such as tumor lysis syndrome (TLS), **cytokine release syndrome** (CRS), **macrophage activation syndrome** (MAS), and neurotoxicity. TLS results from the lysis of large numbers of tumor cells and the accompanying release of nucleic acids and minerals, such as phosphate and potassium, into the circulation. The catabolism of the nucleic acids generates large amounts of uric acid that can cause severe damage to kidneys, as can the deposition of calcium phosphate crystals in the kidney tubules. High levels of potassium can cause life-threatening cardiac irregularities. CRS is a systemic syndrome caused by high levels of inflammatory cytokines, and represents one of the most common toxicities of CAR T-cell therapy. MAS is an uncommon syndrome characterized by uncontrolled activation and proliferation of macrophages and T cells, with laboratory findings of hyperferritinemia (elevated ferritin levels) and hypertriglyceridemia (elevated levels of triglycerides), and clinical findings of hepatosplenomegaly (enlarged liver and/or spleen), liver dysfunction, and pancytopenia (reduction in the numbers of red and white blood cells).

Management of the systemic effects of CAR T-cell therapy.   Toxicities accompanying the use of this therapy are expected and require close management. The large degree of tumor cell lysis results in electrolyte abnormalities, such as high levels of uric acid, phosphate, and potassium. CRS can be treated with glucocorticoids given at, or soon after, the onset of fevers and symptoms. Levels of the inflammatory cytokine IL-6 can be reduced by administration of the anti-IL-6 receptor monoclonal antibody, tocilizumab, which often produces a striking clinical improvement within less than 24 hours. The neurotoxicity associated with CAR T-cell therapy is usually completely reversible with the use of systemic corticosteroids.

CAR T-cell technology remains under active investigation in order to optimize efficacy and reduce potential toxicities. In addition to B-cell ALL, CAR T cells are also showing promise in other B-cell malignancies, including chronic lymphocytic leukemia (CLL) and B-cell non-Hodgkin lymphoma. The CAR T-cell therapies **tisagenlecleucel** and **axicabtagene ciloleucel** were FDA approved for treatment-resistant pediatric B-cell ALL and diffuse large B-cell lymphomas on August 30, 2017, and October 18, 2017, respectively. CAR T cells directed against other antigens on lymphomas and other cancers are also under development.

Take-home points

✓ Acute lymphoblastic leukemia (ALL) is the most common form of cancer in children, accounting for approximately 30% of pediatric cancers.

✓ The majority of cases of pediatric ALL will be cured with chemotherapy, but up to 20% of patients will relapse after standard therapy.

✓ Definitive diagnosis of ALL is made by examination of the morphology, immunophenotype, and cytogenetics of tumor cells obtained from blood and bone marrow.

✓ Treatment of ALL involves multi-drug chemotherapy administered in three phases over the course of 24–36 months, and includes preventative therapy to reduce the risk of leukemic involvement of the central nervous system (CNS).

✓ Treatment requires management of a variety of side effects, including bleeding, thrombosis, tumor lysis syndrome, and infection.

✓ Allogeneic hematopoietic stem cell transplantation (allo-HSCT) is currently the only therapy with the potential to cure relapsed chemotherapy-resistant pediatric ALL. Although, when it is effective, the graft-versus-leukemia (GVL) effects of allo-HSCT can be curative, the graft-versus-host (GVH) response that accompanies this therapy carries high risks.

✓ Chimeric antigen receptor (CAR) T cells are a highly effective adoptive T-cell immunotherapy available for chemotherapy-resistant pediatric B-cell ALL and adult diffuse large B-cell lymphomas. The side effects that accompany this therapy, particularly cytokine release syndrome and neurotoxicity, can be life-threatening, but are usually reversible with diligent management.

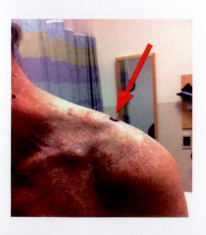

Figure 12.16. Appearance of the left shoulder primary melanoma (*arrow*). (Photo courtesy of Kenneth Tanabe, M.D.)

Discussion questions

1. What findings might lead a primary care physician to suspect acute leukemia, and what recommendation should the doctor make?

2. What findings would provide a definitive diagnosis of B-cell ALL?

3. Comment on the standard course of B-cell ALL therapy and the challenges posed during the course of therapy.

4. Allogeneic HSCT is often employed when other treatments for ALL fail. Why is the powerful approach of HSCT not used at the time of initial diagnosis?

5. What are the management challenges posed by CAR T-cell therapy? How are they currently met?

| Case 12-3 | Metastatic Melanoma |

Topics bearing on this case

Interleukin-2 (IL-2)

CTLA-4 modulation of the cellular immune response

Immune checkpoint inhibitor therapy

Introduction

Melanoma was formerly a disease without many effective therapeutic options in the metastatic setting. Improved molecular understanding of melanoma has led to significant advances in care, as described in Case 4-4. However, long before the advent of kinase inhibitors for this disease, metastatic melanoma was hypothesized to be a sound target for immunotherapies. Advances in immunotherapy coupled with other targeted therapies are providing fresh hope for improved survival of patients with this previously dismal malignancy.

The case of Adam Stephenson, a 66-year-old man with a growing lesion on his left shoulder

Adam Stephenson is a retired landscaper who initially noted a painless, growing bump on his left shoulder (Figure 12.16). The lesion was brown but not uniform in color, rounded, and raised above the level of the surrounding skin. There was no surrounding redness of the skin. Except for the growth, the lesion was not associated with itchiness, pain, or tenderness.

Adam is a white man with brown hair whose skin is not particularly fair. He did not have a history of significant sunburn in childhood, but his work as a landscaper for four decades has included substantial sun exposure. His other medical history includes high cholesterol levels and hypertension. There is no family history of cancer of any kind in his parents or his three siblings.

Adam presented to his primary care physician 3 months after he first noticed the lesion, whereupon he was referred to a dermatologist. In addition to the shoulder lesion, a supraclavicular (above the clavicle, lateral to the neck) lymph node was palpably firm and approximately 1.5 cm in diameter. A shave biopsy of the shoulder lesion was performed, revealing a 1.5-mm-thick, ulcerated melanoma with 14 mitoses/mm^2 (Figure 12.17).

With the possible spread to the supraclavicular lymph node, a PET-CT exam was performed, which revealed intense uptake in the shoulder lesion as well as in two nearby lymph nodes (Figure 12.18). There was no other evidence of spread of the disease.

Because his melanoma was limited to the neck area, Adam underwent a wide local excision and a modified radical neck dissection on the left, 1 month after his presentation. A "neck dissection" is performed to control metastasis to neck lymph nodes, with the aim of removing all lymph nodes to which the cancer may have spread. A "radical neck dissection" removes all lymph nodes on one side of the neck, along with the nearby structures, the spinal accessory nerve, internal jugular vein, and sternocleidomastoid muscle. A "modified radical neck dissection" removes the lymph nodes but spares the nearby, nonlymphatic structures. The pathology revealed a residual invasive melanoma at the shoulder, with ulceration and involvement of seven of the dissected lymph nodes. With extensive involvement of the neck lymph nodes, Adam had a high chance (estimated to be more than 50%) of local recurrence as well as distant spread. He received adjuvant radiation therapy to the neck to reduce the likelihood of recurrence.

At that time of Adam's diagnosis, there was no approved adjuvant systemic therapy. Adam was then monitored for recurrent and distant disease with

Figure 12.17. Pathology of melanoma with mitotic figures in metaphase (*red arrows*). Stain: hematoxylin and eosin. (D. Bogunovic et al., *Proc. Natl. Acad. Sci. U.S.A.* 106:20429–20434, 2009.)

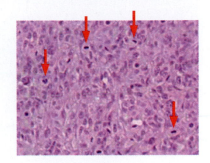

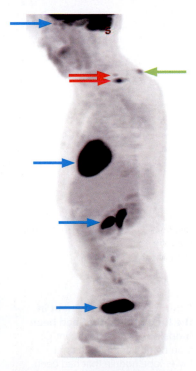

Figure 12.18. PET component of PET-CT scan. Normal physiologic uptake of FDG in the brain, heart, kidneys, and bladder (*blue arrows, top to bottom*). There is abnormal uptake in the primary melanoma and two lymph nodes (*green arrow and red arrows, respectively*).

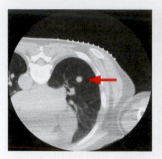
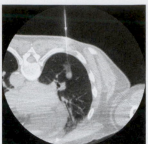
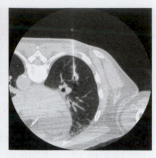

Figure 12.19. CT-guided biopsy of the right lower lobe lung lesion. For orientation purposes, the patient should be imagined to be lying on his stomach; these views are cross-sectional "slices" at the level of the heart, from the vantage point of the patient's feet. Top panel: The arrow indicates the 1.4-cm lung nodule that had grown. Middle and lower panel: The needle is passed under CT guidance to stop at the lung nodule and then sample the tissue.

scans every 3 months. A PET-CT scan performed 6 months after completion of adjuvant radiation therapy revealed new lung lesions in the left upper lobe (1.3 cm in diameter) and right lower lobe (1.4 cm in diameter) that were suspicious for metastatic disease. Adam underwent a CT-guided biopsy of the right lower lobe lesion (Figure 12.19), and pathology confirmed a diagnosis of metastatic melanoma.

The treatment options for metastatic melanoma were discussed by Adam's medical oncologist. Adam's melanoma did not harbor the BRAF mutation (see Chapter 4 for therapies in BRAF-mutant melanoma). Chemotherapy with dacarbazine was briefly discussed but was not recommended. Treatment options were narrowed down to either the monoclonal antibody, ipilimumab, or a clinical trial of the anti-PD-1 antibody, pembrolizumab.

Adam opted for the clinical trial. The trial assigned him to receive 2 mg/kg of pembrolizumab by intravenous infusion every 3 weeks. He tolerated treatment with fatigue, but was able to do all of his normal activities of daily living. After four cycles (12 weeks) of therapy, the lung nodules had decreased in size (Figure 12.20). Adam has had ongoing evidence of response on subsequent imaging, and he has remained on therapy, now for over 2 years.

Immunotherapy and melanoma

Rare cases of spontaneous remission of metastatic melanoma have been described. The mechanism of spontaneous remission is hypothesized to involve an anti-tumor immune response. Numerous immune-based strategies, including cytokines, antibodies, and vaccines, have been developed and tested against metastatic melanoma.

Figure 12.20. Improvement in the lung nodule that had been biopsied. (A) Pre-treatment CT scan demonstrating the right lower lobe nodule that had been biopsied (*red arrow*). For orientation, the patient is rotated through 180 degrees compared with Figure 12.19. (B) CT scan of chest after 12 weeks of pembrolizumab therapy, demonstrating a decrease in the size of the right lower lobe nodule (*green arrow*).

(A)

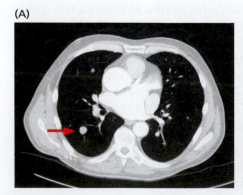

(B)

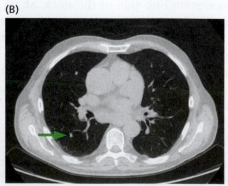

Cytokines.   Cytokines are signaling molecules that are critical in immune responses and stimulation of movement of cells toward sites of inflammation or infection. The cytokine interleukin 2 (IL-2) can induce proliferation of T cells and NK cells, and contributes to the activation of these cells and others. IL-2-stimulated immune cells can lyse autologous tumor cells *in vitro*. In 1998, IL-2 became the first FDA-approved immunotherapy for metastatic melanoma. Approval was based on a study of 270 patients who received "high-dose" recombinant IL-2 (600,000 or 720,000 IU/kg), given as an intravenous infusion every 8 hours for up to 14 doses over 5 days, with a second identical treatment cycle starting 2 weeks after the first infusion. The overall response rate to IL-2 in this study was 16%, with 6% of subjects exhibiting a complete response and 10% showing a partial response. Sustained progression-free survival, including continuous complete remission, was seen in a subset of subjects. Despite the low overall response rate, the possibility of sustained CR, compared with the limited effectiveness of chemotherapy agents at the time (see Case 4-4), led to FDA approval and the possibility of immune-based treatment for other solid tumors. However, high-dose IL-2 is associated with severe multi-organ system toxicity, including capillary leak syndrome, hemodynamic instability, and high risk of infection. In the above trial, 2% of subjects died from adverse events related to **sepsis** (microbial invasion of the body). Because of this toxicity, appropriate IL-2 administration is limited to highly specialized centers that can provide maximal support.

Lower doses of IL-2 have not proved to be as effective as high-dose IL-2. Combinations of IL-2 with cytotoxic chemotherapy drugs, vaccines, or other cytokines have not provided benefit but have increased toxicity compared with high-dose IL-2 as a single agent. IL-2 remains in clinical use, although the advent of effective immune checkpoint inhibitors has diminished its role.

Antibodies.   Ipilimumab is a monoclonal antibody that blocks the interaction between CTLA-4 and CD80/86. By blocking this inhibitory immune checkpoint, ipilimumab enhances the anti-tumor immune response. In metastatic melanoma, anti-CTLA-4 antibodies are associated with a 13% overall response rate. In randomized trials, ipilimumab improved overall survival. However, in some patients it is associated with significant autoimmune toxicity, including but not limited to **enterocolitis** (inflammation of the small intestine and colon) and **hypophysitis** (inflammation of the pituitary gland). In a Phase III study, 60% of subjects experienced an immune-related adverse event, with 10–15% experiencing moderate to severe (grade 3 or 4) events. On March 25, 2011, ipilimumab monotherapy was approved by the FDA for metastatic melanoma.

Antibodies targeting PD-1 and PD-L1 have been developed and evaluated in solid tumors, including metastatic melanoma. Initial reports from 2012 indicated significant overall response rates and reduced toxicity compared with ipilimumab, perhaps because PD-1/PD-L1 interactions occur in peripheral tissues and involve down-regulation of already activated T cells, while CTLA-4/CD80/86 interactions in lymph nodes provide a barrier to the initial activation of T cells. Blockade of CTLA-4 by antibodies such as ipilimumab lowers the threshold for T-cell activation and may allow the activation of more autoreactive T cells, thereby encouraging autoimmune responses. The initial studies of the anti-PD-1 antibody, **nivolumab**, showed overall response rates of 28% for subjects with melanoma, 18% for subjects with non-small-cell lung cancer, and 27% for subjects with renal-cell carcinoma.

Subsequently, other anti-PD-1 antibodies have been developed. Pembrolizumab was studied in a Phase IB trial. For 135 subjects with metastatic melanoma, including some patients who had disease progression on ipilimumab, the overall response rate was 24% for the patients who received the recommended dose of 2 mg/kg by intravenous infusion every 3 weeks. The side effects of pembrolizumab were generally mild, and included fatigue, cough, nausea, pruritus (itchiness), rash, decreased appetite, constipation, arthralgia (joint pain), and diarrhea. Immune-mediated adverse events were not as frequent as with ipilimumab, and included pneumonitis, colitis, hepatitis, and hypophysitis, among others.

The improvement in durable response rates with anti-PD-1 antibodies compared with historic rates with IL-2, as well as improved tolerability due to fewer autoimmune adverse events, led to accelerated FDA approval (on September 4, 2014) of pembrolizumab for the treatment of patients with advanced or unresectable melanoma.

Future prospects and challenges in metastatic melanoma

The management of metastatic melanoma has undergone significant changes since 2011, with approval of immune checkpoint antibodies (ipilimumab, pembrolizumab), BRAF inhibitors (vemurafenib, dabrafenib; see Case 4-4), MEK inhibitor (trametinib; see Case 4-4), and cytokines (peg-interferon-alfa-2b, approved as adjuvant therapy after surgical resection). There are numerous other immune checkpoint inhibitory antibodies in clinical trials that may also be promising. Interestingly, metastatic melanoma is the one of several diseases in solid-tumor oncology in which classic cytotoxic chemotherapy agents have virtually no role in contemporary management.

Because their molecular targets differ, combinations of the above-mentioned agents may offer some promise with regard to improving upon durable response rates and overall survival. Earlier patient introduction to these therapies as adjuvant treatments in localized or locally advanced cases may also improve upon existing disease recurrence and overall survival rates. Exactly how targeted therapies and immune-based approaches will be used in sequence or in combination remains the topic of ongoing and future clinical trials.

Take-home points

✓ Spontaneous remissions occur in rare cases of melanoma.

✓ The most likely mechanism of spontaneous remissions is an anti-tumor immune response.

✓ Immunotherapy strategies that have been FDA approved in melanoma include cytokines and immune checkpoint inhibitors (anti-CTLA-4, anti-PD-1, and anti-PD-L1 antibodies).

✓ High-dose IL-2 has a 16% overall response rate but significant toxicity, including a 2% rate of treatment-related death.

✓ Ipilimumab, an antibody against CTLA-4, prolongs survival in metastatic melanoma but is associated with a significant rate of autoimmune adverse events.

✓ Pembrolizumab and other similar antibodies against PD-1 are associated with higher overall response rates compared with IL-2, and with lower autoimmune adverse events compared with ipilimumab.

✓ Targeting the immune checkpoint in primary or metastatic cancer tissues (PD-1/PD-L1) may result in less autoimmune toxicity compared with immune checkpoint blockade in lymph nodes (CTLA-4/CD80/86), where more nonspecific checkpoint inhibition may occur.

✓ Combinations or rational sequencing of targeted therapy, cytokines, and immune checkpoint inhibitors may lead to enhanced durable response rates and overall survival.

Discussion questions

1. Why are toxicities different among the different immune checkpoint strategies?

2. If immune checkpoint inhibitors are being given to earlier-stage melanoma patients (i.e., with localized or locally advanced disease), when should they be administered? Why is this?

Chapter summary

All of the therapies described in this chapter involve the manipulation of T-cell responses. One of them employs vaccination to prevent HPV infection, which is the cause of most cases of cervical cancer. In addition to prophylactic vaccination to prevent cervical cancer, vaccination against hepatitis B virus (HBV) is also routinely employed to protect against infection by HBV, a major cause of liver cancer. The use of vaccines, exemplified by the case involving prophylactic cervical cancer vaccination, is a deliberate manipulation of the immune system to develop large numbers of antigen-specific T and B cells that are capable of preventing infection by a cancer-causing microbe. Although some progress has been made toward the development of therapeutic vaccines for the treatment of established cancers that are caused and driven by infectious agents, this has proved difficult and awaits the development of approaches that yield more robust and durable effects on disease progression and survival. The other cases in the chapter illustrate strategies that allow the circumvention or disabling of barriers of tolerance and negative regulation that prevent or restrain T-cell responses. Figure 12.21 summarizes the T-cell-based approaches to cancer immunotherapy, and reviews the ways in which antibodies and their derivatives kill targeted cells. The existing therapeutic cancer vaccines and those in development are both intended to increase the intensity of the host's T-cell-mediated responses to existing cancers. The cases presented in this chapter illustrate how an understanding of basic immunology has led to the development of strategies for the rational and targeted treatment of a variety of cancers.

The responses of T cells, which are key regulators and effectors of adaptive immune responses, are negatively regulated by receptor–ligand interactions that establish inhibitory checkpoints. The interruption of these checkpoint interactions by monoclonal antibody blockade releases the inhibition of naturally occurring T-cell anti-tumor responses. The case that describes the use of anti-PD-1 monoclonal antibodies to treat melanoma illustrates the potential of antibody-mediated checkpoint inhibition for therapy. Chimeric antigen receptor-modified T cells (CAR T cells) emerged from a collaboration between monoclonal antibody technology and cytotoxic T-cell biology. The

immunotherapy	description	status	immunotherapy	description	status
tumor antigen / cancer cell / anti-tumor-associated tumor-antigen-specific monoclonal antibody	antibody targets a tumor-associated antigen and effects killing of the targeted cell by a variety of mechanisms	several in clinical use; many in clinical trials	expand cancer-restrictive cells / patient / reinfuse / adoptive T-cell transfer	members of the small populations of autologous anti-cancer CD4 or CD8 anti-cancer T cells are harvested from patients, expanded *in vitro*, and reinfused to attack cancer cells in the patient	in clinical trials
TME-disruptive monoclonal antibody	antibody targets structural elements or cells of the tumor microenvironment that are not neoplastic but support the growth and survival of cancer cells	a few in clinical use; many under investigation	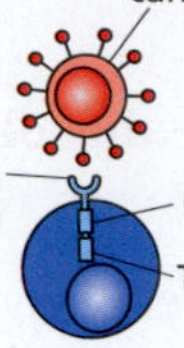 cancer cell / antigen binding / CS / TcR / chimeric antigen receptor T-cell (CAR T-cell)	autologous patient T cells are modified to express antigen receptors specific for a tumor antigen on the patient's cancer cells: the intracellular domains of the antigen receptor provide co-stimulatory signals and T-cell receptor-like signals, bypassing the need for external co-stimulation or conventional antigen presentation	two approved for clinical use; several in clinical trials
cytotoxic drug / cancer cell / monoclonal antibody–drug conjugate (ADC)	antibody targets tumor-specific or tumor-associated antigens and effects killing of the targeted cell by delivery of a highly cytotoxic drug and by a variety of antibody-dependent mechanisms	some in clinical use; many in clinical trials	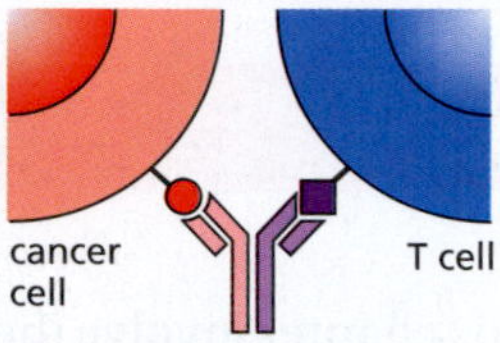 cancer cell / T cell / bispecific antibody (bsAb)	by binding to CD3 of the TCR of T cells and a tumor antigen of a cancer cell, BsAbs can bring T cells together with cancer cells and trigger their T cell-mediated killing	one in clinical use; others in various stages of clinical trials
anti-ligand Ab / inhibitory receptor / inhibitory ligand / anti-receptor Ab / checkpoint-inhibiting monoclonal antibody	relieves inhibition of T-cell responses by interrupting or preventing receptor–ligand interactions between inhibitory receptors and ligands	some in clinical use; many in clinical trials	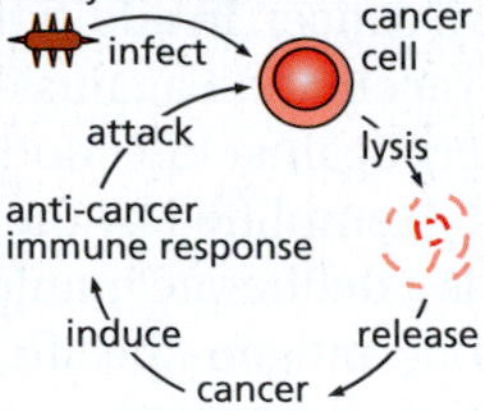 oncolytic virus / infect / cancer cell / attack / lysis / anti-cancer immune response / induce / release / cancer antigens	oncolytic virus kills cancer cell and liberates cancer antigens; cell lysate contains PAMPs DAMPs that promote induction of anti-cancer immune responses	one in clinical use; others in clinical trials
prophylactic cancer vaccine	prevents infection by cancer-causing microbes	two in widespread clinical use; others under investigation			

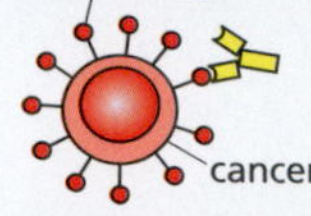

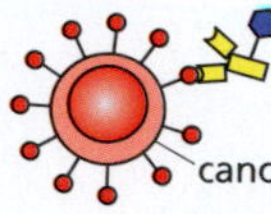

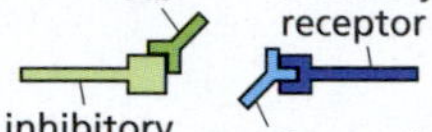

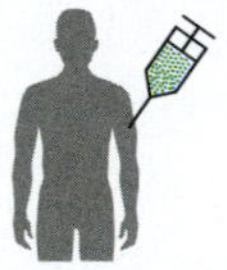

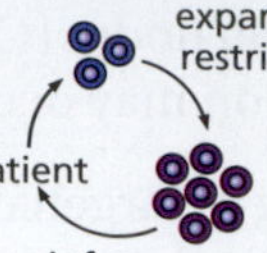

Figure 12.21. Immunotherapeutic strategies for cancer therapy. A wide range of immunotherapeutic approaches have been approved for clinical use or are in advanced stages of development.

therapeutic potential of CAR T cells was illustrated in the acute lymphoblastic leukemia case.

Future efforts will lead to increased understanding of how the growing repertoire of immunotherapies can be integrated into the wider landscape of cancer therapeutics. It is important to emphasize that the immunological strategies discussed in this chapter will often complement rather than replace existing cancer therapies. The future will surely see the clinical use of more monoclonal antibodies and their derivatives (introduced in Chapter 11) against a variety of cancers and cancer-supporting processes. Much more will be learned about the management of CAR T-cell therapy, and there will be intensive exploration of the range of cancers that might be targeted by these engineered cells. NK cells can also be genetically modified to express chimeric antigen receptors, and the resulting CAR NK cells can then be employed for therapy. The spectrum of cytokines secreted by CAR NK cells, although potent, is a somewhat less dangerous one than that of CAR T cells, and consequently may generate less severe side effects. Because they have a shorter half-life than CAR T cells and little immunological memory, CAR NK cells may have shorter lifetimes in patients than do CAR T cells.

The extent to which CAR NK cells challenge CAR T cells will be determined by the comparative clinical performance of these two cell types.

In addition to the immunotherapeutic approaches covered in the three cases presented in this chapter, other approaches, including adoptive transfer of cancer-reactive T-cell clones, oncolytic viruses, and bi-specific antibody-based drugs, are also deployed. Adoptive transfer of cancer-reactive T-cell clones derived from patients has emerged as a powerful and increasingly versatile immunotherapy. In addition to a number of demonstrations of efficacy for the treatment of melanoma, it has shown effectiveness both for the treatment of carcinomas and as a tool for targeting neoantigens generated as a consequence of the extensive mutation and genomic instability that are, to a greater or lesser degree, a hallmark of all cancers. There are now well-documented examples of cancer neoantigens being prospectively identified, and patient T cells capable of attacking tumor cells displaying these often cancer-specific neoantigens expanded in culture and infused into patients. One oncolytic cancer virus is already in clinical use, and others are under active clinical investigation. A bi-specific T-cell engager (BiTE), which is a drug based on bi-specific antibody technology, is in clinical use, and other bi-specific antibodies or bsAb-based drugs are in various stages of development and clinical investigation. All of the approaches mentioned above are summarized in Figure 12.21.

Chapter discussion questions

1. Successful anti-tumor immune responses result in the death of cancer cells. What are the major mechanisms of immune system-mediated cancer cell death? What agents play important roles in each of the mechanisms that you have identified? In addition to the concepts covered in this chapter, your answer should reflect the related content presented in Chapter 11.

2. What are some of the factors that prevent or diminish the immune response to cancer? Why do such apparently counterproductive factors exist?

3. What is allogeneic BMT, and how might it act as an immunotherapy?

Selected references

Introduction

Mahoney KM, Rennert PD & Freeman GJ (2015) Combination cancer immunotherapy and new immunomodulatory targets. *Nat. Rev. Drug Discov.* 14:561–584.

Page DB, Postow MA, Callahan MK et al. (2014) Immune modulation in cancer with antibodies. *Annu. Rev. Med.* 65:185–202.

Palucka K & Banchereau J (2014) SnapShot: cancer vaccines. *Cell* 157:516–516.e1.

Restifo NP, Dudley ME & Rosenberg SA (2012) Adoptive immunotherapy for cancer: harnessing the T cell response. *Nat. Rev. Immunol.* 12:269–281.

Scott AM, Wolchok JD & Old LJ (2012) Antibody therapy of cancer. *Nat. Rev. Cancer* 12:278–287.

Sievers EL & Senter PD (2013) Antibody-drug conjugates in cancer therapy. *Annu. Rev. Med.* 64:15–29.

Tran E, Robbins PF, Lu Y-C et al. (2016) T-cell transfer therapy targeting mutant KRAS in cancer. *N. Engl. J. Med.* 375:2255–2262.

Case 12-1

Doorbar J, Quint W, Banks L et al. (2012) The biology and life-cycle of human papillomaviruses. *Vaccine* 30 (Suppl. 5):F55–F70.

Frazer IH, Leggatt GR & Mattarollo SR (2011) Prevention and treatment of papillomavirus-related cancers through immunization. *Annu. Rev. Immunol.* 29:111–138.

FUTURE II Study Group (2007) Quadrivalent vaccine against human papillomavirus to prevent high-grade cervical lesions. *N. Engl. J. Med.* 356:1915–1927.

Garland SM, Hernandez-Avila M, Wheeler CM et al. (2007) Quadrivalent vaccine against human papillomavirus to prevent anogenital diseases. *N. Engl. J. Med.* 356:1928–1943.

Giuliano AR, Palefsky JM, Goldstone S et al. (2011) Efficacy of quadrivalent HPV vaccine against HPV infection and disease in males. *N. Engl. J. Med.* 364:401–411.

Joura EA, Giuliano AR, Iversen O-E et al. (2015) A 9-valent HPV vaccine against infection and intraepithelial neoplasia in women. *N. Engl. J. Med.* 372:711–723.

Palefsky JM, Giuliano AR, Goldstone S et al. (2011) HPV vaccine against anal HPV infection and anal intraepithelial neoplasia. *N. Engl. J. Med.* 365:1576–1585.

Siegel RL, Miller KD & Jemal A. (2018) Cancer statistics, 2018. *CA Cancer J. Clin.* 68:epub ahead of print.

Trimble CL, Morrow MP, Kraynyak KA et al. (2015) Safety, efficacy, and immunogenicity of VGX-3100, a therapeutic synthetic DNA vaccine targeting human papillomavirus 16 and 18 E6 and E7 proteins for cervical intraepithelial neoplasia 2/3: a randomized, double-blind, placebo-controlled phase 2b trial. *Lancet* 386:2078–2088.

zur Hausen H (2009) Papillomaviruses in the causation of human cancers—a brief historical account. *Virology* 384:260–265.

Case 12-2

Barrett DM, Singh N, Porter DL et al. (2014) Chimeric antigen receptor therapy for cancer. *Annu. Rev. Med.* 65:333–347.

Bhojwani D & Pui CH (2013) Relapsed childhood acute lymphoblastic leukaemia. *Lancet Oncol.* 14:e205–e217.

Davila ML, Riviere I, Wang X et al. (2014) Efficacy and toxicity management of 19-28z CAR T cell therapy in B cell acute lymphoblastic leukemia. *Sci. Transl. Med.* 6:224ra25.

Grupp SA, Kalos M, Barrett D et al. (2013) Chimeric antigen receptor-modified T-cells for acute lymphoid leukemia. *N. Engl. J. Med.* 368:1509–1518.

Locatelli F, Schrappe M, Bernardo ME et al. (2012) How I treat relapsed childhood acute lymphoblastic leukaemia. *Blood* 120:2807–2816.

Case 12-3

Atkins MB, Lotze MT, Dutcher JP et al. (1999) High-dose recombinant interleukin 2 therapy for patients with metastatic melanoma: analysis of 270 patients treated between 1985 and 1993. *J. Clin. Oncol.* 17:2105–2116.

Brahmer JR, Tykodi SS, Chow LQM et al. (2012) Safety and activity of anti-PD-L1 antibody in patients with advanced cancer. *N. Engl. J. Med.* 366:2455–2465.

Chen DS & Mellman I (2013) Oncology meets immunology: the cancer-immunity cycle. *Immunity* 39:1–10.

Hamid O, Robert C, Daud A et al. (2013) Safety and tumor responses with lambrolizumab (anti-PD-1) in melanoma. *N. Engl. J. Med.* 369:134–144.

Hodi FS, O'Day SJ, McDermott DF et al. (2010) Improved survival with ipilimumab in patients with metastatic melanoma. *N. Engl. J. Med.* 363:711–723.

Topalian SL, Hodi FS, Brahmer JR et al. (2012) Safety, activity, and immune correlates of anti-PD-1 antibody in cancer. *N. Engl. J. Med.* 366:2443–2454.

Wolchok JD, Kluger H, Callahan MK et al. (2013) Nivolumab plus ipilimumab in advanced melanoma. *N. Engl. J. Med.* 369:122–133.

Summary

Carter PJ & Lazar GA (2018) Next generation antibody drugs: pursuit of the 'high-hanging fruit.' *Nat. Rev. Drug Discov.* 17:197–223.

Kaufman HL, Kolhapp FJ & Zioza A (2015) Oncolytic viruses: a new class of immunotherapy drugs. *Nat. Rev. Drug Discov.* 14:642–662.

Longo DP & Baden LR (2018) Exploiting viruses to treat diseases. *N. Engl. J. Med.* 379:194–196.

Sadelain M, Riviere I & Riddell S (2017) Therapeutic T cell engineering. *Nature* 545:423–431.

Glossary

A

abiraterone An inhibitor of steroid production that is used in the treatment of prostate cancer. Trade name: Zytiga. (Ch. 6)

acetylation Covalent attachment of an acetyl group to a second molecule such as a protein. (Ch. 8)

acral lentiginous melanoma A form of melanoma that is characterized by black or brown discoloration and typically arises under the nails, on palms of the hands, or on soles of the feet. (Ch. 4)

activated B-cell-like A subtype of diffuse large B-cell lymphoma (DLBCL). (Ch. 11)

active surveillance A management approach reserved for slow-growing cancers that involves closely monitoring a patient's condition but not giving any treatment unless there are changes in test results that indicate that the condition is getting worse. This approach can spare a patient from side effects of treatment and may be offered if the cancer is not expected to shorten the patient's life span. (Ch. 3)

acute lymphoblastic leukemia A hematologic cancer characterized by large numbers of immature lymphoblasts in the bone marrow, circulating blood, lymph nodes, spleen, and other organs. More than 75% of cases in the United States occur in children. (Ch. 11)

acute myelogenous leukemia (AML) A malignant neoplasm of blood-forming tissues characterized by the uncontrolled proliferation of immature granular leukocytes. (Ch. 4)

acute promyelocytic leukemia (APL) A form of acute myelogenous leukemia (AML), characterized by the proliferation of promyelocytes and blast cells with distinctive Auer rods. (Ch. 7)

adaptive immunity Highly specific immunity to antigens mediated by T cells and B cells after initial engagement, with memory of such antigens enduring for subsequent encounters. (Ch. 11)

adenocarcinoma Cancer derived from secretory (gland-forming) epithelial cells. (Ch. 3)

adenomas Premalignant noninvasive growths in various epithelial tissues, which have the potential to progress further to invasive carcinomas. (Ch. 10)

adjuvant In immunology contexts, a substance that enhances the immune response to an antigen with which it is mixed. (Ch. 12)

adjuvant therapy A treatment given following initial therapy to decrease the risk of cancer recurrence. (Ch. 2)

adoptive T-cell therapies Cell-based cancer therapies involving the isolation and *in vitro* expansion of tumor-specific T cells that are then infused into cancer patients to increase the capacity of their immune system to kill tumor cells. (Ch. 12)

adoptive T-cell transfer A procedure in which immune cells are transferred from a donor to a recipient, undertaken with the intention that the donor's cells will be capable of mediating an immune function that the recipient lacks. (Ch. 11)

ado-trastuzumab emtansine An antibody–drug conjugate combining trastuzumab (trade name: Herceptin) with emtansine that is used to treat metastatic HER2-overexpressing breast cancer. *Also called* ado-trastuzumab *or* T-DM1. Trade name: Kadcyla. (Ch. 11)

aggresome pathway An alternative pathway for protein degradation that complements the proteasome pathway. (Ch. 9)

agonist Activating agent; opposite of antagonist. (Ch. 6)

Akt Serine/threonine kinase activated downstream of PI3 kinase with numerous downstream targets involved in cell growth and survival, including activation of the mTOR pathways. *Also called* protein kinase B *or* PKB. (Ch. 5)

alemtuzumab Antibody to CD52 used for lymphocyte depletion, such as for T-cell depletion during bone marrow allografts used in treating chronic myeloid leukemia. Trade name: Campath. (Ch. 8)

ALK Anaplastic lymphoma kinase; a receptor tyrosine kinase encoded by the *ALK* gene. ALK's normal cellular role occurs in the development of the brain. (Ch. 4)

alkylating chemotherapy Treatment with any substance that contains an alkyl radical and is capable of replacing a free hydrogen atom in an organic compound. (Ch. 4)

allogeneic hematopoietic stem cell transplant Transplantation of multipotent hematopoietic stem cells that have come from a donor that is not genetically identical to the recipient. Allogeneic HSC transplantation is most successful when the donor's HLA type is identical to or closely matches the recipient's. (Ch. 4)

all-*trans* retinoic acid (ATRA) An organic compound, also known as tretinoin, that is a potent activator of retinoic acid signaling, used for the treatment of acne and acute promyelocytic leukemia. Trade name: Vesanoid. (Ch. 7)

alopecia A partial or complete lack of hair resulting from normal aging, an endocrine disorder, a drug reaction, an anti-cancer medication, or a skin disease. (Ch. 8)

American Joint Committee on Cancer (AJCC) An organization best known for defining cancer staging standards. (Ch. 3)

anabolic Referring to the constructive phase of metabolism characterized by the conversion of simple substances into the more complex compounds of living matter. (Ch. 5)

anal condyloma Anal wart; a growth of skin caused by a human papillomavirus (HPV) condition that affects the area around and inside the anus. (Ch. 12)

anaplastic large-cell lymphoma A type of non-Hodgkin lymphoma that has a tissue and cellular architecture lacking the differentiated characteristics of an identifiable tissue of origin. (Ch. 4)

androgen Any steroid hormone, such as testosterone, that stimulates development of male characteristics. (Ch. 6)

androgen deprivation therapy (ADT) Reduction of serum testosterone to castration levels, which can be achieved by surgery (removal of testes) or medical means. (Ch. 6)

anemia Deficiency of red blood cells and consequent inadequate blood oxygen-carrying capacity. (Ch. 4)

aneuploidy   A karyotype that deviates from diploid due to an increase or decrease in the numbers of certain chromosomes. (Ch. 2)

angiogenesis   Process by which new blood vessels are formed by sprouting from existing vessels. (Ch. 4)

angiogenic switch   The programmatic shift in a tumor, usually triggered by the inability to obtain adequate amounts of oxygen by diffusion, from a state where it cannot induce neovascularization to a state in which it can. (Ch. 1)

Ann Arbor classification   The staging system for lymphomas, both Hodgkin lymphoma and non-Hodgkin lymphoma. (Ch. 5)

anorexia   A lack or loss of appetite. (Ch. 3)

antagonist   An inactivating agent. (Ch. 6)

antibody   A soluble protein capable of recognizing and binding antigens with high specificity. (Ch. 11)

antibody-dependent cell-mediated cytotoxicity (ADCC)   Killing of antibody-coated target cells by cells with Fc receptors that recognize the constant region of the bound antibody. Most ADCC is mediated by NK cells that have the Fc receptor FcγRIII on their surface. (Ch. 4)

antibody–drug conjugates (ADCs)   A class of anti-cancer drugs that combine the selectivity of monoclonal antibody targeting with the cytotoxic potency of chemotherapy drugs. (Ch. 11)

antigen   A molecule that can bind specifically to an antibody or generate peptide fragments that can be recognized by a T-cell receptor. (Ch. 11)

antigen processing   The intracellular degradation of antigen into peptide fragments that are then bound to MHC molecules and presented on the cell surface. Most T cells recognize antigen only when it is presented in this way. (Ch. 12)

antigen–antibody reactions   Reactions that involve the binding of antibodies to antigens for which they are specific. (Ch. 11)

antigen-presenting cells (APCs)   Highly specialized cells that can process antigens and display their peptide fragments on the cell surface together with other co-stimulatory proteins required for activating naive T cells. The main antigen-presenting cells are dendritic cells, macrophages, and B cells. (Ch. 12)

antisense oligonucleotides   Small pieces of DNA or RNA that can bind to specific molecules of RNA, which may block the ability of the RNA to be translated into a protein or may lead to degradation of the target RNA molecule. (Ch. 8)

apheresis   A procedure in which blood is temporarily withdrawn, one or more components are selectively removed, and the rest of the blood is reinfused into the donor. The process is used in treating various diseases and for obtaining blood elements for the treatment of other patients or for research. *Also called* pheresis. (Ch. 12)

apoptosis   A complex program of cellular self-destruction triggered by a variety of stimuli and involving the activation of caspase enzymes that results in rapid fragmentation of a cell and phagocytosis of resulting cell fragments by neighboring cells. (Ch. 1)

aromatase inhibitors (AIs)   A class of drugs that target the enzymatic activity converting testosterone to estrogen, frequently used in the management of breast cancer. (Ch. 6)

arsenic trioxide (ATO)   An oxidized form of arsenic, used in weed killers and rodenticides. It is also used as an anti-neoplastic in the treatment of acute promyelocytic leukemia. Trade name: Trisenox. (Ch. 7)

ascites   Fluid that accumulates in the peritoneal cavity. In cancer patients, this fluid may contain malignant cells. (Ch. 3)

Ashkenazi Jewish   Pertaining to a Jewish diaspora population comprising the great majority of Northern European Jews, originally recognized as a community by the end of the first millennium in and near the Rhineland area of Germany. Descendants of this community migrated east and proliferated mostly in Poland, western Russia, and adjacent countries. Because the original population was very small and the Ashkenazi Jews followed a strong tradition of marrying only within their ethnoreligious group, deleterious germline mutations imported into the population by in-migration or arising within the population have persisted among them to modern times. (Ch. 9)

assent   A child's affirmative agreement to participate in research. (Ch. 12)

ATM   A serine/threonine protein kinase that is recruited and activated by DNA double-strand breaks. It phosphorylates several key proteins that initiate activation of the DNA damage checkpoint, leading to cell cycle arrest, DNA repair, or apoptosis. (Ch. 4)

Auer rod   An abnormal, needle-shaped or round, pink-staining inclusion in the cytoplasm of myeloblasts and promyelocytes in acute myelogenous, promyelocytic, or myelomonocytic leukemia. These inclusions contain enzymes such as acid phosphatase, peroxidase, and esterase and may represent abnormal derivatives of cytoplasmic granules. (Ch. 7)

autologous   Referring to biological material, usually cells or tissue, that originates in a patient's own body (and may be reintroduced into that patient without expectation of provoking an immune response following some manipulation *ex vivo*). (Ch. 6)

autologous stem cell transplantation   A treatment for advanced or refractory cancers. Hematopoietic stem cells from the bone marrow or blood are withdrawn before the patient receives high-dose chemotherapy with or without radiation that destroys cancer cells as well as hematopoietic stem cells; afterward, the removed cells are reinfused to form a new population of blood cells. (Ch. 3)

autophagy   Process whereby cellular organelles are degraded by engulfment in membranous vesicles, which then fuse with lysosomes, in which degradation occurs. (Ch. 5)

axicabtagene ciloleucel   A CAR T-cell (adoptive cell transfer) therapy targeting CD19 on B cells, employed for treating large B-cell lymphoma. Trade name: Yescarta. (Ch. 12)

axilla   The armpit. (Ch. 6)

axitinib   An oral targeted therapy that inhibits VEGF receptors among other targets. Trade name: Inlyta. (Ch. 5)

5-azacitidine   A water-soluble potent inhibitor of DNA methyltransferase 1 (DNMT1). (Ch. 8)

B

B cell   One of two types of antigen-specific lymphocytes responsible for adaptive immune responses, the other being T cells. The function of the B-cell lineage is to produce antibodies. (Ch. 11)

B-cell aplasia   The absence of B cells due to failure to sustain or generate a B-cell population. (Ch. 12)

B-cell clone   A population of identical B cells. (Ch. 11)

B-cell lymphoma   Any in a large group of non-Hodgkin lymphomas characterized by malignant transformation of B cells. *Also called* B-cell malignancy. (Ch. 11)

B-cell malignancy   *See* **B-cell lymphoma**. (Ch. 11)

B-cell receptor complex (BCR)   A complex comprised of an antigen-binding subunit known as the membrane immunoglobulin (mIg), which is comprised of two immunoglobulin light chains (IgLs) and two immunoglobulin heavy chains (IgHs) as well as two heterodimer subunits of Ig-α and Ig-β. (Ch. 4)

"B" symptoms   Fevers, night sweats, and unexplained weight loss. Seen in patients with lymphoma and other cancers. (Ch. 4)

Bacillus Calmette–Guérin   An attenuated strain of live *Mycobacterium bovis*, the bacterium that causes bovine tuberculosis, used to vaccinate humans against tuberculosis and for the immunotherapy of some forms of bladder cancer. (Ch. 12)

band form   A developing granular (immature) leukocyte in circulating blood, characterized by a curved or indented nucleus. (Ch. 4)

basal cell carcinoma (BCC)   A malignant epithelial cell tumor that begins as a pearly appearing papule and enlarges peripherally, developing a central crater that erodes, crusts, and bleeds. Metastasis is rare, but local invasion destroys underlying and adjacent tissue. (Ch. 4)

base-excision repair (BER)   A form of DNA repair for single-strand DNA breaks, mediated by poly(ADP-ribose) polymerases (PARPs). (Ch. 9)

basophilia   A condition characterized by an excess of basophil cells in the blood. (Ch. 4)

basophils   Granulocytic white blood cells characterized by cytoplasmic granules that stain blue when exposed to a basic dye. (Ch. 4)

BCL-2   A protein belonging to a family of intracellular proteins that includes many members that promote apoptosis and a few that are anti-apoptotic. The *BCL2* gene encoding BCL2 is located at 18q21 on chromosome 18. In some types of non-Hodgkin lymphoma, *BCL2* is translocated to the immunoglobulin heavy-chain promoter region at 14q32, resulting in constitutive expression of BCL2. (Ch. 4)

BCL-6   A transcriptional repressor that opposes differentiation of B cells into plasma cells. (Ch. 5)

BCR–ABL tyrosine kinase   Constitutively active tyrosine kinase fusion protein caused by a chromosomal translocation of the *BCR* with the *ABL* tyrosine kinase genes; associated with chronic myeloid leukemia. (Ch. 4)

bendamustine   An alkylating cytotoxic chemotherapy agent. Trade name: Treanda. (Ch. 4)

benign prostatic hypertrophy (BPH)   A histological diagnosis associated with nonmalignant, noninflammatory enlargement of the prostate, most common among men over 50 years of age. *Also called* benign prostatic hyperplasia. (Ch. 6)

β-2-microglobulin   A protein found on the surface of all nucleated cells that is associated with the heavy chains of major histocompatibility complex (MHC) class I molecules. (Ch. 5)

bevacizumab   A monoclonal antibody that selectively binds to and inhibits activity of human vascular endothelial growth factor (VEGF) to reduce angiogenesis. Trade name: Avastin. (Ch. 5)

bilobed nuclei   Nuclei such as those present in eosinophils that have two connected lobes. (Ch. 7)

bioavailability   The proportion of a substance that is absorbed into circulation to exert biological effects. (Ch. 2)

bioavailable pro-drug   A medication that is readily absorbed from the intestine and converted by the patient's metabolic processes into a form that is pharmacologically active. (Ch. 6)

biochemotherapy   A cancer treatment method that combines chemotherapy and cytokines. (Ch. 4)

Birt–Hogg–Dubé syndrome   A hereditary condition associated with multiple noncancerous (benign) skin tumors, lung cysts, and an increased risk of both benign kidney tumors and kidney cancer.

Affected patients have mutations in *FLCN*, which encodes the protein folliculin. (Ch. 5)

bi-specific antibody (bsAb)   An antibody genetically engineered to possess two structurally distinct binding sites, enabling its simultaneous binding of two structurally different antigens. (Ch. 12)

bi-specific T-cell engager (BiTE)   A fusion protein in which the single-chain variable fragment (scFv) of an antibody specific for an antigen, such as CD3 present on T cells, is genetically engineered to link with the scFv of an antibody specific for an antigen present on the surface of a tumor cell. The resulting protein forces the close engagement of a T cell with the cancer cell, promoting an attack on the cancer cell by the T cell. (Ch. 12)

blastic   In reference to bone metastasis, a shortened term for "osteoblastic," referring to formation of new bone. (Ch. 6)

blinatumomab   A genetically engineered protein that functions as a bi-specific T-cell engager (BiTE) for the treatment of relapsed or refractory acute lymphoblastic leukemia. It targets the B-cell surface antigen CD19 and the T-cell surface protein CD3. Trade name: Blincyto. (Ch. 12)

bortezomib   A targeted therapy that inhibits proteasome activity. Trade name: Velcade. (Ch. 9)

brachytherapy   Placement—either permanent or for a specific period—of radioactive sources, such as seeds, needles, or catheters, in contact with or as implants into tumor tissues to be treated. (Ch. 2)

BRAF   A protein responsible for integrating growth signals from RAS and PI3 kinase (PI3K) to phosphorylate and activate downstream MEK/MAPK signaling. A mutation of BRAF occurs in approximately half of all melanomas. (Ch. 4)

break-apart probe   A probe used in fluorescence *in situ* hybridization (FISH) that indicates when two chromosomal segments that are normally close to each other are separated ("broken apart") due to a chromosomal translocation. (Ch. 4)

brentuximab vedotin   An antibody–drug conjugate used to treat certain Hodgkin and non-Hodgkin lymphomas, composed of an anti-CD30 monoclonal antibody linked to monomethyl auristatin E (MMAE), a cytotoxic chemotherapy that blocks the polymerization of tubulin that is necessary for mitosis. Trade name: Adcetris. (Ch. 11)

Breslow measurement   A measure of the thickness of a melanoma, used to describe the depth of penetration into the skin. (Ch. 4)

bronchioalveolar carcinoma   The less common variant of the two types of adenocarcinoma of the lung, with columnar to cuboidal epithelial cells lining the alveolar septa and projecting into alveolar spaces in branching papillary formations. (Ch. 4)

bronchoscopy   Visual examination of the tracheobronchial tree, using the standard rigid, tubular metal bronchoscope or the narrower, flexible fiber-optic bronchoscope. (Ch. 4)

Bruton's tyrosine kinase (BTK)   A tyrosine kinase important in B-cell receptor signaling. BTK is mutated in the human immunodeficiency disease X-linked agammaglobulinemia. (Ch. 4)

C

cancer-associated fibroblasts (CAFs)   Fibroblast cell type within the tumor microenvironment that promotes tumorigenic features by initiating the remodeling of the extracellular matrix or by secreting cytokines. (Ch. 10)

cancer stem cells   Cells (found within tumors or hematological cancers) that possess characteristics associated with normal stem cells, specifically, the ability to give rise to all cell types found in a particular cancer sample including cancer stem cells. (Ch. 7)

capillary leak syndrome   A condition in which IL-2 induces leakage of plasma and other blood components out of blood vessels and into surrounding tissues, muscle compartments, organs, or body cavities, causing severe hypotension (low blood pressure) that can lead to organ failure and even death. (Ch. 5)

capsid protein   The protein shell of a virus. (Ch. 12)

carcinoembryonic antigen (CEA)   A protein expressed in embryonic development that is present in very small quantities in adult tissue. CEA is overexpressed in some gastrointestinal malignancies. Changes in CEA values may be used to monitor tumor response to treatment. (Ch. 10)

carcinoma   A cancer arising from epithelial cells. (Ch. 1)

carfilzomib   A targeted therapy that inhibits proteasome activity. Trade name: Kyprolis. (Ch. 9)

caspases   A family of cysteine proteases that cleave proteins at aspartic acid residues. They have important roles in apoptosis and in the processing of cytokine pro-polypeptides. (Ch. 9)

castration-resistant prostate cancer   Prostate cancer that keeps growing even when the amount of testosterone in the body is reduced to very low (castration) levels. (Ch. 12)

catabolic   Describing the metabolic process in which complex substances are broken down by living cells into simple compounds. (Ch. 5)

CD5   A protein expressed on the surface of T cells and on a subset of B cells. (Ch. 4)

CD8   A coreceptor that collaborates with T-cell receptors to recognize peptide antigens bound to MHC class I molecules. (Ch. 12)

CD10   A typical protein of the germinal center; also known as common acute lymphoblastic leukemia antigen, or CALLA. (Ch. 5)

CD19   A protein that is a component of the transmembrane B-cell coreceptor, a complex on the B-cell surface that can increase antigen responsiveness when co-ligated with the B-cell receptor. (Ch. 4)

CD20   A protein that is displayed on the surface of many cells of the B-cell lineage. (Ch. 4)

CD23   The low-affinity Fc receptor for IgE. (Ch. 4)

CD28   An activating receptor on T cells that binds members of the B7 family, such as CD80 or CD86, that are co-stimulatory ligands on specialized antigen-presenting cells such as dendritic cells. (Ch. 12)

CD80 or **CD86**   Glycoproteins on antigen-presenting cells that are members of the B7 family and ligands for CD28 co-stimulatory receptors on T cells. (Ch. 12)

cell survival   The span of viability of a cell. (Ch. 5)

centroblasts   Large, rapidly dividing B cells present in the dark zone of germinal centers in follicles or peripheral lymphoid organs. (Ch. 5)

centrocytes   Small B cells that derive from centroblasts in the germinal centers of follicles in lymphoid organs and tissues; they populate the light zone of the germinal center. (Ch. 5)

cerebrospinal fluid   The clear and colorless fluid that flows through and protects the four ventricles of the brain, the subarachnoid spaces, and the spinal canal. (Ch. 12)

Cervarix   A vaccine that protects against cervical cancer development by preventing infection by two of the major oncogenic human papillomaviruses, HPV-16 and HPV-18. (Ch. 12)

cervical intraepithelial neoplasia (CIN)   Abnormal, precancerous changes in the squamous epithelial tissues of the cervix. (Ch. 12)

cervicectomy   Surgical removal of the uterine cervix (the lower portion of the uterus). As the uterine body is preserved, this type of surgery is a fertility-preserving surgical alternative to a radical hysterectomy and may be appropriate for some younger women with early cervical cancer. (Ch. 12)

cetuximab   A monoclonal antibody that inhibits epidermal growth factor receptor (EGFR). This drug is used alone or in combination with chemotherapy for EGFR-expressing cancer. Trade name: Erbitux. (Ch. 10)

checkpoint inhibitors   Antibodies that disrupt receptor–ligand interactions that inhibit T cells' responses, such as those of PD-1 with PD-L1 or CTLA-4 with CD80 or CD86. (Ch. 12)

chemoimmunotherapy   Chemotherapy that is combined with immunotherapy. (Ch. 5)

chemokines   Small chemoattractant proteins that stimulate the migration and activation of cells, especially phagocytic cells and lymphocytes. Chemokines have a central role in inflammatory responses. (Ch. 10)

chemoprevention   The use of natural, synthetic, or biological substances to suppress or prevent a disease such as cancer. (Ch. 9)

chemoradiation   Treatment that combines chemotherapy and radiotherapy. (Ch. 3)

chemotactic factor   Any small molecule that acts as a chemical stimulus along a concentration gradient, attracting macrophages and other cells to a site of inflammation. (Ch. 10)

chemotaxis   Cellular movement occurring in response to chemical signals in the environment. (Ch. 4)

chemotherapy   Medications for the treatment of cancer designed to destroy malignant cells. (Ch. 3)

chimeric antigen receptors (CARs)   Engineered fusion proteins composed of extracellular antigen-specific receptors and intracellular signaling domains that activate and co-stimulate. They are expressed in T cells for use in cancer therapy. (Ch. 12)

chimeric antigen receptor T cells   T cells genetically engineered to express a receptor that incorporates the following elements in a single protein: (1) the antigen-binding domain of an antibody against the targeted tumor antigen; (2) the signaling domain of a chain, usually δ, of the TCR complex; and (3) the intracellular signaling domain of a co-stimulatory molecule such as 4BB1. CAR T cells can recognize and attack target cells without the need for TCR recognition of peptide–MHC complexes or the need for co-stimulation from antigen-presenting cells. (Ch. 11)

choriocarcinoma   Malignancy that arises from trophoblastic tissue, which can derive from the chorionic portion of the products of conception, or from the germ cells in the testis or ovary. (Ch. 3)

chromatin   Complex of DNA, RNA, and proteins that constitutes a chromosome. (Ch. 8)

chromosomal aberration   Any change in the structure or number of any of the chromosomes. (Ch. 2)

chronic lymphocytic leukemia (CLL)   A neoplasm of blood-forming tissues, characterized by a proliferation of small, long-lived lymphocytes, chiefly B cells, in bone marrow, blood, liver, and lymphoid organs. CLL is the rarest type of leukemia and the only leukemia to which there is a possible inheritable genetic predisposition. (Ch. 4)

chronic myelogenous leukemia (CML)   A malignant neoplasm of blood-forming tissues, characterized by a proliferation of granular leukocytes. (Ch. 4)

chronic obstructive pulmonary disease (COPD)   General term for progressive and irreversible conditions characterized by diminished inspiratory and expiratory capacity of the lungs. The condition is aggravated by cigarette smoking and air pollution. Emphysema and chronic bronchitis are types of COPD. (Ch. 4)

circulating tumor cells (CTCs)   Cells that have emerged from a primary tumor cell into the bloodstream. (Ch. 2)

circulating tumor DNA (ctDNA)   DNA in the bloodstream that comes from a tumor and has no explicit association with other cells. (Ch. 2)

cisplatin   A platinum-containing compound in the alkylating agent category of cytotoxic chemotherapy. *Also called* cisplatinum *or cis*-diamminedichloroplatinum (CDDP). Trade name: Platinol. (Ch. 4)

classical Hodgkin lymphoma (CHL)   The more common type of Hodgkin lymphoma (95% of cases), characterized by the presence of Hodgkin Reed–Sternberg (HRS) cells surrounded by inflammatory cells that do not attack the HRS cell. CHL is further divided among four histologic subtypes. (Ch. 11)

clear cell renal cell carcinoma   The most common type of renal cell carcinoma (RCC), which accounts for 60–70% of all RCC cases. Cells are distinguished by clear cytoplasm. (Ch. 5)

clinical oncology   A branch of medicine consisting of three major areas: medical oncology, the treatment of cancer with chemotherapy and other medicines; surgical oncology, the surgical procedures such as biopsy, staging, and resection for cancer treatment; and radiation oncology, the use of radiation for the treatment of cancer. (Ch. 1)

clonal evolution   The accumulation of genetic and epigenetic changes over time in cancer cells that impact cell proliferation and survival. (Ch. 4)

clonality   Derivation from the same cell. (Ch. 4)

clone   Population of identical cells that descend from a common progenitor cell. (Ch. 11)

coagulation   Transforming of the liquid dispersion medium into a gelatinous mass; clotting. (Ch. 7)

coagulopathic   Describing a pathological condition that reduces the ability of the blood to coagulate, resulting in uncontrolled bleeding. (Ch. 7)

combination chemotherapy   Simultaneous use of two or more anti-cancer drugs. (Ch. 2)

co-morbidity   A coexisting medical condition. (Ch. 2)

complement-dependent cytotoxicity (CDC)   Cytotoxicity mediated by the generation of an ensemble of proteins known as the membrane attack complex (MAC) following the activation of complement. The MAC causes cell damage and death by punching holes in the cell membrane and destroying the osmotic integrity of the cell. (Ch. 11)

complement pathway   A system whereby a set of plasma proteins act together as a defense against pathogens in extracellular places. Activation of complement can result in the lysis or phagocytosis of the targeted cell. Complement can be activated by a number of agents, including antigen–antibody complexes and components of some microbial cell walls, such as mannose. (Ch. 11)

complementary DNA (cDNA)   The DNA that is produced on an RNA template by reverse transcriptase. (Ch. 2)

complete remission (CR)   State at which tests, physical exams, and scans show no remaining evidence of cancer. *Also called* no evidence of disease (NED). (Ch. 12)

complete response (CR)   Elimination of all detectable tumor mass following anti-cancer therapy. (Ch. 5)

completion lymph node dissection   A more extensive removal of regional lymph nodes, usually after a sentinel lymph node biopsy revealing spread of cancer to nodes. (Ch. 4)

computed tomography (CT)   A procedure that digitally processes images from successive X-ray scans of a tissue or the entire body to produce visual slices. (Ch. 2)

conformal radiation therapy (CRT)   A type of radiotherapy that uses CT scans to map a cancer in three dimensions. (Ch. 2)

conjunctivitis   Inflammation of the conjunctiva, the mucous membrane that covers the front of the eye and lines the inside of the eyelids, caused by bacterial or viral infection, allergy, or environmental factors. (Ch. 7)

consolidation therapy   A treatment used to kill any cancer cells that may be left in the body after initial "induction" therapy. It may include radiation therapy, a stem cell transplant, or treatment with drugs that kill cancer cells. (Ch. 7)

contralateral   Referring to the opposite side. (Ch. 9)

copy number variations (CNVs)   Genomic differences arising from the generation of repetitions of sections of the genome. (Ch. 2)

co-stimulatory molecules   Cell surface receptors through which cells receive signals that augment those received through antigen receptors. (Ch. 12)

CRAB   Acronym for disease features of multiple myeloma indicating need for therapy: hyper**c**alcemia, **r**enal insufficiency, **a**nemia, **b**one disease. (Ch. 9)

crizotinib   An oral small-molecule inhibitor of ALK. Trade name: Xalkori. (Ch. 4)

CRSPR/Cas9   A gene-editing technology used in cells and organisms to functionally inactivate genes. Abbreviated from **c**lustered **r**egularly **i**nterspersed **s**hort **p**alindromic **r**epeats/**C**RSPR-**as**sociated. *Also termed* CRISPR/Cas9. (Ch. 9)

cryoablation   Process in which a tumor is destroyed by freezing. (Ch. 5)

cryoprecipitate   Any precipitate formed on cooling of a solution. (Ch. 7)

cryptorchidism   A developmental defect in which one or both testicles fail to descend into the scrotum and are retained in the abdomen or inguinal canal. (Ch. 3)

CTLA-4   A high-affinity receptor for B7-family molecules on T cells that inhibits T-cell activation. (Ch. 12)

cutaneous T-cell lymphoma (CTCL)   A general term for T-cell lymphomas that involve the skin. CTCL can also involve the blood, lymph nodes, and internal organs. (Ch. 8)

cyclophosphamide   A DNA alkylating agent that is also used as an immunosuppressive drug. It acts by killing rapidly dividing cells including lymphocytes proliferating in response to antigen. (Ch. 4)

cytarabine   An anti-neoplastic agent that acts through inhibition of DNA polymerase. *Also called* cytosine arabinoside *or* Ara-C. (Ch. 4)

cytochrome P450   A group of enzymes that play important roles in the metabolic alteration and detoxification of drugs. (Ch. 2)

cytogenetic analysis   The study of a cell population's karyotypes to determine whether chromosomal morphology is normal or displays aberrations such as chromosomal breakage, translocation, or duplication. (Ch. 4)

cytogenetic remission   Disappearance of a chromosomal abnormality due to treatment, as measured by fluorescence *in situ* hybridization (FISH); e.g., disappearance of the Philadelphia chromosome in chronic myelogenous leukemia due to therapy. (Ch. 4)

cytokine   Protein made by cells—particularly, cells of the immune system—that affects the behavior of other cells. (Ch. 4)

cytokine release syndrome   A systemic condition caused by a large, rapid release of cytokines into the blood from immune cells affected by some forms of immunotherapy, such as CAR T-cell therapy. (Ch. 12)

cytolytic cells   Cells that can cause the death of targeted cells by a variety of mechanisms. The immune system's repertoire of cells with cytolytic potential includes cytotoxic T cells, NK cells, macrophages, and other myeloid cells including neutrophils. (Ch. 11)

cytopenias   Deficiencies in numbers of the blood cell elements. (Ch. 4)

cytoreduction   The act of reducing the number of tumor cells. (Ch. 8)

cytoreductive nephrectomy   Surgical removal of a kidney harboring kidney cancer, performed even in cases of metastatic kidney cancer. (Ch. 5)

cytotoxic   Referring to the ability of an agent to kill cells; such an agent might be, for example, a drug or another type of cell. (Ch. 3)

cytotoxic chemotherapy   A form of cancer treatment that works by targeting dividing cells and induces cell death (Ch. 2)

cytotoxic T cells   T cells (most often CD8 T cells) that can kill other cells bearing peptide–MHC complexes recognized by their T-cell receptors. CD8 T cells are the principal group responsible for killing virus-infected cells. (Ch. 11)

D

D-dimer   A fibrin degradation product, a small protein fragment present in the blood after a blood clot is degraded by fibrinolysis. (Ch. 7)

dabrafenib   An oral BRAF inhibitor. Trade name: Tafinlar. (Ch. 4)

dacarbazine   *See* **DTIC**. (Ch. 4)

damage-associated molecular patterns (DAMPs)   Molecules released from damaged or dying cells and tissues that are recognized by pattern recognition receptors and are capable of initiating inflammatory responses. (Ch. 11)

dasatinib   A tyrosine kinase inhibitor that inhibits BCR–ABL and Src kinases. Trade name: Sprycel. (Ch. 4)

debulking   Reducing the overall size of a tumor with the intent of removing a great majority of it. (Ch. 2)

dedifferentiation   A change in the structure and orientation of cells, characterized by loss of differentiation and reversion to a more primitive form. (Ch. 7)

deep venous thrombosis   Formation of a blood clot in a deep vein, most commonly in the legs. (Ch. 7)

dendritic cells (DCs)   Key cell type of the immune system that phagocytoses fragments of cells or infectious agents and then presents oligopeptides derived from these phagocytosed particles to helper T cells in the lymph nodes, thereby initiating immune responses. (Ch. 11)

denosumab   A human monoclonal antibody targeting the receptor activator for NF-κB ligand (RANKL) for the treatment of osteoporosis, treatment-induced bone loss, metastases to bone, and giant cell tumors of bone. Trade name for cancer therapy: XGEVA. Trade name for osteoporosis therapy: Prolia. (Ch. 10)

differential   The distribution of different types of white blood cells. (Ch. 4)

differentiation   A process in development in which unspecialized cells or tissues are systemically modified and altered to achieve specific and characteristic physical forms, physiological functions, and chemical properties. (Ch. 7)

differentiation syndrome   A potentially life-threatening complication observed in patients with acute promyelocytic leukemia (APL) treated with all-*trans* retinoic acid (ATRA), caused by the release of inflammatory cytokines from the malignant APL cells in response to ATRA. (Ch. 7)

differentiation therapy   A cancer therapy technique based on the concept that the malignant cell has escaped the normal controls of cell growth and differentiation, and is pathologically arrested at an early stage of differentiation but retains the ability to proliferate. The therapy is thus aimed at inducing normal differentiation of the proliferating cancer cells. (Ch. 7)

diffuse large B-cell lymphoma (DLBCL)   The most common lymphoma and the prototype for aggressive non-Hodgkin lymphoma (NHL). DLBCL is a biologically heterogeneous disease. (Ch. 11)

disseminated intravascular coagulation (DIC)   Blood clotting occurring simultaneously in small vessels throughout the body, which leads to the massive consumption of clotting proteins so that the patient's blood cannot clot appropriately. (Ch. 7)

diuretic   A drug that promotes the formation and excretion of urine. (Ch. 7)

DNA methylation   A process by which methyl groups are added to the DNA molecule. Methylation can change the activity of a DNA segment without changing the sequence. (Ch. 2)

double-strand DNA break   DNA damage in which both strands of the double helix are severed. These types of breaks are particularly hazardous to the cell because they can lead to genome rearrangements or to apoptosis. (Ch. 9)

driver mutations   Mutations that provide a selective advantage to tumor cells or their precursors. (Ch. 1)

DTIC   An intravenous alkylating agent; also known as dacarbazine. (Ch. 4)

dyspareunia   Difficult or painful sexual intercourse. (Ch. 12)

dysplasia   A premalignant tissue composed of abnormal-appearing cells forming a tissue architecture that deviates from normal. (Ch. 8)

dyspnea   Shortness of breath; a distressful subjective sensation of uncomfortable breathing that may be caused by many disorders, including certain heart and respiratory conditions, strenuous exercise, or anxiety. (Ch. 7)

E

E2F family   A family of transcription factors involved in cell cycle regulation. (Ch. 12)

E6   A protein encoded by human papillomavirus that inactivates the p53 protein, a tumor suppressor. (Ch. 12)

E7   A protein encoded by human papillomavirus that inactivates the retinoblastoma protein (pRB), a tumor suppressor. (Ch. 12)

edema   Swelling resulting from an excessive accumulation of serous fluid in the tissues of the body, in various locations depending on the cause. (Ch. 3)

embryonal carcinoma   A malignant neoplasm derived from germinal cells that usually develops in gonads, especially the testes. (Ch. 3)

EML4   Gene encoding a member of the echinoderm microtubule-associated protein-like family. Abnormal fusion of parts of the *EML4* gene with portions of the anaplastic lymphoma kinase gene, which generates *EML4–ALK* fusion transcripts, is one of the primary mutations associated with non-small cell lung cancer. (Ch. 4)

endocrine   Referring to any gland that secretes fluids into the general circulation. (Ch. 6)

endoscopic retrograde cholangiopancreatography (ERCP)   an endoscopic test that provides radiographic visualization of the biliary and pancreatic ducts. (Ch. 3)

endothelial cells   Cells that form the walls of capillaries or lymph ducts. (Ch. 11)

enterocolitis   An inflammation of the large and small intestines. (Ch. 12)

enzalutamide   A nonsteroidal anti-androgen medication that competitively binds the androgen receptor and inhibits testosterone and other androgens from stimulating prostate cancer cell proliferation. Trade name: Xtandi. (Ch. 6)

eosinophils   White blood cell type containing granules that stain with eosin, thought to be important chiefly in defense against parasitic infections but also medically important as effector cells in allergic reactions. (Ch. 4)

epidermal growth factor receptor (EGFR)   The cell surface membrane receptor for epidermal growth factor (EGF), a polypeptide that induces cell proliferation in a variety of cell types. (Ch. 4)

epididymitis   Acute or chronic inflammation of the epididymis, a duct that conveys sperm. (Ch. 3)

epigenetics   The study of changes in gene expression that are not due to alteration in the primary DNA sequences. (Ch. 8)

episome   A genetic element comprised of DNA that exists in a cell, often over extended periods of time, that can replicate independently or can be integrated into a chromosome of the host cell. (Ch. 12)

Epstein–Barr virus (EBV)   A common human herpesvirus that causes infectious mononucleosis and some B-cell lymphomas, and is associated with some nasopharyngeal carcinomas. (Ch. 11)

ERK1/2   Extracellular signal-related kinase 1 and 2, a family of mitogen-activated protein kinases (MAPKs) that lead to cell proliferation in a variety of cell types. (Ch. 4)

erlotinib   A tyrosine kinase inhibitor targeting EGFR. Trade name: Tarceva. (Ch. 4)

erythroderma   An abnormal redness of the skin. (Ch. 8)

etoposide   A cytotoxic chemotherapeutic agent that inhibits topoisomerase II, preventing DNA re-ligation. *Also called* VP-16. (Ch. 4)

everolimus   A targeted therapy inhibiting the mTOR signaling pathway. Trade name: Afinitor. (Ch. 5)

external beam radiation therapy (EBRT)   Treatment by radiation emitted from a source located at a distance from the body. (Ch. 2)

extracellular matrix (ECM)   Mesh of secreted proteins, largely glycoproteins and proteoglycans, that surrounds most cells within tissues and creates structure in the intracellular space. (Ch. 10)

extranodal   Outside the lymph node. (Ch. 5)

F

fibrinogen   A soluble glycoprotein present in blood plasma that is cleaved by the protease thrombin to generate fibrin, a key protein involved in the formation and composition of blood clots. (Ch. 7)

fibroblast   A mesenchymal cell type that is common in connective tissue and in the stromal compartment of epithelial tissues. It is characterized by its secretion of collagen. (Ch. 11)

first-line chemotherapy   Medical therapy when employed as initial treatment of a cancer patient. This term is generally used in the palliative chemotherapy context for incurable cancer. (Ch. 2)

FL international prognostic index (FLIPI)   A clinical tool developed to aid in predicting the prognosis of patients with follicular lymphoma. (Ch. 5)

flow cytometry   A technique that employs lasers, a variety of specialized lenses, and a computer to quantitatively analyze the fluorescence and light-scattering properties of a population of cells as they pass, one by one, through a laser beam. Flow cytometers can determine scattering characteristics and multiple fluorescence parameters for thousands of cells per second. (Ch. 4)

fludarabine   A cytotoxic chemotherapy agent that is a purine analog, inhibiting ribonucleotide reductase and DNA polymerase. (Ch. 4)

fluorescence *in situ* hybridization (FISH)   Procedure in which a sequence-specific DNA or RNA molecule linked to a fluorescent chromophore is annealed to the DNA or RNA of cells that have been immobilized on a microscope slide. (Ch. 2)

fluorodeoxyglucose   A radiopharmaceutical used in positron-emission tomography (PET) imaging; the blood sugar glucose is labeled with fluorine-18 and taken up by metabolically active tissues including certain cancer cells. (Ch. 4)

follicular lymphoma (FL)   The most common indolent lymphoma in the United States. FL is characterized by effacement of the normal lymph node architecture by back-to-back nodules of malignant lymphocytes, known as follicles. (Ch. 5)

fusion PET/CT   Merged CT and PET scans that help with staging a variety of cancers. (Ch. 4)

G

Gardasil   A vaccine that prevents cervical cancer and genital warts by preventing infection by four strains of oncogenic human papillomavirus (HPV-6, -11, -16, and -18). (Ch. 12)

gastrointestinal stromal tumor (GIST)   The most common mesenchymal neoplasm of the gastrointestinal tract, arising in the smooth muscle pacemaker interstitial cells of Cajal, frequently driven by mutations in *KIT*, *PDGFRA*, or *BRAF* genes. (Ch. 4)

gefitinib   A tyrosine kinase inhibitor targeting EGFR. Trade name: Iressa. (Ch. 4)

gene expression arrays   All-inclusive or selective microarrays of representative gene sequences for the identification of expressed genes. (Ch. 6)

genital warts   Small, soft, pink or red swelling of the genital skin that becomes pedunculated and may be painless, caused by a sexually transmitted disease, human papillomavirus (HPV), which accounts for over 50% of all cases of sexually transmitted disease. *Also called* venereal warts *or* condylomata acuminata. (Ch. 12)

genomics   Use of DNA sequences and RNA transcripts to determine the structure and function of genomes. (Ch. 2)

germ cell tumors (GCTs)   Neoplasms derived from germ cells (diploid cells that give rise to gametes). Germ cell tumors can be cancerous or non-cancerous tumors. (Ch. 3)

germinal center   Site of intense B-cell proliferation and differentiation that develops in a lymphoid follicle during an adaptive immune response. Somatic hypermutation and class switching occur in germinal centers. (Ch. 5)

germinal-center B-cell (GCB)-**like**   A subtype of diffuse large B-cell lymphomas (DLBCLs) with cells resembling the phenotypes of rapidly proliferating B cells that develop in lymphoid follicles during an adaptive immune response. (Ch. 11)

germline mutation   Any detectable variation within germ cells (cells that, when fully developed, become sperm and ovum), which is thereby a heritable mutation. (Ch. 9)

Gleason score   The pathology grading system used to determine the aggressiveness of prostate cancer. (Ch. 6)

GNAQ   A gene that encodes guanine nucleotide-binding protein G[q] subunit α. (Ch. 4)

gout   A form of inflammatory arthritis characterized by high uric acid levels in the bloodstream leading to deposition of uric acid crystals in joints, causing significant pain, redness, and swelling. (Ch. 4)

graft-versus-host disease (GVHD)   A rejection response of allogeneic hematopoietic stem cell transplants, in which immunocompetent cells in the donated tissue recognize the recipient's tissues as foreign and attack them. It is commonly associated with inadequate immunosuppressive therapy. (Ch. 4)

graft-versus-leukemia effect   The recognition of and attack on leukemia cells mounted by T cells in allogeneic hematopoietic stem cell transplants. (Ch. 4)

granulocyte-macrophage colony-stimulating factor (GM-CSF)   A cytokine involved in the growth and differentiation of cells of the myeloid lineage, including dendritic cells, monocytes, tissue macrophages, and granulocytes. (Ch. 12)

GTPases   Enzymes whose activity catalyzes the hydrolysis of guanosine triphosphate to guanosine diphosphate and orthophosphate. (Ch. 4)

gynecomastia   An endocrine disorder causing a noncancerous increase in the size of male breast tissue. (Ch. 3)

H

heavy chain (V_H)   The larger of the two types of protein chains in an antibody molecule. (Ch. 12)

helper T cells (T_H)   T cells bearing CD4 coreceptors that promote or "help" antibody production by B cells and promote the activation of naive CD8 T cells. (Ch. 12)

hematogenous   Originating or transported in the blood. (Ch. 6)

hematologic remission   The achievement of a normal complete blood count (CBC). (Ch. 4)

hematopoiesis   Process that results in the formation of all cells in the blood including its red and white cells, the latter including various cells of the immune system. (Ch. 6)

hematuria   Presence of blood in the urine. It is symptomatic of many renal diseases and disorders of the genitourinary system and is detected by microscopic examination of urine sediment. (Ch. 5)

hemoptysis   Coughing up of blood from the respiratory tract. (Ch. 3)

hepatitis B virus (HBV)   A double-stranded DNA virus, a species of the genus *Orthohepadnavirus*, and a member of the *Hepadnaviridae* family of viruses. (Ch. 12)

hepatomegaly   Abnormal enlargement of the liver that is usually a sign of disease, often discovered by percussion and palpation as part of a physical examination. (Ch. 4)

herd immunity   Protection conferred on unvaccinated members of a population by the vaccine-induced protection of vaccinated members of the population. (Ch. 12)

hereditary nonpolyposis colorectal cancer (HNPCC)   An autosomal dominant genetic condition associated with a high risk of colon cancer and other cancers including endometrial cancer, ovarian cancer, stomach cancer, and cancers of the small intestine, hepatobiliary tract, upper urinary tract, brain, and skin. *Also called* Lynch syndrome. (Ch. 10)

herpes simplex virus 1 (HSV-1)   A member of the *Herpesviridae* herpesvirus family that infects humans. HSV-1 (which produces most cold sores and some cases of genital herpes) is ubiquitous and contagious and can be spread when an infected person is producing and shedding the virus. (Ch. 12)

histone   Any of a group of strongly basic, low–molecular-weight proteins that combine with DNA in the nuclei of eukaryotic cells and function in regulating gene activity. (Ch. 8)

histone acetyltransferase (HAT)   An enzyme that transfers acetyl groups from acetyl CoA onto conserved lysine amino acid residues on histone proteins, disrupting the histone–DNA complex and opening the chromatin to allow gene expression. (Ch. 8)

histone deacetylase (HDAC)   An enzyme that removes acetyl groups from an ϵ-*N*-acetyl lysine amino acid residue on a histone, allowing the histones to wrap the DNA more tightly and preventing access for mRNA transcription. (Ch. 8)

Hodgkin Reed–Sternberg (HRS) cell   The common underlying malignant cell of all subtypes of classical Hodgkin lymphoma. The HRS cell is multinucleate and orchestrates a dense immune microenvironment in which the HRS makes up only 1% of the overall tumor cellularity. (Ch. 11)

homeostasis   A relative constancy in the internal environment of the body, naturally maintained by adaptive responses that promote healthy survival. (Ch. 9)

homologous recombination   A type of genetic recombination for double-strand DNA breaks mediated by BRCA1/2, in which nucleotide sequences are exchanged between two similar or identical molecules of DNA. (Ch. 9)

hormone   A complex chemical substance produced in one part or organ of the body that initiates or regulates the activity of an organ or a group of cells in another part. (Ch. 6)

human anti-mouse antibodies (HAMA)   Human antibodies induced in reaction to mouse monoclonal antibodies that have been introduced for therapeutic purposes. (Ch. 11)

human epidermal growth factor receptor 2 (HER2)   A receptor tyrosine kinase overexpressed in many cancers, particularly breast cancer, and related to EGFR. HER2 is the target of trastuzumab therapy. (Ch. 11)

human leukocyte antigen (HLA)   The human version of the major histocompatibility complex (MHC), a gene family occurring in all vertebrate species. (Ch. 12)

human papillomavirus (HPV)   A small DNA virus that is the cause of common warts of the hands and feet, as well as lesions of the mucous membranes of the oral, anal, and genital cavities. It can be transmitted through sexual contact. Specific subtypes of HPV can cause cervical cancer, anal cancer, and head and neck cancer. (Ch. 11)

hydronephrosis   Distension of the pelvis calyces of the kidney by urine that cannot flow past an obstruction in a ureter. (Ch. 5)

hypercalcemia   Presence of elevated concentrations of calcium ions in the blood. (Ch. 9)

hypercellular   Related to or resulting from an abnormal excess of cells. (Ch. 8)

hyperdiploidy   Any increase in chromosome number that involves individual chromosomes rather than entire sets; in humans, this results in more than the normal diploid number of chromosomes (46) in somatic cells. (Ch. 2)

hypermethylation   An increase in the epigenetic methylation of cytosine and adenosine. (Ch. 5)

hypersensitivity   Inflammatory responses, which can be severe, caused by excessive or inappropriate responses of the immune system to an antigen. (Ch. 11)

hypodiploidy   Any decrease in chromosome number that involves individual chromosomes rather than entire sets; in humans, this results in fewer than the normal diploid number of chromosomes (46) in somatic cells. (Ch. 2)

hypogammaglobulinemia   Lower than normal concentration of plasma gamma globulin, usually the result of increased protein catabolism or loss of protein via the urine. It is associated with a decreased resistance to infection. (Ch. 4)

hypomethylating agent   A drug that inhibits DNA methylation. (Ch. 8)

hypophysitis   Inflammation of the pituitary gland. (Ch. 12)

hypothalamus   A portion of the diencephalon of the brain, forming the floor and part of the lateral wall of the third ventricle, that is active in hormonal regulation of various endocrine organs. (Ch. 6)

hypoxia   State of lower than normal oxygen tension. (Ch. 8)

hypoxia-inducible factor (HIF)   Transcription factor that mediates the production of vascular endothelial growth factor (VEGF). (Ch. 5)

hysterectomy   The surgical removal of the uterus. (Ch. 12)

I

ibrutinib   A small-molecule covalent inhibitor of Bruton's tyrosine kinase (BTK). Trade name: Imbruvica. (Ch. 4)

idelalisib   A potent, highly selective targeted inhibitor of PI3K that is effective at decreasing the downstream activity of the PI3K pathway, including decreased activation of AKT and mTOR. Trade name: Zydelig. (Ch. 5)

Ig heavy-chain variable-region genes (IGHV)   Genes encoding the variable regions of immunoglobulin heavy chains. (Ch. 4)

IGH   The immunoglobulin heavy locus, a region on human chromosome 14 that contains a gene cluster encoding the heavy chains of human antibodies (or immunoglobulins). Translocations occur between this locus and *BCL2* in some types of non-Hodgkin lymphoma. (Ch. 5)

imatinib mesylate   An oral tyrosine kinase inhibitor (TKI) that binds to the ATP-binding domain of the BCR–ABL fusion protein, resulting in inhibition of the protein's oncogenic activity. Trade name: Gleevec. (Ch. 4)

imiquimod   A drug known to activate TLR-7, FDA-approved for the topical treatment of basal cell carcinoma and genital warts. Trade names: Aldara, Zyclara. (Ch. 12)

immune checkpoint inhibitor   *See* **checkpoint inhibitors**. (Ch. 11)

immune system   The set of organs, tissues, cells, and molecules involved in innate immunity and adaptive immunity. (Ch. 11)

immunofixation   Process by which antigens in a protein mixture are separated on an electrophoretic gel and identified by the application of labeled antibodies. (Ch. 9)

immunogenic   Capable of provoking an immune response. (Ch. 10)

immunogenic antigen   Any molecule that is able to elicit an adaptive immune response upon introduction into a person or animal. (Ch. 11)

immunoglobulin genes (Ig genes)   Genes that encode the light and heavy chains of immunoglobulins. (Ch. 4)

immunoglobulin heavy- and light-chain genes   Genes that encode the larger and smaller protein subunits that make up antibody molecules. (Ch. 4)

immunological memory   The ability of the adaptive immune system to respond more rapidly and effectively in the future to subsequent exposures to antigens to which it has previously responded.

The technology of vaccination is based on the phenomenon of immunological memory. (Ch. 11)

immunomodulatory   Capable of modifying or regulating an immune response, usually in a beneficial way. (Ch. 8)

immunomodulatory drugs (IMiDs)   A class of drugs containing an imide group that may act by inhibiting production of certain cytokines, co-stimulating immune cells, and increasing production of interferon and IL-2. The IMiD class includes thalidomide and its analogs. (Ch. 10)

immunotherapy   The introduction or induction of agents of the immune system to treat disease. An example of this is the administration of monoclonal antibodies against antigens present on the surface of cancer cells to treat cancer. (Ch. 11)

in vitro   Occurring in tissue culture or in cell lysates or in purified reaction systems outside of a person or animal model. (Ch. 7)

incidence   Number of cases of a condition or disease diagnosed in a population in a defined time period. (Ch. 1)

indels   DNA sequence insertions and deletions. (Ch. 2)

index patient   The first identified case of a disease in a group (such as family members) of patients who have contracted or are at risk of contracting a disease. (Ch. 2)

indoleamine 2,3-dioxygenase (IDO) inhibitors   Molecules that block or inhibit indoleamine 2,3-dioxygenase thereby preventing or relieving its immunosuppressive effects. (Ch. 12)

indolent   Slow to grow. (Ch. 4)

induction chemotherapy   Chemotherapy as the initial treatment for cancer, especially as part of combined modality therapy. This term, in contrast to "first-line chemotherapy," is usually applied to the management of cancer with curative intent. (Ch. 7)

infectious mononucleosis   A usually mild disease with symptoms of fever, malaise, and swollen lymph nodes caused by infection with Epstein–Barr virus. (Ch. 11)

inflammatory bowel disease   General name for a set of inflammatory conditions in the gut, including Crohn's disease and ulcerative colitis. (Ch. 10)

inflammatory cytokines   Cytokines produced in response to infection or injury that induces or increases inflammation. (Ch. 12)

informed consent   Permission to perform a particular test, procedure, or treatment obtained from a patient who is of legal age and mentally fully competent to give consent. (Ch. 12)

inguinal   Pertaining to the groin. (Ch. 5)

innate immunity   The various inherent resistance mechanisms that are first encountered by a pathogen before adaptive immunity is induced. Agents of innate immunity include anatomical barriers, antimicrobial peptides, the complement system, and cells such as macrophages, neutrophils, and NK cells carrying nonspecific pathogen recognition receptors. Innate immunity is present in all individuals at all times and does not increase with repeated exposure to a pathogen, nor is it specific to a particular pathogen. (Ch. 11)

insulin-like growth factor-1 (IGF-1)   A hormone that stimulates protein synthesis; it is the primary mediator of the effects of growth hormone. (Ch. 5)

intensity-modulated radiation therapy (IMRT)   A specialized method of external beam radiation therapy (EBRT) delivering radiation at variable intensities from many different angles to maximally treat the tumor with pinpoint accuracy while sparing much of the surrounding tissue. (Ch. 2)

interferon   A member of one of several closely-related cytokine families originally named for their interference with viral replication. Some interferons have powerful antiviral effects and others display marked effects on immune and inflammatory processes. (Ch. 4)

interleukin-2 (IL-2)   A protein with various immunological functions, including the ability to support proliferation of activated T cells. IL-2 is used in the laboratory to grow T-cell clones with specific helper, cytotoxic, and suppressor functions. (Ch. 4)

International Prognostic Index (IPI)   A prognostic tool composed of five clinical risk factors used in evaluating cases of Hodgkin lymphoma. (Ch. 11)

intravasation   Process of invading a blood or lymphatic vessel from the surrounding tissue. (Ch. 10)

intravenous immune globulin (IVIG)   IgG antibodies purified from the pooled plasma of tens of thousands of donors that is infused intravenously to protect against or treat infections. (Ch. 12)

ipilimumab   Antibody to human CTLA-4, and first checkpoint blockade immunotherapy. Trade name: Yervoy. (Ch. 4)

ipsilateral   Affecting the same side of the body. (Ch. 6)

isoform   A functionally similar form of a protein with similar but nonidentical amino acid sequence; for example, the different forms encoded by different alleles of the same gene. (Ch. 5)

isoform-specific inhibitors   Molecules that selectively inhibit only particular members (isoforms) of a family of proteins that are highly similar. (Ch. 5)

J

jaundice   A yellow discoloration of the skin, mucous membranes, and sclerae of the eyes caused by greater than normal amounts of bilirubin in the blood. (Ch. 3)

JUN   A protein that associates with another protein, FOS, to form AP-1, a transcription factor. (Ch. 4)

K

κ (kappa) or **λ (lambda) light chains**   The two types of smaller protein subunits of an antibody molecule. (Ch. 4)

karyotype   The number and morphology of a cell's chromosomes. (Ch. 2)

karyotyping   The determination of a karyotype used in cytogenetic analysis to detect chromosomal abnormalities. (Ch. 5)

Kit   A receptor tyrosine kinase, mutations of which drive certain tumors including GIST. (Ch. 4)

kyphoplasty   A minimally invasive procedure to treat a vertebral compression fracture, using an inflatable balloon to expand the collapsed vertebral body, followed by injection of bone cement into the created space to restore the height of the vertebral body and reduce pain. In *vertebroplasty*, no balloon is used; cement is injected percutaneously into the collapsed vertebral body. (Ch. 10)

L

lactate dehydrogenase (LDH)   An enzyme that is found in the cytoplasm of almost all body tissues, whose main function is to catalyze the oxidation of lactate to pyruvate. (Ch. 5)

laparoscopic   Relating to laparoscopy, which employs a fiber-optic instrument consisting of an illuminated tube with an optical system to assist in invasive procedures. (Ch. 6)

large-cell carcinoma   A heterogeneous group of undifferentiated malignant neoplasms that lack the cytologic and architectural features of small-cell carcinoma and glandular or squamous differentiation. LCC is categorized as a type of NSCLC (non-small cell lung cancer), which originates from epithelial cells of the lung. (Ch. 4)

lenalidomide   A targeted therapy with multiple effects including inhibition of angiogenesis, cytokine secretion, and proliferation. Used to treat myelodysplastic syndromes and hematologic malignancies including multiple myeloma. Trade name: Revlimid. (Ch. 8)

lentigo maligna melanoma   A neoplasm of melanocytes developing from Hutchinson's freckle on the face or other exposed surfaces of the skin in elderly patients. It is asymptomatic, flat, and tan or brown, with irregular darker spots and frequent hypopigmentation. (Ch. 4)

lentiviral vectors   Genomes of lentiviruses, a family of retroviruses, engineered to deliver genes to cells. (Ch. 12)

leukemias   Malignancies of any of a variety of hematopoietic cell types, including the lineages leading to lymphocytes and granulocytes, in which the tumor cells are predominantly found in the blood and/or bone marrow. (Ch. 1)

leukocytosis   The presence of increased numbers of leukocytes (white blood cells) in the blood. This is commonly seen in acute infection but can also be seen in leukemia. (Ch. 7)

leukopenia   An abnormal decrease in the total peripheral white blood cell count, often associated with chemotherapy or radiation. (Ch. 7)

light chain (V_L)   The smaller of the two types of immunoglobulin molecules comprising an antibody. (Ch. 12)

liquid biopsy   Sampling of blood to determine the presence of circulating cancer cells or circulating DNA of cancer origin. Analysis of such cells or DNA can aid in cancer diagnosis and in monitoring patient responses to some therapies. (Ch. 2)

long noncoding RNA (lncRNA)   RNA transcripts exceeding 200 nucleotides in length that are not translated into polypeptides. Their exact activity in the cell is not well understood but they may impact transcription. (Ch. 8)

loss of heterozygosity (LOH)   A genetic event in which one of two alleles at a heterozygous locus is lost; the lost allele may simply be discarded or may be replaced with a duplicated copy of the surviving allele. (Ch. 5)

lumbar puncture   A diagnostic or therapeutic procedure in which a hollow needle and stylet are introduced into the subarachnoid space of the lumbar part of the spinal canal to obtain cerebrospinal fluid (CSF). Strict aseptic technique is used. *Also called* spinal tap. (Ch. 12)

lymph node dissection   Surgical removal of a lymph node or nodes. (Ch. 4)

lymph nodes   Type of peripheral lymphoid organs present in many locations throughout the body where lymphatic vessels converge, facilitating the activation of lymphocytes present in the nodes by lymphatic fluid-borne antigens for which they are specific. (Ch. 2)

lymphadenopathy   Enlarged lymph nodes. (Ch. 4)

lymphedema   A primary or secondary condition characterized by the accumulation of lymph in soft tissue and the resultant swelling caused by inflammation, obstruction, or removal of lymph channels. (Ch. 4)

lymphocytes   Class of leukocytes that mediate humoral or cellular immunity, encompassing B cells, T cells, and sometimes such innate lymphoid cells as NK cells and some others. (Ch. 4)

lymphocytosis The presence of increased numbers of lymphocytes, as occurs in certain chronic diseases and during convalescence from acute infections. (Ch. 8)

lymphoid cells Mononuclear leukocytes that mediate cellular and humoral immunity. The category includes B cells and T cells and other cell types that lack T- or B-cell receptors, such as NK cells and other innate lymphoid cells as well. (Ch. 4)

lymphomagenesis The growth and development of lymphoma. (Ch. 11)

lymphomas Cancers derived from mature lymphocytes or lymphoid precursors, predominantly found in lymph nodes, other lymphoid organs, or extranodal tissues. (Ch. 1)

Lynch syndrome *See* **hereditary nonpolyposis colorectal cancer**. (Ch. 10)

lytic Referring to disintegration of a cell or tissue; often associated with the potent cytopathic effects of certain viruses on specific host cells. In reference to bone metastasis, a shortened term for "osteolytic," referring to destruction of bone. (Ch. 6)

M

M protein A monoclonal immunoglobulin that can be detected on serum protein electrophoresis with immunofixation and is produced by proliferating myeloma cells. *Also called* paraprotein *or* monoclonal protein. (Ch. 9)

macrophage activation syndrome (MAS) A severe systemic condition caused by uncontrolled activation and proliferation of macrophages and T cells, seen in response to CAR T-cell therapy for cancer patients, as well as several chronic rheumatic diseases of childhood. (Ch. 12)

macrophages Large mononuclear phagocytic cells present in most tissues that have many functions, such as scavenging, pathogen recognition, and production of pro-inflammatory cytokines. Macrophages arise from bone marrow. (Ch. 11)

maintenance therapy A medical therapy that is designed to help a primary treatment succeed. For example, maintenance chemotherapy may be given to people who have a cancer in remission in an attempt to prevent a relapse. (Ch. 7)

major molecular response A PCR-based measure of treatment response based on reduction of a cancer-specific mRNA transcript. In chronic myelogenous leukemia, a complete molecular response is defined as at least a 3 log (1000-fold) reduction of *BCR–ABL* transcripts to <0.1% of baseline after treatment. (Ch. 4)

mastodynia Pain in the breast. (Ch. 3)

median overall survival The length of time—from either the date of diagnosis of a disease or the start of treatment—for which half of a group of patients diagnosed with the disease are still alive. (Ch. 3)

mediastinal Pertaining to a median septum or space between two parts of the body, most commonly referring to the mediastinum. (Ch. 4)

mediastinoscopy An examination of the mediastinum through an incision above the sternum, by using an endoscope with light and lenses. (Ch. 4)

mediastinum A part of the thoracic cavity in the middle of the thorax. (Ch. 3)

melanocytes Cells of neural crest origin that create pigmentation of the skin and iris. (Ch. 4)

melanomas Tumors arising from melanocytes, the pigmented cells of the skin, iris, and retinal pigmented epithelium. (Ch. 1)

membrane attack complex (MAC) A protein complex composed of certain complement components that assemble into a membrane-spanning pore through the cell membrane of pathogens, causing cell lysis. (Ch. 11)

meninges The three membranes enclosing the brain and the spinal cord, comprising the dura mater, the pia mater, and the arachnoid membrane. (Ch. 4)

mesenteric Pertaining to the mesentery, the double layer of peritoneum suspending the intestine from the posterior abdominal wall. (Ch. 5)

MET A single-pass tyrosine kinase receptor essential for embryonic development, organogenesis, and wound healing. (Ch. 4)

meta-analysis A statistical procedure for pooling and integrating the results of a number of distinct experimental, clinical, or epidemiologic studies in order to generate a larger data set and greater statistical significance than that afforded by the data of a single available study. (Ch. 9)

metamyelocytes A stage in the development of the granulocyte series of leukocytes, between the myelocyte stage and the neutrophilic band. (Ch. 4)

methylation The introduction of a methyl group, $-CH_3$, to a chemical compound. (Ch. 8)

MHC molecules Dimeric protein molecules containing one or two highly polymorphic polypeptide subunits. The major histocompatibility complex encodes the polymorphic subunit(s). MHC molecules present peptide antigens to T cells. (Ch. 12)

microarray A collection of sequence-specific DNA molecules that are attached to a solid substrate at specific sites. Microarrays can also incorporate RNA or proteins. (Ch. 2)

microRNAs Endogenously synthesized RNAs transcribed by RNA polymerase II and processed in 21- to 23-nt-long single-strand RNA sequences that interfere with translation of an mRNA or cause its degradation, depending on the degree of complementarity with the mRNA. (Ch. 4)

microsatellite A DNA tract that consists of a succession of repeating units of identical or similar nucleotide sequences. (Ch. 2)

microsatellite instability (MSI) The condition of genetic hypermutability (predisposition to mutation) that results from impaired DNA mismatch repair. In cancers exhibiting the MSI-high phenotype, the number of microsatellites is higher than in normal cells. (Ch. 10)

minimal residual disease (MRD) The name given to small numbers of leukemic cells that remain in a patient during treatment, or after treatment when the patient is in remission. (Ch. 12)

mismatch repair The class of DNA repair processes for proofreading a recently synthesized segment of DNA and removing any misincorporated bases. (Ch. 9)

mitochondrion A rodlike, threadlike, or granular organelle that functions in aerobic respiration and occurs in varying numbers in all eukaryotic cells except mature erythrocytes. (Ch. 5)

mitogen An agent that provokes cell proliferation. (Ch. 10)

mitotic rate The number of cells undergoing division in a specified amount of cancer tissue. (Ch. 4)

molecularly targeted therapies Treatments that use drugs or other substances to target specific molecules involved in the growth, survival, or spread of cancer cells. (Ch. 3)

monoclonal antibodies Antibodies produced by a single clone of B lymphocytes, so that they are all identical. (Ch. 4)

monoclonal B-cell lymphocytosis (MBL)   Presence of fewer than 5000 clonal cells/ml in the absence of lymphadenopathy or splenomegaly. (Ch. 4)

monoclonal gammopathies   A group of abnormal conditions characterized by the presence of high levels of abnormal immunoglobulin proteins in the blood. *Also called* plasma cell dyscrasias. (Ch. 9)

monocytes   Type of white blood cells with bean-shaped nuclei; they are a precursor of tissue macrophages. (Ch. 4)

monomethyl auristatin E (MMAE)   A highly potent toxin that acts as an antimitotic agent by blocking the polymerization of tubulin that is necessary for mitosis. (Ch. 11)

morbidity   The existence of a medical condition or disease. Also, the medical problems caused by a treatment. (Ch. 1)

mTOR   Mechanistic target of rapamycin; a serine/threonine kinase that functions in regulating numerous aspects of cell metabolism and function in complex with the regulatory proteins Raptor and Rictor. (Ch. 5)

mTORC1 and **mTORC2**   Active complexes of mTOR formed with the regulatory proteins Raptor and Rictor, respectively. (Ch. 5)

mucositis   Any inflammation of a mucous membrane, such as the lining of the mouth and throat. (Ch. 7)

MYC   A family of transcription factors that are overexpressed in many cancers and drive cell proliferation. Originally discovered as a homolog of the oncogene v-*myc* carried by an avian myelocytomatosis retrovirus. (Ch. 4).

mycosis fungoides (MF)   A rare chronic lymphomatous skin malignancy resembling eczema or a cutaneous tumor that is followed by microabscesses in the epidermis and can involve lymph nodes, blood, and visceral organs. (Ch. 8)

myeloablative chemotherapy   Therapy involving the elimination of the hematopoietic system, including its components residing in the bone marrow. (Ch. 3)

myeloblasts   Earliest recognizable precursor of the granulocytic leukocytes. The cytoplasm appears light blue, scanty, and nongranular when seen in a stained blood film. (Ch. 4)

myelocytes   The third of the maturation stages of the granulocytic leukocytes normally found in the bone marrow. Granules are visible in the cytoplasm. (Ch. 4)

myelodysplasia   *See* **myelodysplastic syndrome**. (Ch. 11)

myelodysplastic syndrome (MDS)   A hyperproliferative condition of cells of the myeloid lineage in the bone marrow that often progresses to acute myelogenous leukemia. (Ch. 8)

myeloid cells   Cells from the hematopoietic lineage that are not lymphocytes, including erythrocytes, platelets, granulocytes, monocytes, mast cells, and their derivatives. (Ch. 4)

myeloid-derived suppressor cells (MDSCs)   A heterogeneous population of cells in tumors that can inhibit T-cell activation within the tumor. (Ch. 10)

myelomas   Malignancies deriving from plasma cells, the antibody-secreting cells of the B cell lineage. (Ch. 1)

myeloproliferative   Referring to excessive proliferation and resulting elevated levels of one of the several myeloid cell types. (Ch. 4)

myeloproliferative neoplasms   A family of chronic malignant bone marrow and blood diseases caused by mutations that generate clones of myelocytic, erythrocytic, or platelet precursors. (Ch. 4)

N

naive T cells   Lymphocytes that have not encountered their specific antigen under activating or tolerizing conditions. (Ch. 12)

natural killer (NK) **cell**   A type of innate lymphoid cell that is important in innate immunity to viruses and other intracellular pathogens. It is the major cell type responsible for the mediation of antibody-dependent cell-mediated cytotoxicity (ADCC). (Ch. 11)

negative feedback   A regulatory process whereby a stimulatory signal provokes activation of a downstream inhibitory mechanism that proceeds to decrease the initial provoking signal. (Ch. 6)

neo-adjuvant therapy   A treatment given before the main treatment (usually surgery) is applied. (Ch. 2)

neoplastic follicles   Malignant transformation of a normal lymph node component, known as the germinal center, by lymphoma. (Ch. 5)

neural crest   Region of the early embryo that serves as precursor of various specialized tissues and cell types, including certain cells of the peripheral nervous system, bones of the face, melanocytes, and several types of neurosecretory cells. (Ch. 4)

neuroendocrine   Characterized by having features of nervous and endocrine (hormone-producing) cells. (Ch. 4)

neuropathy   Inflammation or degeneration of the peripheral nerves, which can be caused by medications (such as chemotherapy), disease (such as diabetes), or toxins (such as lead poisoning). (Ch. 3)

neutropenia   Deficiency of neutrophils in the blood. (Ch. 3)

neutrophils   An abundant granulocyte in the circulation that expresses Fc receptors and is responsible for recognizing and engulfing various types of infectious agents, notably bacteria. (Ch. 4)

nevi   Atypical moles. (Ch. 4)

next generation sequencing (NGS)   An umbrella term used to describe a set of modern technologies that enable high-throughput sequencing of DNA. (Ch. 2)

NF-κB   A heterodimeric transcription factor composed of p50 and p65 subunits, activated in a variety of ways including by the stimulation of Toll-like receptors and also by antigen receptor signaling. (Ch. 4)

nilotinib   A tyrosine kinase inhibitor that inhibits BCR–ABL and PDGFR. Trade name: Tasigna. (Ch. 4)

nivolumab   A monoclonal antibody that interferes with PD-1, a checkpoint inhibitor that inhibits T cell–mediated immune responses. Trade name: Opdivo. (Ch. 12)

nodular lymphocyte-predominant Hodgkin lymphoma (NLPHL)   A distinct and uncommon category of Hodgkin lymphoma (5% of cases). The malignant cell is the lymphocyte-predominant (LP) cell and widely expresses CD20. (Ch. 11)

nodular melanoma   A subtype of melanoma that is characterized by nodules (lumps) that are uniformly pigmented, usually bluish-black, and sometimes surrounded by an irregular halo of pale, unpigmented skin. The lesion is always raised and may be dome-shaped or polypoid. (Ch. 4)

non-Hodgkin lymphoma (NHL)   Solid tumor of peripheral lymphoid tissue classified by histological features and cell of origin distinct from other lymphomas. (Ch. 5)

non-small-cell lung cancer (NSCLC)   Any of several types of lung cancers with the exception of small-cell lung carcinoma. Subtypes include adenocarcinoma, squamous cell carcinoma, and bronchioalveolar carcinoma. (Ch. 4)

NRAS A member of the RAS family of GTPases. (Ch. 4)

nucleosome A structural subunit of chromatin comprised of a protein octamer (composed of two each of histones H2A, H2B, H3, and H4) around which DNA is wrapped. (Ch. 8)

O

obinutuzumab An anti-CD20 monoclonal antibody. Trade name: Gazyva. (Ch. 4)

ofatumumab A fully human monoclonal antibody specific for CD20, an antigen found on many cells of the B-cell lineage. Trade name: Arzerra. (Ch. 4)

oligometastatic disease A disease state with a small number of new tumors (metastatic tumors) in one or two other parts of the body. Even though metastatic, some cases may still be curable. (Ch. 5)

oligos Collections of DNA oligonucleotide sequences used to create microarrays on a chip. (Ch. 2)

oncogene A cancer-inducing gene or a gene that can transform cells. (Ch. 4)

oncogene addiction The physiological state of a cancer cell in which it is absolutely dependent for its proliferation or survival on the continued function of a certain oncogene. (Ch. 4)

oncolytic virus Virus that causes the lysis of cells it infects, generating inflammatory responses and releasing cellular antigens, some of which may provoke an anti-tumor response. (Ch. 12)

oncomiR A microRNA associated with or causing cell transformation or contributing to tumor progression. (Ch. 8)

oncoprotein A protein product expressed by an oncogene. (Ch. 7)

oral leukoplakia A disease characterize by thickened, white patches inside the mouth. (Ch. 7)

orchitis Inflammation of one or both of the testes, characterized by swelling and pain. (Ch. 3)

osteoblast Mesenchymal cell type related to fibroblasts that constructs mineralized bone through the deposition of a collagenous matrix and apatite crystals. (Ch. 10)

osteoclast Cell type of monocyte origin that functions to degrade and demineralize already assembled bone. (Ch. 9)

osteolytic metastases A type of metastasis to bone in which the tumor has caused bone breakdown or thinning. (Ch. 9)

overall survival (OS) Proportion of patients who are still alive at a certain time following initiation of a treatment, often measured after a certain interval, such as 5 years, following initiation. (Ch. 4)

P

p53 A 53-kDa protein that has a variety of effects on cells as a key tumor suppressor, including roles in apoptosis and genomic stability. *Also called* TP53. (Ch. 12)

pack-year 1 pack-year = the number of cigarettes a 1-pack-per-day smoker smokes in 1 year. (Ch. 4)

palliative Describing a therapy designed to relieve or reduce symptoms but not to produce a cure. (Ch. 2)

pallor An unnatural paleness or absence of color in the skin. (Ch. 12)

pancreatectomy The surgical removal of all or part of the pancreas, performed to excise a tumor. (Ch. 3)

pancreatic intraepithelial neoplasia-1 (PanIN1) A histologically well-defined precursor to invasive ductal adenocarcinoma of the pancreas. (Ch. 3)

pancreatic neuroendocrine cell tumor (PNET) Neuroendocrine neoplasm that arises from cells of the endocrine (hormonal) and nervous system within the pancreas. (Ch. 3)

pancreaticoduodenectomy A surgical procedure in which the head of the pancreas, the entire duodenum, a portion of the jejunum, the distal third of the stomach, and the lower half of the common bile duct are excised, usually to relieve obstruction caused by tumors, often malignant. *Also called* Whipple procedure. (Ch. 3)

panitumumab A monoclonal antibody that inhibits epidermal growth factor receptor (EGFR). Trade name: Vectibix. (Ch. 10)

Papanicolaou screening (Pap smears) A simple smear method of examining stained exfoliative cells, most commonly used to detect precancerous lesions or cancers of the cervix. (Ch. 12)

paratracheal Near the trachea. (Ch. 4)

passenger mutation Mutation that does not confer a selective growth advantage. (Ch. 1)

pathogen-associated molecular patterns (PAMPs) Molecules specifically associated with pathogens that are recognized by cells of the innate immune system. (Ch. 11)

pathologic complete response (pCR) The lack of all signs of cancer in tissue samples removed during surgery or biopsy after treatment with radiation or chemotherapy. (Ch. 9)

pazopanib A potent and selective multi-targeted receptor tyrosine kinase inhibitor that blocks tumor growth and inhibits angiogenesis. Trade name: Votrient. (Ch. 5)

pembrolizumab A monoclonal antibody that interferes with PD-1, a checkpoint inhibitor that inhibits T cell–mediated immune responses. Trade name: Keytruda. (Ch. 4)

penetrance Degree to which or frequency with which an allele of a gene can influence phenotype, e.g., the likelihood that a germline allele will induce a clinical phenotype in a carrier of this allele. (Ch. 6)

peptide–MHC complex A complex of MHC with a peptide derived from the degradation of a protein antigen. Peptide–MHC complexes are the molecular entities recognized by T-cell receptors. (Ch. 12)

percutaneous ablation Procedure for destroying a small tumor by inserting instruments or injecting agents through the skin. (Ch. 5)

percutaneous biopsy A biopsy performed through the skin. (Ch. 4)

performance status Quantification of a patient's general functional status and ability to perform activities of daily life. (Ch. 3)

perioperative Pertaining to the time before or after surgery. (Ch. 3)

peripheral neuropathy *See* **neuropathy**. (Ch. 9)

petechiae Small and often numerous red or purple spots on the skin resulting from tiny hemorrhages from capillaries in the skin. Petechiae are usually flat and do not lose color (blanch) when pressed. (Ch. 12)

pharmacogenomics The study of how the genome determines responses to drugs. (Ch. 2)

pharmacokinetics The effect of the body on an administered drug including its excretion and metabolic transformations. (Ch. 11)

pheochromocytoma Tumor of the neuroectodermal cells of the adrenal glands. (Ch. 5)

Philadelphia chromosome A translocation of the long arm of chromosome 22 with chromosome 9, often seen in chronic myelocytic leukemia. (Ch. 4)

phosphatase An enzyme that removes phosphate groups from a phosphorylated substrate, such as the phosphoamino acid residues in a protein or the phosphorylated inositol of a phospholipid. (Ch. 5)

phosphatidylinositol 3-kinase (PI3K *or* PI3 kinase) A family of enzymes involved in cell functions such as growth, proliferation, differentiation, motility, survival, and intracellular trafficking, which in turn are involved in cancer. (Ch. 5)

phosphorylation Covalent attachment of a phosphate group to a substrate, often a protein. (Ch. 4)

phototherapy Treatment of disorders by the use of light, especially ultraviolet light. (Ch. 8)

PI3K/Akt/mTOR pathway A kinase pathway that plays a central role in regulating the normal function of cells and is one the most frequently dysregulated in human cancer. (Ch. 5)

pituitary gland An endocrine gland suspended beneath the brain in the pituitary fossa of the sphenoid bone, supplying numerous hormones that govern many vital processes. (Ch. 6)

plasma The watery light-yellow fluid part of the lymph and the blood in which leukocytes, erythrocytes, and platelets are suspended. Plasma is made up of water, electrolytes, proteins, glucose, fats, bilirubin, and gases and is essential for carrying the cellular elements of the blood through the circulation, transporting nutrients, maintaining the acid–base balance of the body, and transporting wastes from the tissues. (Ch. 7)

plasma cells Cells of the B-cell lineage that secrete antibodies. (Ch. 11)

plasmacytoma A focal neoplasm containing plasma cells identical to multiple myeloma cells that may develop in the bone marrow, as a *solitary* plasmacytoma of bone; or outside the bone marrow, e.g., in soft tissue, where it is called *extramedullary* plasmacytoma. (Ch. 9)

platelet-derived growth factor-β (PDGF-β) A dimeric protein, composed of two beta subunits, that promotes the growth of cells of mesenchymal origin, such as fibroblasts. (Ch. 5)

platelet-derived growth factor receptors (PDGFRs) Cellular surface tyrosine kinase receptors for members of the PDGF family. (Ch. 4)

platelets Anucleate blood cells, 1 to 3 μm in diameter, involved in clotting. Platelets are formed from bone marrow megakaryocytes. (Ch. 7)

pleural effusion Accumulation of fluid in the space between the lungs and the surrounding pleural membrane. These collections can have benign or malignant causes. (Ch. 4)

pluripotent Referring to the ability of a stem cell to seed progeny that can participate in the formation of all of the tissues of an embryo except the extraembryonic membranes and tissues, such as the placenta. (Ch. 3)

PML–RARA protein A fusion protein consisting of PML and retinoic acid receptor encoded by a novel gene generated by a 15;17 translocation seen in the leukemic cells of nearly all patients with acute promyelocytic leukemia. (Ch. 7)

poly(ADP-ribose) polymerase (PARP) A family of proteins involved in cellular processes including DNA repair, genomic stability, and programmed cell death. (Ch. 9)

polyclonal antibodies Antibodies produced by a population of cells that trace their origins to two or more founding B-cell clones. (Ch. 11)

polymerase chain reaction (PCR) A rapid technique for *in vitro* amplification of specific DNA or RNA sequences, allowing small quantities of short sequences to be analyzed without cloning. (Ch. 8)

polyubiquitination The binding of many ubiquitin molecules to the same target protein. Polyubiquitination of proteins is the triggering signal that leads to degradation of the protein in the proteasome. (Ch. 7)

pomalidomide An IMiD; derivative of thalidomide. Mechanisms of action include inhibition of angiogenesis and immunomodulation. Trade name: Pomalyst. (Ch. 10)

ponatinib An oral tyrosine kinase inhibitor targeting BCR–ABL fusion protein with the T315I mutation. Trade name: Iclusig. (Ch. 4)

pRB A protein that acts as a tumor suppressor by preventing the entry of cells into the G_1 phase of the cell cycle by binding E2F family members. (Ch. 12)

predictive biomarker A measurable indicator associated with response (or lack thereof) to a specific therapy (Ch. 1)

prevalence The number of all (new and old) cases of a disease in a population at a given time. Contrast with *incidence* which, in medical terms, refers to new cases of a disease during a specific period of time. (Ch. 5)

primary mediastinal B-cell lymphoma (PMBCL) A subtype of diffuse large B-cell lymphoma with distinct clinical and biological features, predominantly localized within the mediastinum. (Ch. 11)

primary tumor Tumor growing at the anatomical site where tumor formation began. (Ch. 2)

progenitor cell A cell that, like a stem cell, tends to differentiate into a specific type of cell but, as an early descendant of a stem cell, cannot divide indefinitely and is more limited in the types of cells it can become. (Ch. 8)

programmed cell death protein 1 (PD-1) A receptor expressed on the surface of T cells. It is one of a number of receptors known as checkpoint inhibitors that, when engaged by their ligands, inhibit T-cell responses. (Ch. 12)

programmed death ligand 1 (PD-L1) A protein expressed on macrophages and a variety of other cells including some cancer cells. Its expression by cancer cells can protect them from T cell–mediated immune responses, by binding PD-1 on T cells. (Ch. 12)

programmed death ligand 2 (PD-L2) A protein, related to programmed death ligand 1, that, when expressed by cancer cells, also protects them from cell-mediated immune responses. (Ch. 12)

progression-free survival (PFS) Time elapsed following initiation of treatment during which a clinical condition does not worsen. (Ch. 4)

promyelocyte An immature cell of the myeloid lineage that functions as a precursor of various differentiated granulocyte cell types. (Ch. 7)

prophylactic cancer vaccines Vaccines that prevent cancer by immunization against a pathogen known to cause cancer. (Ch. 11)

proteasomes Protein complexes that degrade unneeded or damaged proteins by proteolysis, a chemical reaction that breaks peptide bonds. (Ch. 8)

protein kinase An enzyme that catalyzes the transfer of a phosphate group from adenosine triphosphate (ATP) to a substrate protein to produce a phosphoprotein. (Ch. 4)

protein kinase B (PKB) *See* **Akt**. (Ch. 5)

protein synthesis The process whereby cells generate new proteins. (Ch. 5)

proteolysis The breakdown of a protein into peptides or its constituent amino acids. (Ch. 12)

proteolytic Used to describe a process, usually mediated by proteases, of cleaving a polypeptide to lower–molecular-weight fragments including individual amino acids. (Ch. 9)

proton beam therapy An alternative form of external beam radiation therapy (EBRT) that uses protons as opposed to photons to deliver energy. (Ch. 2)

pseudotumor cerebri A condition characterized by increased intracranial pressure, headache, blurring of the optic disc margins, vomiting, and papilledema without neurological signs, except palsy of the sixth cranial nerve. (Ch. 7)

pulmonary embolism Blockage of pulmonary circulation by fat, air, tumor tissue, or a thrombus (clot) that usually arises from a peripheral vein (most frequently one of the deep veins of the legs). Predisposing factors include an alteration of blood constituents with increased coagulation, damage to blood vessel walls, and stagnation or immobilization, especially when associated with pregnancy and childbirth, cancer, congestive heart failure, polycythemia, or surgery. (Ch. 7)

pulmonary infiltrates Substances denser than air, such as pus, blood, or protein, that linger within the parenchyma of the lungs. (Ch. 7)

purpura Any of several bleeding disorders characterized by hemorrhage into the tissues, particularly beneath the skin or mucous membranes, producing ecchymosis. Purpura refers to larger lesions than petechiae. (Ch. 12)

Q

quantitative PCR (qPCR) Real-time polymerase chain reaction; the "q" is for quantitative. (Ch. 4)

R

radical nephrectomy The surgical removal of a kidney, usually performed in the treatment of kidney cancer. (Ch. 5)

radio-frequency ablation High-frequency alternating current that is applied to target tissue to raise its temperature and destroy cells. In oncology, this is typically performed percutaneously (through the skin) using imaging guidance. (Ch. 5)

radionuclides Atomic species that undergo radioactive decay, thereby emitting radioactivity (Ch. 2)

radiopharmaceutical A drug that emits radioactivity. In this era of targeted therapies, the radioactivity may be targeted by conjugation of the radiopharmaceutical to an antibody that recognizes a protein on tumor cells or by the differential uptake of the radioactive atom by tumor cells. (Ch. 6)

RAF A protein kinase that is activated by the small GTPase RAS and then activates the MEK1–ERK signaling cascade, often leading to cell proliferation. (Ch. 4)

randomization The random assignment of subjects or objects to an experimental or control group. (Ch. 12)

rapalog An analog of rapamycin. (Ch. 5)

rapamycin An immunosuppressant drug that acts by inhibiting the serine/threonine kinase mTOR, which is essential for the support of some signaling pathways critical for T-cell proliferation and survival. (Ch. 5)

RAS A small GTPase with important roles in intracellular signaling pathways. RAS family members include HRAS, KRAS, and NRAS. For students of history, *RAS* was the first human oncogene cloned. (Ch. 4)

reactive oxygen species Superoxide anion (O_2^-) and hydrogen peroxide (H_2O_2). In the context of immunology, these are produced by phagocytic cells, such as neutrophils and macrophages, to help kill ingested microbes. In the oncology context, increasing ROS by targeted therapies can lead to apoptosis of cancer cells. (Ch. 8)

receptor tyrosine kinases Receptors that have an intrinsic tyrosine kinase activity in their cytoplasmic tails. (Ch. 4)

reconstructive surgery Surgery to heal, restore function, and correct disfigurement or scarring resulting from trauma or acquired or congenital lesions or defects. (Ch. 2)

reduced representation bisulfite sequencing (RRBS) A method using a combination of restriction enzymes and bisulfite-mediated conversion of unmethylated cytosines, but not methylated cytosines, to uracil, for determining sequence of epigenetic modifications in DNA involving methylation of cytosines at CpG sequences. (Ch. 2)

regorafenib An oral multi-kinase inhibitor targeting VEGFR, RAF, KIT, and RET. Trade name: Stivarga. (Ch. 10)

regression A retreat or backward movement in conditions, signs, or symptoms; in oncology, this also describes shrinking of a tumor. (Ch. 5)

regulatory T cell (T_{reg}) Effector CD4 T cell that inhibits T-cell responses and is involved in controlling immune reactions and preventing autoimmunity. In the context of cancer immunotherapy, T_{reg} cells can inhibit immune responses to cancer. (Ch. 11)

relapse Recurrence of a disease state, such as the reappearance of a tumor, after treatment with an initial, ostensibly successful therapy. (Ch. 7)

remission Retreat or disappearance of a disease state with the implied possibility of its eventual reappearance or worsening. (Ch. 7)

renal cell carcinoma (RCC) A malignant neoplasm arising from the tubules of the kidney. (Ch. 5)

resection Removal by surgical excision. (Ch. 2)

retinoids Chemicals with effects similar to vitamin A. (Ch. 7)

retroperitoneal Pertaining to organs closely attached to the posterior abdominal wall and partly covered by peritoneum, rather than suspended by that membrane. (Ch. 5)

retroperitoneal lymph node dissection (RPLND) Surgical removal of lymph nodes bilaterally behind the peritoneum, and the lymph channels and fat around both renal pedicles, the inferior vena cava, and the aorta, including the bifurcation of the aorta. (Ch. 3)

retroperitoneum The space behind the peritoneum. (Ch. 3)

reverse-transcriptase PCR (RT-PCR) Procedure of reverse transcription of mRNA to DNA (called cDNA or complementary DNA) followed by PCR to amplify the resulting cDNA. (Ch. 4)

Richter's transformation Process whereby a patient's CLL will transform into a more aggressive lymphoma. This is an indication that immediate therapy is needed. (Ch. 4)

risk factor A quantitative representation of an individual or group's probability of developing a condition or disease relative to a control population's probability of developing such condition or disease. (Ch. 3)

risk-reducing salpingo-oophorectomy (RRSO) The surgical removal of both fallopian tubes and ovaries as an option for women with *BRCA1* or *BRCA2* mutations not thought to have cancer, but who have a high lifetime risk of developing ovarian, fallopian tube, and breast cancers. (Ch. 9)

rituximab A chimeric anti-CD20 monoclonal antibody bearing a murine antigen-combining (variable) domain and a human constant domain. Trade name: Rituxan. (Ch. 5)

RNA-induced silencing complex (RISC) A multiprotein complex, specifically, a ribonucleoprotein, that incorporates one strand of a single-stranded RNA (ssRNA) fragment, such as microRNA (miRNA), or double-stranded small interfering RNA (siRNA). RISC binding to the target mRNA can result in its degradation or inhibition of its translation. (Ch. 8)

RNA interference-mediated knockdown Reduction in the level of a specific messenger RNA by small interfering RNA-mediated cleavage, resulting in a reduction in the synthesis of the protein encoded by that mRNA. (Ch. 9)

RNA sequencing (RNA-seq) Whole transcriptome sequencing to determine the nature and quantity of RNA transcripts in a sample. (Ch. 2)

romidepsin An intravenous anti-cancer agent targeting HDAC, currently used in cutaneous T-cell lymphoma (CTCL) and other peripheral T-cell lymphomas (PTCLs). Trade name: Istodax. (Ch. 8)

S

salvage chemotherapy Therapy administered when previous therapies have failed and the disease has recurred. (Ch. 2)

sarcomas Tumors derived from mesenchymal cells, usually those constituting various connective tissue cell types, including fibroblasts, osteoblasts, endothelial cell precursors, and chondrocytes. (Ch. 1)

satiety A state of being satisfied, as in the feeling of being full after eating. (Ch. 3)

secondary cancer prevention Interventions such as screening and early detection to discover and control cancerous or precancerous processes while they are localized. (Ch. 10)

secondary malignancies Treatment-related cancers that occur in patients as the result of prior radiation therapy or chemotherapy. (Ch. 2)

secondary tumor A tumor that forms when tumor cells spread from the primary tumor site to form a tumor at a new location. *Also called* metastasis. (Ch. 2)

selective estrogen receptor down-regulator (SERD) A compound that binds to the estrogen receptor and promotes its degradation. (Ch. 6)

selective estrogen receptor modulator (SERM) An agent that exhibits agonist activity for estrogen receptors in some tissues but antagonist activity for estrogen receptors in other tissues. (Ch. 6)

seminoma A malignant tumor of the testis. It is the most common testicular tumor and is believed to arise from the seminiferous epithelium of the mature or maturing testis. (Ch. 3)

sensitivity Ability of a test to identify persons with the disease or condition of interest. In statistical terms, this is the "true positive rate" and often refers to how well a specific test identifies patients *with* a disease—the fraction of patients who have a disease that test positive. (Ch. 10)

sentinel lymph node (SNL) biopsy Surgical removal and dissection of the first lymph node in the chain of lymph nodes that might drain lymph and metastatic cells from a primary tumor. (Ch. 4)

sepsis A potentially life-threatening, systemic inflammation as a result of microbial infection. (Ch. 12)

sequencing The definitive methodology of genomics, as it enables the determination of the precise sequence of nucleotides composing any or all of a genome or RNA molecules transcribed from it. (Ch. 2)

serine/threonine kinase inhibitor A molecule capable of inhibiting the enzymatic activity of protein kinases that phosphorylate serine or threonine residues on their substrate proteins. (Ch. 4)

serine/threonine protein kinase A protein kinase that catalyzes the transfer of phosphate from ATP to the hydroxyl of specific serine or threonine residues of its target protein or proteins. (Ch. 5)

serum protein electrophoresis (SPEP) A procedure that separates and can be used to identify and quantify specific proteins in serum.

It can be used to diagnose some diseases and is frequently used to evaluate immunoglobulins when monoclonal gammopathies are suspected. (Ch. 9)

Sézary syndrome (SS) A leukemic version of mycosis fungoides characterized by circulating Sézary cells, diffuse erythroderma, and an aggressive clinical course. (Ch. 8)

single-chain variable fragment (scFv) A fusion protein in which the variable region domains of the heavy and light chains of an antibody molecule are joined to each other by a highly flexible peptide linker. (Ch. 12)

single-molecule sequencing Use of technologies allowing direct sequencing of long stretches of individual DNA molecules, the long reads reducing the ambiguities introduced by repetitive stretches of DNA and facilitating the assembly of genomic sequences. (Ch. 2)

single-strand DNA breaks (SSBs) DNA damage in which only one strand of the DNA duplex is severed. They are typically corrected by base-excision repair pathways. (Ch. 9)

small-cell lung cancer (SCLC) A lung cancer of specialized cells having neurosecretory properties. (Ch. 4)

small lymphocytic lymphoma (SLL) A type of lymphoma in which the CLL cells involve predominantly lymph nodes or other solid tissues (as opposed to primarily being found circulating in the blood). (Ch. 4)

somatic mutation Mutation that strikes the genome of a cell that is not a germline cell. Such a mutation cannot, by definition, be transmitted to a person's offspring. (Ch. 4)

sorafenib An oral kinase inhibitor targeting RAF and other kinases. Trade name: Nexavar. (Ch. 5)

specificity Ability of a test to distinguish between patients with or without a disease or condition of interest. In statistical terms, this is the "true negative rate" and often refers to how well a test identifies patients *without* a disease—the fraction of patients without a disease who test negative (as opposed to "false negative rate"—the fraction of patients *with* a disease who test negative). (Ch. 10)

spiculated Spiked. (Ch. 4)

splenomegaly Enlargement of the spleen, which might cause feelings of abdominal fullness, left upper quadrant abdominal discomfort, referred pain felt in the left shoulder, or early satiety, which describes fullness before finishing a normal-sized meal. (Ch. 4)

squamous cell carcinoma A slow-growing malignant tumor of squamous epithelium, frequently found in the lungs and skin, and occurring also in the anus, cervix, larynx, nose, and bladder. (Ch. 4)

STAT Signal transducers and activators of transcription. A family of transcription factors activated by many cytokine and growth factor receptors. (Ch. 4)

stereotactic radiosurgery (SRS) A specialized external beam radiation therapy technique delivering photon beams at high concentration to a precise target in the brain in either a single or a few treatments. When SRS is used to treat other areas of the body, it is termed "stereotactic body radiotherapy" (SBRT). (Ch. 2)

steroidogenesis The biological synthesis of steroid hormones. (Ch. 6)

subcarinal Pertaining to the area below the split of the trachea into left and right bronchi. (Ch. 4)

sunitinib An oral kinase inhibitor targeting VEGFR, mutated KIT, and other kinases. Trade name: Sutent. (Ch. 5)

superficial spreading melanoma The most common melanoma, which grows outward, spreading over the surface of the affected organ or tissue. The lesion is typically raised and palpable, unevenly pigmented, and irregularly shaped and has an unclear border. (Ch. 4)

supraclavicular Pertaining to the area above the clavicle, or collarbone. (Ch. 4)

Surveillance, Epidemiology, and End Results The "SEER" database, which provides cancer statistics in the United States. (Ch. 4)

survivorship The state of being a survivor; for oncology patients, this state includes management of long-term cancer- and treatment-related physical, psychiatric, social, and economic consequences. (Ch. 2)

synthetic lethal A cell phenotype susceptible to lethality resulting from the combined effect of deficient expression of two genes or inhibition of their protein products where deficient expression of just one of the two genes would not carry the lethal risk. (Ch. 9)

T

T315I mutation A mutation in the BCR–ABL protein in which a threonine at position 315 in the amino acid sequence is replaced by an isoleucine, resulting in a kinase that is resistant to imatinib, a powerful inhibitor of unmutated BCR–ABL. (Ch. 4)

T-cell-dependent bi-specific antibodies (TBD) Full-length monoclonal antibodies, usually human, in which one chain is specific for a protein of the T-cell receptor complex and the other for an antigen associated with the intended cancer cell target. Such antibodies will bring cancer cell targets together with T cells, thereby enabling attack by T cells even though they may have no specificity for the targeted cancer cell. (Ch. 12)

T-cell receptor (TCR) A heterodimer of α and β or γ and δ polypeptide chains that functions as the T-cell antigen binding receptor. (Ch. 12)

T cells Category of lymphocytes that develop mainly in the thymus, which includes T_H cells, T_C cells, and T_{reg} cells. (Ch. 11)

talimogene laherparepvec (T-Vec) An oncolytic, genetically modified herpes simplex virus 1 that is used as immunotherapy for certain melanomas. Trade name: Imlygic. (Ch. 12)

targeted therapies Drugs or approaches that eliminate or block the growth and spread of cancer by interfering with specific molecules or pathways involved in the growth, progression, survival, or spread of cancer. (Ch. 1)

Tdt Terminal deoxynucleotidyl transferase, a specialized DNA polymerase expressed in lymphoid tissue that inserts nontemplated nucleotides into the V, D, and J exons during antibody gene recombination, enabling junctional diversity. Tdt expression is a marker of immature lymphoid cells and some lymphomas. (Ch. 12)

telomerase An enzyme specialized to extend telomeric DNA, typically characterized by an RNA subunit and a reverse transcriptase. (Ch. 1)

telomeres Protective nucleoprotein structure at the end of a eukaryotic chromosome that protects this end from degradation and from fusion with other chromosomes. (Ch. 1)

temsirolimus An intravenously administered targeted inhibitor of mTOR signaling. Trade name: Torisel. (Ch. 5)

teratogenicity The property or capability of producing congenital malformations. (Ch. 7)

teratoma Benign tumor formed by embryonic system stem cells in which a wide variety of differentiated stem cells may be formed. (Ch. 3)

thalidomide A drug originally marketed as a treatment for nausea and morning sickness but found to be teratogenic; later found to be

an effective cancer therapy via mechanisms of action including inhibition of angiogenesis and immunomodulation. Trade name: Thalomid. (Ch. 10)

The Cancer Genome Atlas (TCGA) A project of cataloging mutations found in a wide and broadly diverse variety of cancers but not in their paired normal tissue counterparts. (Ch. 2)

therapeutic cancer vaccines Immunizations intended to generate immune responses that treat cancers. (Ch. 11)

thoracentesis Perforation of the chest wall and pleural space with a needle to aspirate fluid for diagnostic or therapeutic purposes or to remove a specimen for biopsy. The procedure is usually performed using local anesthesia, with the patient in an upright position. (Ch. 4)

threonine/tyrosine kinase inhibitor A substance that inhibits the activity of dual specificity kinases that can phosphorylate the hydroxyls of either specific threonine or tyrosine residues of their target protein. (Ch. 4)

thrombocytopenia A platelet count below the lower limit of the reference interval, usually 150,000/mm³. (Ch. 4)

thromboembolism A condition in which a blood vessel is obstructed by a blood clot (thrombus) carried in the bloodstream from its original site of formation. *See also* **venous thromboembolic event**. (Ch. 10)

thrombosis Abnormal condition in which a clot (thrombus) develops within a blood vessel. (Ch. 4)

tinnitus A subjective noise sensation, often described as ringing, heard in one or both ears. (Ch. 3)

tisagenlecleucel A CAR T-cell therapy targeting CD19 on B cells, employed for treating large B-cell lymphoma. Trade name: Kymriah. (Ch. 12)

tocilizumab Humanized anti-IL-6 receptor antibody used in treating rheumatologic conditions such as rheumatoid arthritis and giant-cell arteritis, as well as cytokine release syndrome following CAR T-cell therapy. Trade name: Actemra. (Ch. 12)

tolerance State in which the immune system shows a lack of reactivity toward certain antigens, notably those that are expressed by normal cells and tissues. (Ch. 12)

TP53 Gene that encodes the tumor suppressor p53. (Ch. 4)

trametinib An oral small-molecule inhibitor of the two forms of MEK, MEK1 and MEK2. Trade name: Mekinist. (Ch. 4)

transformation Process of converting a normal cell into a cell having some or many of the attributes of a cancer cell. (Ch. 7)

transforming growth factor-α (TGF-α) A mitogenic polypeptide that is a ligand for the epidermal growth factor receptor, EGFR. (Ch. 5)

transforming growth factor-β pathway (TGF-β pathway) A signaling pathway involved in many cellular processes in both the adult organism and the developing embryo including cell growth, cell differentiation, apoptosis, cellular homeostasis, and immunosuppression. (Ch. 12)

translocation Rearrangement of chromosomes that results in the fusion of two chromosomal segments that are not normally attached to one another. (Ch. 2)

triple-negative Term describing a breast cancer that does not express estrogen receptor, progesterone receptor, or HER2. (Ch. 9)

TSC1 (and TSC2) The GTPase activating proteins (GAPs) of the mTOR pathway. (Ch. 5)

tumor-associated antigens (TAAs) Antigens that are expressed by cancer cells but can also be found on normal cells of the same or related tissue. (Ch. 11)

tumor-associated macrophages (TAMs) Class of immune cells present in high numbers in the microenvironment of solid tumors, subpopulations of which might mount cytotoxic immune responses while others might suppress immune responses. (Ch. 10)

tumor-infiltrating lymphocytes (TILs) Lymphocytes that have moved from the blood into a tumor. The presence of TILs may indicate that the immune system is responding to and perhaps attempting an attack on the cancer. (Ch. 12)

tumor lysis syndrome (TLS) A condition due to a large number of cancer cells dying in a short period of time, resulting in release of cell contents into the bloodstream and consequent metabolic abnormalities including high uric acid, potassium, and phosphate levels and acute kidney injury. (Ch. 12)

tumor marker A measurable biomarker found at higher than normal levels in the blood, urine, or body tissue of some people with certain types of cancer. (Ch. 3)

tumor microenvironment (TME) The combination of malignant cells and normal cells, such as fibroblasts, cells of the immune system, and bone marrow–derived inflammatory cells, that comprise the tumor and the supporting vasculature within and outside the tumor. (Ch. 1)

tumor-specific antigens (TSAs) Antigens produced by a particular type of tumor that are absent in or on normal cells of the tissue in which the tumor developed. (Ch. 11)

tumor suppressor A gene product whose absence or partial or complete inactivation leads to an increased likelihood of cancer development. (Ch. 1)

tyrosine kinase inhibitor (TKI) A targeted therapy that inhibits tyrosine kinase activity. (Ch. 4)

U

ubiquitin A small protein that can attach to other proteins and can function as a protein interaction module, or can target proteins for degradation by the proteasome. (Ch. 8)

ubiquitination The process of attachment of one or many subunits of ubiquitin to a target protein, which can mediate either degradation by the proteasome or formation of scaffolds used for signaling, depending on the nature of the linkages. (Ch. 8)

ubiquitin–proteasome pathway (UPP) The principal mechanism for protein catabolism in the mammalian cytoplasm and nucleus. (Ch. 9)

up-regulation Increase in expression of a gene; in the narrowest sense, that in which transcription of a specific mRNA is increased, but also used more broadly to refer to an increase in mRNA levels for a particular gene from any cause, such as increased stability of the specific mRNA. (Ch. 6)

ureter One of a pair of tubes, about 30 cm long, that carries urine from the kidney into the bladder. (Ch. 5)

uvea The vascular, pigmented middle coat of the eye. (Ch. 4)

V

vaccination The deliberate induction of adaptive immunity to a pathogen by injecting a dead or non-pathogenic live form of the pathogen or its antigens (a vaccine). (Ch. 11)

vascular endothelial growth factor (VEGF) A protein factor that stimulates the proliferation of cells of the endothelium of blood vessels. It promotes tissue vascularization and is important in blood vessel formation in tumors. (Ch. 5)

vasculature Network of blood vessels. (Ch. 10)

vemurafenib A targeted inhibitor of activated BRAF. Trade name: Zelboraf. (Ch. 4)

venous thromboembolic event A blood clot that starts in a vein and can spread via the bloodstream to obstruct circulation (embolize) at another site, most commonly the lungs. (Ch. 3)

virus-like particles (VLPs) Multiprotein structures that resemble viruses but lack the viral genome and are thus non-infectious. (Ch. 12)

Von Hippel–Lindau syndrome A hereditary syndrome characterized by congenital tumor-like vascular nodules in the retina; hemangioblastomas of the cerebellar hemispheres and similar spinal cord lesions; cysts of the pancreas, kidneys, and other viscera; kidney cancers; and seizures. Mental retardation may be present. (Ch. 5)

W

Whipple procedure *See* **pancreaticoduodenectomy**. (Ch. 3)

whole exome sequencing (WES) A technique used to determine the sequence of all the protein-coding genes in an individual's genome. (Ch. 2)

whole genome sequencing (WGS) A process used to determine the DNA sequence of an individual's entire genome. (Ch. 2)

X

X-linked agammaglobulinemia A genetic disorder in which B-cell development is arrested at the pre-B-cell stage and no mature B cells or antibodies are formed. The disease is due to a defect in the gene encoding protein tyrosine kinase BTK, which is encoded on the X chromosome. (Ch. 4)

Y

yolk sac tumor A rare malignant tumor of cells that line the yolk sac of the embryo. These cells normally become ovaries or testes; however, the tumor can also occur in areas such as the brain or chest. (Ch. 3)

Z

zoledronic acid A bisphosphonate inhibitor of osteoclastic bone resorption used for the treatment of hypercalcemia of malignancy, multiple myeloma, and osteoporosis. It is administered intravenously. Trade name for cancer therapy: Zometa. Trade name for osteoporosis: Reclast. (Ch. 10)

Index

A

ABC-like lymphoma, 250–51, *250*, 255
ABCDE warning signs (melanoma), *90*
abiraterone, 135, 154, *155*
ABL1 gene, Philadelphia chromosome, 66–67, *66*, 69, *69*
ABVD, *40*, 259–60, *261*
AC → T, *40*
accelerated phase, chronic myelogenous leukemia (CML), 70, 73, *73*
accessory cells, 232
acetylation, 172
acidic fibroblast growth factor (aFGF), *218*
acquired endocrine resistance, 143
acral lentiginous melanoma, 91, *91*
activated B-cell (ABC)-like lymphoma, 250–51, *250*, 255
activated partial thromboplastin time (aPTT), 164
active surveillance, 45, 149–50
acute erythroid leukemia, *161*
acute leukemia, 70
acute lymphoblastic leukemia (ALL), 21, *273*, 280, 281, 291–96
  case study, 294–96, *295*
  diagnosis, 292, 298
  epidemiology, 292, 298
  genomics, 293
  histology, 292–93, *293*
  management, 293–94, 298
    with CAR T cells, 296–98, *296*
    immunosuppressive therapy, 294, 296–98, *296*
  symptoms, 292
acute megakaryocytic leukemia, *161*
acute monocytic leukemia, *161*
acute myelogenous leukemia (AML), *40*, 73, 161
  diagnosis, 164, *165*
  epidemiology, 163
  etiology, 163
  prognostic factors, *162*
  subtypes, 161, *161*
  symptoms, 164
acute myelomonocytic leukemia, *161*
acute promyelocytic leukemia (APL), 161–68
  case study, 166–67
  diagnosis, 164, *165*
  epidemiology, 163
  etiology, 163
  genomics, 162–63, *162*, *163*
  management, 165–66, 167
  prognostication, *164*, 166
  symptoms, 164
adaptive immunity, 239, 270–71, *271*
ADCC. *See* antibody-dependent cell-mediated cytotoxicity
Adcetris, *244*
ADCs (antibody-drug conjugates), 240, 246, 247, *248*, *304*
adenocarcinoma, pancreatic, 50–58
adenomas, 219
adenomatous polyposis coli, *33*
adjuvant therapy, 39, 223, 289

ado-trastuzumab emtansine (T-DM1), *244*, 265–66, 267
adoptive T-cell therapies, 269, 270–77, *271*, *273–76*, 282, *304*
adoptive T-cell transfer, 240, 305
adriamycin, 140, 230, 259
ADT. *See* androgen deprivation therapy
adult T-cell leukemia/lymphoma, *118*, 249
afatinib, *64*
aFGF (acidic fibroblast growth factor), *218*
Afinitor, *64*
aggresome pathway, 196, *196*
aggressive lymphomas, *118*, 119, *249*
Aiolos, 234, 235
AIs (aromatase inhibitors), 41, 135, 141, 143, 144
AKT (protein kinase B, PKB), 103, 108
AKT1 gene, *33*
ALCL. *See* anaplastic large-cell lymphoma
aldosterone, 134–35
Alecensa, *64*
alectinib, *64*
alemtuzumab, 187, *244*
ALK, function of, 87, *87*
ALK inhibitors, 88, 100
ALK rearrangements, *28*, 87, 89
alkylating agents, 20, *20*
alkylating chemotherapy, 78
ALL. *See* acute lymphoblastic leukemia
all-*trans* retinoic acid (ATRA), 165–66
allogeneic hematopoietic stem cell transplantation (allo-HSCT), 73, 74, 80, 182, 208, 277, 298
alopecia, 184
alpha fetoprotein (AFP), *28*, 43, *43*
American Joint Committee on Cancer (AJCC), 52
AML. *See* acute myelogenous leukemia
amyloid light-chain (AL) amyloidosis, 205
anabolic pathways, 105
anal condyloma, 290
anal sphincter, 223
anaplastic large-cell lymphoma (ALCL), *118*, 183, *249*
anastrazole, 135, 143
androgen deprivation therapy (ADT), 151, 152, 155, 156
androgen receptor (AR), 135, 153–54, 156
androgens, 131, 151
anemia, in chronic myelogenous leukemia, 68
aneuploidy, 24
angiogenesis, 227
  cancer and, 5–6, 61, 217
  in tumor microenvironment (TME), 217–18, *218*
angiogenesis inhibitors, 230
angiogenic switch, 5–6, 217
angioimmunoblastic T-cell lymphoma, *118*, 183, *249*
angiopoietin, *218*
Ann Arbor classification, 121, *121*
anorexia, 51
anthracyclines, *20*, 140, 142
anti-androgens, 135–36, 153, 156
anti-apoptotic proteins, 5
anti-CD4 monoclonal antibodies, 237

anti-CD20 monoclonal antibodies, 123
anti-HBV vaccine, 303
anti-HPV vaccine, *286*, 288–90, *289*
anti-tumor antibiotics, *20*
antibodies, 239, 240, *241*, 246, 301
antibody-dependent cell-mediated cytotoxicity (ADCC), 78, 243
antibody-drug conjugates (ADCs), 240, 246, 247, *248*, *304*
antifolates, *20*
antigen–antibody reactions, 240
antigen-presenting cells (APCs), 270
antigen processing, 270
antigens, 239
antimetabolites, *20*, 142
antisense oligonucleotides, 191
apalutamide, 136
APC gene, *33*, 220
APC mutations, 16
APCs (antigen-presenting cells), 270
APEX study, 211
APL. *See* acute promyelocytic leukemia
apoptosis, 5, 165
aPTT. *See* activated partial thromboplastin time
AR (androgen receptor), 135, 153–54
Arimidex, 143
Aromasin, 143
aromatase, 135
aromatase inhibitors (AIs), 41, 135, 141, 143, 144
arsenic trioxide (ATO), 165, 166
Arzerra, *244*
ascites, 51
asparaginase, 295
ASPIRE trial, 211
atezolizumab, *244*
ATM, 75
ATM gene, *198*
ATO (arsenic trioxide), 165, 166
ATP, 62
ATP-binding pockets, 62, *63*
ATRA (all-*trans* retinoic acid), 165–66
autologous cellular immunotherapy, 155
autologous stem cell transplant, 47, 73, 166, 208
autophagy, 105–6
Avastin, *244*
axitinib, *64*, 116, 227
5-azacitidine, 179, 181, 182

B

B-cell-activating factor (BAFF), 232
B-cell aplasia, 297
B-cell clone, 240–41
B-cell lymphomas, 118, *118*, 183, 189, 243, 248–49, *249*, 298
B-cell prolymphocytic leukemia, *118*, 249
B-cell receptor complex (BCR), 75, 81
B cells, 231, 239, 240
B-Raf proto-oncogene, *33*
"B" symptoms, 76, 251, 258
Bacillus Calmette-Guérin (BCG), 277, 281
band forms, 68, *68*
basal cell carcinoma (BCC), 90
BASC. *See* BRCA1-associated genome surveillance complex
base-excision repair (BER), 198
basic fibroblast growth factor (bFGF), *218*
basophils, 66
BCB-ABL translocation, 293
BCC. *See* basal cell carcinoma
BCG (Bacillus Calmette-Guérin), 277, 281
BCL-2, 5, 75

BCL-2 inhibitors, 81
BCL-6, 120
BCL2 translocation, 119–21, *119*, 250
BCL6 gene, 249
BCR (B-cell receptor complex), 75, 81
BCR-ABL fusion gene, 23, *28*, 67
BCR gene, Philadelphia chromosome, 66–67, *66*, 69, *69*
BCR signaling cascade, *80*, 81, 82
belinostat, 189
bendamustine, 78, 123, 124–25
benign prostatic hypertrophy (BPH), 146
BEP chemotherapy, *40*, 46
BER. *See* base-excision repair
β human chorionic gonadotropin (βhCG), *28*, 43, *43*
β-2-microglobulin, *28*, 122
bevacizumab, 85, 116, 228, *244*
bexarotene, 187
bFGF (basic fibroblast growth factor), *218*
bi-specific antibodies (bsAb), 279–80, *304*, 305
bi-specific T-cell engager (BiTE), 280
bicalutamide, 136
bile duct carcinoma, *273*
bilirubin, 51
bilobed nuclei, 167
bioavailability, 21
bioavailable prodrug, 154
The Biology of Cancer (Weinberg), 71
biomarkers, 10
biopsy, 13
    liquid biopsy, 35, 37
    percutaneous biopsy, 83
    sentinel lymph node (SLN) biopsy, 92
    transrectal ultrasound-guided biopsy, *147*
Birt-Hogg-Dubé syndrome, 109, *110*
bisphosphonates, 208, 233, *233*, 235
BiTE (bi-specific T-cell engager), 280
bladder cancer, management, 281
blast phase, chronic myelogenous leukemia, 70, 73, *73*
blastic metastases, 149
bleomycin, *20*, 47, 259
blinatumomab, 280
BMSCs, 232, 233
BMT. *See* bone marrow transplant
bone marrow microenvironment
    immunomodulatory drugs in, 233–34
    in multiple myeloma (MM), 231–32, *232*, 235
bone marrow transplant (BMT), 277
bone metabolism, multiple myeloma (MM), 232–33
bone scan, staging with, *9*
bortezomib, 209–11, *211*, 212, 213, 229, 230, 231
Bosulif, *64*
bosutinib, *64*, 73
BRACAnalysis CDx, 204
brachytherapy, 17, 19, 150–51
BRAF gene, 23, *28*, *33*, 94–95
BRAF inhibition, 96–97, 98, 99
BRAF/MEK/MAPK pathway, 95, *95*
BRAF mutations, 23, 94–95
brain lesions, 16–17, 21
BRCA-associated breast cancers, 197–204
    epidemiology, 200
    management, 203–4
    recommendations for *BRCA1/2* asymptomatic carriers, 200–201
    risk factors, *200*
BRCA1-associated genome surveillance complex (BASC), 197
BRCA1 gene, 136, 197, *198*, *198–200*, 200, 204
BRCA2 gene, 136, 197, *198*, *198–200*, 200, 204
BRCA2 protein, 197

breast, draining lymph nodes, 14, *14*
breast cancer, 136–44
    about, 263
    case studies, 136, 142–43, *142*, 201–3, *201–3*, 213, 264–66
    diagnosis, 137–39
    epidemiology, 132, 136, 144, 197
    etiology, 136–37
    future prospects and challenges, 143–44, 266
    genomics, 23, 136–37, 139–40, 143
    management
        BRCA-related breast cancers, 204
        combination chemotherapy, *40*, 140, 144
        GnRH agonists, 133
        history of surgical management, 14–15, *15*
        hormone therapy, 133, 143, 144
        localized, 140–41
        metastatic, 141–42, 143
        monoclonal antibodies, 139, 140–41, 263–67, *263–65*
        neoadjuvant therapy, 203
        radiation therapy, 141
        reconstructive surgery, 15
    metastatic, 141–42, 143, 156
    prophylactic mastectomy, 16, 203
    risk factors, *136*, 137
    screening, 137, 144, 204, 213
    staging, *138*, 139
    subtypes, 139–40, *139*, 144
    "triple negative" breast cancer, 139, *139*, 144, 204
breast-conserving surgery, 15
breast mammography, 137, *137*, 144
brentuximab vedotin, 189, *244*, 260, 261, *261*, 268
Breslow measurement, 92
bronchoscopy, 83
Bruton, Ogden, 80
Bruton's tyrosine kinase (BTK), 80, 127, 254
BsAb. *See* bi-specific antibodies
Burkitt lymphoma, *118*, *249*

C

c-KIT mutations, 95
CA1-3/CA27.29, *28*
CA19-9, *28*
CA125, *28*
cabazitaxel, 155, *155*
cabozantinib, *64*
CAFs (cancer-associated fibroblasts), 6, 7, 214
calcitonin, *28*
Campath, *244*
cancer
    adaptive immune response to, 270–71, *271*
    biomarkers, 10
    categorization, 1
    cell growth and, 21, 23
    co-morbidities, 16
    deaths from, 2–3, *2*, *3*, *41*
    genetic heterogeneity within, 7
    genetics, 4, 6
    genomic profiling, 65, 66
    hallmarks of, 4–6, *4*
    herd immunity, 290
    incidence, 2, *2*, 11
    "index patient," 16
    primary tumors, 13, 14
    protein kinases and, 61–62
    screening tests, 8, 10
    secondary malignancies, 17
    secondary tumors, 13, 14

    staging, 8, *9*
    statistics by type and gender, *2*
    tumor microenvironment (TME), 6–8, *6*, 216, *304*
    viruses implicated in causation, 282, *282*
    See also cancer cells; diagnosis; screening tests
cancer-associated fibroblasts (CAFs), 6, 7, 215
cancer cells
    angiogenesis, 5–6
    apoptosis, 5
    fast cell division, 21, 23
    glycolysis, 6
    mutations and, 193
    proliferative signaling in, 4–5
The Cancer Genome Atlas (TCGA), 31
cancer stem cells, 156
cancer vaccines. *See* vaccination
capecitabine, 54, 142
capillary leak syndrome, 112, 301
CAPN. *See* The Solid Tumor Targeted Cancer Gene Panel by Next
        Generation Sequencing
Caprelsa, *64*
capsid proteins, 285
CAR (chimeric antigen receptor), 274
CAR T cells, 240, *273*, 275–77, *275*, *276*, 296–98, *296*, 303, 304, *304*
carboplatin, *20*, 287
carcinoembryonic antigen (CEA), *28*, 222
carcinomas, 1
carfilzomib, 209, 211
case studies
    acute lymphoblastic leukemia (ALL), 294–96, *295*
    acute promyelocytic leukemia (APL), 166–67
    breast cancer, 136, 142–43, *142*, 201–3, *201–3*, 213, 264–66
    cervical cancer, 288
    chronic lymphocytic leukemia, 79–80
    chronic myelogenous leukemia, 71–73
    of classical Hodgkin lymphoma (CHL), 260
    colorectal cancer (CRC), 225–26, *225*, *226*
    cutaneous T-cell lymphoma (CTCL), 187–88
    diffuse large B-cell lymphoma (DLBCL), 253–54
    hereditary breast cancer, 201–3, *201–3*, 213
    Hodgkin lymphoma (HL), 260
    lung cancer, 86–87, *86*
    melanoma, 95–96, *96*
    metastatic melanoma, 297–98
    multiple myeloma (MM), 208–9, 230–31
    myelodysplastic syndromes (MDSs), 178–80
    non-Hodgkin lymphoma (NHL), 124–25
    pancreatic cancer, 55–56
    prostate cancer, 136, 152–53, *152*
    renal cell carcinoma (RCC), 114–15, *116*, *117*
    Sézary syndrome, 187–88
    testicular cancer, 47–48
caspase (cysteine-aspartic protease), 173
castration-resistant prostate cancer (mCRPC), 153–54, *155*, 156, 272
catabolic pathways, 105
CCL2, 216
CCR2, 216
CCR4, 216
CD4 T cells, 270
CD8 T cells, 270
CD10, 120, 292
CD19, 292
CD20, *28*, 120, 127, 243
CD28, 275
CD80, 277
CD86, 277
CDC (complement-dependent cytotoxicity), 243
CDH1 gene, 137, *198*

CDRs (complementarity-determining regions), *241*
CEA (carcinoembryonic antigen), *28*, 222
cell division, as hallmark of cancer, 4–5
cellular homeostasis, 193, 194
centroblasts, 120
centrocytes, 120
cereblon (CRBN), 234, *234*, 235
cereblon E3 ubiquitin ligase complex, *234*
cerebrospinal fluid, 292
Cervarix, 288, 291
cervical cancer, 282–91
  about, 282–83
  adoptive cell transfer, *273*
  case study, 288
  causation, 282, *282*, 291
  diagnosis, 284–85
  epidemiology, 283–84
  etiology, 283–84, *284*
  HPV vaccination, 288–90, *289*
  human papillomavirus (HPV), 243, 272, 282, *282*, 284, *284*,
      285–87, *286*, 291
  management, 287
  screening, 283, 284
  staging, 285, *285*
  subtypes, 284
  symptoms, 284, *284*
cervical intraepithelial neoplasia (CIN), 289
cervicectomy, 287
cetuximab, 227, *244*
checkpoint inhibitors, 240, 269, 277, 282, *304*
CHEK2 gene, *198*
chemoimmunotherapy, 125
chemokine receptors, 216
chemokines, 216, 237, 280
chemoradiation, for pancreatic cancer, 54, 58
chemotactic factors, 216
chemotaxis, 81, 216
chemotherapy, 19–22, *19*, 39–41, *41*, 47, 59
  adjuvant, 39
  benefits of, 39–41, *41*
  bioavailability, 21
  "cycle" of, 22
  first-line chemotherapy, 22
  history, 19–20
  induction chemotherapy, 165, 295
  mechanism of action, 20, *20*, 40
  molecular targeted therapies, 39
  multi-drug combinations, 22, 40
  myeloablative chemotherapy, 47
  neoadjuvant chemotherapy, 19, 39, 202
  palliative, 39
  for pancreatic cancer, 54–55, 57
  pathologic complete response (pCR), 202
  perioperative, 39
  pharmacogenomics, 35, 36, 37
  routes of administration, 21, *21*
  salvage chemotherapy, 22
  second-line chemotherapy, 22
  side effects, 21–22, *22*
  single-agent chemotherapy, 21
  for testicular cancer, 45, 46–47, 48
  understanding failures, 40
chemotherapy agents, 39–41, *41*
  categories of, 20, *20*
  clinical trials, 40
  drug development, 11, *11*
  history of, 19–20
  mechanism of action, 20, *20*, 40
  routes of administration, 21, *21*

  side effects, 21–22, *22*
  See also case studies; combination chemotherapy
chimeric antigen receptor (CAR), 274
chimeric antigen receptor (CAR) T cells, 240, *273*, 275–77, *275*, *276*,
      296–98, *296*, 303
"chip," 25
CHL. *See* classical Hodgkin lymphoma
chlorambucil, 78, 79
cholangiocarcinoma, *273*
CHOP, 253, 254, 256
choriocarcinomas, 44
chromatin, *172*
chromogranin A, *28*
chromophobe renal cell cancer, 108, *108*, *109*
chromosomal aberrations, 23–24
chromosomal translocation, 119–20, *119*
chromosome 13q14, 78
chromosomes
  aneuploidy, 24
  karyotype, 24, *24*, 69, *69*
  Philadelphia chromosome, 66–67, *66*, 69, *69*
  translocation, 23–24
  trisomy 12, 78
  See also genomics/genetics
chronic lymphocytic leukemia (CLL), 75–82, *118*, 249
  case study, 79–80
  diagnosis, 76–77, *76*, *77*, 82
  epidemiology, 75–76
  future prospects and challenges, 81–82
  genetics, 75, 82
  management, 78–79
    adoptive cell transfer, *273*
    CAR T cell therapy, 298
    relapsed CLL, 80–81, *81*
  Richter's transformation, 78
  staging, 77
  symptoms, 76, 82
  white blood cells in, 76, *76*
chronic myelogenous leukemia (CML), 23, 64, 66–74, *70*
  case study, chronic phase, 71–73
  clinical phases, 66, 69–70
    accelerated phase, 70, 73, *73*
    blast phase, 70, 73, *73*
    chronic phase, 69, 70–73
    WHO criteria for, *73*
  diagnosis, 67–69, *67*, *68*
  epidemiology, 67
  evaluation, 67–70
  future prospects and challenges, 74
  genetic hallmark, 66
  management, 74
    accelerated phase, 73
    blast phase, 73
    chronic phase, 70–73
    when initial TKI therapy fails, 74
  mutations, 74
  Philadelphia chromosome, 66–67, *66*, 69, *69*, 74
  symptoms, 67, 74
chronic myelomonocytic leukemia (CMML), 177
chronic obstructive pulmonary disease (COPD), 83–84
CIN (cervical intraepithelial neoplasia), 289
circulating tumor cells (CTCs), 34–35, *34*, 37
circulating tumor DNA (ctDNA), 34–35, *34*, 37
cisplatin, *20*, 54, 85, 203, 287
classical Hodgkin lymphoma (CHL), 256
  case study, 260
  diagnosis, 258–59
  epidemiology, 258
  future prospects and challenges, 262

incidence of death, 259, *259*
management, 259–60, 261, 262
risk factors, 258
staging, 258
clear cell renal cell cancer, 108, *108, 109,* 113
Clinical Lung Cancer Genome Project (CLCGP), 36
CLL. *See* chronic lymphocytic leukemia
clodronic acid, 233
clonal evolution, 70
clonality, 76–77
clones, 240–41, 243
CML. *See* chronic myelogenous leukemia
CMML. *See* chronic myelomonocytic leukemia
comorbidities, 16, 147
co-stimulatory molecules, 270
coagulation tests, 164
Cobas 4800 BRAF V600 mutation test, 98
cobimetinib, *64*
Cologuard set, *28*
colon, anatomy, 223
colon cancer
    combination chemotherapy regimen, *40*
    prophylactic colectomy, 16
    See also colorectal cancer
colon polyps, 220
colony-stimulating factor-1 (CSF-1), 237
colorectal cancer (CRC), 219–28
    case study, 225–26, *225, 226*
    diagnosis, 222
    EGFR/RAS/RAF signaling in, 220, *220*
    epidemiology, 219–20, 221, 228
    etiology, 221
    evaluation, 222–23
    future prospects and challenges, 227–28
    genomics, 23, 220–21
    inherited syndromes, 221, *221,* 228
    management
        localized, 223–24, 228
        metastatic, 225, 226–28
        surgery, 15–16, *15*
        targeted therapy, 226–27
    metastatic CRC, 222–23, *223,* 225, 226–28
    microsatellite instability (MSI), 220
    risk factors, 221, *222*
    staging, 223, *224*
    symptoms, 222
    tumorigenesis, *219,* 220
colostomy, 223
colposcopy, 285
combination chemotherapy, 21, 40
    for breast cancer, *40,* 140, 144
    for chronic lymphocytic leukemia, 78–79
    for chronic myelogenous leukemia, 70, 71
    for colon cancer, *40*
    for melanoma, 98–99
    for non-Hodgkin lymphoma, *40,* 123
    for pancreatic cancer, 57
    resistance to, 88, 89
    for testicular chemotherapy, *40*
Cometriq, *64*
complement-dependent cytotoxicity (CDC), 243
complement pathway, 243
complementarity-determining regions (CDRs), *241*
complete lymph node dissection, 93
complete remission, 293
computed tomography (CT)
    lung cancer evaluation, 84
    pancreatic cancer evaluation, 51, *52, 53*
    for radiation therapy, 18
staging with, 9
    testicular cancer evaluation, 43, *43*
computer-aided radiation therapy, 17–18
conformal radiation therapy (CRT), 18
consolidation therapy, 165
copanlisib, 126, 127
COPD. *See* chronic obstructive pulmonary disease
Cotellic, *64*
CRAB criteria, multiple myeloma (MM), 205, 206, 212
CRBN (cereblon), 234, *234,* 235
CRBN–CRL4 ubiquitin ligase, 234, *234,* 235
CRC. *See* colorectal cancer
CRd, 209, 211
crizotinib, *64, 86–88,* 87–89, 88, *88,* 89
Crohn's disease, 221
CRISPR/Cas9. *See* CRSPR/Cas9
CRS (cytokine release syndrome), 297
CRSPR/Cas9-mediated knockouts, 213
CRT. *See* conformal radiation therapy
cryoablation, 111
cryotherapy, 150
cryptorchidism, 42
CSF-1 (colony-stimulating factor-1), 237
CSF-1 receptor, 237
CT. *See* computed tomography
CT colonography, 222
CTCL. *See* cutaneous T-cell lymphoma
CTCs. *See* circulating tumor cells
ctDNA. *See* circulating tumor DNA
CTLA-4, 277
Cullin-4A (CUL4A), 234
curative radiation therapy, 17
Curie, Marie, 17
cutaneous T-cell lymphoma (CTCL), 183–90
    case study, 187–88
    diagnosis, 184–85, *184, 185*
    epidemiology, 184
    future prospects and challenges, 189
    management, 186–87, *187,* 189
    staging, 185–86, *186*
CXCL-12, 215, 216
CyberKnife, 18
cyclophosphamide, *20,* 78, 123, 140, 202, 203
CYP17, 135
Cyramza, *244*
cysteine-aspartic protease (caspase), 173
cytarabine, *20,* 70, 71, 165
cytochrome P450-bearing enzymes, 35
cytogenetic analysis
    acute lymphoblastic leukemia, 293
    chronic lymphocytic leukemia, 78
    chronic myelogenous leukemia, 68–69
cytogenetic approaches, 23–25, *24*
cytokeratin fragments 21-1, *28*
cytokine release syndrome (CRS), 297
cytokine therapy, *19,* 237, 280, 301
cytolytic cells, 239
cytopenia, 76
cytoreduction, 180
cytoreductive nephrectomy, 113
cytoreductive surgery, 16, 17
cytotoxic chemotherapy. *See* chemotherapy
cytotoxic T cells, 7, 239, 270

D

dabrafenib, *64,* 96, 97, *97,* 98
dacarbazine, *20,* 259
damage-associated molecular patterns (DAMPs), 239–40

damaged DNA binding protein 1 (DDB1), 234
dasatinib, *64*, 71, 73
daunorubicin, *20*, 165, 295
DCs (dendritic cells), 240
DDB1 (damaged DNA binding protein 1), 234
death rates
    from cancer, 2–3, *2*, *3*, *41*
    leading causes of death in U.S., *41*
debulking, 16
decitabine, 181
dedifferentiation process, 159
dendritic cells (DCs), 240, 270
denosumab, 208, 233, 235
deubiquitinases (DUBs), 213
dexamethasone, 166, 209, 229, 230, 231, 234, 295
diagnosis, 8
    genomics for, 23, 32–33, 37
    surgery for, 13–14, *14*
diagnostic tests, FDA-approved, 27, *28–29*
DIC. *See* disseminated intravascular coagulation
differentiation syndrome, 165, 166, 167
differentiation therapy, 159–61, *161*, 165–67, 168
diffuse large B-cell lymphoma (DLBCL), *118*, 249–55
    case study, 253–54
    diagnosis, 251–53, *252*
    epidemiology, 251
    future prospects and challenges, 254–55, *255*
    genomics, 249–51, *250*, 255–56
    histology, 251–52, *251*
    management, 253, 254, 256
    risk factors, 253
    staging, 252
diphtheria, prophylactic vaccination for, *283*
disseminated intravascular coagulation (DIC), 164
distal pancreatectomy, 53
DLBCL. *See* diffuse large B-cell lymphoma
DNA
    circulating tumor DNA (ctDNA), 34–35, *34*, 37
    karyotypes, 24–25, *24*, 69, *69*
    methylation, 30, 37, 171, 173–74, 181
    microarrays, 25–26, *26*, 30, 32, 35, 37
    oligonucleotide sequences (oligos), 25
DNA methyltransferases (DNMTs), 173, 182
DNA repair, 193, 197
    base-excision repair (BER), 198
    BRCA genes and, 197
    double-strand DNA break repair, 197, *199*
    hereditary breast cancer, 197–98
    mechanisms, 193–97
    mismatch repair (MMR), 197, 221
    single-strand DNA breaks, 198
DNMT inhibitors, 181
DNMTs (DNA methyltransferases), 173, 182
docetaxel, *20*, 54, 142, 152
double-strand DNA break repair, 197, *199*
doxorubicin, *20*, 123, 140, 142, 187, 202, 203, 230
driver mutations, 4, 23, 32, 61–62, 66
DUBs. *See* deubiquitinases
durvalumab, *244*
duvelisib, 189
dyspareunia, 284

E

E6 HPV oncoprotein, 286, *286*
E7 HPV oncoprotein, 286, *286*
E255K/V mutation, 74
E317L/V/I/C mutation, 74
Early Breast Cancer Trialists' Collaborative Group, 140

Eastern Cooperative Oncology Group Performance Status scale
    (ECOG PS scale), *54*, 66
EBRT. *See* external beam radiation therapy
EBV (Epstein-Barr virus), 258
ECM (extracellular matrix), 215, 219
ECOG PS scale. *See* Eastern Cooperative Oncology Group
    Performance Status scale
ECP. *See* extracorporeal photopheresis
edema, 51
EGF-R antibodies, 227
EGF receptor (epidermal growth factor receptor), *33*, 56
EGFR gene, *33*
EGFR mutation analysis, *28*
embryonal carcinomas, 44, *44*
EML4-ALK fusion, 87, *87*
EML4-ALK translocation-positive NSCLC, *86–88*, 87–89
emtansine, 139
endocrine glands, 131
endogenous T cells, 273–74, *273*, *274*
endoscopic retrograde cholangiopancreatography (ERCP),
    56, *57*
endothelial cells, 7, 216, 232
enterocolitis, 301
enteropathy-associated T-cell lymphoma, *118*, 249
enzalutamide, *152*, 153–54, *155*
eosinophils, 66
epidermal growth factor (EGF) receptor, *33*, 56
epididymitis, 42
epigenetic therapy, 171–91
    about, 171–72, 190–91
    for cutaneous T-cell lymphoma (CTCL), 188–89, *189*
    DNA methylation, 171, 173–74
    histone modification, 171–73, *172*, *173*
    long noncoding RNAs (lncRNAs), 171, 175
    microRNA (miRNA) targeting, 171, 174–75, *174*
    for myelodysplastic syndromes (MDSs), 179–82
epigenetics, 171
episome, 287
Epstein-Barr virus (EBV), 258, *282*
ER (estrogen receptor), *28*, 135
ErbB2, 139
Erbitux, *244*
ERCP. *See* endoscopic retrograde cholangiopancreatography
eribulin, 142
erlotinib, 57, *64*
erythroderma, 183
erythroleukemic leukemia, *161*
estrogen, 136, 141
estrogen receptor (ER), *28*, 135, 144
etoposide, *20*, 85
everolimus, *64*, 116, 129, 144
exemestane, 135, 143, 144, 201
external beam radiation therapy (EBRT), 17, 150, *150*, 151
extracellular matrix (ECM), 215, 219
extracorporeal photopheresis (ECP), 187
extranodal NK T-cell lymphoma, *118*, 249
EZH2 gene, 250

F

F359V/C/I mutation, 74
FAB (French American British) classification, of myelodysplastic
    syndromes, 176–77
familial adenomatous polyposis (FAP), 16, *221*
FCR, 78
females, cancers statistics by type and gender, *2*
Femara, 143
FGF (fibroblast growth factor), 215, 232
fibrin/fibrinogen, *28*

fibrinogen, 164
fibroblast growth factor (FGF), 215, 232
first-line chemotherapy, 22
FKBP12, 116
FL. *See* follicular lymphoma
FL international prognostic index (FLIPI), 122, *122*
flow cytometry, 76, *77*
fludarabine, 78
fluorescence *in situ* hybridization (FISH), 24–25, *24*
    in chronic myelogenous leukemia, 69, *69*, 164
    in myeloid myeloma, 206–7
fluorodeoxyglucose-avid lymph nodes, 84
fluorodeoxyglucose (FDG), as tracer, *9*
5-fluorouracil, *20*, 54, 223, 225, 226, 227
flutamide, 136
FOLFIRI chemotherapy, *40*
FOLFIRINOX chemotherapy, 54, 55, 58
FOLFOX chemotherapy, *40*, 223–24, 225, 226
follicle-stimulating hormone (FSH), 131, *132*
follicular lymphoma (FL), *118*, 119–27, *249*
    chromosomal rearrangement, 119–20, *119*, 121
    future prospects and challenges, 126–27
    genetics, 119–20, *119*, 121
    PI3-kinase inhibition, 125–26, *125*, 127
    staging, 121–22, *121*
FoundationOne CDx diagnostic test, *29*
French American British (FAB) classification, of myelodysplastic
        syndromes, 176–77
FSH (follicle-stimulating hormone), 131, *132*
full-body CT scan, 10, 122
fulvestrant, 141
fused PET/CT scan, *9*
fusion PET/CT, 84, *84*

G

Gamma Knife, 18
Gardasil, 288, 290, 291
gastrointestinal stromal tumor (GIST), 64
Gazyva, *244*
GCTs. *See* germ cell tumors
gefitinib, *64*
gemcitabine, *20*, 54, 55, 56, 58, 142, 187
gender, cancers statistics by type and gender, *2*
gene amplification, 23
gene expression, 171, 172
    See also epigenetic therapy; epigenetics
gene expression arrays, breast cancer, 139–40
gene repression, 172
genital warts, 290
genomic markers, 27, *28–29*
genomic profiling, 65, 66
genomics/genetics, 4, 6, 23–35, 35–36, *37*
    of acute lymphoblastic leukemia (ALL), 293
    of acute promyelocytic leukemia (APL), 162–63, *162*, *163*
    of breast cancer, 23, 136–37, 139–40, 143
    The Cancer Genome Atlas (TCGA), 31
    chromosomal aberrations, 23–24
    of chronic lymphocytic leukemia, 75, 82
    of chronic myelogenous leukemia, 66–67, *66*, 69, *69*, 74
    circulating tumor cells (CTCs), 34–35, *34*, 37
    circulating tumor DNA (ctDNA), 34–35, *34*, 37
    clinical use of, 23
    of colorectal cancer (CRC), 23, 220–21
    cytogenetic approaches, 23–25, *24*
    defined, 23
    for diagnosis, 23, 32–33, 37
    of diffuse large B-cell lymphoma (DLBCL), 249–51, *250*, 255–56
    driver mutations, 4, 23, 32, 61–62, 66

genome instability and mutation, 4, 6
    of hereditary breast cancer, 197–98, *198*, *199*
    of hereditary renal cell carcinoma, 109, *109*
    heterogeneity within cancers, 7
    of human papillomavirus (HPV), 285–86
    inherited mutations, 16
    karyotype, 24, *24*, 69, *69*
    of lung cancer, 86, *86*, 89
    of melanoma, 23, 94–95, 99
    microarrays, 25–26, *26*, 30, 32, 35, 37
    of multiple myeloma (MM), 206–7, *207*
    of myelodysplastic syndromes (MDSs), 176–77, *177*, 181
    of non-Hodgkin lymphoma (NHL), 119–20, *119*, 121
    passenger mutations, 4, 62
    pharmacogenomics, 35, 36, 37
    Philadelphia chromosome, 66–67, *66*, 69, *69*
    of prostate cancer, 145–46
    protein kinases, 61–62, 65
    of renal cell carcinoma (RCC), 109, *109*, 113
    sequencing, 27, 29–32, *29–32*, 35–36
    The Solid Tumor Targeted Cancer Gene Panel by Next Generation
        Sequencing (CAPN), 33–34, *33*
    synthetic lethality, 193–94, *194*, *199*, 212, 214
    translocation, 23–24
    See also epigenetic therapy; mutations
germ cell tumors (GCTs), 41, 44
germ cells, 41
germline mutation, 75
germinal-center B-cell (GCB)-like lymphoma, 250–61, *250*
Gilman, Alfred, 20
Gilotrif, *64*
GIST. *See* gastrointestinal stromal tumor
Gleason grade, 148–49, *148*, *160*
Gleevec, 64, *64*
glioblastoma, mutations in, 23
gliomas, about, 1
glycolysis, 6
GM-CSF (granulocyte-macrophage colony-stimulating
        factor), 272
GNAQ gene, *33*
GNAQ mutations, 95
GnRH (gonadotropin-releasing hormone), 131, *132*
GnRH agonists, 133, *133*, 136
Goodman, Louis S., 20
goserelin, 133, 141
gout, 67
graft-versus-host disease (GVHD), 80, 277
graft-versus-leukemia effect (GVL), 73, 277, 294, 298
granulocyte-macrophage colony-stimulating factor (GM-CSF), 272,
        280
Greenberg, Peter L., 178
GTPases, 95
guanine nucleotide binding protein Q, *33*
GVHD. *See* graft-versus-host-disease
GVL. *See* graft-versus-leukemia effect
gynecomastia, 42

H

hairy cell leukemia, *118*, *249*
HAMAs (human anti-mouse antibodies), 246
Hanahan, Douglas, 4
HATs (histone acetyltransferases), 172, *173*
HBV (hepatitis B virus), 272, *282*, 303
HCV (hepatitis C virus), *282*
HDAC inhibitors, 173, 196
    for cutaneous T-cell lymphoma (CTCL), 187, 188–89, 190–91
    mechanism of action, *189*
HDACs (histone deacetylases), 172–73, *173*, 196, 197

HE4, *28*

heart disease, death rates from, 3, *3*

helper T cells, 270

hematopoiesis, 148

hematopoietic cancers, 1

hematopoietic stem cell transplantation (HSCT), 179–80

hematuria, 109

hemicolectomy, 223

hemoptysis, 42, 83

Heng criteria, 112

hepatitis B virus (HBV), 272, *282*, 303

hepatitis C virus (HCV), *282*

hepatocyte growth factor (HGF) receptor, *33*

hepatomegaly, 77

hepatosplenic T-cell lymphoma, *118*, *249*

HER-2/neu gene, 23, *28*, 263

HER2 (human epidermal growth factor 2), 10, 139, 263

HER2-blocking drug, 140–41, 142

HER2 breast cancer, *139*, 140, 144, 263, 265

Herceptin, 139, *244*, 263–66

herd immunity, 290

hereditary breast cancer, 197–204

  about, 197

  case study, 201–3, *201–3*, 213

  genomics, 197–98, *198*, *199*

  management

    localized, 203

    neoadjuvant therapy, 202–3

  prophylactic mastectomy, 203

  symptoms, 201

hereditary leiomyomatosis, *109*

hereditary non-polyposis colon cancer (HNPCC), 221, *221*

hereditary renal cell carcinoma, 109, *109*, 117

herpes simplex virus 1 (HSV1), 279

HGF receptor (hepatocyte growth factor receptor), *33*

HHV-8 (human herpesvirus-8), *282*

HIF (hypoxia-inducible factor), 113, 116

high grade B-cell lymphoma, *118*

high-intensity focused ultrasound (HIFU), 150

high-throughput screening, 213, 214

histone acetylation, 172

histone acetyltransferases (HATs), 172, *173*

histone deacetylases (HDACs), 172–73, *173*, 196, 197

histone methylation, 172

histone modification, 171–73, *172*, *173*

HLA (human leukocyte antigen), 294

HNPCC. *See* hereditary non-polyposis colon cancer

Hodgkin lymphoma (HL), 256–62

  about, 256–58, 262

  case study, 260

  diagnosis, 258–59

  epidemiology, 258

  future prospects and challenges, 262

  histology, 2662

  incidence of death, 259, *259*

  management, *40*, 259–60, 261, 262

  risk factors, 258

  staging, 258

Hodgkin Reed–Sternberg (HRS) cell, 256, 257, 261, 262

homologous recombination, 197, 198

hormone-refractory prostate cancer, 153

hormone therapy, *19*, 136–56

  acquired endocrine resistance, 143

  for breast cancer, 133, 143

  mechanism of action, *133*, 136

    inhibition of conversion of steroids in peripheral tissues, *133*, *134*, 135

    inhibition of gonadotropin release, 133–34, *133*, *134*

    steroid hormone receptor antagonists, *133*, *134*, 135–36

    steroidogenesis inhibition, *133*, 134–35, *134*, 156

  for prostate cancer, 133, 135, 151, 153–54, 155, *155*

  resistance to, 143, 144

  types, 133

hormones, 131

HPV (human papillomavirus), 243, 272, 282, *282*, 284, *284*, 285–87, *286*, 291

  anti-HPV vaccine, *286*, 288–90, *289*

HRAS gene, *33*

HRS cell (Hodgkin Reed-Sternberg cell), 256, 257, 261, 262

HSCT. *See* hematopoietic stem cell transplantation

HSV1 (herpes simplex virus 1), 279

HTLV-1 (human T-cell leukemia virus-1), *282*

Huggins, Charles, 151

human anti-mouse antibodies (HAMAs), 246

human epidermal growth factor 2 (HER2), 10, 139, 263

human herpesvirus–8 (HHV-8), *282*

human leukocyte antigen (HLA), 294

human papillomavirus (HPV), 243, 272, 282, *282*, 284, *284*, 285–87, *286*, 291

  anti-HPV vaccine, *286*, 288–90, *289*

human T-cell leukemia virus–1 (HTLV-1), *282*

hydronephrosis, 124

hydroxyurea, *20*, 166

hypercalcemia, 205, *205*

hyperlipidemia, 231

hypersensitivity, 246

hypogammaglobulinemia, 76

hypomethylating agents, 179, 181, 182

hypophysitis, 301

hypothalamic/pituitary/gonadal axis, 131–32, *132*

hypoxia-inducible factor (HIF), 113, 116

hysterectomy, 287

I

ibritumomab, *244*

ibrutinib, *64*, 80, *80*, 81, *81*, 82, 254

Iclusig, *64*

idelalisib, 80, 81, 125, 126, *126*, 127, 129

IDH1 gene, *33*

IDH2 gene, *33*

IDO inhibitors, 281

IFL, 225, 227

IFN (interferon), 112, 280

Ig heavy-chain variable-region (IGHV) genes, 75

IGF-1 (insulin-like growth factor-1), 107, 232

Ikaros, 234, 235

IKZF1, 234

IL-2 (interleukin-2), 96, 112, 117, 280, 301, 302

IL-6 (interleukin-6), 215, 232

imatinib mesylate, 23, 64–65, *64*, 70–71, *70*, 73, 100

Imbruvica, *64*

Imfinzi, *244*

imiquimod, 277, 280–81

immortalization, *242*

immune checkpoint inhibitors, 240, 269, 277–78, *278*, 282, *304*

immune responses, 5

immune system, 239, 281, 290

immunocytes, 215

immunogenic antigen, 240

immunoglobulin (Ig) genes, 75

immunoglobulins, *28*

immunological memory, 239

immunomodulatory drugs (IMiDs), 181, 229, 303

  in bone marrow microenvironment, 233–34

  chemical structure, *233*

intravenous immune globulin (IVIG), 296
  for multiple myeloma, 229, 230–31, 235
  See also immunotherapy
immunosuppressive microenvironment, 217, 219
immunotherapy, 239, 240, *240*, 269–72, 303
  acute lymphoblastic leukemia, 294, 296–98, *296*
  adoptive T-cell therapies, 269, 270–71, *271*, 272, 273–77, *273–76*, 282, *304*
  adoptive T-cell transfer, 240, 305
  Bacillus Calmette-Guérin (BCG), 277, 281
  bi-specific antibodies (bsAb), 279–80, *304*, 305
  cancer vaccines, *19*, 240, 269, 271–73, 282
  CAR T cells, 240, *273*, 275–77, *275*, *276*, 296–98, *296*, 303, 304, *304*
  checkpoint inhibitors, 240, 269, 277, 282, *304*
  cytokines, *19*, 237, 280
  future challenges, 304
  imiquimod, 277, 280–81
  indoleamine 2,3-dioxygenase (IDO) inhibitors, 281
  intravenous immune globulin (IVIG), 296
  mechanisms of action, 303–5, *304*
  metastatic melanoma, 299, 300–302
  oncolytic viruses, 279, 305
  for renal cell carcinoma, 112, 117
  thalidomide, 229, 230, 233, 235, 277, 280
  tolerance, 20
  transforming growth factor-β (TGF-β), 281
  See also immunomodulatory drugs; monoclonal antibodies; monoclonal antibody therapy
implantable chemotherapy administration, *21*
IMRT. *See* intensity-modulated radiation therapy
"index patient," 16
indoleamine 2,3-dioxygenase (IDO) inhibitors, 281
indolent disease, 75, 82
indolent lymphomas, 118, *118*, 119, 249
  See also follicular lymphoma
induction chemotherapy, 165, 295
infectious mononucleosis, 258
inflammation, tumor-promoting effects of, 5
inflammatory bowel disease, 221
inflammatory cytokines, 297
Inlyta, *64*
innate immunity, 239, *240*
insulin-like growth factor–1 (IGF–1), 107, 232
intensity-modulated radiation therapy (IMRT), 18, *18*, 151
interferon (IFN), for renal cell carcinoma (RCC), 112
interferon alpha, in chronic myelogenous leukemia, 70, 71, 280
interferons, 187, *187*
interleukin-2 (IL-2), 96, 112, 117, 280, 301, 302
interleukin-6 (IL-6), 215, 232
International Germ Cell Consensus Classification, 44, *45*
International Prognostic Index (IPI), 251
International Prognostic Scoring System (IPSS), 178
International Staging System (ISS), 208
interstitial pulmonary fibrosis, 47
intra-arterial chemotherapy, *21*
intramuscular chemotherapy, *21*
intraperitoneal chemotherapy, *21*
intrathecal chemotherapy, *21*
intravasation, 218
intravenous chemotherapy, *21*
intravenous immune globulin (IVIG), 296
intravenous radionuclides, 17, 19
intravesicular chemotherapy, *21*
IPI (International Prognostic Index), 251
ipilimumab, *244*, 300, 301, 302
IPSS. *See* International Prognostic Scoring System
Iressa, *64*
irinotecan, *20*, 54, 225, 227

islet cell carcinomas, 51
isocitrate dehydrogenase 1 and, 2, *33*
isoform-specific inhibitors, 105
isoforms, 105
ISS. *See* International Staging System
ixabepilone, 142
ixazomib, 211

J

JAK-STAT pathway, 257
JAK2 gene, *33*
JAK3 gene, *33*
Jakafi, *64*
Janus kinase 1 and 2, *33*
jaundice, 51, 58

K

Kadcyla, 139, *244*
Kaplan–Meier curve, *201*
kappa light chains, 77
Karnofsky Performance Status (KPS) scale, 55, *55*
karyotype, 24, 69, *69*
  defined, *69*
  normal human karyotype, *24*
  Philadelphia chromosome, 66–67, *66*, 69, *69*
ketoconazole, 135
ketorolac, 123
Keytruda, *244*
kidney cancer
  cytoreductive surgery, 17
  renal cell carcinoma (RCC), 106–17
  treatment, 17
kinase inhibitors. *See* protein kinase inhibitors
KIT (stem cell factor receptor), *28*, 49
KPS scale. *See* Karnofsky Performance Status (KPS) scale
KRAS gene, *33*
KRAS mutation analysis, *28*, 56
kyphoplasty, 231

L

L1 protein, 285
L2 protein, 285
lactate dehydrogenase (LDH), *28*, 43, *43*, 122
lambda light chains, 77
laparoscopic prostatectomy, 150, *150*
laparoscopy, 16
lapatinib, *64*, 114, 139, 142
LDH. *See* lactate dehydrogenase (LDH)
lenalidomide, 180, 181, 182, 189, 209, 211, 212, 219, 229, 230, 231, 233, 234, 235, 280
lentigo maligna melanoma, 91, *91*
lentiviral vectors, 297
letrozole, 135, 143, 144
leucovorin, 224, 225, 226
leukemia
  about, 1
  classification, 281–82
  defined, 66, 291
  graft-versus-leukemia effect, 73
  stem cell transplant, 73, 80, 166
  See also individual forms of leukemia
leukocytosis, 164
leuprolide, 133, 141, 152
LH (luteinizing hormone), 131, *132*
liquid biopsy, 35, 37

LKB1 gene, 137, *198*

long noncoding RNAs (lncRNAs), 171, 175, 191

loss of heterozygosity (LOH), 113

low-penetrance genes, 137

LP cell (lymphocyte-predominant cell), 256

lumbar puncture, 292

luminal A breast cancer, 139, *139*, 140

luminal B breast cancer, 139, *139*, 140

lumpectomy, 15, *15*

lung cancer, 82–89

  case study, 86–87, *86*

  diagnosis, 83, *86*

  EML4-ALK translocation-positive NSCLC, *86–88*, 87–89

  epidemiology, 82–83

  future prospects and challenges, 89

  genomics, 86, *86*, 89

  management, 85, 89

  metastatic, 85

  mortality, 83

  mutations and, 31

  non-small-cell lung cancer (NSCLC), 84, *84*, 85, 86, *86*, 89

  screening, 83

  small-cell lung cancer (SCLC), 84, 85, 89

  smoking and, 82–83, 85, 89

  staging, 84, *84*, 89

  symptoms, 83, *83*, 89

  tissue biopsy, 83–84

luteinizing hormone (LH), 131, *132*

lymph node dissection, 92

lymph node draining, in breast, 14, *14*

lymphadenopathy, 67, 292

lymphatic system, 14

lymphedema, 92, 140

lymphocyte-predominant (LP) cell, 256

lymphocytes, 66, 76–77

lymphocytosis, 76, *76*, 185

lymphoid cells, 66, 232

lymphomagenesis, 249

lymphomas

  about, 1

  adoptive cell transfer treatment, *273*

  indolent lymphomas, 118, *118*, 119

  nitrogen mustard therapy, 20

  subtypes, *118*

  See also Hodgkin lymphoma; non-Hodgkin lymphoma

lymphoplasmacytic lymphoma, *118*, 249

Lynch syndrome, *221*

Lynparza, 204

M

M protein (monoclonal protein), 205

MAbs. *See* monoclonal antibodies

MAC (membrane attack complex), 243

macrophage activation syndrome (MAS), 297

macrophages, 237, 239

magnetic resonance imaging (MRI), staging with, *9*

maintenance therapy, 165

major molecular response, 71

males, cancers statistics by type and gender, *2*

malignant cell transformation, 159

MammaPrint diagnostic test, *29*

mammography, 137, *137*, 144

mantle cell lymphoma, *118*, 249

marginal zone lymphoma, *118*

MAS (macrophage activation syndrome), 297

mastectomy, modified radical, 14–15, *15*, 16, 203

mastodynia, 42

matrix metalloproteinase-9 (MMP-9), 218, *218*

Mayo Clinic Solid Tumor Targeted Cancer Gene Panel, 32–33, *33*

MBL. *See* monoclonal B-cell lymphocytosis

MCCV (Merkel cell carcinoma virus), *282*

MCP-1 (monocyte chemotactic protein-1), 216

mCRPC. *See* castration-resistant prostate cancer

MDS-U. *See* myelodysplastic syndrome unclassified

MDSCs (myeloid-derived suppressor cells), in TME, 216, 217

MDSs. *See* myelodysplastic syndromes

mechanism-based targeted therapy, 7

mechanistic target of rapamycin (mTOR), 103, 105–6, 108

mechlorethamine, 20, *20*, 259

mediastinal carcinoma, 44, *44*

mediastinoscopy, 13–14, *14*, 16, 83

mediastinum, 44

medical cancer therapies, 7, 13, 19, *19*

  neo-adjuvant therapy, 17

  perioperative treatment, 17

  pharmacogenomics, 35, 36, 37

  See also chemotherapy; hormone therapy; immunotherapy; monoclonal antibody therapy; radiation therapy; surgery; targeted therapy; vaccination

medical castration, 133, 151

megakaryocytic leukemia, *161*

MEK inhibition, 98

MEK/MAPK signaling, 98

Mekinist, *64*

melanocytes, 90

melanoma, 90–99

  about, 1

  acral lentiginous melanoma, 91, *91*

  BRAF gene in, 23, 94–95

  case study, 95–96, *96*

  diagnosis, 90

  epidemiology, 90, 99

  evaluation, 92

  future prospects and challenges, 99, 302

  genetic alterations, 23, 94–95, 99

  imaging, 93

  lentigo maligna melanoma, 91, *91*

  lymph node involvement, 92

  management

    adoptive cell transfer, 273, *273*, 274

    cytoreductive surgery, 17

    IDO inhibitors, 281

    immunotherapy, 299, 300–302

    kinase inhibition, 96–99, *97*

    localized melanoma, 93–94

    metastatic melanoma, 94

    oncolytic virus therapy, 279

  metastatic melanoma, 93, 94, 99, 274, 297–302

  mutations in, 23, 31

  nodular melanoma, 91, *91*

  recommended surgical margins, 92, *92*

  staging, 91, 92, *92*

  subtypes, 91, *91*

  superficial spreading melanoma, 91, *91*

  uveal melanoma, 90, 95

  warning signs, 90, *90*

melphalan and prednisone (MP), 229

membrane attack complex (MAC), 243

meninges, 90

Merkel cell carcinoma virus (MCCV), *282*

mesenteric, 124

mesorectum, *15*

MET gene, *33*

meta-analysis, 210

metalloproteinases, 218, *218*

metamyelocytes, 68

metastasis, 5

blastic metastases, 149
malignant cell transformation, 159
metastatic breast cancer, 141–42, 143, 156, 203–4
metastatic castration-resistant prostate cancer (mCRPC), 151–52, 156
metastatic colorectal cancer (CRC), 222–23, 223, 225, 226–28
metastatic lung cancer, 85
metastatic melanoma, 93, 94, 99, 274, 297–302
metastatic prostate cancer, 148–49, 149, 151–52, 155, 156
metastatic renal cell carcinoma (RCC), 111–13, 112, 115–16, 117
methotrexate, 20, 187
methylated cytosines, identification of, 30
methylation, 172
methylene diphosphonate (MDP), 9
myeloproliferative neoplasm, 66
MF. See mycosis fungoides
MGUS. See monoclonal gammopathy of undetermined significance
MHC-I, 270
MHC-II, 270
MHC molecules, 270
microarrays, 25–26, 26, 30, 32, 35, 37
microRNA (miRNA) targeting, 171, 174–75, 174, 191
microRNAs, 75
microsatellite instability (MSI), 220
microtubule inhibitors, 142, 260
minimal residual disease (MRD), 293
mismatch repair (MMR), 197, 221
mitochondrion, 120
mitogen, 226
mitomycin-C, 20
mitotic rate, 92
MLL translocation, 293
MM. See multiple myeloma
MMAE (monomethyl auristatin E), 260, 261, 262, 267
MMP-9 (macrophage metalloprotein-9), 218, 218
MMR (mismatch repair), 197, 221
modified radical mastectomy, 15, 15
molecular biomarkers. See biomarkers
molecular targeted therapies, 39
moles, 90
monoclonal antibodies (MAbs), 62, 239–48
about, 241, 242, 243, 247
checkpoint inhibition in, 278, 278, 304
mechanism of action, 304
methods of killing or inhibiting growth of cancer cells, 243–45, 245, 246–47
nomenclature, 245–46, 246
pharmacokinetics, 247
sources of, 241, 242
structure, 246, 246
tumor microenvironment (TME) and, 244–45
monoclonal antibody therapy, 19, 247, 269
for breast cancer, 139, 140–41, 263–67, 263–65
for chronic lymphocytic leukemia, 78, 79, 80
for cutaneous T-cell lymphoma, 187
for Hodgkin lymphoma, 267–68
mechanism, 243–45, 245, 246–47
for non-Hodgkin lymphoma, 123
monoclonal B-cell lymphocytosis (MBL), 77
monoclonal gammopathies, 205
monoclonal gammopathy of undetermined significance (MGUS), 205–6
monoclonal protein (M protein), 205
monocyte chemotactic protein-1 (MCP-1), 216
monocytes, 66, 243
monocytic leukemia, 161
monomethyl auristatin E (MMAE), 260, 261, 262, 267
mononucleosis, 258
MOPP, 40, 259
MP (melphalan and prednisone), 229

MRD (minimal residual disease), 293
MRI. See magnetic resonance imaging
MSI (microsatellite instability), 220
mTOR (mechanistic target of rapamycin), 103, 105–6, 108, 116
mTORC1, rapalogs and, 106, 116
mucositis, 161
multi-drug resistance (MDR) transporter, 236
multiple myeloma (MM), 205–12, 228–35
about, 205, 212
bone marrow microenvironment in, 231–32, 232, 235
bone metabolism, 232–33
case studies, 208–9, 230–31
CRAB criteria, 205, 206, 212
diagnosis, 206, 207
epidemiology, 205–6, 212, 228
etiology, 206, 206
future prospects and challenges, 211–12, 235
genomics, 206–7, 207
management, 208, 212, 228–29, 235
immunomodulatory drugs (IMiDs), 229, 230–31, 233–34, 235, 280
proteasome inhibition, 209–11, 211, 212, 213, 235
transplant-ineligible patients with newly diagnosed MM, 229–30
proteasome inhibition, 209–11, 211, 212, 213
risk factors, 207
staging, 208, 208
symptoms, 232
mural cells, 216
mustard gas, 19–20
mutations, 31–32, 32
APC mutations, 16
BRAF mutations, 23, 94–95
c-KIT mutations, 95
cancer cells from, 193
The Cancer Genome Atlas, 31
chronic myelogenous leukemia, 74
driver mutations, 4, 23, 32, 61–62, 66
germ-line mutation, 75, 202
GNAQ mutations, 95
hereditary breast cancer, 198
incidence in different types of cancer, 31–32, 32
inherited mutations, 16
NRAS mutations, 95
passenger mutations, 4, 62
somatic mutation, 75
See also genomics/genetics; translocation
MYC gene, 250
Mycobacterium bovis, 281
mycosis fungoides (MF), 118, 183–84, 190, 249
diagnosis, 184–85, 184, 185
epidemiology, 184
future prospects and challenges, 189
management, 186–87, 187, 188–89, 190
prognosis, 190
TNMB classification, 185, 186
myeloablative chemotherapy, 47
myeloblastic leukemia, 161
myelocytes, 68, 68
myelodysplasia, 259
myelodysplastic syndrome unclassified (MDS-U), 177
myelodysplastic syndrome with isolated del(5q), 177, 177, 181–82
myelodysplastic syndromes (MDSs), 175–82
case study, 178–80
classification, 176–77, 177
diagnosis, 176–77, 176, 177
epidemiology, 176
genomics, 176–77, 177, 181
karyotypic analysis, 177–78, 178
management, 179–82

myelodysplastic syndromes (MDSs) (*continued*)
  prognosis, 178, 182
myelogenous leukemia. *See* acute myelogenous leukemia; chronic myelogenous leukemia
myeloid cells, 66, 161, 218, 232, 237
myeloid-derived suppressor cells (MDSCs), in TME, 216, 217
myeloma cells, 205
myelomas, about, 1
myelomonocytic leukemia, *161*
myeloproliferative neoplasms, 64

N

naive T cells, 270
nanoalbumin-bound paclitaxel (nab-paclitaxel), 237
natural killer cells (NK cells), 239, 243, 277, 304
negative feedback, 131
neoadjuvant chemotherapy, 19, 39, 202
neoplastic follicles, 120
nephrectomy, 111, 113
neratinib, *64*
Nerlynx, *64*
Network Genomic Medicine (NGM), 36
Neu, 139
neural crest, 90
neuroblastomas, about, 1
neuroectodermal cancers, about, 1
neuroendocrine cancers, 84
neutropenia, 55
neutrophils, 66, *68*
Nexavar, *64*
next generation sequencing (NGS), 27, 30, 31–32, *31*, *32*, 36, 37
NGM. *See* Network Genomic Medicine
NGS. *See* next generation sequencing
NHL. *See* non-Hodgkin lymphoma
nilotinib, *64*, 71, 73, 74
nilutamide, 136
nitrogen mustard, 20
nivolumab, *244*, 301
NK cells (natural killer cells), 239, 243, 304
nodular lymphocyte-predominant Hodgkin lymphoma (NLPHL), 256
nodular melanoma, 91, *91*
non-Hodgkin lymphoma (NHL), 118–27, 183, 243, 248–56
  about, 118–19, 127, 248–49
  case study, 124–25
  diagnosis, 119–20, 251–52, *252*
  epidemiology, 120, 251
  future prospects and challenges, 126–27
  genomics, 119–20, *119*, 121
  histology, 119, *119*, 120–21, *121*
  imaging, *119*
  management, 122–23, 127, 280
  prognostication, 122, *122*
  staging, 121–22, *121*
  subtypes, 183, *183*, 248–49, *249*
  symptoms, 119, 251
non-small-cell lung cancer (NSCLC), 84, *84*, 85, 86, *86*
nonseminomas, 46, 47
NOTCH1, 75
NRAS gene, *33*
NRAS mutations, 95
NSCLC. *See* non-small-cell lung cancer
nuclear matrix protein 22, *28*
nucleoside analogs, 78

O

obinutuzumab, 79, 123, *244*
ofatumumab, 80, *244*

olaparib, 203, 204
oligometastatic disease, 113
oligonucleotide sequences (oligos), 25
oncocytic renal cell cancer, 108, *108*, *109*
oncogene addiction, 62
oncogenes, 61
oncogenesis, protein kinases and, 61–62
oncology, 3–6, 11
oncolytic virus therapy, 279, 305
oncomiRs, 174
oncoprotein, 165
Oncotype DX, *29*
Oncovin, 259
Opdivo, *244*
OPG (osteoprotegerin), 233
oprozomib, 211
oral chemotherapy administration, *21*
oral leukoplakia, differentiation therapy, 160–61, *161*
orchiectomy, 43, 44
orchiopexy, 42
orchitis, 42
OS. *See* overall survival
osteoblasts, 232, *232*, 233
osteoclasts, 205, 232, *232*, 233, 235
osteolytic bone lesions, 232
osteolytic metastases, 205, *206*
osteoprotegerin (OPG), 233
OVA1 diagnostic test, 29
ovarian cancer
  cytoreductive surgery, 17
  prophylactic salpingo-oophorectomy, 16
ovaries, estrogens and, 135
overall survival (OS), 88
oxaliplatin, 54, 224, 226

P

p53, 5
p53 gene, 137
paclitaxel, *20*, 140, 142, 202, 203, 237, 287
paired-sample analysis, 27, 29, *29*
PALB2 gene, 136
palbociclib, 144
palliative chemotherapy, 39
palliative surgery, 16
pallor, 292
pamidronate, 233, *233*
PAMPs (pathogen-associated molecular patterns), 239
pancreas, anatomy, *51*
pancreatectomy, 53
pancreatic cancer, 50–58
  case study, 55–56
  clinical trials, 58
  death rate, 50
  diagnosis, 51–52
  evaluation, 52, *52*
  incidence, 50
  management, 53–58
  prognosis, 50, 53, 58
  risk factors, 50, *50*, 58
  staging, 52–53, *52*
  subcategories, 50–51
  survival, 50, 53, 54, 58
  symptoms, 51–52, *51*, 58
  targeted therapy, 56–57, 58
pancreatic intraepithelial neoplasia-1 (PanIn1), 56
pancreatic neuroendocrine cell tumors (PNETs), 51
panitumumab, 227, *244*
PAP (prostatic acid phosphatase), 272–73

Pap smear, 283, 284
papillary renal cell cancer, 108, *108*, *109*
paraprotein, 205
PARP inhibitors, 204, 213
PARPs [poly(ADP-ribose) polymerases], 198, 213
partial mastectomy, 15, *15*
passenger mutations, 4, 62
pathogen-associated molecular patterns (PAMPs), 239
pathologic complete response (pCR), 202
pazopanib, *64*, 115, 116, 227
PCR. *See* polymerase chain reaction
PD-1 (programmed cell death protein-1), 278
PD-L1 (programmed death ligand 1), 278
PD-L2 (programmed death ligand 2), 278
PDGF (platelet-derived growth factor), 215
PDGF-β (platelet-derived growth factor-β), 113, *114*
PDGFR (platelet-derived growth factor receptor), 64, 106–7
pembrolizumab, *244*, 281, 300, 301–2
pemetrexed, *20*, 85
pentostatin, 187
peptide–MHC complexes, 270
percutaneous ablation, 111
percutaneous biopsy, 83
performance status, *54*, 55, *55*
pericytes, 7, 216
perioperative treatment, 17, 39
peripheral neuropathy, 210–11
peripheral T-cell lymphoma, *118*, 183, *249*
Perjeta, 139
pertuzumab, 139
PET. *See* positron-emission tomography
petechiae, 292
PFS. *See* progression-free survival
PGF (placenta growth factor), 227
pharmacogenomics, 35, 36, 37
pharmacokinetics, 247
pheochromocytomas, 113
Philadelphia chromosome, 66–67, *66*, 69, *69*
phosphatidylinositol 3,4,5-trisphosphate (PIP3), 103
phosphatidylinositol 3-kinase (PI3K), 103, 104, 107
phosphatidylinositol 4,5-bisphosphate (PIP2), 103, 104
phosphorylation, 61, 65
phototherapy, 186
PI3 kinase inhibitors, 81, 107
PI3K (phosphatidylinositol 3-kinase), 103, 104–5, *106*, 107
PI3K/Akt/mTOR pathway, 103–29, *128*
   about, 103–4
   inhibiting components of, 107
   summary of, *104*
   tumorigenesis and, *106*
PI3K inhibitor, 125, 127, 144
PIK3CA mutations, 143
PIN (prostatic intraepithelial neoplasia), 145
PIP2 (phosphatidylinositol 4,5-bisphosphate), 103, 104
PIP3 (phosphatidylinositol 3,4,5-trisphosphate), 103
pituitary gland, 131
PKB (protein kinase B), *33*, 103
placenta growth factor (PGF), 227
plasma cells, 231, 240
plasmacytoma, 205
plasminogen activator inhibitor (PAI-1), diagnostic test, *29*
platelet-derived growth factor (PDGF), 215
platelet-derived growth factor-β (PDGF-β), 113, *114*
platelet-derived growth factor receptor (PDGFR), 64, 106–7
platinum analogs, *20*
pleural effusion, 84
pluripotent cells, 41, 44, 159
PLX4032, 96–97
PMBCL. *See* primary mediastinal B-cell lymphoma

PML–RARα fusion protein, 162, 163
PNETs. *See* pancreatic neuroendocrine cell tumors
polio, prophylactic vaccination for, *273*
poly(ADP-ribose) polymerases (PARPs), 198, 213
polyclonal antibodies, 241
polymerase chain reaction (PCR), 164, 185
polyubiquitination, 165, 195, *195*
pomalidomide, 233, 234, 235
ponatinib, *64*, 72–73, *74*
positron-emission tomography (PET), staging with, *9*
PR (progesterone receptor), *28*, 144
pralatrexate, 187
precursor B-cell lymphoblastic lymphoma, *118*, *249*
precursor T-cell lymphoblastic lymphoma, *118*, *249*
predictive biomarkers, 10
prednisone, 123, 259
preventative surgery, 16
primary mediastinal B-cell lymphoma (PMBCL), 250, 251
primary tumors, 13, 14
procarbazine, 259
progesterone receptor (PR), *28*, 144
progestogens, 131
programmed cell death protein-1 (PD-1), 278
programmed death ligand 1 (PD-L1), 278
programmed death ligand 2 (PD-L2), 278
progression-free survival (PFS), 88
promyelocytic leukemia, *161*
prophylactic surgery, 16, 203
prophylactic vaccination, *272*, *272*, 283, *304*
prostate cancer, 145–56
   case study, 136, 152–53, *152*
   castration-resistant prostate cancer (mCRPC), 151–54, *155*, 156,
      272
   diagnosis, 146, 147–48, *147*
   epidemiology, 132, 136, 145, *145*, 155
   future prospects and challenges, 155
   genomics, 145–46
   hormone-refractory prostate cancer, 153
   intensity-modulated radiation therapy (IMRT), 18
   management
      active surveillance, 149–50
      androgen deprivation therapy (ADT), 151, 152, 155, 156
      hormone therapy, 133, 135, 151, 155
      localized, 149–51, *150*, 155, 156
      metastatic, 151–52
      radiation therapy, 150–51, *150*
      surgery, 150, *150*
      targeted therapy, 153–54
   medical castration, 133, 151
   metastatic, 148–49, *149*, 151–52, 155, 156
   risk factors, *145*
   screening, 10, *28*, 145, *145*, 146–47
   staging, 148–49, *148*, 155, 160
   symptoms, 146, *146*
prostate gland
   anatomy, 146, *146*, 160
   benign prostatic hypertrophy (BPH), 146
   biopsy, 147–48, *147*
prostate-specific antigen (PSA) test, 10, *28*, 145, *145*, 146, 155
prostatectomy, 150, *150*
prostatic acid phosphatase (PAP), 272–73
prostatic intraepithelial neoplasia (PIN), 145
proteasomal degradation, 172
proteasome inhibition, 209–11, *211*, 212, 213, 229, 235
proteasomes, 195, *195*, 196
   ubiquitin-proteasome pathway (UPP), 194–96, *195*, 212, 213
protein degradation, 193–94
   aggresome pathway, 196, *196*
protein kinase B (PKB), *33*, 103, 108

protein kinase inhibitors/inhibition, 61–101, 107
   about, 100
   approved for cancer therapy, 62, 64–65, *64*, 100–101
   in chronic myelogenous leukemia (CML)
      in accelerated phase, 70, 73, *73*
      in blast phase, 70, 73, *73*
      chronic phase, 70–73
   clinical use, 62, 64–65, *64*
   decision tree for use, 100–101, *101*
   in melanoma, 96–99, *97*
   resistance to, 65, 100
protein kinases, action, 61–62, *63*, 65
5-protein signature (OVA1), diagnostic test, 29
prothrombin time (PT), 164
proton beam therapy, 18, 151
PSA screening test. *See* prostate-specific antigen (PSA) test
pseudotumor cerebri, 166
psoralen, 186
PTEN gene, 137, *198*
purpura, 292
PUVA, 186, 187

Q

quantitative PCR (qPCR), 69

R

R-bendamustine, 123
R-CHOP, *40*, 123, 253, 254, 255
R-CVP, 123
RA. *See* refractory anemia
RAD51, 197
radiation therapy, 17–19
   brachytherapy, 17, 19
   for breast cancer, 141
   computer-aided targeting, 17–18
   conformal radiation therapy (CRT), 18
   curative, 17
   external beam radiation therapy (EBRT), 17
   for Hodgkin lymphoma, 259
   intensity-modulated radiation therapy (IMRT), 18, *18*
   intravenous radionuclides, 17, 19
   for lung cancer, 85
   for non-Hodgkin lymphoma, 123
   for prostate cancer, 150–51, *150*
   proton beam therapy, 18
   sensitivity/insensitivity to, 17
   stereotactic radiosurgery (SRS), 18
   for testicular cancer, 45
radical inguinal orchiectomy, 43
radical mastectomy, 14–15, *15*
radical nephrectomy, 111
radical prostatectomy, 150
radio-frequency ablation, 111
radionuclides, 17, 19
radiopharmaceuticals, 155
radium-223, 19, 155, *155*
RAEB. *See* refractory anemia with excess blasts
Rai staging system, 77
ramucirumab, *244*
randomization, 289
RANK ligand, 233
RANK ligand inhibitors, 208
rapalogs, 106, 116
rapamycin, 106
RARα, 162
RARα–RXR complex, 162

RARS. *See* refractory anemia with ringed sideroblasts
RAS proto-oncogenes, *33*
RCMD. *See* refractory cytopenia with multilineage dysplasia
RD regimen, 229
reactive oxygen species, 173
receptor tyrosine kinase (RTK), 62, 106
recombinase, 197
reconstructive surgery, 15
rectal cancer
   surgical management, 15–16, *15*
   See also colorectal cancer
reduced representation bisulfite sequencing (RRBS), 30, 37
refractory anemia (RA), 176
refractory anemia with excess blasts (RAEB), 177, *177*
refractory anemia with ringed sideroblasts (RARS), 177, *177*
refractory cytopenia with multilineage dysplasia (RCMD), *177*
regorafenib, *64*, 227
regulator of cullin-1 (ROC1), 234
regulatory T cells, 7, 257, 270
renal cell carcinoma (RCC), 106–17
   case study, 114–15, *116*, *117*
   diagnosis, 109, 117
   epidemiology, 108
   evaluation, 109–10, *110*
   future prospects and challenges, 116–17
   genomics, 109, *109*, 113
   hereditary syndromes, 109, *109*, 117
   imaging, 109–10, *110*, *115*, *116*
   management
      immunotherapy, 112, 117
      localized, 111, 117
      metastatic, 111–13, *112*, 115–16
      targeted therapies, 115–16, 117
   metastatic, 111–13, *112*, 115–16, 117
   risk factors, 108–9
   signaling pathways in, 113, *114*
   staging, 110, *110*, *111*, 117
   subtypes, 108, *108*, *109*
   symptoms, 117
resection, 13, 14
resistance
   to AR pathway agents, 154
   to BRAF inhibition, 98
   to hormone therapy, 143, 144, 154
   to protein kinase inhibitors, 65, 100
   to targeted therapy, 88, 89
retinoblastoma protein (RB), 5
retinoid X receptors (RXRs), 162, 187
retinoids, 161, 187
retroperitoneal lymph node dissection (RPLND), 46
retroperitoneum, 43, 124, 148
reverse-transcriptase PCR (RT-PCR), 69
Revlimid, 209, 230
Richter's transformation, 78
RISC (RNA-induced silencing complex), 174
risk-reducing salpingo-oophorectomy (RRSO), 200, *201*
Rituxan, *244*
rituximab, 78, 123, 124–25, 127, *244*, 253, 254, 256
RNA-induced silencing complex (RISC), 174
RNA interference-mediated knockdowns, 213
RNA sequencing (RNA-seq), 30, 35, 37
ROC1 (regulator of cullin-1), 234
romidepsin, 188, 189
Röntgen, Wilhelm, 17
RPLND. *See* retroperitoneal lymph node dissection
RRBS. *See* reduced representation bisulfite sequencing
RRSO. *See* risk-reducing salpingo-oophorectomy
RT-PCR. *See* reverse-transcriptase PCR

RTK (receptor tyrosine kinase), 62, 106
ruxolitinib, *64*
RVD, 209, 230
RVD lite, 229, 231
RXRs (retinoid X receptors), 162, 187

S

salpingo-oophorectomy, prophylactic, 16
salvage chemotherapy, 22
samarium (^{135}Sm) isotopes, 19
sarcomas, about, 1
SCC. *See* squamous cell carcinoma
scFv (single-chain variable fragment), 276
schwannomas, 1
SCLC. *See* small-cell lung cancer
screening tests, 8, 10
    biomarkers, 10
    BRACAnalysis CDx, 204
    for breast cancer, 137, 144, 204, 213
    for cervical cancer, 283, 284
    full-body CT scans, 10
    for lung cancer, 83
    mammography, 137, *137*
    Pap smears, 283, 284
    for prostate cancer, 10, *28*, 145, *145*, 146–47, 155
    PSA test, 10, *28*, 145, *145*, 146, 155
    for testicular cancer, 42
    visual inspection with acetic acid (VIA), 283, 284–85
    See also diagnostic tests
SDF-1 (stromal cell-derived factor-1), 216, 232
second-line chemotherapy, 22
secondary malignancies, 17
secondary tumors, 13, 14
SEER database (Surveillance, Epidemiology, and End Results database), 229
selective estrogen receptor down-regulator (SERD), 141
selective estrogen receptor modulator (SERM), 141, 142
seminomas, 44, 45
sentinel lymph node (SLN) biopsy, 92, 140
sentinel lymph nodes, evaluation of, 15, 140
sepsis, 301
sequencing, 27, 29–32, *29–32*, 35–36
    next generation sequencing (NGS), 27, 30, 31–32, *31*, *32*, 36, 37
    paired-sample analysis, 27, 29, *29*
    reduced representation bisulfite sequencing (RRBS), 30, 37
    RNA sequencing (RNA-seq), 30, 35, 37
    strengths and weaknesses of, 30–31
    whole exome sequencing (WES), 29–30, 32–33, 35, 37
    whole genome sequencing (WGS), 29, 30, 32–33, 35, 37
SERD (selective estrogen receptor down-regulator), 141
serine/threonine kinase inhibitors, 96
serine/threonine kinase mTOR, 103, 105
serine/threonine kinases, 61
SERM (selective estrogen receptor modulator), 141, 142
serum protein electrophoresis (SPEP), 207
70-gene signature, diagnostic test, *29*
sex steroids, 131, 132, *132*, 156
Sézary syndrome (SS), 183–84, 190
    case study, 187–88
    diagnosis, 184, 185, *185*
    epidemiology, 184
    future prospects and challenges, 189
    management, 187, *187*, 188–89, 190
    prognosis, 190
    TNMB classification, 186, *186*
SF3B1, 75
single-agent chemotherapy, 21

single-cell analysis, 36
single-chain variable fragment (scFv), 276
single-strand DNA breaks, 198
sipuleucel-T, 155, *155*, 272
skin
    cutaneous T-cell lymphoma, 183–90
    normal differentiation of, 160, *160*
skin cancer, 90
    See also basal cell carcinoma; melanoma; squamous cell carcinoma
SLL. *See* small lymphocytic lymphoma
SLN biopsy. *See* sentinel lymph node (SLN) biopsy
small-cell lung cancer (SCLC), 84, 85
small lymphocytic lymphoma (SLL), 75, 77, *118*, 249
smoking, lung cancer and, 82–83, 85, 89
The Solid Tumor Targeted Cancer Gene Panel by Next Generation Sequencing (CAPN), 33–34, *33*
somatic mutation, 75
sonography. *See* ultrasound
sorafenib, *64*, 96, *97*, 116
SPEP. *See* serum protein electrophoresis
spiculated mass, 83, *83*
spinal tap, 292
splenectomy, 53
splenomegaly, 67, *67*
Sprycel, *64*
squamous cell carcinoma (SCC), 90, 97, 99, 284
SRS. *See* stereotactic radiosurgery
SS. *See* Sézary syndrome
staging, 8, *9*
    with AJCC staging classification, *138*, 139
    with Ann Arbor classification, 121, *121*
    of breast cancer, *138*, 139
    of cervical cancer, 285, *285*
    of chronic lymphocytic leukemia, 77
    of colorectal cancer (CRC), 223, *224*
    of cutaneous T-cell lymphoma (CTCL), 185–86, *186*
    of diffuse large B-cell lymphoma (DLBCL), 252
    Gleason grade, 148–49, *148*, *160*
    of Hodgkin lymphoma (HL), 258
    with International Germ Cell Consensus Classification, 44, *45*
    with International Staging System (ISS), 208
    of melanoma, 91, 92, *92*, 93, *93*
    of multiple myeloma, 208, *208*
    of non-Hodgkin lymphoma, 121–22, *121*
    of non-small-cell lung cancer, 84, *84*, 89
    of pancreatic cancer, 52–53, *52*
    of prostate cancer, 148–49, *148*, 155, 160
    with Rai system, 77
    of renal cell carcinoma, 110, *110*, *111*, 117
    of testicular cancer, 45
    with TNM staging, 52–53, *52*, 92, 93, *93*
    TNMB system, 185–86, *186*
    with ultrasound, *9*
stem cell factor receptor (KIT), *28*, 49
stem cell transplant
    allogeneic hematopoietic stem cell transplant, 73, 74, 80, 182, 208, 298
    autologous stem cell transplant, 47, 73, 166, 208
    for chronic lymphocytic leukemia, 80
    for chronic myelogenous leukemia, 73, 166
    graft-versus-host disease, 80
    hematopoietic stem cell transplantation (HSCT), 179–80
    for multiple myeloma, 208
    for myelodysplastic syndromes (MDSs), 180, 182
stereotactic radiosurgery (SRS), 18
steroid hormone receptor antagonists, *133*, *134*, 135–36
steroidogenesis, 131, *132*, *133*, 134–35, *134*, 156

Stivarga, *64*
STK11 gene, 137, *198*
stromal cell-derived factor-1 (SDF-1), 216, 232
strontium (89Sr) isotopes, 19
subcutaneous chemotherapy administration, *21*
subcutaneous panniculitis-like T-cell lymphoma, *118*, *249*
succinate dehydrogenase-associated familial cancer, *109*
SUMMIT trial, 210
sunitinib, *64*, 115, 116, 227
superficial spreading melanoma, 91, *91*
surgery, 13–17
    for breast cancer, 14–15, *15*
    curative procedures, 14
    cytoreductive, 16, 17
    debulking and palliative operations, 16
    for diagnosis, 13–14, *14*
    laparoscopic, 16
    for pancreatic cancer, 53, 58
    perioperative treatment, 17
    preventative surgery, 16
    prophylactic, 16
    for prostate cancer, 150, *150*
    reconstructive surgery, 15
    for rectal cancer, 15–16, *15*
    resection, 13, 14
    for testicular cancer, 42, 43, 44, 48
surveillance, 45
Surveillance, Epidemiology, and End Results (SEER) database, 229
survivorship, 15
Sutent, *64*
SVC syndrome, 258, *258*
synthetic lethality, 193–94, *194*, *199*, 212, 214

T

T-cell-dependent bi-specific antibodies (TBDs), 280
T-cell large granular lymphocyte leukemia, *118*, *249*
T-cell lymphoblastic lymphoma, *118*
T-cell lymphomas, *118*, 183, 190, *249*
T-cell prolymphocytic leukemia, *249*
T-cell receptors (TCRs), 270
T cells, 239, 269, 270, 393
    CAR T cells, 240, *273*, 275–77, *275*, *276*, 296–98, *296*, 303, 304, *304*
    CD4 T cells, 270
    CD8 T cells, 270
    endogenous T cells, 273–74, *273*, *274*
    helper T cells, 270
    naive T cells, 270
    regulatory T cells, 7, 257
T-DM1 (ado-trastuzumab emtansine), *244*, 265–66, 267
T-Vec (Talimogene laherparepvec), 279
T315A mutation, 74
T315I mutation, 72, 74
TAAs (tumor-associated antigens), 243, *244*, 245, 274
Tafinlar, *64*
Talimogene laherparepvec (T-Vec), 279
tamoxifen, 141, 142, 143, 201
TAMs (tumor-associated macrophages), 217, 218
TANs (tumor-associated neutrophils), 218
Tarceva, *64*
targeted therapy, 7, *19*, 39, 128
    for lung cancer, 88
    mechanism-based targeted therapy, 7
    for melanoma, 99
    for metastatic colorectal cancer, 226–27
    for metastatic renal cell carcinoma, 115–16, 117
    for non-Hodgkin lymphoma, 127
    for pancreatic cancer, 56–57, 58
    for prostate cancer, 153–54
    resistance to, 88, 89
    for testicular cancer, 48–49
    See also protein kinase inhibitors/inhibition
Tasigna, *64*
taxanes, *20*, 140, 142, 152
Taxol, 140
TBDs (T-cell-dependent bi-specific antibodies), 280
TCGA. *See* The Cancer Genome Atlas
TCRs (T-cell receptors), 270
Tecentriq, *244*
telomerase, 5
telomeres, 5
temsirolimus, *64*, 106, 115, 116, *116*, 129
teratogenicity, 161
teratomas, 41–42, 44
testicular cancer, 41–49
    case study, 47–48
    combination chemotherapy, *40*
    diagnosis, 42
    evaluation, 43–44, *43*, *44*
    fertility, 46
    incidence, 42
    management, 45–49
    metastatic disease
        management, 46–47
        prognosis, 44–45, *45*
    recurrence, 42
    risk factors, 42, *42*
    screening, 42
    staging, 45
    subcategories, 44
    symptoms, 42
    tumor markers, 43, 44–45, *45*
    tumor types, 44
testicular torsion, 42
testosterone, 136
TGF-α (transforming growth factor-α), 113, *114*
TGF-β (transforming growth factor-β), 281
thalidomide, 229, 230, 233, 235, 277, 280
therapeutic vaccines, 272–73
thoracentesis, 84
thrombocytopenia, 77, 210, 292
thromboembolism, 229
thrombosis, 67
thyroglobulin, diagnostic test, *29*
TILs (tumor-infiltrating lymphocytes), 273
TKIs. *See* tyrosine kinase inhibitors
TLS (tumor lysis syndrome), 294
TME. *See* tumor microenvironment
TMPRSS translocations, *145*
TNF-α (tumor necrosis factor-α), 232
TNM staging
    melanoma, 92, 93, *93*
    pancreatic cancer, 52–53, *52*
TNMB staging, 185–86, *186*
tocilizumab, 295, 297
tolerance, 20
topical chemotherapy administration, *21*
topoisomerase inhibitors, *20*
topotecan, 85
Torisel, *64*
total mesorectal excision, 15, *15*
TP53, 75
TP53 gene, *33*, *198*
tracers, for imaging, *9*
trachelectomy, 287

trametinib, *64*, 96, *97*, 98
transforming growth factor-α (TGF-α), 113, *114*
transforming growth factor-β (TGF-β), 281
translocation, 23–24
    Philadelphia chromosome, 66–67, *66*, 69, *69*
    See also genomics/genetics
transrectal ultrasound-guided biopsy, *147*
trastuzumab, 10, 23, 139, 140–41, 142, *244*, 263–66
trastuzumab–emtansine, 139
"triple negative" breast cancer, 139, *139*, 144, 204
trisomy 12, 78
TSAs (tumor-specific antigens), 243, *244*, 245, 274
tuberous sclerosis complex, *109*
tumor-associated antigens (TAAs), 243, *244*, 245, 274
tumor-associated macrophages (TAMs), 217, 218
tumor-associated neutrophils (TANs), 218
tumor-infiltrating lymphocytes (TILs), 273
tumor lysis syndrome (TLS), 294
tumor markers
    testicular cancer, 43, 44–45, *45*
    See also biomarkers; genomic markers
tumor microenvironment (TME), 6–8, *6*, 215–19, *216*, 236
    angiogenesis, 217–18, *218*
    extracellular matrix, 237
    immunosuppressive TME, 217, 219, *304*
    monoclonal antibodies and, 244–45
    targeting and disturbing, 218–19
    therapeutic approaches to modifying, 236–37, *237*
tumor necrosis factor-α (TNF-α), 232
tumor protein p53, *33*
tumor-specific antigens (TSAs), 243, *244*, 245, 274
tumor suppressors, 5, 75
tumors
    germ cell tumors (GCTs), 41, 44
    pluripotency, 41
    protein kinases and, 61–62
21-gene signature (Oncotype DX), diagnostic test, *29*
Tykerb, *64*, 139
tyrosine kinase inhibitors (TKIs), 65, 95
    for breast cancer, 139
    for chronic myelogenous leukemia, 64–65, *64*, 70–74
    for melanoma, 95, 99
    second-generation drugs, 71
tyrosine kinases, 61

U

ubiquitin, 195
ubiquitin groups, 172
ubiquitin-proteasome pathway (UPP), 194–96, *195*, 212, 213
ubiquitination, 172
ulcerative colitis, 221
ultrasound (US), staging with, *9*
ultraviolet light therapy, 186
United States, leading causes of death in, *41*
UPP. *See* ubiquitin-proteasome pathway
urokinase plasminogen activator (uPa), *29*
US. *See* ultrasound
uvea, 90
uveal melanoma, 95

V

V299L mutation, 74
V600E, 97, 98
vaccination, *19*, 240, 269, 271–73, 282
    anti-HBV vaccine, 303
    anti-HPV vaccine, *286*, 288–90, *289*

for polio and diphtheria, *283*
    prophylactic vaccination, *272*, *272*, 283, *304*
    therapeutic vaccination, 271–73
vandetanib, *64*
vascular endothelial growth factor (VEGF), 49, 113, 115, 215, 218, *218*, *227*, 232
Vectibix, *244*
VEGF antibody, 228
VEGF inhibitors, 226
VEGF-R, 227
VEGF receptors, 115
VEGF signaling, testicular cancer, 49
Velcade, 209, 230
vemurafenib, 23, *64*, 96, 96–98, *97*
venetoclax, 80, 81
VHL tumor suppressor, 113
VIA (visual inspection with acetic acid), 283, 284–85
vinblastine, *20*, 259
vinca alkaloid inhibitors, *20*
vinca alkaloids, *20*, 142
vincristine, *20*, 123, 230, 259, 295
vinorelbine, 142
VIP chemotherapy, testicular cancer, 46–47, 48
virus-like particles (VLPs), 286
viruses, implicated in cancer causation, 282, *282*
visual inspection with acetic acid (VIA), 283, 284–85
von Hippel–Lindau syndrome, 109, *109*, 113
vorinostat, 189
Votrient, *64*
VP-16, 47

W

Waldenström macroglobulinemia, *118*, 205, *249*
Weinberg, Robert A., 4, 71
Whipple procedure, 53
white blood cells
    in chronic lymphocytic leukemia, 76, *76*
    in chronic myelogenous leukemia, 67–68, *68*
    precursors, 68, *68*
    types, 66
    See also leukemia
WHO classification, of myelodysplastic syndromes (MDSs), 177, *177*
whole exome sequencing (WES), 29–30, 32–33, 35, 37
whole genome sequencing (WGS), 29, 30, 32–33, 35, 37
World Health Organization (WHO) criteria, *73*

X

X-linked agammaglobulinemia, 80–81
Xalkori, *64*

Y

Y253H mutation, 74
Yervoy, *244*
yolk sac tumors, 44

Z

Zelboraf, *64*
Zevalin, *244*
zinc finger proteins, 234
zoledronic acid, 208, 231, 233, *233*